# SPECTRAL EVOLUTION OF GALAXIES

# ASTROPHYSICS AND SPACE SCIENCE LIBRARY

A SERIES OF BOOKS ON THE RECENT DEVELOPMENTS
OF SPACE SCIENCE AND OF GENERAL GEOPHYSICS AND ASTROPHYSICS
PUBLISHED IN CONNECTION WITH THE JOURNAL
SPACE SCIENCE REVIEWS

*Editorial Board*

R.L.F. BOYD, *University College, London, England*

W. B. BURTON, *Sterrewacht, Leiden, The Netherlands*

L. GOLDBERG, *Kitt Peak National Observatory, Tucson, Ariz., U.S.A.*

C. DE JAGER, *University of Utrecht, The Netherlands*

J. KLECZEK, *Czechoslovak Academy of Sciences, Ondřejov, Czechoslovakia*

Z. KOPAL, *University of Manchester, England*

R. LÜST, *European Space Agency, Paris, France*

L. I. SEDOV, *Academy of Sciences of the U.S.S.R., Moscow, U.S.S.R.*

Z. ŠVESTKA, *Laboratory for Space Research, Utrecht, The Netherlands*

VOLUME 122
PROCEEDINGS

# SPECTRAL EVOLUTION OF GALAXIES

PROCEEDINGS OF THE FOURTH WORKSHOP
OF THE ADVANCED SCHOOL OF ASTRONOMY
OF THE "ETTORE MAJORANA" CENTRE
FOR SCIENTIFIC CULTURE, ERICE, ITALY,
MARCH 12–22, 1985

Edited by

CESARE CHIOSI

*Institute of Astronomy, University of Padova, Italy*

and

ALVIO RENZINI

*Department of Astronomy, University of Bologna, Italy*

D. REIDEL PUBLISHING COMPANY

A MEMBER OF THE KLUWER  ACADEMIC PUBLISHERS GROUP

DORDRECHT / BOSTON / LANCASTER / TOKYO

Library of Congress Cataloging in Publication Data

Main entry under title:

Spectral evolution of galaxies.

(Astrophysics and space science library ; v. 122)
  1. Galaxies – Evolution – Congresses.  2. Stars – Congresses.
3. Astronomical spectroscopy – Congresses.  I. Chiosi, C. (Cesare)
II. Renzini. Alvio.  III. Ettore Majorana International Centre for
Scientific Culture. Advanced School of Astronomy.  IV.  Series.
QB856.S64  1986     523.1'12     85-30065
ISBN 90-277-2187-4

Published by D. Reidel Publishing Company,
P.O. Box 17, 3300 AA Dordrecht, Holland.

Sold and distributed in the U.S.A. and Canada
by Kluwer Academic Publishers,
190 Old Derby Street, Hingham, MA 02043, U.S.A.

In all other countries, sold and distributed
by Kluwer Academic Publishers Group,
P.O. Box 322, 3300 AH Dordrecht, Holland.

All Rights Reserved
© 1986 by D. Reidel Publishing Company, Dordrecht, Holland
No part of the material protected by this copyright notice may be reproduced or
utilized in any form or by any means, electronic or mechanical
including photocopying, recording or by any information storage and
retrieval system, without written permission from the copyright owner

Printed in The Netherlands

TABLE OF CONTENTS

PREFACE ix

LIST OF PARTICIPANTS xi

INTRODUCTORY LECTURE

J. SILK
Galaxy evolution: some issues and questions 3

I.    GALAXY AND STAR FORMATION PROCESSES

J. SILK
Towards a theory of galaxy formation 15

J. SILK
Physical processes in star formation 31

A. DI FAZIO, R. CAPUZZO DOLCETTA
Theoretical and observational evidence for the existence
of a single fragmentation law, and galactic evolution 47

J. LEQUEUX
Star formation tracers in galaxies 57

I. TARRAB
Age and processes of star formation in very
young LMC associations 75

G. COMTE
On the relationship between neutral hydrogen mass,
luminosity and color in irregular galaxies 81

D. KUNTH
IZW18 and the search for very metal poor galaxies 87

C. J. LONSDALE
Bolometric luminosity evolution 91

B. F. MADORE
Galaxy encounters and the Holmberg effect 97

J. M. VAN DER HULST, E. HUMMEL, W. C. KEEL, R. C. KENNICUTT JR
The effects of galaxy-galaxy interactions on nuclear activity ... 103

T. DE JONG
Galaxies in the infrared ... 111

K. BRINK
IRAS far-infrared observations of interacting galaxies ... 127


II. STELLAR POPULATIONS IN THE LOCAL GROUP

J. MOULD
The stellar populations of galaxies in the local group ... 133

J. A. FROGEL
M giants in the galactic nuclear bulge ... 143

A. E. WHITFORD
Age and mass of M giants in the galactic bulge ... 157

M. AARONSON, K. H. COOK, J. NORRIS
The AGB population of nearby galaxies ... 171

W. L. FREEDMAN
M33: radial distributions and a comparison of its
global luminosity function with other nearby galaxies ... 183


III. TOWARDS MODELLING THE SPECTRAL EVOLUTION OF GALAXIES

A. RENZINI, A. BUZZONI
Global properties of stellar populations and the
spectral evolution of galaxies ... 195

C. CHIOSI
Advancements in the stellar evolution theory: the role of
convective overshooting all across the HR diagram ... 237

G. BRUZUAL A.
Spectral evolution of galaxies ... 263

G. BARBARO, F. M. OLIVI
Spectrophotometric models of galaxies ... 283

N. ARIMOTO, Y. YOSHII
Photometric evolution of elliptical galaxies in the color –
magnitude diagram ... 309

## IV. EMPIRICAL POPULATION SYNTHESIS

R. W. O'CONNELL
Analysis of stellar populations at large lookbacks ... 321

A. PICKLES
Population synthesis and epochs of star formation in NGC 1316 (Fornax A) ... 345

B. ROCCA-VOLMERANGE, B. GUIDERDONI
Far-UV stellar populations of S0 galaxies ... 357

F. BERTOLA
The UV energy distribution of elliptical galaxies ... 363

R. NESCI
UV spectra of normal ellipticals ... 369

## V. GALAXIES AT LARGE LOOKBACK TIMES

A. DRESSLER
Studies of cluster galaxies at large lookback times ... 375

D. HAMILTON
Observational tests for galaxy evolution ... 391

P. R. M. EISENHARDT
Colors of 3CR and first-ranked high redshift galaxies ... 403

B. GUIDERDONI, B. ROCCA-VOLMERANGE
Evolution of disk galaxies in high-redshift clusters ... 411

D. C. KOO
Quests for primeval galaxies: a review of optical surveys ... 419

P. R. F. STEVENSON, T. SHANKS, R. FONG
New observations of galaxy number counts ... 439

## VI. CHEMICAL EVOLUTION OF GALAXIES AND MISCELLANEA

R. GÜSTEN
The chemical evolution of galaxies ... 449

M. TOSI, A. I. DIAZ
Nitrogen and oxygen evolution in nearby spiral galaxies ... 469

C. FORIERI
**Effects of metal-dependent stellar models on the yield of nitrogen**     473

M. A. SHAW, G. F. GILMORE
**Surface brightness distributions in two edge-on spiral galaxies**     477

**SUBJECT INDEX**     485

PREFACE

As it was said by one of the participants to this workshop " In our attempts to understand the spectral evolution of galaxies, we are fortunate indeed to have the ability to look back in time and observe galaxies as they were billions of years ago. Perhaps in no other discipline is it possible to gain such a direct view to history. The galaxies we seek to study are remote, their light faint, and thus only recently has it become technically feasible to sample the spectra of normal luminosity galaxies at lookback times of five billion years or more".... or, perhaps, even to see galaxies in the process of their formation, or shortly afterwards.
This fourth workshop organized by the "Advanced School of Astronomy was indeed centered on the "Spectral Evolution of Galaxies", on reviewing and discussing the relevant astrophysical processes and on assessing our current ability to model and understand the evolution of stellar populations.
Following an opening session dealing with some outstanding questions of galaxy evolution, Session I addressed the specific problems of galaxy and star formation processes, topics of uncertainty and controversy to which IRAS observations may give novel perspectives.
The properties of stellar populations in the local group of galaxies formed the basis of Session II.
Session III dealt with the fundaments of the theory of spectral and photometrical evolution of stellar populations, and with recent developments in the theory of stellar structure, a necessary step to model and understand galactic evolution.
Session IV was primarily concerned with empirical population syntheses and problems related to the UV spectra of elliptical and S0 galaxies.
The properties of galaxies at large lookback times and the search of primeval objects were the subject of Session V.
Finally, Session VI intended to focus on the interrelation between chemical evolution of galaxies and how this would affect the spectrophotometric properties of these.
The most recent observational and theoretical results were presented and extensively discussed by the participants that weather conditions helped to bring together during the ten days of the meeting. At the end, there was a general feeling of satisfaction about this educational experience, and anxious expectation for the exciting developments that the next generations of space and ground based telescopes will certainly promote.
The workshop was held in Erice (Sicily), at the "Ettore Majorana Centre for Scientific Culture", from March 12 through 22, 1985, ,and was organized about a sequence of review lectures, each followed by contributed talks and discussions, most of which are to be found in these proceedings.

The Italian National Research Concil (CNR-GNA) and the Centre "E. Majorana" are gratefully acknowledged for having provided financial assistance to several participants. Finally, we would like to thank the local staff of the Centre "E. Majorana", and in particular Mrs. Pinola Savalli and Dr. Alberto Gabriele for their efficient organization of the logistic aspects of the meeting.

                            Cesare Chiosi    and    Alvio Renzini

LIST OF PARTICIPANTS


AARONSON, M., Steward Observatory, Tucson, U.S.A.
ARIMOTO, N., Observatoire de Paris Meudon, Meudon, France
BALKOWSKI, C., Observatoire de Paris Meudon, Meudon, France
BARBARO, G., Istituto di Astronomia, Padova, Italy
BERTOLA, F., Istituto di Astronomia, Padova, Italy

BOULADE, O., C.E.N./IRF-DPHG, Gif sur Yvette Cedex, France
BRESSAN, A., S.I.S.S.A., Trieste, Italy
BRINK, K., Astronomical Institute, Amsterdam, The Netherlands
BROADHURST, T., Department of Physics, Durham, U. K.
BRUZUAL, G., C.I.D.A., Merida, Venezuela

BUSON, L., Osservatorio Astronomico, Padova, Italy
BUZZONI, A., Dipartimento di Astronomia, Bologna, Italy
CAPUZZO DOLCETTA, R., Istituto Astronomico, Roma, Italy
CHIOSI, C., Istituto di Astronomia, Padova, Italy
COMTE, G., Observatoire de Marseille, Marseille, France

O'CONNEL, R., University of Virginia, Charlottesville, U.S.A.
DRESSLER, A., Mt. Wilson-Las Campanas Observatories, Pasadena, U.S.A.
EISENHARDT, P., Nat. Opt. Astron. Observatories, Tucson, U.S.A.
DI FAZIO, A., Osservatorio Astronomico, Roma, Italy
FLIN, P., Jagiellonian University Observatory, Krakow, Poland

FOCARDI, P., Dipartimento di Astronomia, Bologna, Italy
FORIERI, C., S. I. S. S. A., Trieste, Italy
FREEDMAN, W., Mt. Wilson-Las Campanas Observatories, Pasadena, U.S.A.
FROGEL, J. A., C. T. I. A. O., Tucson, U.S.A.
GREGGIO, L., Dipartimento di Astronomia, Bologna, Italy

GUSTEN, R., M. P. f. Radioastronomie, Bonn, F. R. Germany
HAMILTON, D., Nat. Opt. Astron. Observatories, Tucson, U.S.A.
HELD, E., Istituto di Astronomia, Padova, Italy
VAN DER HULST, J. M., Found. for Radio Astronomy, Dwingeloo, The Netherlands
DE JONG, T., Astronomical Institute, Amsterdam, The Netherlands

KOO, D. C., Space Telescope Science Institute, Baltimore, U.S.A.
KUNTH, D., Institut Astrophysique, Paris, France
LEQUEUX, J., Observatoire de Marseille, Marseille, France
LONSDALE PERSSON, C., JPL/CALTECH, Pasadena, U.S.A.
MACLAREN, I., Department of Physics, Durham, U. K.

MADORE, B., CALTECH, Pasadena, U.S.A.
MOULD, J., Palomar Observatory, CALTECH, Pasadena, U.S.A.
NESCI, R., Istituto Astronomico, Roma, Italy
NEWBERRY, M. V., University of Michigan, Ann Arbor, U.S.A.
OCCHIONERO, F., Istituto Astronomico, Roma, Italy

OLIVI, F., S. I. S. S. A., Trieste, Italy
ORIO, M., Department of Physics, Technion, Haifa, Israel
PICKLES, A., Kaptein Laboratorium, Groningen, The Netherlands
PIGATTO, L., Osservatorio Astronomico, Padova, Italy
RAMPAZZO, R., Istituto di Astronomia, Padova, Italy

RENZINI, A., Dipartimento di Astronomia, Bologna, Italy
ROCCA-VOLMERANGE, B., Institut Astrophysique, Paris, France
SALUCCI, P., S. I. S. S. A., Trieste, Italy
SCHAREIN, R., University of British Columbia, Vancouver, Canada
SHAW, M., Department of Astronomy, Edinburgh, Scotland

SILK, J., Department of Astronomy, Berkeley, U.S.A.
DE SOUZA, R., E. S. O., Garching-Munchen, F. R. Germany
STEVENSON, F., Department of Physics, Durham, U. K.
TARRAB, I., Institut Astrophysique, Paris, France
TOSI, M., Dipartimento di Astronomia, Bologna, Italy

WHITFORD, A. E., Lick Observatory, Santa Cruz, U.S.A.

INTRODUCTORY LECTURE


J. SILK
**Galaxy evolution: some issues and questions**

# GALAXY EVOLUTION: SOME ISSUES AND QUESTIONS

Joseph Silk
Department of Astronomy
University of California
Berkeley, CA   94720, USA

In this introductory talk, I am going to list in a very subjective fashion some outstanding questions that will certainly be addressed, if not resolved, at this workshop on spectral evolution of galaxies.

## I. GAS SOURCES AND SINKS

A galaxy is, or certainly was, a mixture of gas and stars. Even ellipticals often contain gas, as evidenced by the presence of dust lanes. At the present epoch, stellar mass loss provides perhaps the major source of gas. Infall of gas may also occur: one probable instance of this is at the center of cooling flows around cD galaxies in clusters. Some of the gas must be primordial, in the sense that it has not been processed through stars. However, much of the gas both in spirals and in the intracluster medium has been highly enriched by stellar ejecta. The intracluster gas may be partially stripped from galaxies by ram pressure of the hot ambient medium, and partially fuelled by winds driven by the collective interaction of many supernova remnants. Indeed it has been argued that galactic winds are necessary to clear some ellipticals out of their accumulating gas reservoir. We would like to know the sources and sinks of the gas in a galaxy to be better able to ascertain when and for how long the bulk of its star formation occurred.

## II. TRIGGERING THE GAS TO FORM STARS

The gas in galaxies accumulates in clouds, presumably as in our own galaxy. But what triggers cloud collapse and star formation? There are several different possibilities, including interaction with HII regions expanding around massive OB stars and cloud collisions, driven by some combination of spiral density waves, gravitational instability, or magnetic Rayleigh-Taylor instability. The efficiency at which stars form under any of these situations is not known, but is crucial to understanding their early evolution. The best one can do, again, is to look at interstellar clouds, where efficiencies are known, but for a particular type of galaxy at a mature stage of its life.

One can, perhaps more fruitfully, look at other galaxies in search for possible triggers of star formation. One such trigger is galaxy interactions, and de Jong will be discussing IRAS evidence for star formation in interacting systems. It has long been known that excessively blue colors are associated with close interactions, and these are also integrated as evidence for enhanced star formation. Finally, radio jets have recently been shown to trigger star formation, as in Minkowski's Object.

## III. THE MASS FUNCTION OF NEWLY FORMED STARS

The knowledge of the initial mass function (imf) is essential for inferring chemical yields and enrichment. Observational evidence is consistent with a universal imf. This is based on comparison of the field and cluster imfs in our galaxy, and on data on imfs in nearby galaxies. However, the uncertainties remain large, and it is difficult to rule out spatial variations. Some evidence suggests that one can form exclusively low mass stars, for example in T associations and possibly in accretion flows in rich clusters. The latter argument is very indirect, however, being based on a presumed lack of heating in the cooling flows, and there is no theoretical justification for the presumption that a high pressure environment suppresses OB star formation. Indeed, Minkowski's object provides a nice counterexample, where the interaction of a high pressure radio jet with a gas clouds appear to result in massive star formation. If star formation is bimodal, however, it seems clear that massive star formation is accompanied by lower mass star formation, although the converse may not necessarily be true.

The time evolution of the imf represents another area of uncertainty. The overabundance of oxygen in old population II stars is suggestive of enhancement of massive stars in the imf. However, a normal imf could also explain this if the oldest stars formed after the first generation of massive stars had produced oxygen over $\sim 10^7 yr$ but before the rather longer time scale $\sim 10^9 yr$ required to make Fe in lower mass stars. Perhaps a potentially stronger case for massive star enhancement may be made for intracluster gas, which appears to be enriched in oxygen but not in Fe, Si or S in the core of the M87 cluster of galaxies. There is so much enriched material required, as inferred from the near-solar abundance of Fe in many clusters where the amount of gas is comparable to the luminous mass in galaxies, that recourse to an epoch of early star formation, enrichment, and mass loss in the cluster galaxies seems necessary.

## IV. CHEMICAL EVOLUTION

In any closed system of gas and stars, continuing mass loss and star formation results in the chemical abundances approaching the chemical yield, which, for a standard imf, means that stellar abundances will be solar (or somewhat larger) after a few billion years. Many galaxies, especially dwarfs, have abundances well below solar, and this is usually understood to mean that they have suffered extensive gas loss during an earlier phrase of their evolution. Evidently chemical and dynamical evolution are intimately coupled together. Stripping and catastrophic mass loss cannot be the entire answer, since the presence of carbon stars in nearby dwarf spheroidals shows that some gas is being retained and star production is occuring over several billion years.

Study of relative abundances of different elements provides another probe on early chemical evolution. For example, nitrogen is predominantly a secondary product of nucleosynthesis, requiring prior production of carbon and oxygen in an earlier generation of stars. Examination of the range in metallicity over which N/O is constant allows an estimate of the duration of the secondary enrichment phase. The presence of some N in the oldest stars can be explained by dredge-up by convective mixing, and in this primary production regime, one might expect N to be independent of O.

It is by no means evident that mixing of enriched stellar ejecta would be extremely rapid, so that the usual instantaneous enrichment approximation would be valid. There are indications that $^3He$ is variable between different galactic HII regions: if this is taken at face value, rather slow mixing over $\sim 10^9 yr$ or more is inferred . This slow time scale is not at all improbable: for example, turbulent diffusion at a velocity $10 km\ s^{-1}$ and with mean free path $\sim 100 pc$ (corresponding to characteristic interstellar cloud parameters) yields a turbulent diffusion coefficient $10^{26} cm^2\ s^{-1}$ and a chemical homogenization scale $\sim (Dt)^{\frac{1}{2}} \sim 1 kpc (t/10^9 yr)^{\frac{1}{2}}$.

## V. ROLE OF GALAXY MORPHOLOGY AND STRUCTURE

There is an old and ongoing debate about whether galaxies form by nature or by nurture, that is to say, by intrinsic properties determined by initial conditions of the parent protogalactic clouds or by interaction with their environments. Ellipticals are found in dense regions, spirals in low density regions, and there are correlations between luminosity, metallicity and dynamical and structural parameters of galaxies over a wide range. Yet whether these should be ascribed to environmentally induced effects such as galaxy collisions or infall, or to initial angular momentum or density in individual protogalaxies, is not established.

Indeed, morphology provides an increasingly unusable classification scheme for distant galaxies. How can one distinguish an elliptical undergoing a burst of star formation from a spiral galaxy with continuing star formation?

Studying starbursters offers an intriguing prospect of probing the galaxy formation process in nearby systems. Objects such as $1Zw18$, the most metal poor emission line galaxy known, must be relatively young systems. A particularly important and still not resolved issue is whether such blue compact galaxies have underlying populations of old stars.

## VI. GALAXY FORMATION

A major and even central topic at this workshop will be the appearance of a very young galaxy. But what exactly is a protogalaxy? It is not at all obvious that one should define it to be a parent, actively starforming cloud containing the mass of the mature galaxy. For example, a protogalaxy could be assembled in dribs and drabs, each drib (or drab) being responsible for some star formation and enrichment. In this way, there might not be a single very luminous phase when population II metals were produced, but a more extended period of activity that would be relatively hard to trace in deep searches.

Cosmology plays an important role in predicting when galaxies formed. This prediction arises from N-body simulations of the large scale matter distribution, with initial conditions being very tightly constrained by the observed isotropy of the microwave background radiation, by the nucleosynthesis of the light elements early in the big bang, and by the apparent dominance of dark matter on scales in excess of about $10kpc$. Two rival models have emerged, depending on whether $30eV$ neutrinos (hot dark matter) or massive weakly interacting particles (cold dark matter) conspire to produce a closure density of dark matter. Massive neutrinos are temporarily out of favor, primarily because a very recent epoch ($z \sim 1$) of galaxy formation is predicted owing to the suppression of all primordial galactic and subgalactic density fluctuations. Galaxies and clusters form simultaneously at low redshift in this theory. The fact that some objects (quasars, very luminous galaxies) are seen at $z \sim 2$ creates a serious difficulty for a neutrino-dominated universe.

While there is no question that very luminous galaxies must have an old population being seen in one or two systems at $z > 1$ it is by no means obvious that the average galaxy ($L \sim L_*$) could not have formed at $z \sim 1$. Deep redshift surveys have not yet probed the presence of such galaxies to a redshift greater than 0.5 or so, and already, at this redshift, there are some preliminary indications of the presence of a strongly evolving population of blue galaxies.

The cold dark matter theory makes some very specific predictions about the epoch and rate of galaxy formation. Luminous galaxies must necessarily form from rare fluctuations, otherwise excessive amounts of clustering occur, at a rate that declines sharply with epoch, and cuts off approximately at a redshift of 2 or 3. The common fluctuations, which collapse later, are assumed to not form massive galaxies, and are likely to provide a continuing reservoir of hot intergalactic gas. that can form dwarf galaxies at a relatively recent epoch as well as be a source of infall. This offers a rich choice of possibilities for models of galactic evolution, and poses a challenge to observers and theoreticians alike. This workshop comes at a timely moment, and its proceedings should provide a very useful and lively assessment of the degree to which any theoretical deductions about galactic evolution can presently be drawn.

Whitford: 1. What Salpeter exponent x does Scalo find for the upper end of his IMF?

2. Can you comment on the dichotomy urged by Wirth and Gallagher between luminous, high surface brightness E-galaxies (following the $r^{1/4}$ law and having very low current SFR) and diffuse dwarf spheroidals like NGC 205 and other still less luminous? And does this imply that only the former have a continuing gas–ejection mechanism, such as a stellar wind?

Silk: 1. Scalo's favored index for the massive star end of the IMF is -1.7($\pm$0.3); the Salpeter value is 1.35.

2. The very low surface brightness dwarfs must have been stripped early on, probably by stellar winds. The high surface brightness dwarfs are only recently over the past $10^9 yr$ being cleared out of gas. Why the difference? I suspect initial conditions, namely the high surface brightness dwarfs formed from dense, and hence less easily strippable, clouds.

Renzini: 1. Accretion rates in CD's have been estimated assuming that the x-ray gas is just cooling, i.e. that there are no heating sources. What happens if heating is present?

2. From an estimated low value of the Jeans mass in cooling flows, it has been inferred that massive stars ($M > M\odot$) should not form at all. Could you comment about that?

Silk: 1. If the flows are being heated, then it is meaningless to estimate flow rates by taking the ratio of gas mass to cooling time. In fact, heating by massive stars forming with a standard IMF suffices to reduce the inferred mass flow rates by up to an order of magnitude in the central galaxies.

2. The Jeans mass should only be interpreted as providing a lower bound on the masses of forming stars. While high ambient pressure as in the cooling flows, reduces the minimum Jeans mass. It can also drive non-linear interaction of fragments, which could raise the minimum surviving fragment scale.

Occhionero: 1. If future observations will push down the upper limits on $\delta T/T$ by one or two orders of magnitude, then dark matter will not be able to explain the puzzle. If so, we need reionization? But, then will not our need for dark matter be reduced and consequently weaken our "theoretical" evidence for it?

2. If $3\sigma$ - fluctuations correlate with ellipticals and $1\sigma$ - fluctuations with spirals, shouldn't there be a similar correlation with low and high angular momentum? If so, why?

Silk: 1. I agree.

2. It has been argued that the rarer $3\sigma$ fluctuations are also more highly correlated. However, they suffer lower tidal torques and acquire lower angular momentum than $1\sigma$ and this may largely cancel out the former effect.

Aaronson: 1. Reply to Silk's question on M/L in dwarfs: the situation with regard to dark matter in the spheroidals is a bit up in the air. It is going to take several more years to sort out the effects of binary stars and atmospheric motions from the true velocity dispersion.

2. Reply to de Jong question about evidence of younger stars in dwarf spheroidals: First, a number of these systems contain luminous AGB carbon stars. Second, in several systems deep color–magnitude diagrams have been obtained which reach the main sequence and show directly a younger turnoff.

3. Reply to Renzini: I disagree with your statement that the dwarf spheroidals all lost their gas at the same time. We know now that in some systems, such as Fornax, star formation continued for a much longer period of time than in other systems, such as Draco and Ursa Minor. The ratio of the intermediate age to old population in these various systems is clearly very different.

Dressler: You showed the result of Scalo (also found by Gilmore) that the luminosity function for faint stars turns down. Arcadio Poveda has presented circumstantial evidence that these stars are radiating gravitational energy for $\lesssim 10^9$ years but will fade to brown dwarfs if nuclear burning does not ignite. If so, the luminosity function will be a poor representation of the mass function, which may continue to rise with decreasing mass.

Di Fazio: An alternative scenario on the evolution (or, anyway, on the interpretation of the nowadays structure) of clusters of galaxies is that of chemical pollution coming from primordial protogalactic explosions in the protocluster medium (see e.g. Di Fazio, Vagnetti, Wilson, 1980*). The new-formed galaxies violently eject from 10 - 15% of their mass, enriched with metals, and with a substantial stellar component, that in its turn, continues to pollute the intracluster medium. After the first protocluster free-fall time has passed, fragmentation into galaxies stops at an efficiency $\mu \lesssim 50\%$, where $\mu = M_{gas}/M_{tot}$. The fragmentation halt induced by partial gas deflection and heating of the gas is thus indicating that roughly more than 50% of the initial cloud mass is converted into galaxies. I think that, with the evidence we have** available on Fe abundances, x luminosities, estimates on the fraction of the total mass in gas, etc., we cannot exclude the mentioned alternative scenario. Do you agree?

Silk: I agree.

Di Fazio: If we consider the various mass functions (of very different astrophysical self-gravitating objects) that can be computed from the available observational counts (about single stars in the SN*, binaries in the SN, open clusters of our galaxy, globular clusters of various galaxies, galaxies themselves, clusters of galaxies) we see a very confusing remarkable similarity (if not equality) of the shape of these functions. The ranges are different, but, with small differences in the slopes probably due to internal evolution of the objects subsequent to their birth via fragmentation, the mass functions, once normalized, and brought to dimensionless units $(M/M_{max})$, are the same under statistical significance comparison tests. Moreover, they all show a turnover at low masses, which in almost all cases was shown not to be due to selection effects. (See, e.g. in the stellar case, the discussion and evidence for turnover given by Miller and Scalo, Hughes, Probst and O'Connell. Also the cluster mass function, which I will show, deduced from Dressler's sample in collaboration with Dr. P. Flin, has a well clear-of-selection-effects low mass turn over. The same can be said for the globular cluster mass function. When we use mass functions, we are interested in them mainly because we want to understand how all those different objects formed. So, when we discuss the "completeness" we must first understand what we define by "complete". For example, binary stars are born together, so the counts for mass function should count only the total mass of a binary system. Also, in understanding how the fragmentation to stars

---

* *Astrophys. Spa. Sci.*, L72, 204.

went on, when we count stars, if we could see planets, we would exclude from the count the planets, as they are probably the result of a sub–fragmentation of the protostellar clouds. Analogously, in counting galaxies, maybe we have at least 2 families (or generations): one is of the various "dwarfs", with masses from some $10^7 M_\odot$, to some $10^9 M_\odot$ that almost always belong dynamically to the halos or "bigger" galaxies, and another is that of larger galaxies. We should probably cut a sample of galaxies, "complete" according just to total number, into two parts, according to the understanding that dwarf galaxies in halos ($R > 100 = 150 kpc$) are born probably as debris of the mother central protogalaxy, as A. Renzini was proposing, or as sub–fragmentation of the mother cloud. This careful way to decide what is a meaningful definition of completeness is very important. In fact, if we make a mass function mixing together different (also in time sequence) classes, or families, or generations of objects, we will obtain an "overcomplete" count, which is not telling us true information on the birth process. Can you please comment?

(1) Miller, G. E., and Scalo, J. M. 1979, *Astrophys. J. Suppl.*, **41**, 513.

(2) Hughes, D. W. 1980, *Astron. Astrophys.*, **87**, 136.

(3) Probst, R. G., and O'Connell, R. W. 1982, *Astrophys. J. Lett.*, **252**, L69.

Silk: I agree especially with regard to the interpretation of the field IMF for stars: the role of binaries is very uncertain. The physics is so different of star galaxy, and cluster formation, however, that, any resemblance between IMF's must be either illusory, due to sampling biases, or coincidental.

Kunth: You still find objects and metals at very early epochs. At least quasars are seen up to $z = 3.8$ and metal absorption lines are found in at very high $z$. How do you account for their formation?

Silk: Quasars are still rare objects even at $z = 3$. I think that a small number of objects could well have formed prior to the epoch of "average" galaxy formation, whenever that was. This applies equally to luminous galaxies for example, Spinrad's object at $z = 1.8$.

I.  GALAXY AND STAR FORMATION PROCESSES

J. SILK
Towards a theory of galaxy formation

J. SILK
Physical processes in star formation

A. DI FAZIO, R. CAPUZZO DOLCETTA
Theoretical and observational evidence for the existence of a single fragmentation law, and galactic evolution

J. LEQUEUX
Star formation tracers in galaxies

I. TARRAB
Age and processes of star formation in very young LMC associations

G. COMTE
On the relationship between neutral hydrogen mass, luminosity and color in irregular galaxies

D. KUNTH
IZW18 and the search for very metal poor galaxies

C. J. LONSDALE
Bolometric luminosity evolution

B. F. MADORE
Galaxy encounters and the Holmberg effect

J. M. VAN DER HULST, E. HUMMEL, W. C. KEEL, R. C. KENNICUTT JR
The effects of galaxy-galaxy interactions on nuclear activity

T. DE JONG
Galaxies in the infrared

K. BRINK
IRAS far-infrared observations of interacting galaxies

# TOWARDS A THEORY OF GALAXY FORMATION

Joseph Silk
Department of Astronomy
University of California
Berkeley, CA  94720, USA

## I. INTRODUCTION

The theories of star and galaxy formation are intimately related. Unfortunately, there is no accepted theory of star formation, despite the fact that laboratories where stars are currently forming exist within 200 pc of the sun. This may suggest that attempting to understand how galaxies formed would be a futile endeavor. However such a prospect has never deterred theoreticians, and this lecture is devoted to describing some aspects of how star formation processes are relevant to galaxy formation. I commence by discussing the formation of the first stars in the collapsing protogalactic gas cloud. I then examine the characteristic parameters of the old spheroidal components of galaxies, and show how important a role star formation processes must have played in their development and evolution.

## II. THE FIRST STARS

The principal distinction between a primordial cloud and one in the present interstellar medium is the lack of heavy elements. Provided that the metallicity is $[Z] \lesssim 10^{-3}$, or less than 0.1 percent of solar, low temperature gas cooling is predominantly via $H$, $He$, or $H_2$ excitations. Since the critical scale which determines whether fragmentation can occur, the Jeans mass, depends sensitively on temperature, one might anticipate that in the absence of fine-structure cooling by C, Si, Fe *etc.*, to temperatures below 1000K, and even below 100K, the minimum fragment masses would be very large in the primordial environment. In fact, this is not the case, at least so long as $H_2$ molecules can form.

Although no dust grains are present in the primordial environment, $H_2$ molecules can form via the intermediary step of negative hydrogen ion formation. There is a residual electron abundance remaining after recombination, as high as $n_e/n_H \sim 10^{-3}$ if the baryon density is low. The sequence of reactions

$$H + e^- \to H^- + h\nu,$$
$$H^- + H \to H_2 + e^-,$$

commences at a redshift $z \lesssim 100$, when the radiation field first allows $H^-$ survival. An alternative channel is via $H_2^+$ formation. Rapid formation and destruction ensures that a collapsing cloud attains an equilibrium $H_2$ abundance of $n_{H_2}/n_H \approx 10^{-3}$, and the $H_2$ rotational cooling suffices to maintain the temperature near $1000K$. Only at very high density ($n_H \gtrsim 10^9$ cm$^{-3}$) does the physics change. Two effects intervene. Three-body formation ($3H \to H_2 + H$) becomes rapid and drives the fractional $H_2$ abundance to nearly unity. Line opacity also becomes important at high density, and the cooling efficiency per $H_2$ molecule is reduced. However the temperature rises very slowly with increasing density, and collisional dissociation of $H_2$ only sets in above $\sim 5000K$. At this stage, the density may exceed $\sim 10^{15}$ cm$^{-3}$, depending on the optical depth, or equivalently, the cloud mass. The minimum Jeans mass attained is $\sim 0.1~M_\odot$ (Palla et al. 1983).

Now the Jeans mass criterion is necessary but not sufficient for star formation: for example, coagulation may result in much larger minimum mass fragments. However, a similar Jeans scale is found in the present interstellar medium, and this leads one to suspect that the initial stellar mass function in a primordial cloud should not differ greatly from that at the present epoch. This weak sensitivity on physical conditions arises because the minimum Jeans mass is of order $0.01~T^{1/4}~M_\odot$ for fragments undergoing free-fall collapse that are of unit optical depth (Rees 1977; Silk 1977).

A stable protostar only forms when the free-fall collapse of a fragment is halted, and the opacity has built up sufficiently that quasi-static contraction commences. Prior to this stage, any fragment in excess of the minimum Jeans mass is liable to further subfragmentation, dynamical disruption, or coalescence. The condition for quasistatic contraction is equivalent to requiring that the fragment mass exceed $0.01~T^{5/6}~(\kappa/\kappa_{es})^{1/3}~M_\odot$, where T now denotes the central temperature, $\kappa$ is the opacity, and $\kappa_{es}$ is the Thomson scattering opacity (Silk 1977). This criterion is very sensitive to $T$: in fact, there is a considerable difference in the following situations. If $T$ is several hundred degrees Kelvin, as would be the case for formation of protostars in the present interstellar medium, or $T \sim 10^3~K$ as is appropriate in clouds containing a substantial abundance of $H_2$, then one can show that with appropriate opacities the typical protostellar mass is a few tenths of a solar mass. In the absence of $H_2$, however, $T \sim 10^4 K$, and the protostellar mass would be $\sim 20\ M_\odot$.

This extreme sensitivity to the presence of $H_2$ has the following implications. At the outset, star formation in a primordial cloud should result in an initial stellar mass function that does not differ appreciably from the solar neighborhood IMF. However, once massive stars form, then it is likely that ionization fronts, stellar winds, and supernova remnants will drive strong shocks throughout much of the cloud. At a density in excess of $\sim 100$ cm$^{-3}$ in a primordial cloud, once $H_2$ is destroyed, it is very difficult to reform it (Silk 1985). This is because in the absence of heavy element coolants, the temperature is $10^4\ K$ or larger, and the formation rate of $H_2$ does not compete with collisional dissociation at high density (n $\gtrsim$ 100cm$^{-3}$) and temperature ($T \gtrsim 5000\ K$). Photodissociation is less of a problem, because $H_2$ self-shields itself efficiently. Now to some extent, this also occurs in ordinary interstellar clouds: heating of ambient gas by massive stars tends to suppress formation of low mass stars, as fragmentation becomes inhibited. However, because of the vulnerability of the $H_2$ molecule, I expect this tendency to be greatly enhanced in primordial clouds. Thus the final IMF is likely to contain stars ranging in mass from $\sim 0.1$ to $\sim 100\ M_\odot$, but should have an excessive number of massive stars compared to the present day IMF.

There are possible observational indications that oxygen, a characteristic nucleosynthetic signature of massive stars, may have been overproduced at early stages of galactic evolution, relative to the present yield. Oxygen in intracluster gas may be overabundant relative to iron by a factor of 5 (Canizares *et al.* 1982). Moreover, from known mass loss rates, we are reasonably confident that most of the observed intracluster heavy elements must have been produced by stellar mass loss in the early phase of evolution of the cluster galaxies. Oxygen is also found to be enhanced in old, metal-poor halo stars: however, this effect may be interpreted as due to the fact that the chemical yield contribution from lower mass stars had not yet reached a steady state when the old halo stars formed.

## III. GALAXY FORMATION

There are three principal pathways towards a theory of galaxy formation. One approach is to search for traces of the linear regime, including both the anisotropy in the microwave background and the fluctuations in the large-scale mass distribution. This has been fruitful, but has not yet led to very definitive conclusions. A second approach has been to search for protogalaxies, the link

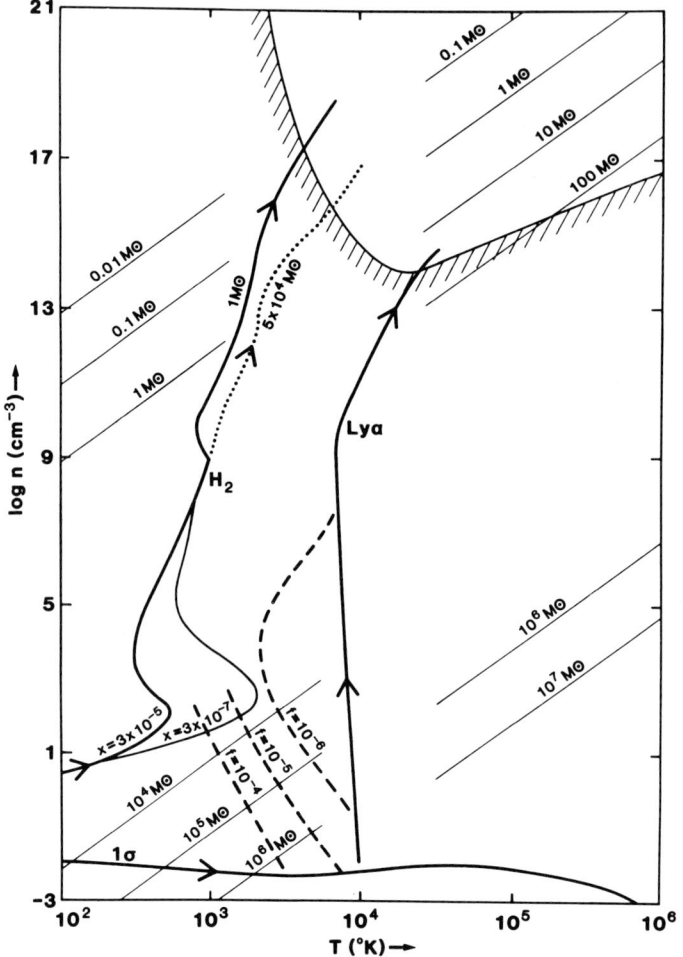

**Figure 1.** Evolution of collapsing cloud in the (n, T) plane. Curves marked $H_2$ and $Ly\alpha$ describe evolution of a spherical cloud in free-fall collapse with primordial abundance of heavy elements for two cases, corresponding to whether $H_2$ formation occurs or is suppressed by photodissociation or shocks. Evolution curves (solid lines) for $H_2$ cooling correspond to residual ionization fractions after recombination $x = 3 \times 10^{-5}$ and $3 \times 10^{-7}$ and for a 1 $M_\odot$ fragment; the dotted line indicates the effects of increased opacity for a $5 \times 10^4$ $M_\odot$ cloud (adapted from Palla et al. 1983). The dashed lines indicate cooling for a given fractional $H_2$ abundance, equal to $f = 10^{-4}$, $10^{-5}$, or $10^{-6}$. The hatched line demarcates the region where quasi-static contraction of a protostar commences. The diagonal lines marked in $M_\odot$ are contours of constant Jeans mass. The $1\sigma$ curve denotes the initial baryon density (taken to be 0.1 of the total density) of structures that have collapsed dissipationlessly from a scale-invariant cold dark matter primordial fluctuation spectrum, normalized to have unit variance on a scale of $8h^{-1}$ Mpc.

between density fluctuations and the mature galaxies that surround us. Despite intensive searches, there is still no identification of a new-born primeval galaxy at high redshift. A third technique that seems rather promising but has not been very vigorously pursued, is to regard galaxies as fossils, and to try to infer their formation history by reproducing their observed characteristics. The hope here is that there is a sufficient number of parameters to be explained that a reasonably unique model can be developed.

The input data consists of the following. Focussing on spheroidal components, as being the oldest and most primitive objects to study, even a brief overview would take note of the surface brightness profiles, the correlations between luminosity, velocity dispersion and metallicity, the metallicity gradients, the rotation or lack thereof, and the high central surface brightnesses of many systems. All of these properties have changed slowly, if at all, over the past $10^{10}$ yr in the life of the average galaxy. Both dynamical and chemical evolution time-scales for spheroidal systems are exceedingly slow, unless a rare merger or tidal interaction between galaxies occurs.

Theoretical discussions of galaxy formation have usually emphasized one of two extreme possibilities. These are that galaxies form by the hydrodynamic dissipative collapse of a parent gas cloud, which subsequently fragments and forms stars, or by the collapse and violent relaxation of a gravitationally unstable system of pre-existing stars. Both approaches have had certain successes, but neither one alone is capable of producing an acceptable model of galaxy formation. Dynamical N-body collapse studies yield elegant explanations of the Hubble-de Vaucouleurs light profiles, and of the low rotation of luminous ellipticals, while failing to result in sufficiently high central densities and metallicity gradients (van Albada 1982). Dissipative collapses seem ideal for metallicity gradients and for obtaining the observed correlations between ellipticals, but seem a somewhat contrived route to the density profiles (Larson 1975).

Evidently, one needs a hybrid model, involving aspects both of dissipative collapse and dynamical relaxation. Motivation for such a model comes from the cosmological arguments. The alternative hot and cold dark matter schemes do not yield galaxy mass clouds, but rather, both theories lead to an agglomeration of small gas clouds of mass $10^6 - 10^8 M_\odot$ that must somehow merge to form a giant galaxy. Additional clues come from observational arguments. The most promising candidates for young galaxies are the least chemically evolved gas-rich systems, of which a prototypical example is I Zw 18. High resolution HI studies show that this galaxy, which is just now undergoing a vigorous burst of star formation for the first time (Kunth and Sargent 1985), is embedded in an extended clumpy cloud of atomic hydrogen. Recently, interacting galaxies have been found by the IRAS satellite studies to be undergoing bursts of star formation, from which we infer that gravitational interaction is responsible for initiating substantial amounts of star formation (Icke 1985; de Jong 1986). Studies of molecular gas and of massive OB star associations in spiral galaxies suggest a similar sequence of events. Preexisting, often molecular, clouds are triggered by spiral density waves or by the magnetic Rayleigh-Taylor instability to undergo collisions and coalescence or possibly assumption. This destabilizes the clouds, which grow in mass, are compressed, collapse and form stars, thereby accounting for the enhanced star formation rates seen in the gas density maxima in spiral arms (Elmegreen 1985).

One is led to a working hypothesis for conditions at the onset of protogalactic collapse, namely that a galaxy forms from an ensemble of many small gas clouds. These clouds can interact dynamically and violently relax during the initial collapse. Cloud collisions will lead to substantial dissipation, and one might reasonably expect such a model to combine the best features of dissipational and dynamical models. If one assumes that cloud collisions trigger or at least enhance the local star formation rate, then a collision-dominated, high surface brightness core will develop, surrounded by a halo where direct encounters between clouds occur rarely (relative to typical orbital times). Attempts to analyze such a model have been made by Silk and Norman (1981) and Carlberg (1984), and it can plausibly account for most of the correlations observed for spheroidal systems.

Whether one believes in the details of such models is a matter of faith, and it is accordingly more effective to give rather less specific arguments. The cooling of gas clouds interacting in a galactic potential well determines whether or not effective dissipation can occur (Silk 1983). The gas interacts at the virial temperature, whether it fills the potential well or is in the form of discrete colliding clouds. In order for effective cooling to occur behind a planar shock, the preshock cloud column density must exceed a critical value

$$\Sigma_{cr} = \rho \sigma t_{cool},$$

where $\sigma$ is the velocity dispersion of the galaxy and $t_{cool}$ is the cooling time-scale of the shocked gas. Since $T \propto \sigma^2$ and $t_{cool} \propto \rho^{-1} f(T)$, one sees that $\Sigma_{cr}$ is a function only of the post-shock temperature, or equivalently of $\sigma$. In order for there to be effective cooling, the cloud column density must exceed $\Sigma_{cr}(\sigma)$. Essentially the identical argument applies for cooling of a collapsing uniform spherical gas cloud (Rees and Ostriker 1977; Silk 1977).

Strong dissipation will result in formation of a dense layer of shocked gas, which, if the compression occurs more or less isotropically, will result in gravitational instability and fragmentation. Cloud coagulation should be effective between clouds of unequal mass, and dominate over cloud disruption (Chieze and Lazareff 1980). Certainly, loss of cloud kinetic energy occurs, and the gas will settle into the core of the potential well. The net effect of a short dissipation time-scale should surely be to strongly enhance the rate of star formation. Once stars form, the dissipation ceases for the stellar component, which freezes out dynamically while the gas continues to dissipate.

One major uncertainty lies in the initial mass function of the newly formed stars. Opinions differ as to whether stars that form out of a cloud of primordial abundance should be very massive (Tohline 1980), or have very low masses, or have a normal mass range appropriate to the solar neighborhood (Palla *et al.* 1983; Silk 1983). Opinions favoring each of these possibilities may be found in the literature. The safest assumption would simply be that the mass range is normal. Indeed the paucity of very low metallicity solar mass stars, which is consistent with $dN/dZ \approx$ constant (Beers *et al.* 1985), favors a formation efficiency of primordial stars with normal masses which is proportional to the mass fraction in heavy elements. Pursuing the implications of this result, one concludes that some massive stars will be among the first generation of stars to form. These will become supernovae, and the enriched mass loss and energy input will affect the ambient primordial gas.

There clearly must be chemical feedback. That there also is dynamical feedback may be inferred from the facts that metal-poor dwarf galaxies must have incurred considerable mass loss and that the intergalactic medium in rich galaxy clusters is highly enriched in metals that must have been ejected from galaxies. Supernova-driven winds are the most likely causes of the enriched ejecta and mass loss. Also, by analogy with our own interstellar medium, stellar energy input must be invoked to heat and energize molecular clouds, thereby preventing them from collapsing in a free-fall time.

Suppose that supernova energy input self-regulates the protogalaxy contraction rate (Silk 1984, 1985). Initially, the mean cooling time will exceed the free-fall time. Stars will form in clouds that can cool, and star formation can continue for up to $(\lambda/\sigma^2)$ free-fall times, where $\lambda \sim 10^{16}$ erg g$^{-1}$ is the energy input rate per unit star formation rate. After a free-fall time has elapsed, a mass fraction $\sim \sigma^2/\lambda$ of the initial cloud has been converted into stars. This implies that if star formation were arrested at this stage, the metallicity of the existing population would vary as $\sigma^2$, and in fact be equal to $\sim y\sigma^2/\lambda$, where $y$ is the yield.

In fact, there is an excellent reason to believe that this is precisely what happens. After a dynamical time elapses, the cooling time for the overall system first becomes equal to the free-fall time. This heralds the onset of greatly enhanced dissipation, and it seems plausible to argue that a burst of star formation is initiated. This will strip most of the remaining gas, driving a wind out of the galaxy, providing that the potential well is not too deep.

Exactly how much gas remains to form stars depends on the cooling efficiency. The higher the heavy element abundance, the more efficient is the cooling, and consequently the greater the resulting dissipative collapse and star formation rates. This in turn produces more supernovae, thereby self-regulating the density of gas that can remain and continue to form stars. If all of the gas at this critical density is converted into stars with a metallicity $y\sigma^2/\lambda$, appropriate to the enrichment at the onset of the vigorous star formation phase, one ends up with a density

$$n \approx 30\sigma_{100}^{0.4} \text{cm}^{-3}$$

where $\sigma_{100} = \sigma/100 \text{km s}^{-1}$, and the total stellar mass in the galaxy satisfies

$$M = 10^{11} \sigma_{100}^{3.8} \ M_\odot.$$

As for the very low mass galaxies, stripping by winds is so effective that essentially no gas is retained to make stars once the winds develop, apart from some gas that might be in dense clouds. One finds that supernova remnants overlap before deceleration and cooling dominate, leaving enough thermal energy in the hot shocked gas in the remnant interiors to drive a wind, if (Dekel and Silk 1985)

$$\sigma \lesssim 65 n^{\frac{1}{22}} E_{51}^{\frac{4}{11}} \text{ km s}^{-1},$$

where $E_{51} \equiv E/10^{51} erg$ is the total energy initially in a supernova explosion, and an efficiency of one supernova per $100 M_\odot$ of gas forming stars has been adopted. It follows that galaxies with $M_B$ fainter than about $-18$ should have been stripped in their protogalactic evolutionary phase, while more luminous galaxies will more nearly approximate closed systems that can undergo continuous self-enrichment. The stripped systems should also have low surface brightness compared to more luminous galaxies.

There are several trends apparent in the observational data that suggest these theoretical speculations may be on the right track (Figure 2). Luminous galaxies have approximately constant density within their half-light radii and satisfy a relation of the form $L \propto \sigma^4$. The predicted normalization agrees reasonably well with the data. More generally, galaxies lie in a regime in the $(n, \sigma)$ plane where cooling and dissipation have been efficient during the formation phase, while galaxy clusters and groups lie outside the cooling regime. The Hubble types are ordered in the $(n, \sigma)$ plane, when spheroids are compared: the luminous, high $\sigma$ spheroids of early type galaxies are distinct from the lower $\sigma$ and surface brightness spheroids of later type systems. The dwarfs form a distinct low density and surface brightness sequence relative to that of normal galaxies. Finally, comparison of oxygen abundances and the ratio of nitrogen to sulphur abundances with galaxy luminosities displays evidence that the low luminosity galaxies have undergone significant stripping (Wyse and Silk 1985). Both of these chemical indicators are constant for low luminosity ($L_B \lesssim 10^9 L_\odot$) disk galaxies, but rise for more luminous galaxies.

How is all of this to be interpreted? The chemical evolution is straightforward. Luminous, massive galaxies undergo self-enrichment by successive stellar generations. Hence nitrogen, which is produced by mass–losing intermediate mass stars ($\lesssim 10 M_\odot$), is mostly produced in this latter stage of gas recycling, while oxygen and sulphur, produced by massive stars, are synthesized whenever star formation occurs. The metal-poor dwarfs produce some N, O, and S before stripping occurs, but only the luminous galaxies can systematically self enrich and thereby develop the observed correlations between metallicity indicators and stellar luminosity. The dwarf surface brightnesses require some dark matter to be present, otherwise the stripping process could be catastrophic. In the context of a cold dark matter theory for formation of galaxies (Blumenthal et al. 1984), the combination of two conditions, that cooling occurs within a dynamical collapse time and that stripping occurs below $\sigma \sim 70$ km s$^{-1}$, leads to the explanation of the location of the dwarfs in the $(n, \sigma)$ diagram as being formed from the typical density peaks in the primordial fluctuation spectrum (Dekel and Silk 1985). The rare peaks collapse earlier, are denser, and many occur at higher velocity dispersion: this means that they can form the "normal", high surface brightness galaxies.

The Hubble sequence segregation that shows up in the $(n, \sigma)$ diagram is highly suggestive that primordial conditions, either before or during protogalactic evolution, rather than environmental influences, are responsible for galaxy morphology. It should be possible, with an improved theoretical framework, to draw evolutionary tracks for protogalaxies, connecting primordial fluctuations with the observed locus of different galaxy types. Whether the discriminants between galaxy types arise at the level of the primordial fluctuations or via wind-stripping

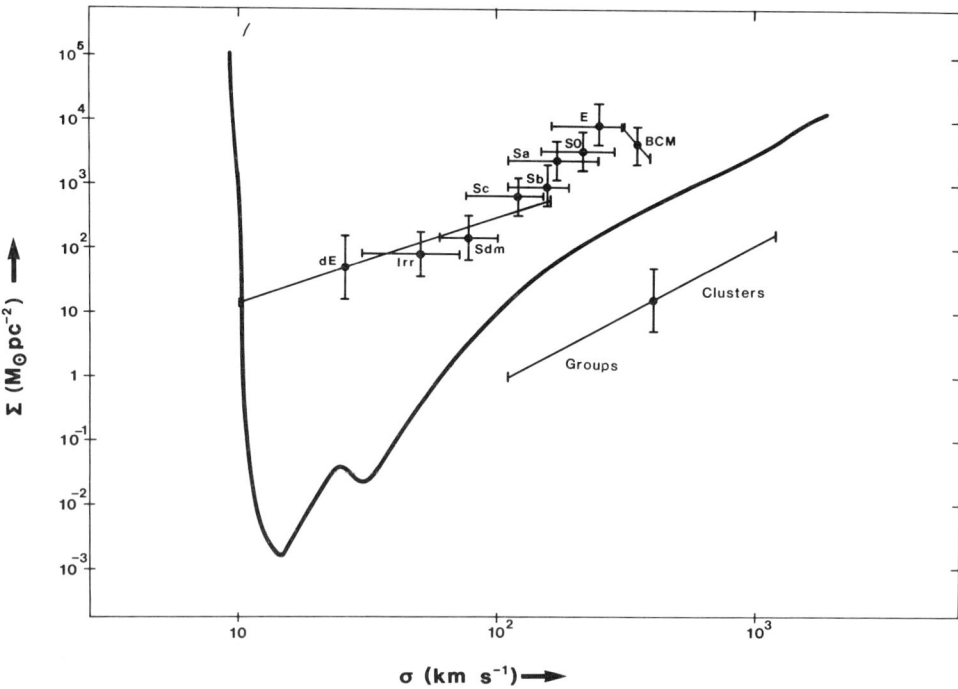

**Figure 2.** Surface density $\Sigma(M_\odot pc^{-2})$ versus velocity dispersion $\sigma(\mathrm{km\ s}^{-1})$ for different morphological types of galaxies and for galaxy groups and clusters. Plotted data is based on statistical correlations between luminosity L and $\sigma$ (for ellipticals and brightest cluster members), and L and maximum rotational velocity (for spirals, SO's, irregulars), L and half-light radius (dwarf ellipticals), and number counts within the central megaparsec radius cores for groups and clusters. The velocity dispersion is the line-of-sight central velocity dispersion for the ellipticals, and the one-dimensional velocity dispersion of a presumed isotropic velocity dispersion for the other systems. All correlations are converted to a $(\Sigma, \sigma)$ relation by assuming that the systems predominantly consist of stars out to the half-light radii, with appropriate mass-to-light ratios. The cooling curve shown is calculated for a collapsing cloud of primordial abundance. From Silk (1985).

during the protogalactic phase is not well understood. While the dwarf-luminous galaxy separation may reflect initial conditions, the more complex differences between luminous galaxies, including bulge-to-disk ratios and specific angular momenta may arise as protogalaxies develop and interact with neighboring protogalaxies in groups of clusters, or with ambient gas that accumulates in their vicinity. For example, continued gas infall can occur more readily for isolated systems, thereby forming disks. It does seem probable that the explanation of the most prominent characteristics of typical galaxies lies more than $10^{10}$ years ago, during the early gas-rich phase of evolution. There are exceptions, notably with the rare mergers occuring today and with instances of cD formation by cannibalism, but one really needs the malleability of a gas-rich system to account for typical galaxy characteristics.

If the cold dark matter scenario is taken seriously, then dwarf galaxies contain the key to the true mass distribution in the universe. For only the dwarfs are the unbiased representatives of the average peaks in the primordial density fluctuation spectrum. However it is finally worth re-emphasizing that, despite the motivation from inflationary cosmology, all is not well in a cold dark matter-dominated universe. The sucesses are several: low $\delta T/T$ (Bond and Efstathiou 1984; Vittorio and Silk 1984), consistency with primordial nucleosynthesis if $\Omega_b \approx 0.1$, and a viable biasing scheme that effectively hides the bulk of the cold dark matter from astronomical measurement, and yields reasonable peculiar velocities for the galaxies and an acceptable galaxy correlation function (Davis et al. 1985). The prime difficulty is that a biased cold dark matter scenario predicts too smooth a universe on large scales. It is difficult to explain the motion of the Virgo supercluster relative to the background radiation frame (Vittorio and Silk 1985), it is difficult to account for the largest voids and superclusters, and it is difficult to account for the enhanced cluster-cluster correlations.

However the reality of all of these effects has been challenged. One would like to verify some of these effects, for example by understanding the flow field in the local supercluster, a crucial step in separating the supercluster motion from the dipole anisotropy, or by searching for evidence of baryons, such as hot gas in the voids, or for the failed galaxies, to verify the reality of the voids. For now, it seems that inflationary cosmology provides an attractive framework for the observable universe.

REFERENCES

Albada van, T., 1982, *M.N.R.A.S.*, **201**, 939.
Beers, T. C., Preston, G.W. and Shectman, S. A., 1985, preprint.
Blumenthal, G., Faber, S., Primack, J. and Rees, M., 1984, *Nature*, **311**, 517.
Bond, J. R. and Efstathiou, G., 1984, *Ap. J. Letters* **285**, L45.
Canizares, C., Clark, G., Jernigan, G., and Markert, T. 1982, *Ap. J.*, **262**, 33.
Carlberg, R., 1984, *Ap. J.*, **286**, 403.
Chieze, J. and Lazareff, B., 1980, *Astr. Ap.*, **91**, 290.
Davis, M., Efstathiou, G., Frenk, C. S., and White, S. D. M., 1985, *Ap J.*, (in press).

Dekel, A. and Silk, J., 1985, *Ap. J.* (in press).
Elmegreen, B. G., 1985, in *Birth and Infancy of Stars*, ed. R. Lucas, A. Omont and R. Stora (North-Holland: Amsterdam, in press).
Icke, V. 1985, *Astr. Ap.*, **144**, 115.
Jong de, T., 1986, *Proc. Erice School on Spectral Evolution of Galaxies.*
Kunth, D. and Sargent, W. L. W., 1985, *Ap. J.* (in press).
Larson, R., 1975, *M.N.R.A.S.*, **173**, 671.
Palla, F., Stahler, S., and Salpeter, E. E., 1983, *Ap. J.*, **271**, 632.
Rees, M.J. 1976,*M.N.R.A.S.*, **176**, 483.
Rees, M. J. and Ostriker, J. P., 1977, *M.N.R.A.S.* **179**, 5411
Silk, J., 1977, *Ap. J.*, **211**, 638.
Silk, J., 1983, *Nature*, **301**, 574.
Silk, J., 1983, *M.N.R.A.S.*, **205**, 705.
Silk, J., 1984, *The Big Bang and Georges Lemaître*, ed. A. Berger (D. Reidel: Dordrecht), p. 279.
Silk, J., 1985, *Ap. J.* (in press).
Silk, J. and Norman, C., 1981, *Ap. J.*, **247**, 59.
Tohline, J., 1980, *Ap. J.*, **239**, 417.
Vittorio, N., and Silk, J., 1984, *Ap. J. Letters*, **285**, L39.
Vittorio, N., and Silk, J., 1985, *Ap. J. Letters*, (in press).
Wyse, R. F. G. and Silk, J., 1985, *Ap. J. Letters*, (in press).

Koo: 1. How do you distinguish between globulars and dwarf ellipticals in your scheme for galaxy formation?

2. What would primeval galaxies, if they exist, look like?

Silk: 1. Globular clusters have core relaxation time–scales that area small fraction of the Hubble time. Hence it is not possible to infer with any confidence the parameters of precursor clouds from which globulars formed. Dwarf ellipticals, on the other hand, are really fossils, whose observed properties were frozen in at formation. Perhaps both could originate from similar clouds; alternatively, globulars could be a secondary phenomenon.

2. More or less like anything you can imagine. For example, today, I can imagine plausible models in which primeval galaxies were exceedingly luminous, and other equally plausible models in which there was no single luminous phase, but rather a whole sequence of inconspicuous mini–starbursts in which the first stars formed.

Aaronson: Suppose it turned out that there was not dark matter after all. How would your galaxy formation picture change?

Silk: Dwarf spheroidals might be hard to explain. Low abundances suggest considerable mass loss: without stabilization by a dark halo it is not clear that very much of a galaxy would survive. However, only a modest amount of dark matter is needed: at least an amount roughly equal to the mean luminous mass.

Dressler: Your picture seems to favor nature over nurture for the generation of the Hubble sequence, since the sequence of spheroids for the different types is already established before the disks are added.

Freedman: If some of the gas is in the form of dense clouds, then it won't be heated and lost within one free fall time.

Silk: I agree.

Dressler: Is there an estimate of the amount of baryonic matter in $Ly\alpha$ clouds of the early universe? Could all of the "failed galaxies" be represented by these clouds, or are these just the next lower group of strong correlations?

Silk: The observed $Ly\alpha$ clouds represent a rather small contribution to $\Omega$, probably around $10^{-4}$. I suspect that these clouds are very fragile and easily destroyed by ionizing radiation from the quasars, since their density contrast is very low relative to the mean density at that time. This suggests that the $Ly\alpha$ clouds have just begun to collapse: the "failed" galaxies must already have formed.

Tarrab: What is the difference in the time scale in forming globular clusters and galaxies?

Silk: This is given, to a zeroth approximation, by comparison of the dynamical time–scales: $10^6 yr$ as opposed to $10^8 yr$. In fact, these are only lower bounds since collapse could have occured much more slowly than free–fall. Also, formation of globulars could be secondary and need not have commenced until after galaxies had begun to form.

Frogel: Will formation of OB stars in the SSPSF mode inhibit the subsequent formation of low mass stars?

Silk: The best guess seems to be that whenever OB stars form, so also do lower mass stars form. However, as in our galaxy, there may have been an extended preceding period of low mass star formation. This would be disrupted by vigorous OB star formation.

Kunth: Is there a way to quantify a bit the self–propagation of massive OB star formation so as to convince oneself that blown away gas will result in a collapse rather than simply disrupt?

Silk: One can show with a linear stability analysis that if the swept–up shell column sensity exceeds a certain critical value, then it will be gravitationally unstable to breaking–up and collapse.

Frogel: Carbon stars in dwarf spheroidals imply star formation over a significant fraction of their lifetime. What are the implications of this for the dark matter requirement in dwarfs?

Silk: Dark matter helps. It provides a gravitational reservoir where primordial gas and gas lost in stellar evolution can accumulate, eventually cool and fall in.

Madore: Since the interaction of star formation with the gas is a critical phenomenon undergoing self–regulation, is it not also possible that if the burst is strong and a great deal of gas is expelled that gas could then return at some later epoch and prolong the star formation phase, possibly allowing carbon stars to form in such systems as Carina?

Silk: Gas accumulation in a halo and later infall is likely for isolated galaxies and for galaxies in loose groups. Galaxy interactions in clusters are likely to heat up the gas and supress late infall. For Carina–like systems, any loosely bound gas would fall in very slowly since the outer potential well is so shallow. This would indeed provide a mechanism for continuing star formation until the gas was finally removed by the interaction with a nearby large galaxy. In the Local Group the latter time–scale is very long and so could allow carbon stars to have formed in dwarf spheroidals.

Mould: The biased galaxy formation picture has a lot of surplus baryonic matter. What form is this in?

Silk: Whatever you cannot rule out! Plausible possibilities include low surface brightness galaxies and hot, x-ray emitting gas.

Mould: We need continuous gaseous infall to keep galactic disks alive. According to Bothun's recent observations, fairly strict limits can be put on envelopes of HI attached to spirals (the suggestion of Larson, Caldwell and Tinsley). I wonder if the left over material in the biased galaxy formation picture could provide this flow of fuel?

Silk: This is an interesting possibility.

PHYSICAL PROCESSES IN STAR FORMATION

Joseph Silk
Department of Astronomy
University of California
Berkeley, CA  94720, USA

I. INTRODUCTION

Star formation occurs close at hand. There are active sites of low mass star formation, and regions of vigorous OB star formation within 500 pc of the sun. Yet star formation processes have not yielded all their secrets. Indeed, not a single genuine protostar has been discovered, despite intensive infrared searches.

In this brief review of star formation processes, I will emphasize aspects that are most relevant for constructing models of galaxy evolution. In particular, I will discuss constraints from observations of molecular clouds, and the initial mass function, large-scale star formation processes, and cloud support and fragmentation.

II. CONSTRAINTS FROM MOLECULAR CLOUD OBSERVATIONS

Cloud properties are summarized in Table 1. Structure is seen over a range from $10^6 M\odot$ down to $1 M\odot$ or less. The smallest scale structure is presently unresolved. Much of the evidence for clumpiness is indirect, and for example is derived from densities required for excitation of $NH_3$ lines. Large clumps can be mapped in nearby molecular clouds, and there is some evidence that they may be rotationally supported. Only for the smallest, coldest clumps do the observed line widths approach the thermal line widths, and even for these, there is significant suprathermal broadening. However for large clumps, especially those in clouds near HII regions, line widths are very broad, with widths amounting up to an order of magnitude larger than the thermal line width. Typical values are $\Delta u =$ 2-3 km s$^{-1}$ in warm clouds, and $\Delta u \sim 1$ km s$^{-1}$ in cold clouds, with the coldest clumps having $\Delta u \sim 0.1$ kms$^{-1}$. For comparison, thermal broadening yields

$$\Delta u = 0.2(T/20K)^{1/2} \text{ kms}^{-1}.$$

Low luminosity embedded infrared sources are seen in many cold clouds, although some clumps appear to contain no internal sources. Luminous infrared sources are embedded in the warm clouds. Bipolar flows are usually associated with these central infrared objects, and one of the principal issues to be discussed below will be whether these flows are capable of energizing the clouds and accounting for the observed line widths.

TABLE 1

Cloud Properties

|  | $Mass(M\odot)$ | $Scale(pc)$ | $Density(cm^{-3})$ | $T(°K)$ |
|---|---|---|---|---|
| HI superclouds | $10^7$ | $10^3$ | 1 | 70 |
| Giant molecular clouds | $10^5 - 10^6$ | 20 - 100 | $10^2$ | 30 |
| Warm molecular clouds | $10^2 - 10^3$ | 1 - 10 | $10^2 - 10^3$ | 30 |
| Cold molecular clouds | $10 - 10^3$ | 0.05 - 20 | $10^3 - 10^4$ | 15 - 20 |
| Clumps | 1 - 10 | $\lesssim 0.01$ | $10^5 - 10^6$ | 10 |

## III. LARGE–SCALE STAR FORMATION PROCESSES

The formation of giant molecular clouds is the initial hurdle that any theory of star formation in the galaxy must first tackle. Such molecular complexes are sites of vigorous star formation, the associated OB associations forming the "beads on a string" that delineate spiral arms.

There are three theoretical approaches to the problem of large–scale star formation (Elmegreen 1985a). Coagulation of small clouds orbiting in the galaxy potential is a slow process; from the observed cloud number per kpc and random motions, one estimates a time–scale of $\sim 2 \times 10^8$ yr for substantial coagulation to occur. However, this process is enhanced by the spiral density wave pattern. The cloud density enhancement stimulates additional cloud collisions, which result in orbital energy loss and in further density enhancement. Tomisaka (1984) finds that up to fifty percent of the molecular gas is concentrated into giant molecular clouds within $\sim 4 \times 10^7$ yr at the solar radius. Star formation is prominent in spiral arms primarily because the cloud density is higher. Grand design spiral arms are presumed to form by this type of mechanism.

A second and not entirely independent theory commences with the magnetic Rayleigh–Taylor (or Parker) instability. The light cosmic ray fluid couples to the galactic magnetic field parallel to the galactic plane, and that configuration is unstable to rippling of the field in the gravitational field of the galaxy. Interstellar gas accumulates between the field undulations over a free–fall time scale ($\sim 2 \times 10^7$ yr) and forms superclouds ($\sim 10^7$ M⊙) that are separated by a distance $2\pi H \sim 1$ kpc, where H is the HI scale height. The density contrast achieved in an initially uniform medium is very modest (less than an order of magnitude), and it is necessary to combine the Parker instability with a cloud coagulation scheme in order to build up giant molecular clouds.

A final mechanism appeals to cumulative effects of many stellar winds in a massive OB association which sweep out a dense shell of ambient gas. The swept–up shell is gravitationally unstable when its column density exceeds a critical value, of order

$$(p/G)^{1/2} \sim 4.10^{-3} n^{1/2} T_4^{1/2} \text{ g cm}^{-2},$$

where the temperature $T_4 \equiv T/10^4 K$ and density n refer to the ambient gas. The critical column density corresponds, coincidentally, to $A_v \approx 1$ in a medium with ambient pressure $p/k = nT \approx 10^4$ cm$^{-3}$ K. The shell fragments are supposed to form more OB stars, thereby providing a means of understanding the sequential formation of OB associations. This also provides a physical basis for the concept of stochastic, self-propagating star formation, a theory which does especially well in accounting for the visual appearance of the more chaotic spirals or irregular galaxies.

## IV. EVIDENCE FROM OPEN CLUSTERS AND ASSOCIATIONS

The study of sites of recent star formation provides a useful probe of conditions during the formation epoch. For example, the color–magnitude diagram for the Pleiades can be used to infer the age spread of cluster members. The nuclear age is inferred, from the main sequence turn–off, to be about 70 million years. The locations of the faintest stars, however, appear to lie slightly above the (poorly known) lower main sequence, and must have formed at least 100 million years ago. This suggestion of an age spread is supported by the fact that some of the faint stars are rapid rotators, and hence must be recent arrivals on the main sequence. We infer that star formation must have continued for at least 30 million years in the Pleiades (Stauffer 1984).

To learn something about the masses of forming stars, we must examine an association in which star formation is still occurring at present. One example of this is NGC 2264, an OB assocation for which isochrone fitting suggests that low mass and massive stars form contemporaneously (Stahler 1985). This conclusion differs from that of earlier analyses, which underestimated stellar ages by not allowing for the fact that the more massive stars are already on the main sequence. Nor is there any strong evidence that the star formation rate rises substantially with time. However, in other OB associations, there is evidence that the most massive stars are $(1-2) \times 10^7$ yr younger than the lower mass stars (Doom et al. 1985).

Study of T-associations, which contain exclusively low mass stars, tells a somewhat different story. The star formation rate allegedly increases with time (Cohen and Kuhi 1979), and over a period up to $\sim 10$ million years, it is apparent that no massive stars have formed. The Pleiades, however, are not deficient in low mass stars. Proper motion surveys have shown that the outer regions of the Pleiades contain many low mass stars, and the initial stellar mass function is similar to the field star IMF (van Leeuwen 1980).

One concludes from these results that low mass stars form exclusively for an extended period, which may last for as long as 10-30 million years. Once massive stars form, low mass stars continue to form as well, but cloud disruption and cessation of star formation occurs after another 10-20 million years. Disruption requires formation of a sufficient number of OB stars which have produced energetic winds and supernovae. Indirect evidence for the duration of the massive star formation phase comes from interpretation of the galactic CO distribution in terms of clouds on ballistic trajectories that are eventually disrupted by OB star formation (Bash 1979).

A recent study of the field star initial mass function also concludes that there is evidence for bimodality (Scalo 1985). There is a feature in the IMF above $1 M_\odot$ which can be best explained if low mass stars form independently of massive stars.

Stellar densities in open clusters and associations yield information on the parent molecular cloud. One of the densest clusters, the Trapezium, has a star density exceeding 300 pc$^{-3}$. This is equivalent to a gas density of about $10^4$ H atoms cm$^{-3}$, comparable to that of molecular cloud cores. Blaauw (1978) was able to use proper motions of stars in expanding sub-associations in Orion to infer the size and shape of the parent molecular cloud. According to Mathieu (1983), the efficiency must have been at least 30 percent in order not to overexpand or even unbind the newly formed star clusters.

Direct estimates of star formation efficiency come from combining optical and infrared data on embedded stars with sensitive CO surveys (Wilking and Lada 1985). Observed efficiencies are as low as 0.3 percent in the $\lambda$ Orionis complex, and range up to 25 percent in the $\rho$ Ophiuchi core and 50 percent in NGC 7023. One can use these efficiency measurements to infer cloud ages, if we assume that star formation proceeds everywhere at the mean rate per unit mass of gas in the solar neighborhood. The local star formation rate is $4 \times 10^{-9}$ M$\odot$ pc$^{-2}$, and the local gas density is 2 M$\odot$ pc$^{-2}$. Hence the age of a cloud is $5 \times 10^8 \, \varepsilon$ yr, where $\varepsilon$ is the observed star formation efficiency. The inferred ages are only $\sim 10^6$ yr for $\lambda$ Ori and are of order $10^8$ yr for NGC 2068, $\rho$Oph, and NGC 7023. The timescales inferred for giant molecular clouds from morphological considerations favor an age of $\sim 10^7$ yr, consistent with the observed efficiencies of a few percent. However, one cannot exclude the possibility that dark clouds have been quiescently forming stars for $3 \times 10^7$ or even $10^8$ yr, nor that the giant molecular complexes are still more long-lived, provided that the efficiency of star formation is sufficiently great. However the lack of giant molecular clouds in the interarm regions does constrain their ages to not exceed $(1-2) \times 10^8$ yr (Elmegreen 1985a).

## TABLE 2

### STAR FORMATION EFFICIENCY AND CLOUD AGES

| Dark cloud Complex | Star Density (pc$^{-3}$) | Efficiency ($\rho_*/\rho_{H2}$) | $t_{cl}$(yr) |
|---|---|---|---|
| $\lambda$ Ori | 0.03 | 0.3 % | $10^6$ |
| Tau-Aur | 0.5 | 2 | $10^7$ |
| NGC 2023 | 15 | 3 | $1.5 \times 10^7$ |
| Tau-Aur clumps | 30 | 10 | $5 \times 10^7$ |
| Chamaeleon | 8 | 12 | $6 \times 10^7$ |
| NGC 2068 | 70 | 20 | $1 \times 10^8$ |
| $\rho$ Oph | 145 | 25 | $1.2 \times 10^8$ |
| NGC 7023 | 14 | 50 | $2.5 \times 10^8$ |

Notes to Table 2:

1. Cloud ages assume universal star formation rate per unit gas mass of $(5 \times 10^8 \text{ yr})^{-1}$.

2. Data from Wilking and Lada (1985).

An alternative approach to inferring statistical ages for molecular clouds comes from counting clouds with and without star formation (Elmegreen 1985b). Since there is about a factor 10 more mass in cold clouds than in warm-centered clouds containing luminous embedded stars (Rowan-Robinson 1979), one infers that the age of the cold clouds must be between $3\times10^7$ and $10^8$ yr. The ages of the star clusters in the starforming clouds lie in the range $3\times10^6$-$10^7$ yr.

The time-scales for molecule formation and chemical evolution of molecular clouds also provide measures of age, in principle (Stahler 1983). The difficulty in applying this approach is that processes of accretion of complex molecules onto grains and of their ejection from grain surfaces are poorly understood.

## V. MOLECULAR FLOWS AND CLOUD SUPPORT

If molecular clouds were in free-fall collapse, the rate of star formation, given the observed efficiencies, would vastly exceed the observed rate. Moreover, clouds are not in pressure balance: the pressure in a molecular cloud core exceeds $p/k \sim 10^5$ cm$^{-3}$ K, far greater than that of the diffuse interstellar medium. Hence dispersal would occur in $(L/1\text{pc})$ $(1\text{km/s}/\Delta u) \sim 10^6$ yr unless molecular clouds were gravitationally bound.

The role of gravitational support is confirmed by the well-studied correlations (Larson 1981, Myers 1983) between line width and cloud scale: $\Delta u \propto R^{1/2}$, and density and scale: $n \propto R^{-1}$. Together, these correlatons imply that $(\Delta u)^2 \sim GM/R$, confirming that clouds are in approximate virial equilibrium. Possible sources of support against collapse include rotation and magnetic fields, although the former is probably too ordered to fully account for cloud longevity. Nevertheless, there is accumulating evidence for rotation in cloud cores (Vogel 1985). The situation with regard to magnetic support is much more uncertain, since there are few measurements of magnetic fields in dense regions, and the measurements are restricted primarily to Zeeman splitting of the 21 cm line of atomic hydrogen, at best a very minor constitutent. The rotating cloud cores are unlikely to be magnetically supported, otherwise Alfven wave propagation would have removed most of the angular momentum, although diffuse cloud envelopes may well be supported by magnetic stresses. The ionization is probably so low in cloud cores that magnetic fields will have undergone ambipolar diffusion: however they will remain dynamically significant in cloud envelopes.

The key to cloud support and longevity undoubtedly lies in understanding the origin of the suprathermal line widths. These yield enough dynamical pressure to give support against collapse. Yet supersonic turbulence would dissipate very rapidly, within a cooling time, or in much less than the collapse time-scale. A promising source of internal energy input that may be capable of keeping clouds from collapse consists of premain sequence objects generating strong winds (Norman and Silk 1980). This phenomenon manifests itself as high velocity often bipolar, molecular gas flows around the embedded objects, themselves usually seen only in the far infrared. More than 50 such flows have now been discovered within 2 kpc of the sun. Table 3 summarizes the relevant data for those flows with adequate maps and central sources, and gives the rate of momentum input as well as the formation rate, as a function of the bolometric luminosity of the central source.

## TABLE 3

### MOLECULAR FLOWS

| (1) | (2) | (3) | (4) | (5) | (6) | (7) |
|---|---|---|---|---|---|---|
| $L_*$ | N | $<D>$ | $<\dot{M}V>$ | $\pi f$ | $\dot{N}f$ | $<\dot{M}V>t_{cl}f$ |
| $1 \leq L_* \leq 10$ | 5 | 0.3 | 1.6(-4) | 11 | 4(-4) | 1.8(4) |
| $10 \leq L_* \leq 10^2$ | 8 | 0.38 | 12.8(-4) | 17 | 3.5(-4) | 1.8(4) |
| $10^2 \leq L_* \leq 10^3$ | 7 | 0.62 | 9.1(-3) | 2.6 | 3.6(-5) | 2.4(5) |
| $10^3 \leq L_* \leq 10^4$ | 10 | 0.97 | 5.8(-2) | 1.1 | 2.3(-5) | 6.3(5) |
| $10^4 \leq L_* \leq 10^5$ | 6 | 1.1 | 1.1(-1) | 1 | 7.1(-5) | 1.1(6) |

Notes for Table 3

Column (1): Bolometric luminosity of central source in $L_\odot$
Column (2): Number of mapped flows
Column (3): Mean distance in kpc
Column (4): Mean momentum flow rate in $M_\odot yr^{-1} km\, s^{-1}$
Column (5): Incompleteness factor out to 1kpc ($f \equiv <D>^{-2}$)
Column (6): Birth rate in $yr^{-1}\, kpc^{-2}$
Column (7): Momentum ($M_\odot\, km\, s^{-1}\, kpc^{-2}$) injected over cloud lifetime, with $t_{cl}=10^7 yr$.

Data sources: Lada (1985); Edwards and Snell (1984); Levreault (1984).

The formation rate of the high velocity flows is $\sim 10^{-3}\, yr^{-1}\, kpc^{-2}$. This is comparable to the solar neighborhood star formation rate inferred from stellar associations, and suggests that essentially all newly forming stars produce molecular outflows. The momentum input over the $\sim 10^7$ yr lifetime of cold dark clouds amounts to $\sim 2 \times 10^6\, M_\odot\, km\, s^{-1}\, kpc^{-2}$. Comparing this with the local gas density of 2x10⁶ M☉ kpc⁻², we infer that the molecular flows can maintain $<\Delta u> \sim 1$ km s⁻¹$(t_{cl}/(10^7$ yr$)$ per unit cloud mass. This is probably sufficient to account for observed linewidths, which amount to

$$\Delta u_{obs} \approx 0.7(R/1pc)^{1/2} \text{ km} s^{-1}$$
$$\approx 0.5(t_{cr}/10^6 yr) \text{ km} s^{-1}$$

where $t_{cr}$ is the eddy crossing time $(R/\Delta u_{obs})$. Hence if clouds can avoid internal dissipation for $\sim 5 \, t_{cr}$, the energy input from the flows suffices to stir up the moleuclar clouds.

Now dissipation over $t_{cr}$ occurs only if the geometrical clump cross-section is relevant for deceleration. The most likely means of achieving the required degree of elasticity in cloud internal motions is to boost the effective cross-section via appeal to magnetic fields. According to Clifford and Elmegreen (1983), entangling of magnetic flux tubes increases the effective collision cross-section between magnetized clumps over the geometric cross-section by a factor $F \equiv (u_A n_c / \bar{n} \, \Delta u)^{2/3}$ for internal clump density $n_c$, mean cloud density $\bar{n}_c$, internal Alfven velocity $u_A$, and clump velocity $\Delta u$. Magnetic fields act to enhance the momentum transfer between clumps, and thereby increase the effective collision cross-section. Now the interstellar magnetic field is consistent with an approximate scaling as $n^{1/2}$, and a normalization $u_A = 6$ km s⁻¹ that should be approximately valid in clouds. Hence

$$F = 15(\delta/10\Delta u_1)^{2/3},$$

where $\delta$ is the clump density contrast relative to the cloud and $\Delta u_1 \equiv \Delta u/1 \text{km s}^{-1}$. Thus one seems to have available the necessary boost in efficiency of momentum transfer to allow the observed molecular flows to energize clouds and account for molecular line widths, and hence for cloud support against free-fall collapse.

## VI. CLOUD EVOLUTION AND STAR FORMATION

The characteristic scale of energy input associated with protostellar outflows is given by momentum balance: a spherical wind flowing at rate $\dot{M}$ and velocity $V_w$ into a uniform medium of pressure $p_o$ sweeps out a shell of radius

$$R = (\dot{M}V_w/4\pi p_o)^{1/2} \approx 0.1 pc$$

for typical parameters ($\dot{M}=10^{-6}$ M$\odot$ yr$^{-1}$, $V_w=300$ kms$^{-1}$, $p_o/k=10^5$ cm$^{-3}$K). In order for the shell velocities to be supersonic, the collision time between adjacent shells must be less than the duration of the wind phase. Now the effective collision velocity between decelerated shells is given by the relative velocities of the central protostars, and the condition for supersonic stirring reduces to a lower bound on the density of outflows:

$$n_w \gtrsim 3V_{*,1}V_{w,300}n_3\left(\frac{0.1}{\delta M_*/M_*}\right) \text{pc}^{-3},$$

where $V_{*,1} \equiv V_*/1\text{km s}^{-1}$ is the stellar velocity dispersion in the cloud, $V_{w,300} \equiv V_w/300$ km s$^1$ is the wind velocity, $n_3 \equiv n/10^3$ cm$^{-3}$ is the ambient density, and $\delta M_*/M_*$ is the mass fraction of the exciting protostar that is expelled in the outflow.

For comparison, the number density of T-Tauri stars in the Taurus-Auriga clumps may exceed 30 pc$^{-3}$, while in the $\rho$ Oph core, the frequency of premain–sequence objects is $\sim 150$ pc$^{-3}$. Only a few percent of these objects need be actively driving winds, and the inferred cloud lifetime is $(\rho/\rho_{*,w})t_w \sim 100\ t_w$, where $\rho_{*,w}$ is the mass density of stars with winds, $\rho$ is the total mass density available for forming stars, and $t_w$ ($\sim 10^6$yr) is the duration of the wind phase.

The wind–driven shells may be dynamically or gravitationally unstable prior to intersection. Certainly, intersection should result in clump formation. It is likely there will be a wide range of clump masses, and the overall evolution of the clumps is necessarily rather uncertain. Supersonically moving clumps will collide and sweep up ambient gas, as well as expand because of the inefficiency of ram pressure confinement. Typical clump masses are in the range 0.1-1M$\odot$. If clump coagulation rather than disruption dominates, then it is only a matter of time before clumps become gravitationally unstable. For a given external pressure $p_o$ the critical mass above which collapse of an isothermal clump occurs is

$$M_{cr} = 4.7(T/20K)^2\left(\frac{p_o}{k}/10^5\text{cm}^{-3}\text{K}\right)^{1/2} M\odot.$$

If the ambient pressure increases sufficiently, even low mass clumps are induced to collapse. This increase in pressure will be inevitable as more and more stars form.

Suppose that the dominant process in triggering instability is pressure enhancement rather than the relatively uncertain competition between clump coagulation and leakage or disruption. Now $p_o$ is proportional to the momentum input rate per wind, to the number density of winds ($n_w$), and to the mean separation between winds ($\propto n_w^{-1/3}$). Hence $p_o \propto n_w^{2/3}$, whence $M_{cr} \propto n_w^{-1/3}$. It follows that more active protostar formation induces higher pressure, which in turn results in lower mass clumps being induced to collapse. Therefore low mass star formation should be a stable, self-sustaining process. This is envisaged to be the case for cold dark clouds.

Next, consider a situation in which clump coagulation is suddenly enhanced. This could be the primary effect of passage of a shock wave, due to propagation of a nearby ionization front or collision with another cloud, for example. Direct evidence for enhanced clump motions near HII regions comes from a recent study by Wilson (1985): such motions will enhance the clump collision rate. Collisions between map out clumps of unequal masses are likely to result in coagulation, and some very massive clumps should accumulate. Again, these will be gravitationally unstable, and form massive stars. Such stars are capable of triggering clump collapse both by dynamical interaction, such as radiatively driven implosion, or by their cumulative heating of molecular gas throughout the cloud. Once stars form that are capable of significantly enhancing the local radiation field, the dust and gas heating substantially raises the critically unstable mass–scale, which is proportional to $T^2$ at fixed pressure or optical depth. This means that once massive stars begin to form, there is sufficient feedback into thermal pressure gradients within clumps that massive star formation should continue. Indeed, a simple version of this argument yields an initial stellar mass spectrum proportional to $M^{-2.3}$ (Silk 1977).

One can even pursue this scheme further: if the critical clump mass spectrum is proportional to $m^{-\nu}$ with $\nu > 2$, most of the mass reservoir is in low mass clumps, and coagulation results in a steady rate of formation of clumps of critical mass which form massive stars. However if $\nu < 2$, the clump mass distribution is dominated by the most massive clumps, most of the mass reservoir is rapidly exhausted, and a burst of massive star formation ensues. Needless to say, an interesting test of this scheme will eventually be to map out the clump mass distribution with high resolution maps of molecular clouds.

## VII. FRAGMENTATION

In the course of this brief review, I have chosen to say rather little about one topic which has been the focus of many previous discussions of star formation, namely hierarchical fragmentation during cloud collapse. I have taken the viewpoint that fragmentation and star formation is largely a secondary process, driven by prior generations of protostars and mass outflows. Of course, this neatly sidesteps the non-trivial question of how the first stars were formed. Fragmentation during cloud collapse provides a possible resolution of this issue. The analytic theory is mostly based on the hierarchical fragmentation scheme of Hoyle (1953), and follows fragmentation during isothermal collapse to successively smaller scales. The objections of Layzer (1964) to this scheme are still considered valid, and the principal utility of later discussions has been to set a lower bound on the fragmentation scale. Neglecting pressure and assuming spherical symmetry, this turns out to be about 0.01 M$\odot$, although more realistic treatments suggest this value should be increased by an order of magnitude (Silk 1980).

Three dimensional numerical hydrodynamical simulations of collapsing clouds include pressure and rotation, but have inadequate spatial resolution to follow much more than one generation of fragmentation. Provided that the initial fluctuation amplitudes are large enough, multiple fragmentation occurs during the dynamical collapse phase despite the inhibiting presence of gradients (Rozyczka 1983). Most of the fragment growth takes place when the collapse is slowed by pressure gradients or rotation. Within the limited dynamical range followed, the fragments are found to contain many Jeans masses and low specific angular momentum (Bodenheimer et al. 1980).

Extension of the analytic theory of fragmentation to primordial clouds, necessary to properly address the issue of how the first stars formed, has led to conflicting viewpoints. On the one hand, Tohline (1980) concludes that primordial stars were extremely massive. If the preexisting fluctuations on stellar mass–scales were of order $\delta$, then the minimum Jeans mass of $M_J^{min}$ in spherical collapse of a warm cloud is enhanced by a factor of order $\delta^{-1}$ relative to that in a cold cloud. This is because linear fluctuation growth is retarded because of thermal pressure gradients until the initial Jeans mass of the cloud is sufficiently small. Silk (1982) showed that anisotropic collapse would be more favorable, collapse to thin sheets or pancakes resulting in an enhancement of $M_J^{min}$ by only a factor of $\delta^{-1/2}$. One might expect that fluctuations would be present on all scales once even a few massive stars had formed; hence this issue of enhancement of $M_J^{min}$ by finite pressure effects is likely to be irrelevant.

A more serious concern is the actual value of $M_J^{min}$ attained in collapse of a primordial cloud. Early estimates suggested that $M_J^{min}$ should scale as $T^{1/4}$, and hence be an order of magnitude larger in primordial clouds than in clouds of solar abundance (Rees 1976, Silk 1977). However incorporation of molecular hydrogen cooling at very high density, where $H_2$ is formed by three–body combination of H atoms ($3H \rightarrow H_2 + H$), has shown that $M_J^{min}$ is of order 0.01 M$\odot$ in primordial cloud collapse, and hence indistinguishable from its value in solar abundance clouds despite the very different opacities and cooling rates (Palla et al. 1983). There are some differences: thermal instability may play a role (Silk

1983), and if $H_2$ is destroyed, for example in shocks, the ensuing Lyman alpha cooling will ensure much larger values of $M_J^{min}$ (Silk 1985). There are possible indications of a phase of enhanced massive star formation during early galactic evolution (e.g. Larson 1985), but the uncertainties in theoretical modelling are such that one could derive equally viable scenarios with solar neighborhood initial mass function that applied even in primordial clouds. These issues are addressed in more detail in my following lecture.

## REFERENCES

Bash, F. 1979 *Ap.J.*, **233**, 524.
Blaauw, A. 1978, *Problems of Physics and Evolution of the Universe* (Yerevan: Armenian Academy of Sciences), (p. 101).
Bodenheimer, P., Tohline, J.E. and Black, D.C. 1980, *Ap.J.*, **242**, 209.
Clifford, P. and Elmegreen, B.G. 1983, *M.N.R.A.S.*, **202**, 629.
Cohen, M. and Kuhi, L. 1979, *Ap.J. Suppl.*, **41**, 743.
Doom, C., de Greve, J.P. and de Loore, C. 1985, *Ap.J.*, **290**, 185.
Edwards, S. and Snell, R. 1984, *Ap.J.*, **281**, 237.
Elmegreen, B.G. 1985a, in *Protostars and Planets II*, ed. D. Black and M. Mathews (University of Arizona Press: Tucson) (in press).
Elmegreen, B.G. 1985b, in *Birth and Infancy of Stars*, eds. R. Lucas, A. Omont, R. Storer (Amsterdam: North–Holland) (in press).
Hoyle, F. 1953, *Ap.J*, **118**, 513.
Klein, R.I., Sandford, M.T. and Whitaker, R.W. 1983, *Ap.J. Letters*, **271**, L69.
Lada, C.J. 1985, preprint.
Larson, R. 1985, *M.N.R.A.S.* (in press).
Larson, R.B. 1981, *M.N.R.A.S.* **194**, 809.
Layzer, D. 1963, *Ap.J.* **137**, 351.
Levreault, R. M. 1984, *Ap.J.*, **277**, 634.
Mathieu, R. 1983, *Ap.J. Letters*, **267**, L97.
Myers, P. 1983, *Ap.J.* **270**, 105.
Norman, C. and Silk, J. 1980, *Ap.J.*, **238**, 158.
Palla, F., Stahler, S. and Salpeter, E.E. 1983, *Ap.J.*, **271**, 632.
Rees, M.J. 1976, *M.N.R.A.S.*, **176**, 483.
Rowan-Robinson, M. 1979, *Ap.J.*, **234**, 111.
Rozyczka, M. 1983, *Astr. Ap.*, **125**, 45.
Scalo, J. 1985, preprint.
Silk, J. 1982, *Ap.J*, **256**, 514.
Silk, J. 1977, *Ap.J.*, **211**, 638.

Silk, J. 1980, in *Star Formation*, eds. A. Maeder and L. Martinet (Geneva: Geneva Observatory), p 133.
Silk, J. 1983, *M.N.R.A.S.*, **205**, 705.
Silk, J. 1985, *Ap.J.*, (in press).
Stahler, S. 1983, *Ap.J.*, **274**, 822.
Stahler, S. 1985, *Ap.J.*, **293**, 207.
Stauffer, J. 1984, *Ap.J.*, **280**, 189.
Tohline, J. 1980, *Ap.J.*, **239**, 417.
Tomisaka, K. 1984, *P.A.S.J*, **36**, 457.
van Leeuwen, F. 1980, in *Star Clusters*, ed. J.E. Hesser (D. Reidel: Dordrecht), p. 157.
Vogel, S. 1985, preprint.
Wilking, B.A. and Lada, C.J. 1983, *Ap.J*, **274**, 698.
Wilking,B.A. and Lada, C.J. 1985, *Protostars and Planets II:* ed. D. Black D. and M. Mathews (University of Arizona Press: Tucson), (in press).
Wilson, T. 1985, preprint.

Eisenhardt: Is the time of formation of low and high mass stars in your model consistent with observations of NGC 2264?

Silk: The reanalysis of the NGC 2264 H-R diagram, with isochrone fitting that allows for the premain sequence contraction phase indeed suggests that this cluster did not experience an extended period of exclusively low mass star formation. One caveat is that this result presumes a coeval origin of the contraction phase. Certainly, however, over the past $10^6 yr$, it has been forming both massive and low mass stars. A successful model should allow both this possibility and the formation of exclusively low mass stellar associations.

Pickles: With respect to the onset of high mass star formation and disruption of molecular clouds...Is it possible for some other process to lead to cloud disruption, thus resulting in lower mass star formation only?

Silk: I cannot think of any obvious candidate, apart from simple exhaustion of the gas supply.

Eisenhardt: It's interesting to note that Rieke et al. (1980) in a study of M 82 find a mass function weighted towards high mass stars in this starburst galaxy.

Silk: One problem in interpreting a flat massive star–dominated IMF is that my simple theoretical argument about star formation rates relies on the mass distribution of gas fragments that are destined to become stars. Since I do not know with any precision the fraction of fragments that make either no stars or more than one star, it is not easy to draw any quantitative inference about the origin of the IMF slope.

Mould: A few years ago it seemed that the subject of the IMF would be observationally driven by star counts in young LMC globular clusters. Only a few clusters were studied by Freeman and the data were not published. But a survey of young clusters IMFs could lead to correlations with mass, density, etc.

Silk: This would be very interesting to pursue, especially the dependence of cluster IMF on kinematics and metallicity.

Koo: Theoretically, at what densities would dust form?

Silk: Dust forms wherever there is substantial mass loss from cool evolved stars, M giants or supergiants. However, dust has no effect on the cooling of primordial clouds if the heavy element abundance is sufficiently low, below about $10^3 Z_\odot$.

de Jong: Your discussion of the slope of the mass spectrum of fragments in a cloud suggests that in galaxies or clouds experiencing a burst of star formation the mass spectrum should have a flatter slope.

Pickles: What is the effect of chemical inhomogeneities on fragmentation in contracting protostellar clouds?

Silk: Chemical inhomogeneities may play a role in driving fluctuations by thermal/chemical instability.

Di Fazio: Many authors (like, e.g., Palla, Salpeter and Stahler, 1984), in their attempt to predict the minimum mass produced in the first star formation process, devote their efforts to the thermodynamics (cooling the molecules and thus enhancing the cooling), while they simply assume a given collapse of the mother cloud, so obtaining a time–increasing density. Now, we must stress that, to be able to follow meaningfully the Jeans mass in time, we should not push the computations too far ahead in time, that is, we should stop at $t \approx$ Jeans time. In fact, we know that gravitational instability boosts up the fragmentation rate when the Jeans mass is falling (and this is obtained, e.g., by a very good cooling, inducing isothermality, plus a time–increasing density). But now, if the fragmentation rate is increasing, we know that, at about one Jeans time from the beginning of the collapse of the unstable mother cloud, the gas is exhausted. From this time on (actually, from a little before) the Jeans mass cannot be calculated letting simply: not $\rho$ = total density as now we have but instead $\rho$ = total density minus density in fragments, and this is almost zero (or is tending to zero). Thus, even with exceptionally good cooling inducing isothermality, even a free–fall (increasing total density) eventually shows a stop in the $M_J$ decrease (with subsequent reincrease) due *not* to thermodynamical reasons, but to gas exhaustion. So, if one has no time dependent fragmentation rate function available, the Jeans mass time–evolution should not be pushed over one Jeans time, otherwise the predictions will be meaningless. Can you comment about this, please? Also, incidentally, where are your predicted zero–metal small mass Pop III stars? They should be all around us as stars of high velocity (halo component), like the Pop II stars. And, to make an example, a $0.8 M_\odot$ or $1 M\odot$ Pop III star should still be alive from halo formation times, and it should be detectable just like a good ol' honest Pop II star.

Silk: 1. I think that the fragmentation efficiency must be strongly limited by non-linear fragment interactions. Hence I see no reason why there should not always be a continuing reservoir of material of increasing density, capable in principle of producing fragments of decreasing mass, until opacity effects inhibit continued cooling and thereby stabilize the collapse against fragmentation.

2. If Population III stars are defined as being those which produced the observed extreme population heavy elements, the number required is very small, only amounting to $dN/dZ \approx$ constant. Of course this assumes several generations as Z builds up from zero to its old Pop II value. Even if population III had a Salpeter IMF, this is consistent with the observed numbers of very low or zero metallicity stars.

Occhionero: Can we play in star formation theory the dark matter game of galaxy formation? If 1 GeV photinos were around, then solar size condensation seeds would be available. Are they useful in triggering collapse or coagulation?

Silk: The average dark matter density is so low in the disk of our galaxy compared to the density of a molecular cloud that I cannot imagine how clumpiness of the dark matter could have any significant impact.

Pickles: What is the global efficiency of star formation?

Silk: Looking at giant molecular cloud complexes, I would guess around one or two percent. But allowance for low mass star formation, seen to occur in cold cloud cores, would increase this to 10-20 percent.

Eisenhardt: Isn't a critical parameter in the stochastic self propagating SF models the probability of inducing SF in a neighboring region? Spiral structure results only from a narrow range in this parameter.

Dressler: Isn't it true that the greatest challenge for SSPSF is to explain the great symmetry of the spiral pattern in some galaxies? This seems to point to a global mechanism as well.

Silk: I absolutely agree. The spiral density wave induced by tidal interactions with a neighbor combined with a cloud coagulation scheme or collision scheme for triggering star formation probably provides the best mechanism to account for the "grand design" spirals.

# THEORETICAL AND OBSERVATIONAL EVIDENCE FOR THE EXISTENCE OF A SINGLE FRAGMENTATION LAW, AND GALACTIC EVOLUTION.

A. Di Fazio: Osservatorio Astronomico di Roma, Viale del Parco Mellini 84, 00136 - Roma, Italy
R. Capuzzo Dolcetta: Istituto Astronomico, Università 'La Sapienza', via G. Lancisi 29, 00161- Roma, Italy

ABSTRACT. Mass functions for various objects are elaborated from observational data. Their shapes are compared among themselves quantitatively. A striking similarity suggest the hypothesis that all the different kinds of astrophysical objects considered were formed by essentially the same process. Theoretical mass functions, based on the hypothesis of fragmentation due to gravitational instability (Di Fazio, 1985), are compared with the observations, yielding a remarkably good agreement.

I. Theoretical modeling of fragmentation.

Following Di Fazio (1985) and taking the gravitational instability as the relevant process for the fragmentation of a self-gravitating gas cloud, the following expression for the differential of the number of perturbations of wavevector k is obtained:

$$dN = |4\pi k^2 dk|/(2\pi)^3 dVdt/\tau(k) \tag{1}$$

where $\tau(k) = 2\pi/|\omega(k)|$ is the time scale for the growth of mode k. $\omega(k)$ comes from the dispersion relation (in absence of angular momentum and magnetic field):

$$\omega^2(k) \equiv k^2 c_s^2 - 4\pi G \rho < 0, \tag{2}$$

where this inequality is the condition for the growth of the perturbation (see Jeans, 1902, 1928; Weinberg, 1972). Proceeding as in Di Fazio (1985), one obtains a time dependent mass function describing the spectrum of objects created up to time t by gravitational instability only. This corresponds to the strongest definition of IMF and it is:

$$\phi(M,t)dM \equiv dn = AdM \int_{t_0}^{t} M^{-2} \rho^{3/2} \sqrt{1-(M_J/M)^{2/3}} \, dt' \qquad (3)$$

$A = \pi^{3/2}/36$ is a constant, n is the number density of fragments existing at time t, with mass between M and M+dM, created by gravitational instability alone. A fragmentation rate spectrum can be obtained:

$$(\partial \rho_{frag}/\partial t)_{M,M+\Delta M} \equiv \frac{\partial}{\partial t} \int_{M}^{M+\Delta M} M\phi(M,t)dM = 3A\sqrt{G}\, \rho^{3/2} \times$$

$$\times \left[ \sqrt{1-(M_J/M)^{2/3}} - \sqrt{1-(M_J/(M+\Delta M))^{2/3}} - \log \left| \frac{1+\sqrt{1-(M_J/M)^{2/3}}}{(M_J/M)^{1/3}} \right| \times \right.$$

$$\left. \times \frac{(M_J/(M+\Delta M))^{1/3}}{1+\sqrt{1-(M_J/(M+\Delta M))^{2/3}}} \right] \qquad (4)$$

(see Di Fazio, 1985). The total fragmentation rate, i.e. $(\partial \rho_{frag}/\partial t)_{M_J,M_{gas}}$, can be compared with Schmidt's (1959) empirical law. We have: $(\partial \rho_{frag}/\partial t)_{M_J,M_{gas}} = 3A\sqrt{G}\, \rho^{3/2} \times$

$$\times \left[ \log \left| \frac{1+\sqrt{1-(M_J/M_{gas})^{2/3}}}{(M_J/M_{gas})^{1/3}} \right| - \sqrt{1-(M_J/M_{gas})^{2/3}} \right] \qquad (5).$$

The explicit density dependence of the fragmentation rate is $\rho^{3/2}$, thus, already lower than Schmidt's (1959) $\rho^2$. Moreover, the function in brackets, depending on mass and Jeans mass, shows a cut-off at $M_{gas} \leq M_J$. Forcing a $\rho^\alpha$ dependence to simulate the actual fragmentation rate, taking into account the time-average values of the function in brackets, one obtains $0.4 \leq \alpha \leq 1.5$ depending on the evolution models, i.e. substantially lower than Schmidt's (1959) empirical value 2. This is in agreement with more recent observational determinations of the density dependence of the fragmentation rate (see e.g. Madore, 1977, Miller and Scalo, 1979, and Lequeux, 1980). The mass dependence of the integrand in (4) is shown in Fig. 1. Fig.2 shows the ratio of the Di Fazio fragmentation rate to Schmidt's, divided by $\rho^{-1/2}$, vs. $x \equiv M_J/M_{gas}$.

II. Evolution models and comparison with observations.

A dynamical, radiative, 2-fluids evolutionary model is needed in order to have the time evolutions of the state variables to be used in the initial mass function (IMF) (3).

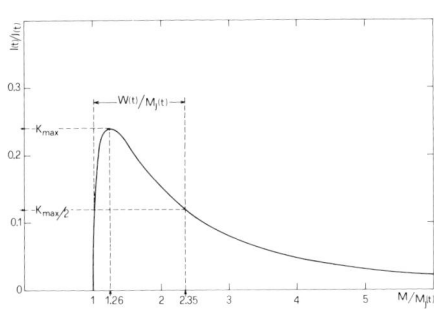

Fig. 1
The integrand of the IMF (3) in dimensinless units. $J(t) = \rho^{3/2} M_J^{-2}$.

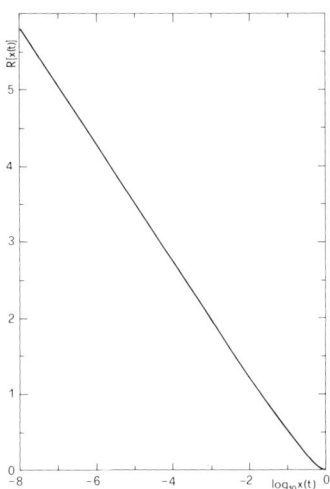

Fig. 2
The ratio of the Di Fazio fragmentation rate to Schmidt's, in units of $\rho^{-1/2}$.

This model, presented by Di Fazio (1982) and then furtherly developed (Di Fazio , 1985), treats the evolution of a self-gravitating object made out of two phases, namely, gas and fragments. The two fluids are coupled via energy, mass, momentum conservations and through the fragmentation law (4). The radiative treatment uses the same scheme described in Di Fazio and Palla (1981). The chosen initial conditions correspond to an initially pure gas cloud, of mass $M = 10^{11} M_\odot$, temperature $T = 100°K$, radius $R_0 = 50$ Kpc, i.e. typical of an average unstable protogalactic cloud. The evolution is followed until the gas exhaustion induced by the fragmentation process causes the Jeans mass to start rising again (notwithstanding a substantially isothermal behaviour) up to the point of equivalence $M_J = M_{gas}$. At that point, in fact, the fragmentation process stops (as can be seen from (5)) and the attained mass spectrum represents the first fragment generation. Starting from this point, one can consider the subsequent evolution of the new-born fragments by taking the most abundant representative one, with a mass corresponding to the peak of the mass function. The new initial conditions can be assigned as the final conditions of the previous mother-cloud evolution and so on, up to the attainment of a generation of fragments that is stable against further fragmentation.

Figs. 3 to 6 show the four generations obtained.

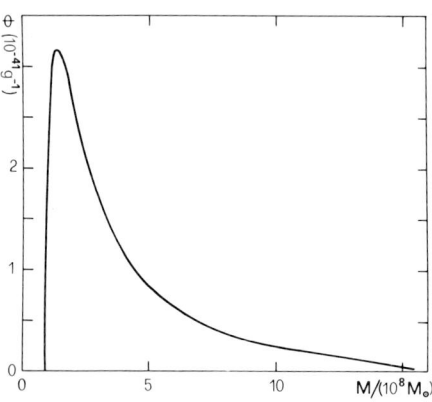

Fig. 3
The first generation mass function per unit mass. In abscissa is the mass.

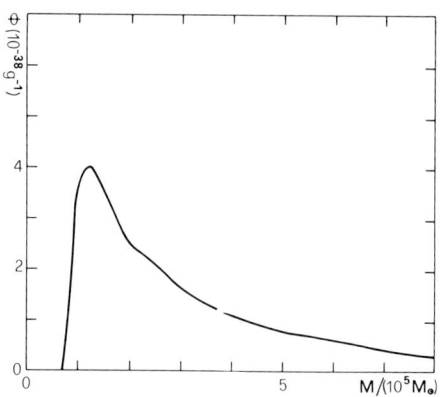

Fig. 4
The second generation mass function. It peaks at a typical globular cluster value.

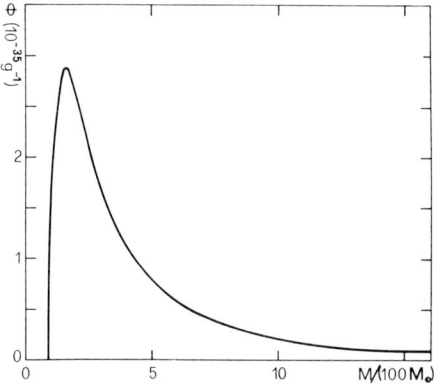

Fig. 5
The third generation mass function. In its range fall the first stars.

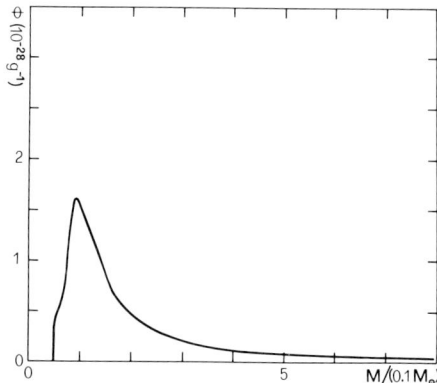

Fig. 6
The 4th generation mass function, representative of the first metal enriched stars.

They represent, respectively, a family of giant gas clouds, a family of protoglobular clusters, a family of objects that could be interpreted as pop.III giant nuclear burners, and "normal" stars. As an example of the time behaviour of the state variables of one of these evolutions, we display vs. time in Figs. 7 ÷ 9, respectively, i) the total radii, ii) the gas temperature and the velocity dispersion of the fragments,

iii) the fragmentation rate and the Jeans mass.

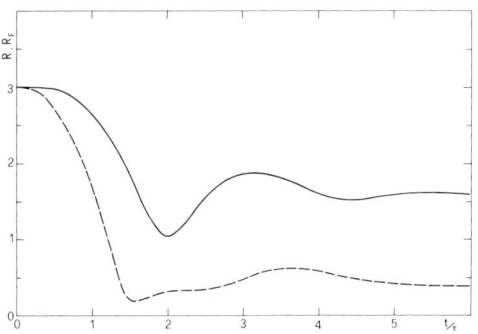 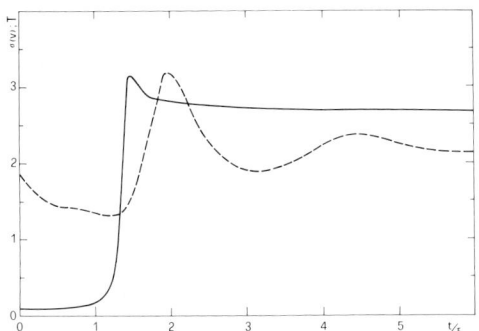

Fig. 7 The time evolution of the radii of the gas cloud (dashed line) and of the cloud of fragments (continuous line) in units $10^{20}$ cm.

Fig. 8 Gas temperature T/1000°K (cont. line) and velocity dispersion $\sigma/0.5$ Kms$^{-1}$ of the fragments (dash. line) vs. time.

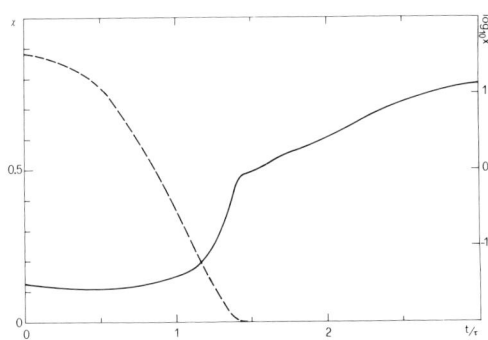

Fig. 9 Time evolution of $x = M_J/M_{gas}$ (cont. line) and of $\chi = (\partial \rho_{F,tot}/\partial t)/$Schmidt's SFR (dash. line)

In Figs. 7 ÷ 9 the time is in units of the free fall time $\tau$.

Let us now go back to the process of formation of fragments and make some considerations on how the Jeans mass evolutions are frequently calculated in the literature. The crucial point is that even if the collapsing cloud undergoes a very efficient cooling, such as to keep a quasi-isothermal state, the Jeans mass will stop decreasing and will eventually rise when the gas density will attain sufficiently low values . The latter happens (notwithstanding the collapse) due to fragmentation, which exhausts the available gas. Unfortunately, many authors attempting to predict the typical values of the masses of the first stars, assume a given collapse law and use the total density instead of the ambient gas density (as it would be correct) in the Jeans mass expressions (see e.g. Palla et al. , 1983 or other similar dynamical approximations, e.g. Silk, 1980). The approximations of letting $\rho_{gas} = \rho_{tot}$ in $M_J$ yields meaningless predictions when it is extended to times of the order or greater than the Jeans time, i.e. the time at which an overwhelming fraction of the gas is converted into fragments.

III. Comparing the various mass functions: a single fragmentation law?

The mass spectra of various self-gravitating astrophysical objects (open clusters, binary systems, globular clusters, galaxies, clusters of galaxies, molecular clouds) were elaborated by the authors on the basis of observational data collected by many observers (also for different purposes than that of obtaining mass functions) with the exception of the case of the single stars of the solar neighbourhood (SN), obtained by Scalo (1985) and shown by J. Silk during one of his talks in this Workshop. Frequency-mass histograms (see Fig. 10) were obtained, showing a general common trend: i) a cut-off at low masses, ii) a maximum displaced at the very left side of the mass range, iii) a slow decrease in the high mass tail, with similar slopes in all the considered cases.

Quantitative comparisons performed by the application of Kolmogorov-Smirnov test confirmed a striking similarity, with levels of confidence ranging from 86% to more than 99%. These results suggest us to propose a strong version of Reddish's (1978) hypothesis that all these objects, notwithstanding the different evolutionary hystories (probably responsible for the slight differences in the mass spectra) were formed by essentially the same process.

Recalling the theoretical evolution model and mass function (Di Fazio, 1985) previously described, Kolmogorov-Smirnov tests between some observed mass spectra (the globular clusters of the Galaxy, of M31, and the single stars of the SN (Scalo, 1985), and the corresponding theoretical mass functions show an agreement at levels of confidence, respectively, 99%, 83%, 70%.

More details about the described comparisons will be given in a forthcoming paper.

In Fig. 10 frequency vs. mass hystograms for various objects are reported. Sources of data are described in Di Fazio (1985) and in Di Fazio and Capuzzo Dolcetta (1985). For the Large and Small Magellanic cloud globular clusters we show two histograms obtained by different methods (see Capuzzo Dolcetta and Battinelli, 1985, and Di Fazio, 1985). Clusters of galaxies 1, 2 are obtained from Dressler's Catalogue, with two different attribution of masses to the single galaxies components of the clusters. The choice of selecting two sub samples for galaxies in depen dence of the radial velocity was made in order to take into account the selection effect due to the distance. $V_R$ is in $Kms^{-1}$.

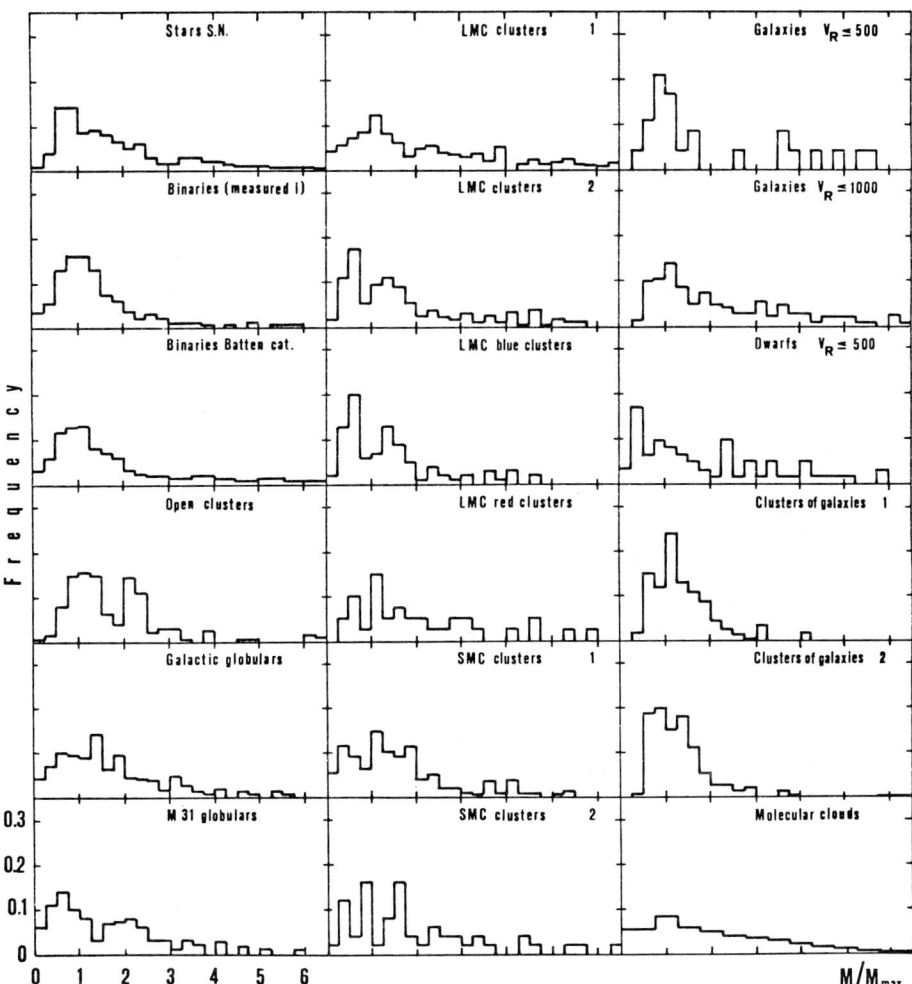

Fig. 10 Frequency-mass hystograms for various objects. Mass is in units of $M/M_{max}$, i.e. the peaking mass of a smoothed curve fitting the distribution.

## REFERENCES

Capuzzo Dolcetta, R., Battinelli, P.: 1985, 'A population synthesis approach to the Magellanic Clouds clusters', in press

Di Fazio, A.: 1982, 'Fragmentation by gravitational instability and star formation in the protohalo', Frascati Workshop 1982, The first stellar generations, Mem. S. A. It., 54, n.1, 243

Di Fazio, A.: 1985, 'Analytical time dependent mass function and fragmentation rate due to gravitational instability', Astron. Astrophys., in press

Di Fazio, A., Capuzzo Dolcetta, R.: 1985, 'Evidence for a single fragmentation law by gravitational instability', in preparation
Di Fazio, A., Palla, F.: 1981, 'Thermodynamical and dynamical evolution of a self-gravitating object and the influence of the ionization', Astrophys. and Space Sci., 76, 391
Jeans, J.: 1902, Phyl. Trans. Roy. Soc., 199A, 49
Jeans, J.: 1928, in Astronomy and Cosmogony, Cambridge University Press
Lequeux, J.: 1980, in Star Formation, 10th Advanced Course Swiss Soc. Astron. Astrophys., Eds. Maeder, L. Martinet
Madore, B.: 1977, Monthly Notices Roy. Astron. Soc., 178, 1
Miller, G.E., Scalo, J.M.: 1979, Astrophys. J. Suppl., 41, 513
Palla, F., Salpeter, E.E., Stahler, S.W.: 1983, Astrophys. J., 271, 632
Reddish, V.C.: 1978, in Stellar Formation, Publ. Pergamon Press, p. 81
Salpeter, E.E.: 1955, Astrophys. J., 121, 161
Scalo, J.M.: 1985, in Protostars and Planets.II., in press
Silk, J.: 1980, in Star Formation, 10th Advanced Course Swiss Soc. Astron. Astrophys., Eds. Maeder, L. Martinet
Weinberg, S.: 1972, Gravitation and Cosmology, Publ. J. Wiley and Sons, Inc.

## DISCUSSION

Silk: There is no observational necessity for invoking an IMF for the precursor of population II that was dominated by very massive stars. The required number of massive stars is very small, so that even with a Salpeter IMF, one does not predict an excess of very low Z low mass stars.

Capuzzo Dolcetta: I do not agree with Joe Silk. In fact there are at leats two observational necessities for invoking an IMF for the precursors of Pop. II that was dominated by massive objects. One is that no zero-metal stars (i.e. primordial low-mass stars) have ever been detected, even in the zone nearby the Sun. So, I think this can suggest that the first generation of stars was mainly composed by massive stars. Moreover, the IMF concept is more general than to be applied just to stars. In fact, to be able to account for the globular clusters formation, we need an IMF, some time before the birth of Pop. II, peaking at masses high enough ($\gtrsim 10^5 M_\odot$) and of the right shape (see our data on the observed mass functions), in particular, with a cut-off at low masses, i.e. not expressible as power laws.

Di Fazio: I would like to add that Salpeter's IMF (for the stars of the SN), even though "it does not imply an excess of very low Z low mass stars", simply does not have the observed shape of the IMF for the SN stars. In fact, Joe Silk has just shown Scalo's (1985) IMF for these stars. We saw that it cannot be expressible as a power law, and that it has a turnover and a cut-off at masses lower than about 0.4 $M_\odot$. The shape of an IMF is a very important observational evidence of the modes of formation of stars (or other astrophysical objects). Thus, as a criterion on the IMF's we cannot take into account only the amount of heavy elements predicted.

We should also remeber that Salpeter (1966) used Luyten (1939) luminosity function, and that the power law came fro a fit restricted to the mass range 0.4 — 10 $M_\odot$. Finally, since Joe says that using Salpeter's IMF is "more economical", I note that even ignoring the disagreement of Salpeter mass function with the most recent observations (Luyten, 1968; Probst and O'Connel, 1983; Miller and Scalo, 1979; Scalo, 1985) it is much more "economical" to use a single, compact theory (such as ours) to explain the formation of many different classes of objects (stars, open clusters, globular clusters, molecular clouds, dust clouds, galaxies, clusters of galaxies, etc.) than having to resort to empirical, ad hoc mechanisms for each single very particular family of objects.

De Jong: I would like to point out that what you call the Schmidt's law has lately lost some of its popularity. Observations of CO in our own and other galaxies have shown that there is a direct proprtionality between the number of newly formed stars and the density of the gas from which they form and not a quadratic dependence.

Di Fazio: In fact, if you followed closely what I showed, you could check that, indeed, my fragmentation rate is very different from Schmidt's empirical "law", in that: i) it produces a cut-off at high temperature and/or at low density, when $M_J \gtrsim M_{gas}$, ii) it depends on temperature, iii) it depends on the chemical composition of the gas, iv) if also the temperature-dependent part was forced to be expressed as a power law of the density, since my SFR implies long periods of star formation inhibition, we would obtain a power n sensibly smaller than 2, i.e. $0.4 \lesssim n \lesssim 1.5$, compatibly with the observations (e.g. Lequeux, 1980; Miller and Scalo, 1979).

STAR FORMATION TRACERS IN GALAXIES

James LEQUEUX

Observatoire de Marseille, 13248 Marseille Cedex 04, France

I. INTRODUCTION

This paper deals in a critical way with various ways of determining the present rate of star formation and the history of star formation in galaxies, with emphasis on quantitative aspects and on irregular galaxies. Irregular galaxies are apparently the simplest systems where active star formation is presently occuring, and one may hope that understanding them will help to understand the more complex spiral galaxies. As to elliptical galaxies, they are mainly fossils of a past evolution and they will not be discussed here, as well as lenticular galaxies.

Under the general name of irregular galaxies, one finds a wide variety of objects, namely:
1) Small systems with incoherent star formation : these are the irregular galaxies *stricto sensu*.
2) Interacting systems, sometimes amorphous. Mutual interactions can distort galaxies so much than their previous morphology is lost : they may be classified as irregular galaxies - some are amorphous. A recently studied example of such a system where a direct collision is seen is Mk 171 = NGC 3690 + IC 694, where an active burst of star formation covers the whole optical image (Gehrz et al., 1983; Bushouse and Gallagher, 1984; Augarde and Lequeux, 1985).
3) Isolated (?) anomalous systems. An interesting such galaxy is the amorphous object NGC 1800 (Gallagher et al., 1981).
4) Blue compact galaxies. French (1980) distinguishes two classes in this category : objects with relatively low luminosity and low metal abundance (e.g. I Zw 18) and objects with high luminosity and high metal abundance (ex. B234: French and Miller, 1981). There are intermediate cases like B272 (Downes and Margon, 1981). All these objects have an extremely active present star formation extending over a small region.

I will mainly concentrate on the first category of irregular galaxies, but the methods which will be described below can be extended to other galaxies. It is worth remembering some general properties of the "normal" irregulars:

a) In spite of looking chaotic, they usually have an underlying exponential disk like spiral galaxies, albeit of fainter surface brightness (see e.g. Hunter and Gallagher, 1985).

b) They have high interstellar hydrogen-mass to luminosity ratios. The hydrogen is generally distributed in a smooth regular way.

c) They have blue colors. Fainter irregulars look bluer but this may be a selection effect.

d) Almost all irregulars for which good 21-cm line observations exist (in particular made with the Very Large Array) seem to be supported by rotation rather than random motions. This is true for even very small and faint systems (Viallefond et al., 1985 ; Carignan, 1985). However some compact galaxies have a chaotic 21-cm structure and are not supported by rotation (Lequeux and Viallefond, 1980 ; Viallefond and Thuan, 1983).

## II. DETERMINATION OF THE PRESENT-DAY STAR FORMATION RATE (PDSFR)

### 1. Star counts

The most direct way to measure the PDSFR is obviously to count the brightest stars, which are also the most massive and young ones. In principle, one could thus determine both the PDSFR and the initial mass function (IMF). Photometric data exist for most nearby irregular galaxies down to M $\sim$ 10 $M_\odot$, and also for several spiral galaxies (**Freedman**, 1984). However there are several problems about obtaining and interpreting those data.

Observational problems: a) Since the photometry is limited to some magnitude in the visible (U,B or photographic), there is a strong selection effect against detection of very blue or very red stars which have large bolometric corrections. As the evolution of massive stars is nearly at constant $M_{bol}$, one sees only the most massive of the O-type blue main-sequence stars. Even when these stars are observed, it is almost impossible to classify them from photometry alone, and their $M_{bol}$ and $T_{eff}$ are very uncertain. b) Crowding is a problem for relatively faint stars, and the samples are increasingly incomplete and photometry less accurate when going to fainter magnitudes.

Problems in the interpretation. a) Interpretation of star counts in terms of IMF and PDSFR is very sensitive to the models of stellar evolution, which are still uncertain for massive stars (see the lectures by Chiosi, this volume). In particular the $T_{eff}$ extent of the main sequence is poorly known and might be much larger than predicted by conventional models, up to A0 supergiants (Meylan and Maeder, 1982; Bisiacchi et al., 1983; Bertelli et al., 1984). This main-sequence widening may be required to explain the presence of many "blue" stars with colors as red as B-V $\sim$ 0, U-V $\sim$ 0 in the HR diagrams of irregular galaxies. b) There may be age effects if star formation is irregular in time. These effects are visible in some cases and are more severe for smaller systems where fluctuations in the SFR and IMF are relatively larger.

A good, complete upper HR diagram in $M_{bol}$, $T_{eff}$ coordinates exists only for the solar neighbourhood (SN) in our Galaxy: Bisiacchi et al., 1983; Humphreys and Mc Elroy, 1984. The latter authors have discussed carefully the incompleteness selection effects in the SN and in the Magellanic Clouds and have been able to compare the bolometric luminosity functions and IMFs for these three systems. They are similar and a relative PDSFR can be derived from their data. This is given in Table 1, from their Table 7, down to $M_{bol} = -7.5$, with (arbitrary) normalization to the LMC. The V luminosity function for the SN is shown in Fig. 1 from their Fig. 6 down to the completeness limit of $M_V = -5$. These new results supersede previous attempts e.g. by Vangioni-Flam et al., 1980.

Upper HR diagrams now exist for 13 nearby irregular galaxies. These galaxies **form** certainly the most unbiased sample for which the PDSFR **can** be determined. There are several problems with these data, however. The brightest stars are lacking in the smallest systems, and LSG 3, the faintest galaxy of the sample, does not seem to have supergiants at all and its brightest stars are asymptotic giant branch stars if this object is at a distance similar to M31 or M33 (Christian and Tully, 1983). The lack of massive stars in dwarf systems may be a statistical small-number effect (except perhaps for LSG 3), as shown by Schild and Maeder (1983); see also Sandage, 1984. Age effects are sometimes visible in the relative numbers of blue and red stars, e.g. in Ho I vs Ho II (Hoessel and Danielson, 1984). The distances of some of these galaxies (in particular Leo A) have been determined from the magnitude of the brightest stars and are somewhat problematic in view of the above effects.

In spite of these difficulties, it is worthwhile to build from the published HR diagrams a V luminosity function in what can be estimated as the complete part of the stellar photometry [1]. The results are shown in Fig. 1 which compares the integral V luminosity functions of 10 galaxies with that of the SN (1 kpc$^2$). It is gratifying to see that these luminosity functions are not significantly different from each other (see Freedman, 1984 for a similar study of a sample of spiral and irregular galaxies, that yields the same result). It thus appears justified to use the difference in ordinates of these luminosity functions in Fig. 1 to derive relative PDSFRs for the galaxies of the sample. The results of this exercise is given Table 1 after normalization to the LMC. It is seen that the PDSFR <u>per unit mass of gas</u> is approximately constant for all galaxies within a factor 2-3 (note that the errors probably

---

[1] Due to the possible, poorly known, extension of the main-sequence to the red, the transformation of these luminosity functions into mass functions needs much care and is not attempted here

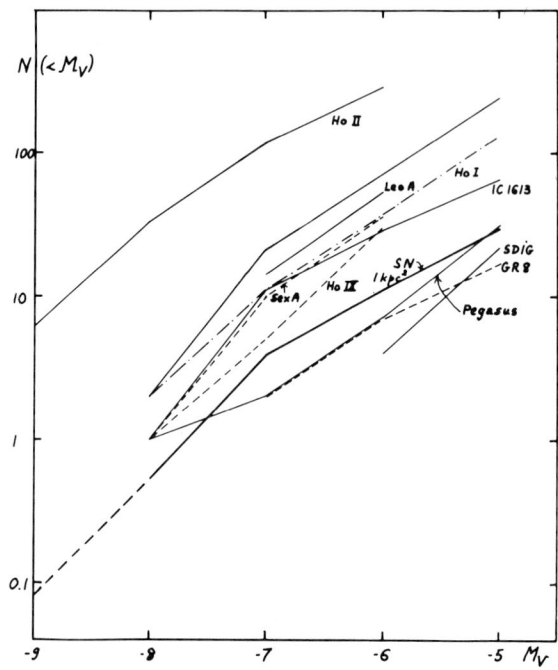

Fig. 1. Cumulative V-luminosity functions for nearby irregular galaxies. The adopted distances are as in Table 1. References: SN = Solar neighbourhood, 1 kpc$^2$, from Humphreys and McElroy (1984) Fig. 6; IC 1613, from Humphreys (1980), numbers × 4/3 to take into account incomplete coverage; NGC 6822, from Humphreys (1980), neglecting stars with 0.5 < B-V < 1.65, probably some contamination by red galactic stars; SDIG, from Lequeux and West (1981), provisional photometry; Pegasus, from Hoessel and Mould (1982), the survey may be somewhat incomplete; Sextans A, from Sandage and Carlson (1982) with magnitude sequence and distance revised by Sandage (1984); GR 8, from Hoessel and Danielson (1983); Leo A, from Demers et al. (1984); Holmberg I and Holmberg II, from Hoessel and Danielson (1984); Holmberg II, from Sandage (1984) but using the same distance modulus $(m-M)_o$ = 27.5 as Holmberg I and II, from the previous reference.

amount already to a factor 2). This is not true for the PDSFR <u>per unit luminosity</u> which may increase systematically for fainter systems. Note that some galaxies are apparently somewhat lazy in forming stars (SMC, Sextans, LSG 3). Ho II is active for its luminosity but not for its gas content which is large. There is no relation between the PDSFR and color, indicating that the luminosity is not in general dominated by young, massive stars.

The above study of a relatively unbiased sample of irregular galaxies shows that the gas content seems to determine the present day formation rate of massive stars. We know <u>nothing</u> on the rate of formation of low-mass stars which are likely to dominate the luminosity. However it is clear that the faintest systems, which have a large PDSFR/$L_B$ ratio, either are not forming many low-mass stars at present, or if they do are young systems which have not been forming stars for very long in the past. Chemical abundance determinations, which unfortunately are not yet available for the faintest galaxies, may help to solve this dilemma.

Table 1. Relative Present-day formation rates of massive stars in nearby irregular galaxies

| Galaxy | Adopted $(m-M)_o$ | $M_{B_o}$ [1] | $(B-V)_o$ [1] | $(U-B)_o$ [1] | $M_{gas}$ [2] ($M_\odot$) | PDSFR[3] / $M_{gas}$ | PDSFR[3] / $L_B$ |
|---|---|---|---|---|---|---|---|
| The Galaxy (SN, 1kpc$^2$) | – | – | – | – | $6.0\ 10^6$ | 4.6 | – |
| LMC | 18.5 | -18.1 | 0.48 | – | $7.6\ 10^8$ | 1 | 1 |
| Ho II (DDO 50) | 27.5 | -16.4 | 0.30 | -0.35 | $1.3\ 10^9$ | 0.7 | 5.1 |
| SMC | 19.0 | -16.3 | 0.46 | – | $5.1\ 10^8$ | 0.3 | 1.1 |
| NGC 6822 | 23.2 | -14.9 | – | – | $1.6\ 10^8$ | 1.1 | 4.4 |
| Ho I (DDO 63) | 27.5 | -14.5 | -0.01 | -0.11: | $1.5\ 10^8$ | 0.6 | 3.0 |
| IC 1613 | 24.3 | -14.4 | 0.58 | – | $5.1\ 10^7$ | 1.4 | 2.9 |
| Sextans A (DDO 75) | 26.2 | -14.4 | 0.38 | – | $1.6\ 10^8$ | 0.5 | 3.3 |
| Leo A (DDO 69) | 26.8: | -14.2: | 0.19 | -0.18 | $1.2\ 10^8$ | 0.9 | 5.3 |
| Pegasus (DDO 216) | 26.1 | -14.0: | 0.62 | 0.14 | $2.3\ 10^7$ | 1.0 | 1.3 |
| Ho IX (DDO 66) | 27.5 | -13.3: | 0.67 | -0.50 | $2.7\ 10^8$: | 0.2: | 6.1: |
| SDIG | 27.1 | -11.1 | 0.64 | -0.32 | $1.3\ 10^7$ | 1.5 | 16 |
| GR 8 (DDO 155) | 25.6: | -10.9 | 0.30 | -0.51 | $4.9\ 10^6$ | 3.4 | 16 |
| LSG 3 | 24.6: | -9.2: | – | – | $3.1\ 10^5$ | 0 | 0 |

[1] From $B_T$, B-V and U-B of de Vaucouleurs et al. (1981) and if not available, de Vaucouleurs et al. (1976). For SDIG, Laustsen et al. (1977) and for LSG 3, Christian and Tully (1983). Corrected from galactic extinction only using the E(B-V) of Burstein and Heiles (1982).

[2] Solar Neighbourhood: Vangioni-Flam et al. (1980). LMC: Mc Gee and Milton (1966). SMC: Hindman (1967). Ho II, Ho I, Sextans A, Leo A, Pegasus, GR 8: Fisher and Tully (1981). NGC 6822: Roberts (1975). IC 1613: Huchtmeier et al. (1981). Ho IX: Appleton et al. (1981), very uncertain because of confusion with M 81. SDIG: Cesarsky et al. (1977). LSG 3, Thuan and Martin (1979). The mass of HI (or HI + $H_2$) has been multiplied by 1.4 to take helium and heavy elements into account.

[3] Normalized to the LMC. The Present Day Star Formation Rate (PDSFR) had first be normalized to the Solar Neighbourhood SN using Fig. 1 ; the LMC and SMC have been then compared to the SN inside 3 kpc using the bolometric luminosity function of Table 7 of Humphreys and Mc Elroy (1984).

## 2. Lyman continuum flux

Lyman continuum photons are produced only by O and early B stars. They ionize the interstellar gas which emits recombination lines and a free-free continuum. The intensity of the thermal radio continuum or of the Balmer (or Paschen etc.) lines of hydrogen are proportional to the number of Lyman continuum photons absorbed by the gas (for numerical relations see e.g. Lequeux et al., 1981). Problems when transforming the corresponding observed fluxes into a PDSFR are :

- On the observational side : extinction corrections when using Hα or Hβ. This problem is far from straightforward (Caplan and Deharveng, 1985)
- On the side of interpretation: a) separation between thermal and non-thermal components when using radio fluxes. Apparently the non thermal flux is still important at centimeter wavelengths (see e.g. the discussion of blue compact galaxies by Klein et al., 1984). b) Absorption of Lyman continuum photons by dust (Smith et al., 1978, Appendix A). c) The interpretation is <u>very sensitive</u> to the adopted IMF and upper mass limit, and sensitive to the evolutionary tracks. d) The derived number of photons has to be divided by an average lifetime of HII regions to obtain a rate, and this is somewhat arbitrary : usually one takes $6 \; 10^6$ years for this lifetime.

This method has recently been applied by Hunter et al.(1982), Gallagher et al.(1984) and Hunter and Gallagher (1985) to irregular galaxies. They find a relation between the PDSFR per unit luminosity and color, which is not surprising as their sample contain galaxies with strong bursts which dominate the light, and they find no clear correlation between PDSFR and global gas and abundance parameters. They also obtain a more or less constant PDSFR per unit B luminosity. All this seems contradictory with the results of Table 1. The discrepancy may be more apparent than real however, since their samples are rather heterogeneous and contain many (rare) galaxies with bursts of star formation : for these galaxies, it is clear that the light is dominated by the burst itself and that there is a tendancy for getting a uniform PDSFR per unit B luminosity. However more work on well-defined samples (as for example the irregulars in the Virgo cluster) is required to clear up that point.

Kennicutt (1983) and Kennicutt et al.(1984) have done a similar study for spirals in the Virgo and more distant clusters. What they actually measure the Hα equivalent width which gives a PDSFR per unit luminosity. When all galaxy types are mixed, they find a good correlation of this quantity with U-B or B-V which might just in fact illustrate the fact that later-type galaxies form more stars per unit luminosity than earlier-type ones and also have bluer colors. More interesting is the lack of correlation they find between the PDSFR and the gas content : HI-poor <u>spirals</u> may form stars actively while anemic or Hα-weak galaxies are not necessarily HI-poor. This did not appear in our sample of irregular galaxies discussed above. Their sample of irregulars is too small to check if there is a discrepancy for this type of galaxies. To end this section, I would like to warn the users of Balmer or radio data of galaxies that any attempt to calculate gas consumption rates from such data is premature as we do not know whether or not low-mass stars are formed together with the observed O stars, and if the answer is yes, whether the ratio of massive to low-mass stars is constant or not.

## 3. Far-IR emission

This emission results from re-radiation of stellar radiation absorbed by interstellar dust. In a galaxy which does not form gas actively, the ratio $L_{IR}/L_B$ is of the order of 0.1 and the dust is rather cold (de Jong, this book). Any galaxy in which $L_{IR}/L_B \gg 0.1$ and which has hot dust is likely to experience active star formation. Recent results on the PDSFR obtained in this way from observations with the IRAS satellite are discussed by de Jong in this book : I refer the reader to his paper.

## 4. Far-UV emission

Except for early types, the far-UV emission of galaxies is dominated by young main-sequence stars with masses $\gtrsim 3\ M_\odot$. It may thus be used to derive a PDSFR. However the corresponding results, although not as sensitive to the IMF as those obtained from the Lyman continuum photons, are somewhat sensitive to metallicity and are critically affected by interstellar extinction which is very high in the far-UV, and perhaps even different for massive stars (Mochkovitch and Rocca-Volmerange, 1984). However, this method has been used for the Magellanic Clouds by Vangioni-Flam et al.(1980), revised in Lequeux (1984), with results consistent with those from star counts. Recently, Donas and Deharveng (1984) have used OAO 2 data at 1910 A to determine the PDSFR for those (relatively nearby) galaxies detected by this satellite. They find a good correlation (within a factor $\simeq 3$) between this PDSFR and the mass of gas, in disagreement with the findings of Kennicutt (1983) and Kennicutt et al.(1984) using H$\alpha$ photometry, but in agreement with our results on the nearby irregulars. They also find a correlation between the PDSFR per unit blue luminosity and color which may just, as discussed in section II 3, reflect the run of color with galaxy type. However there is a large spread in PDSFR per unit luminosity inside galaxies of Sab to Sbc types, which reduces for types Sc to Im.

## 5. Other tracers

Lequeux (1979) has discussed other tracers for the PDSFR which might be useful qualitatively. The cepheids are particularly interesting since they can be dated, as indeed are to some extent the brighter supergiants. They have been used to show how the sites of star formation have moved over the face of the LMC during the last few $10^8$ years (Payne-Gaposchkin,

1972; Isserstedt, 1984). Another tracer of PDSFR which has been used
occasionally is the radio <u>non-thermal</u> emission (see e.g. Gehrz et al.,
1983), under the assumption that it can be considered as the superimposition of the emission of supernova remnants. However this assumption is
most probably wrong even if all relativistic electrons are accelerated
in supernova remnants, and seem to yield PDSFRs that look excessive by
other standards (see e.g. Augarde and Lequeux, 1985; Jenkins, 1984).

III. DETERMINATION OF THE HISTORY OF STAR FORMATION IN GALAXIES

We are entering here a very complex and difficult subject onto which
only a few glances are given in the present paper: its purpose is only
to illustrate a few methods for attacking this subject.

1. Stellar studies

For the moment, such studies have been published only for our Galaxy,
the Magellanic Clouds and dwarf spheroidals. However they are rapidly
extending to more remote systems (Mould, this book), and will extend
even more in the future with the advent of the Space Telescope.

In our Galaxy, a number of studies have been made trying to fix
limits to the past average SFR. For example Mayor and Martinet (1977)
conclude from a general review that if the SFR has decreased with time
as $\exp(-t/T)$, T can be infinite (uniform SFR) and must in any case be
smaller than $5 \; 10^9$ years: this corresponds to an average past SFR
smaller than 3 times the present one.

In the Magellanic Clouds, deep star counts in several regions have
been used to infer something about the past star formation: see the
reviews by Stryker (1984) and Frogel (1984). The results of the fit of
theoretical models to the observed luminosity functions in the LMC are
model-dependent, however it appears that any good fit requires a strong
intermediate-age stellar component formed $3-5 \; 10^9$ years ago. The situation for the SMC is somewhat more complex, although a $3 \; 10^9$ year-old
component appears in the halo. The existence of an intermediate-age
component in the LMC is confirmed by the large number of carbon stars
found by Blanco and Mc Carthy (1983): these stars are believed to be
$3-5 \; 10^9$ years old. Finally the position in a $(J-K)_o$, $K_o$ diagram of
M AGB stars in clusters and the Bar West field of the LMC give another
confirmation of this event (Frogel and Blanco, 1983; Frogel, 1984). The
case for a recent burst which would have occured some $10^8$ years ago is
much less compelling.

2. Colors

The use of colors to trace back the evolution of galaxies has first been
discussed extensively by Searle, Sargent and Bagnuolo (1973). They notice that star clusters evolve in the (U-B), (B-V) diagram following a
well-defined track. If a galaxy can be considered as the superimposition
of star clusters of various ages (this assumes explicitely that the IMF
is invariable) its position in the two-color diagram is characteristic,

Fig. 2. Evolution of a galaxy in the two-color diagram. This figure, built from Searle et al. (1973) explains how the colors of a galaxy can be used to infer their past history. It assumes normal metallicity and a Salpeter initial mass function: changes in these parameters would affect the location of the loci in the figure. The heavy line is the locus of the **colors** of a star formation burst having occured at the indicated ages in the past. The lighter line corresponds to a uniform star formation having started at the indicated ages in the past. Any galaxy $10^{10}$ years old with a uniform or decreasing SFR has colors which lie on the dashed line connecting the $10^{10}$ year points of the two previous loci. Another dashed line indicates the locus of the colors of galaxies $10^9$ years old, etc. The effect of a recent burst occuring in a $10^{10}$ years old galaxy with a previously uniform SFR is indicated as the dotted line as a function of its strength. The hatched region corresponds to the colors of most elliptical and spiral galaxies which might all be around $10^{10}$ years old but have experienced different decreasing SFRs. Some irregular galaxies and other objects have bluer colors and may be interpreted as either younger galaxies or galaxies dominated by recent SF bursts. Colors alone cannot help for chosing between these two alternatives.

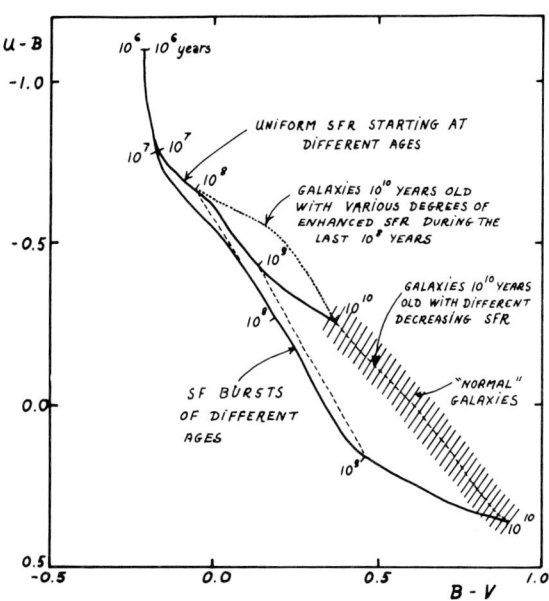

although not unambiguously, of its past history (Fig. 2).
They remark that actual spiral and elliptical galaxies lie between the position of a cluster $\sim 10^{10}$ years old (implying that all star formation has taken place in such a remote past) and that of continuous uniform star formation since this time. Some irregular, blue compact and other galaxies lie higher in the diagram, implying that their light is strongly contributed or even dominated by recent star formation. Later, Larson and Tinsley (1978) demonstrated that the location of a galaxy in the (U-B), (B-V) diagram depends only (for a given invariable IMF and metallicity) on the ratio :

$$H = \int_0^{now} SFR(t)\,dt/PDSFR \quad Gyr,$$

the PDSFR being an average over the last $\simeq 10^8$ years. Deviations with respect to the mean (U-B), (B-V) locus indicate a recent burst; however the characteristics of such a burst cannot be unambiguously determined in this way. More recently, Rocca-Volmerange et al. (1981) showed that the conclusions of Larson and Tinsley can be extended to any combination of colors, from the far-UV to the near IR.

For any application, it should be remembered that an IMF has to be chosen, more or less arbitrarily. One such application is to the Magellanic Clouds, for which Rocca-Volmerange et al. (1981) succeeded in matching the whole spectrum from 1690 A to 6250 A; they derive the following values of H corresponding to the plausible range of IMFs:
H(LMC) = 7 to 10 Gyr
H(SMC) = 9 to 20 Gyr

This can be interpreted in several ways, since only the average past SFR is determined by the PDSFR and R: in particular it is not incompatible with the active period of star formation 3-5 $10^9$ years ago discussed in the previous section. One also notes that the PDSFR in the SMC is lower than in the LMC, a result already obtained from the direct study of stellar populations (see Table 1).

Another interesting problem which can be attacked through the study of colors concerns the past history of star formation in blue compact galaxies. In many such objects, an extended red structure underlying the burst can be seen, and there is no doubt that star formation has taken place before the present burst (for a recent example concerning object ESO 338-IG 04, see Bergvall, 1985). This is not so clear for example for IZw 18 and IZw 36 for which no such underlying structure has been detected. These two objects have extreme colors. After correcting for the emission lines but not for extinction. I find from the data of Huchra (1977), Sargent and Searle (1970) and Viallefond and Thuan (1983): for IZw 18: V = 16.21, B-V = 0.19, U-B = -0.88, and for IZw 36: V = 14.76, B-V = 0.54, U-B = -0.68. These colors are too red in B-V to correspond to a burst, and this suggests an underlying red population. However they are also deviating from the predictions of the whole range of models by Larson and Tinsley (1978). Thuan (1983) performed IR photometry for blue compact galaxies including the two objects above and concludes from a comparison of (U-B) and (V-K) colors that they contain red giants whose progenitors must have been formed before the present, very young burst (see also Thuan, 1985). However this conclusion, which rests on models, is disputed by Gondhalekar et al. (1984) who compare directly colors of extragalactic HII regions for which the underlying old population should make a negligible contribution with those of blue compact galaxies corrected from line emission (and of course extinction): they conclude that some of these galaxies have the same colors as HII regions and burst models and hence are truly young galaxies. Unfortunately their sample contains neither IZw 18 nor IZw 36. This problem is clearly worth further studies.

## 3. Spectral synthesis

Spectral synthesis is clearly a higher spectral-resolution, in principle more powerful, version of the previous studies. For applications, see the papers by Barbaro, Bruzual and O'Connell in this volume and also Bushouse and Gallagher (1984).

## 4. Chemical abundances

Heavy element abundances are clearly telling us something about the past evolution of galaxies. However their interpretation is not unambiguous, because galaxies probably do not evolve as closed boxes, but may experience infall of primordial material, stripping and winds. For a discussion, see Matteucci and Chiosi (1983). However the very low abundances in IZw 18 clearly show that the present burst is probably the first one of such importance in this galaxy (Lequeux and Viallefond, 1980). Heavy-element ratios are less subject to contamination and might prove useful if the corresponding nucleosynthesis is well understood. For example, the C/O ratio in the Magellanic Clouds is noticeably weaker than in our Galaxy (Dufour et al., 1982); since carbon appears to be produced in lower-mass stars than oxygen, a smaller C/O ratio may be interpreted as resulting from a smaller integrated death-rate of low-mass stars compared to high-mass stars in the Clouds. This in turn implies time changes in the IMF. However this result does not seem to be confirmed by the rather high abundance of iron (see a discussion in Lequeux, 1983).

## 5. Total mass, mass of gas

Here we clearly enter the realm of very dangerous tracers. Although the ratio $M_{gas}/M_{total}$ is in principle an indicator of the degree of evolution of a galaxy, it can be affected by infall, stripping and winds. The total mass is sometimes poorly known (not better than a factor 3 for the SMC). Dark material can be present, even in very small systems, e.g. under the form of dark halos around irregular galaxies. In this context, one should remember that the virial mass is about 10 times the mass of HI in IZw 18 (Lequeux and Viallefond, 1980) and 5 to 7 times in IZw 36 (Viallefond and Thuan, 1983), while these systems look otherwise very unevolved, especially IZw 18. Hence the use of $M_{gas}/M_{tot}$ or mass/luminosity ratios is likely to produce spurious results. Gallagher et al.(1984) have however somewhat bravely tried to use total mass as a measure of integrated past SFR since $15 \ 10^9$ years, while the B light is supposed to measure the integrated SFR since $3 \ 10^9$ years and the Hα or radio flux is used to determine the PDSFR. Assuming an unvariable IMF, they find that the SFR should have been roughly constant in the past for low-mass ($M < 10^{10} M_\odot$) irregulars while higher-mass irregular and spiral galaxies have experienced a decreasing SFR. Alternatively, this might simply imply that the latter systems contain more dark matter than the former.

IV. CONCLUSION

Our present knowledge of present-day and past star formation rates in galaxies is still very primitive. The methods exist but must be applied with a critical eye and, if correlations are searched for, on unbiased samples of galaxies. Although indirect tracers are interesting, they will never substitute for direct studies of the stellar populations requiring deep photometry and classification: these studies, although limited to nearby galaxies, remain the first priority for the future since they will be the basis upon which a sound understanding of star formation in galaxies has to rest.

REFERENCES

Augarde, R., Lequeux, J.: 1985, Astron. Astrophys., in press
Appleton, P.N., Davies, R.D., Stephenson, R.J.: 1981, Monthly Not. Roy. Astron. Soc., 195, 327
Bergvall, N.: 1985, Astron. Astrophys., in press
Bertelli, G., Bressan, A.G., Chiosi, C.: 1984, Astron. Astrophys. 130, 279
Bisiacchi, G.F., Firmani, C., Sarmiento, A.F.: 1983, Astron. Astrophys. 119, 167
Blanco, V.M., McCarthy, M.F.: 1983, Astron. J., 88, 1442
Burstein, D., Heiles, C.: 1982, Astron. J., 87, 1165
Bushouse, H.A., Gallagher, J.S.: 1984, Publ. Astron. Soc. Pacific, 96, 273
Caplan, J., Deharveng, L.: 1985, Astron. Astrophys., in press
Carignan, C.: 1985, in preparation
Cesarsky, D.A., Falgarone, E., Lequeux, J.: 1977, Astron. Astrophys. 59, L 5
Christian, C.A., Tully, R.B.: 1983, Astron. J., 88, 934
Demers, S., Kibblewhite, E.J., Irwin, M.J., Bunclark, P.S., Bridgeland, M.T.: 1984, Astron. J., 89, 1160
de Vaucouleurs, G., de Vaucouleurs, A., Corwin, H.G.: 1976, Second Reference Catalogue of Bright Galaxies, University of Texas
de Vaucouleurs, G., de Vaucouleurs, A., Buta, R.: 1981, Astron. J. 86, 1429
Donas, J., Deharveng, J.M.: 1984, Astron. Astrophys., 140, 325
Downes, R.A., Margon, B.: 1981, Astron. J., 86, 19
Dufour, R.J., Shields, G.A., Talbot, R.J. Jr.: 1982, Astrophys. J. 252, 461
Fisher, J.R., Tully, R.B.: 1981, Astrophys. J. Suppl., 47, 139
Freedman, W.L.: 1984, Ph. D. Thesis, University of Toronto
French, H.B.: 1980, Astrophys. J., 240, 41
French, H.B., Miller, J.S.: 1981, Astrophys. J., 248, 468
Frogel, J.A.: 1984, Publ. Astron. Soc. Pacific, 96, 856

Frogel, J.A., Blanco, V.M.: 1983, Astrophys. J. Lett., 274, L 57
Gallagher, J.S., Hunter, D.A., Knapp, G.R.: 1981, Astron. J., 86, 344
Gallagher, J.S.III, Hunter, D.A., Tutukov, A.V.: 1984, Astrophys. J., 284, 544
Gehrz, R.D., Sramek, R.A., Weedman, D.W.: 1983, Astrophys. J., 267, 551
Gondhalekar, P.M., Morgan, D.H., Dopita, M., Phillips, A.P.: 1984, Monthly Not. Roy. Astron. Soc., 209, 59
Hindman, J.V.: 1967, Australian J. Phys., 20, 147
Hoessel, J.G., Danielson, G.E.: 1983, Astrophys. J., 271, 65
Hoessel, J.G., Danielson, G.E.: 1984, Astrophys. J., 286, 159
Hoessel, J.G., Mould, J.R.: 1982, Astrophys. J., 254, 38
Huchra, J.P.: 1977, Astrophys. J. Suppl., 35, 161
Huchtmeier, W.K., Seiradakis, J.H., Materne, J.: 1981, Astron. Astrophys. 102, 134
Humphreys, R.M.: 1980, Astrophys. J., 238, 65
Humphreys, R.M., Mc Elroy, D.B.: 1984, Astrophys. J., 284, 565
Hunter, D.A., Gallagher, J.S., Rautenkranz, D.: 1982, Astrophys. J. Suppl., 49, 53
Hunter, D.A., Gallagher, J.S.III : 1985, Astrophys. J. Suppl., in press
Isserstedt, J.: 1984, Astron. Astrophys., 131, 347
Jenkins, C.R.: 1984, Astrophys. J., 277, 501
Kennicutt, R.C.Jr.: 1983, Astron. J., 88, 483
Kennicutt, R.C.Jr., Bothun, G.D., Schommer, R.A.: 1984, Astron. J., 89, 1279
Klein, U., Wielebinski, R., Thuan, T.X.: 1984, Astron. Astrophys., 141, 241
Larson, R.B., Tinsley, B.M.: 1978, Astrophys. J., 219, 46
Laustsen, S., Richter, W., van der Lans, J., West, R.M., Wilson, R.N.: 1977, Astron. Astrophys., 54, 539
Lequeux, J.: 1979, Astron. Astrophys., 79, 1
Lequeux, J.: 1984, in Structure and Evolution of the Magellanic Clouds, S. van den Bergh and K.S. de Boer, eds., Reidel, Dordrecht, p. 67
Lequeux, J., Viallefond, F.: 1980, Astron. Astrophys., 91, 269
Lequeux, J., West, R.M.: 1981, Astron. Astrophys., 103, 319
Lequeux, J., Maucherat-Joubert, M., Deharveng, J.M., Kunth, D.: 1981, Astron. Astrophys., 103, 305
Matteucci, F., Chiosi, C.: 1983, Astron. Astrophys., 123, 121
Mayor, M., Martinet, L.: 1977, Astron. Astrophys., 55, 221
Mc Gee, R.X., Milton, J.A.: 1966, Australian J. Phys., 19, 343
Meylan, G., Maeder, A.: 1982, Astron. Astrophys., 108, 148
Mochkovitch, R., Rocca-Volmerange, B.: 1984, Astron. Astrophys., 137, 298
Payne-Gaposchkin, C.: 1972, in IAU Coll. No.17 "L'âge des étoiles", G. Cayrel de Strobel and A.M. Delplace, eds., Meudon Observatory, p. III 1
Roberts, M.S.: 1975, in Stars and Stellar Systems, Vol.IX, ed. Sandage A., Sandage, M., Kristian, J., The University of Chicago Press, p. 309
Rocca-Volmerange, B., Lequeux, J., Maucherat-Joubert, M.: 1981, Astron. Astrophys., 104, 177

Sandage, A.R.: 1984a, Bull. Amer. Astron. Soc., 16, 880
Sandage, A.: 1984b, Astron. J., 89, 621
Sandage, A.R., Carlson, G.: 1982, Astrophys. J., 258, 439
Sargent, W.L.W., Searle, L.: 1970, Astrophys. J. Lett., 162, L 155
Schild, H., Maeder, A.: 1983, Astron. Astrophys., 127, 238
Searle, L., Sargent, W.L.W., Bagnuolo, W.G.: 1973, Astrophys. J., 179, 427
Smith, L.F., Biermann, P., Mezger, P.G.: 1978, Astron. Astrophys., 66, 65
Stryker, L.L.: 1984, in Structure and Evolution of the Magellanic Clouds, S. van den Bergh and K.S. de Boer, eds., Reidel, Dordrecht, p. 79
Thuan, Trinh X.: 1983, Astrophys. J., 268, 667
Thuan, T.X.: 1985, preprint
Thuan, T.X., Martin, G.E.: 1979, Astrophys. J. Lett., 232, L 11
Vangioni-Flam, E., Lequeux, J., Maucherat-Joubert, M., Rocca-Volmerange, B.: 1980, Astron. Astrophys., 90, 73
Viallefond, F., Trinh X. Thuan : 1983, Astrophys. J., 269, 444
Viallefond, F., Comte, G., Lequeux, J.: 1985, in preparation


DISCUSSION

The discussion is reorganized to follows the plan of the paper.

R. Capuzzo-Dolcetta. A general comment about the determination of the SFR of stars of different masses (the present day IMF): the flux of a galaxy at different wavelengths cannot give unambiguous information on this quantity. The dominant contribution to the integrated light of a galaxy comes from a relatively narrow range of masses whatever the wavelength, and this range is too narrow for allowing a good determination of the IMF using colors or spectrophotometry alone. Little can be said in particular about low-mass stars.

R. O'Connell, Question: Could you say something more about the extended main sequences observed in galactic clusters and associations? How would they affect estimates of the PDSFR?
        Answer: Massive stars spend about 90% of their lifetimes on the MS and evolve at roughly constant $M_{bol}$. Thus in a $M_{bol}$, $T_{eff}$ diagram the MS ends on a line such that 90% of the stars are at the left of it. When looking at the observed HR diagrams for the Galaxy, this line is clearly in the region of the A-F supergiants, to the right of the end of the MS as given by conventional theoretical models. Reconciling theory and observation requires more internal mixing than conventionally assumed, and this is an active subject for current investigations. It is clear that the use of conventional models yields incorrect results for the IMF, the PDSFR etc. The actual quantitative importance of such effects has yet to be investigated. However the effect should be small for relative estimates of the PDSFR as done in the present

study, if the extent does not differ from galaxy to galaxy, as it is apparently the case for our Galaxy, the LMC and the SMC (Humphreys and Mc Elroy, 1984).

T. de Jong. Q.: In the relation you show between the massive star PDSFR and the gas mass for irregular galaxies (Table 1), I guess that you are referring to the mass of atomic hydrogen. What about molecular hydrogen?
A.: There seems to be a consensus mainly based on the weakness or absence of the CO line in irregular galaxies that they contain relatively little molecular hydrogen. This is far from certain, however. A better argument is that the gamma-ray emission is the external parts of our Galaxy, which are deficient in heavy elements like irregular galaxies, can be accounted for by the interaction of cosmic rays with atomic hydrogen alone (Bloemen et al., 1984, Astrophys. J. 279, 136).

B. Madore. Q.: Is it not true that the observers who have published color-magnitude diagrams for irregular galaxies did not make complete counts? Could you comment on the corrections you applied for this incompleteness?
A.: I have checked that except for IC 1613 the counts have been made over the whole extent of the galaxy. I plot in Fig. 1 only the part of the luminosity function that I estimate to be complete.

A. Dressler. Q.: Could dust be masking a correlation between PDSFR and other parameters?
A.: This is hard to say.

C. Lonsdale. I have studied far-IR emission compared to H-alpha and find a large amount of scatter which implies that a large fraction of the H-alpha may be lost due to extinction.

A. Dressler. Gunn and I have produced a similar diagram to Kennicutt's using [HII] vs. (B-V) color, published in our papers I (1982, Astrophys. J. 263, 533) and III (in press) on the study of galaxy populations in distant clusters. Those data show that even for a sample of narrow morphological range Sbc - Sd the correlation of color and star formation rate is quite good.

M. Aaronson. Blanco and Mc Carthy (1983, Astron. J., 88, 1142) have estimated that the contribution of carbon stars to the total bolometric luminosity in the Large Magellanic Cloud is only 3-4%. However it ranges from 25 to 75% in the intermediate-age clusters. This suggests to me that the strength of the so-called intermediate-age burst in this galaxy was not all that large.

A. Renzini. (concluding a discussion on Magellanic Clouds and dwarf spheroidal galaxies after the previous remark). Even if a majority of people likes to think at dwarf spheroidals as objects which evolved by themselves, i.e. as isolated systems, still I think it is worth mentioning the minority's viewpoint according to which they may just be the remnants of tidal interactions among satellites of our Milky Way (i.e.

the Magellanic Clouds and perhaps one third major object now surviving as the Fornax dwarf spheroidal). I think that a hint for an intermediate age component in all the seven dwarfs is present, as either AGB carbon stars or "anomalous cepheids".

D. Kunth. Q.: When deriving the masses of IZw 18 and IZw 36, how sensitive does they depend upon the assumption that these systems are relaxed?

A.: Of course the authors have assumed virial equilibrium. These systems are completely isolated, with no hydrogen cloud or galaxy possibly at the same distance within almost half a degree. One does not expect in such a situation large deviations from virial equilibrium.

G. Comte, to D. Kunth. Q.: What about the colors of IZw 18 if it is genuinely a young galaxy? Is an eventual main-sequence accompanying the blue stars of the burst able to match the colors, or is it necessary to have red giants?

A.: From near-IR photometric studies of blue compact galaxies Thuan (1983) finds that the IZw 18 colors are best matched with a burst of duration $3 \, 10^7$ years but also require a contribution from red giants. Since there is not enough time for stars formed in the present burst to have evolved there must have been previous SF. On the other hand, unpublished CCD pictures of IZw 18 obtained by Laurent Vigroux at the CFH Telescope indicate that the V and I images are coeval so that star formation would only have occured at the very same place in this galaxy. I also would like to stress that the SF history presented here to match both the oxygen abundance and the $H\beta$ flux does not rule out the possibility of a bivariate SF mode: low-mass stars could have been formed on a slower process mode (Lequeux and Viallefond, 1980).

M. Aaronson, to J. Lequeux. Q.: How old do you mean when you say IIZw 40 and similar objects have an "old" population?

A.: I mean stars less than about 2 solar masses, so stars at least a few Gyr old.

J. Silk. Q.: Cannot abundance ratios be a useful probe of the IMF in irregular galaxies?
T. de Jong. Q.: You are warning us against using chemical abundances as a tracer of star formation. Could abundances not be used in a meaningful way for trying to discriminate between the possibilities left open by the interpretation of the colors (constant SFR over lifetime of the galaxy versus strong burst $\simeq 5 \, 10^9$ years ago)?

A.: (to both). Absolute abundances of a single element e.g. oxygen are of relatively little use except e.g. to fix a limit to the integrated past SFR or number of bursts etc. However abundance <u>ratios</u> are useful. I gave an example in the lecture. Another example concerns the very first stellar nucleosynthesis after the Big Bang: the relatively high O/Fe ratio in extreme Population II stars indicates that their enrichment was due to massive stars only (see e.g. Barbuy, 1983, Astron. Astrophys. <u>123</u>, 1). I have some doubts however that we can answer the question of de Jong from abundance ratios alone.

D. Koo. Q.: How can the conclusions of Gallagher, Hunter and Tutukov that the SFR has been uniform for extremely blue irregulars with U-B < 0.5 be reconciled with the redder U-B colors predicted by Sargent, Searle and Bagnuolo (1973) for constant SFR?

A.: The relatively high past SFR they find may in fact mostly reflect the presence of dark matter, as discussed in the paper.

Comte. In irregular galaxies, it is especially dangerous to use the total mass as an indicator of past SF because the fraction of dark matter is highly variable from one object to the other (global M/L ratio varying from 0.5 to 15).

Occhionero. Q.: If dwarf galaxies and irregulars are sunk in cold dark matter halos as globular clusters might be, why are they irregular in shape rather than nicely spherical?

A.: Irregulars are only irregular at first glance. The background disk is usually very regular as well as the overall HI distribution. They are probably flat rather than spherical as the disk is supported by rotation.

# AGE AND PROCESSES OF STAR FORMATION IN VERY YOUNG LMC ASSOCIATIONS

Irene Tarrab
Institut d'Astrophysique
98 bis Bld. Arago
75014 Paris - France

Abstract.

The simultaneous use of several tracers of young stars allows a more reliable determination of the present star formation rate and an age estimation for extragalactic HII regions and associations in the neighboring galaxies.

Theoretical time evolution models of the UV-flux, the number of Lyman continuum photons, the effective temperature of the exciting stars and of the equivalent width of H$\beta$ have been computed for three star formation rates (burst, constant and decreasing rate), considering that they should lead to differences in the observed global parameters. In these calculations the slope x and the upper mass limit of the initial mass function (IMF) were keep variable.

This method was applied to some OB-associations in the Large Magellanic Cloud (LMC); 30 Dor and the neighboring regions. Coherent results were obtained for a burst of star formation and for a decreasing rate. A constant star formation is not consistent with observational data.

Introduction

In our galaxy the star forming regions and associations are very absorbed and look too extended. That means a big difficulty to include the entire regions to get global parameters which allow us to use them as tracers of young stars.

However this kind of study is easier to do it for extragalactic star forming regions. Presently, photometric data, UV, H$\alpha$, H$\beta$ and radio fluxes are available for some OB-associations, HII regions and HI regions.

The analysis of five OB-associations in the LMC (LH100, LH101, LH104, LH105, LH113) has been done in this work where you can find in the following sections the data compilations, the description of the theoretical models, the interpretation and finally some conclusions.

## Data Compilations

From the Lucke and Hodge (1970) list of OB stellar associations, five regions (30 Dor and 4 neighboring HII regions) were observed by Israël and Koorneef (1979) in several UV passbands by the Astronomical Netherlands Satellite (ANS). In this work only the passband centered at 1550 A was considered.

In order to correct the UV fluxes from extinction, H$\alpha$ and H$\beta$ fluxes observed by Caplan and Deharveng (1983) were used to get the color excess E(B-V) of each region.

The correction for extinction was ruled from the next relation;

$$I_{int} = I_{obs} 10^{0.238 + 4.44 E(B-V)_{LMC}}.$$

where the interstellar extinction for the LMC, and for the galaxy were taken respectively from Nandy et al (1980) and Nandy et al (1975), and $E(B-V)_{LMC}$ represents the derived color excess of the individual regions.

The H$\beta$ line emission come from the rate of ionized gas to continuum, emitted by stars contained in the HII region. Then the H$\beta$ equivalent width can be a good age tracer. The observed equivalent width of H$\beta$ line emission come from Dottori and Bica (1981).

Another parameter which allows to get information about massive stars contained in the region, is the radio continuum flux. If this one is purely thermal, it can be used to infer the Lyman continuum photon flux, in this way the 6cm radio continuum observations by McGee et al (1972) were used.

Finally the mean effective temperature of the exciting stars of the HII regions (Teff) is a very helpful parameter to make an age estimation for hot regions, then very young. The Teff for the LMC HII regions have been computed, as suggested by Stasinska (1980), from the ratio [OIII]/H$\beta$, once the chemical composition (O/H) is fixed. Ionization theoretical models by Stasinska (1982) have been used. For the chemical composition, previous determination from Aller et al (1974), Faulkner and Aller (1965) and Peimbert and Torres-Peimbert (1974) have been used. Table 1 shows; the color excess, the intrinsic UV-flux, the H$\beta$ equivalent width, the Lyman continuum photon flux, the Teff and the ratio R defined as:
$R \equiv Log(N_{Ly-c}/L_{UV})$.

## Theoretical Models.

In order to compute the time evolution of the parameters used as tracers of young stars, an assumed star formation law as well as a given initial mass function are needed. Three models were computed; a single burst population, a continuous star formation and a decreasing star formation rate. Two different IMF were considered; with x=1.5 and x=2.

The selected theoretical evolutionary stellar tracks were those by Maeder (1980) with a metallicity z=0.008 and mass loss case

2K(BC) for stars of initial mass from 30 M☉ to 180 M☉. For less massive stars 15 and 9 M☉, case B (Maeder 1981) and for lighter stars from 5 to 1.25 M☉ the Maeder and Mermilliod (1981) tracks with $\alpha=0.5$ were used.

Then a far-UV flux was assigned to every couple of values (Log L, Log Teff), using the ratio r = UV-flux/total-luminosity, corresponding to a given Teff. This ratio was obtained from TD1 measures by Nandy et al (1976), using intrinsic colors $(U_\lambda-V)_o$ as a function of the spectral type and luminosity class. This ratio is defined as; $r \equiv 10^{-0.04((U_\lambda-V)_o-BC+9.49)}$ where the bolometric correction were assigned as a function of the Teff from Code et al (1976), and the constant 9.49 comes from the normalization of the bolometric magnitudes (Kurucz,1979).

A similar calculation has been undertaken on the basis of the Mihalas NLTE model (1972) for the synthetic time evolution of the Lyman continuum flux and the effective temperature by Viallefond (private communication).

The Hβ equivalent width ($W_{H\beta}$) gives us information about the number of A and B stars which dominate the continuum in Hβ to the number of O stars responsable for the ionization, and hence for the Hβ emission line. $W_{H\beta}$ was computed by:

$$W_{H\beta}(Å) = 4.785 \cdot 10^{-13} N_{Ly-c}(\text{ph s}^{-1})/I_{4862,\text{cont}}(\text{erg s}^{-1}\text{Å}^{-1})$$

The final step is to compute the total flux emitted by a star cluster by adding member intensities. This has been done weighting the tracks by two initial mass functions normalized to 1M☉, then the emission produced by 1 solar mass was calculated.

The initial mass function is defined as the number of stars of mass M per logarithmic interval and the cut off masses are the following:

$$\Phi(M) = CM^{-x} \quad \text{for } 1.8\ M_\odot < M < 100\ M_\odot \text{ and } x = 1.5, 2$$
$$\phantom{\Phi(M)} = CM^{-0.6} \quad \text{for } 0.007\ M_\odot < M < 1.8\ M_\odot$$

For the decreasing case, the considered star formation rate (SFR) is: SFR = $e^{-t/\tau}$ where t represents the time and $\tau$ the decreasing factor (both expresed in the same units).

The computed Lyman continuum flux decreases faster with time than the UV-flux does; it disappears after $6 \cdot 10^6$ yrs. The reason is that most of Lyman flux comes from massive stars. which have a small life. However even if the UV flux decreases slowly the contribution of small mass stars (M<2 M☉) is not very important. It should be notice that the ratio $R \equiv Log(N_{Ly-c}/L_{UV})$ as a function of the slope x and of the upper mass limit Mu, doesn't change considerably with x nor with Mu, when the upper masses are not very small. This ratio R will be used as an age tracer instead of using separatly $N_{Ly-c}$ and $L_{UV}$.

For a burst model, the Teff decrease rapidly with time, whereas for a continuous star formation model, it rises to an asymptotic

## References.

- Aller, L.H., Czyzak, S.J., Keyes, C.D. and Boeshaarg, 1974, Proc. Acad. Sci. USA, Vol 71, N°11, p.4496.
- Caplan, J., and Deharveng, L., 1985, Astron. Astrophys. in press.
- Code, A.D., Davis, J., Bless, R.C., Hanbury-Brown, R., 1976, Astrophys.J. 203, 417.
- Dottori, W.A., Bica, E, E.L.D., 1981, Astron. Astrophys., 102,245
- Faulkner, D.J., and Aller, L.H., 1965, M.N.R.A.S., 130, N°6.
- Israël, F.P., and Koornneef, J., 1979, Ap.J., 230, 390.
- Kurucz, L.R., 1979, Ap.J. Suppl. Series, 40, 1.
- Lucke, P.B., 1974, Ap.J. Suppl. N°255, 28.
- Lucke, P.B., and Hodge, P.W., 1970, Astron. Journ., 75, 171.
- Maeder, A., 1980, Astron. Astrophys., 92, 101.
- Maeder, A., 1981, Astron. Astrophys., 102,101.
- Maeder, A., and Mermilliod, J.C., 1981, Astron. Astrophys., 93, 136.
- Mc Gee, R.X., and Lynette M. Newton., 1972, Autralian Journ. of Phys., 25, 5, 581.
- Mihalas, D., 1972, Non-LTE Model Atmospheres for B and O stars NCAR, Boulder, Colorado.
- Nandy, K., Schmidt, Edward, G., 1975, Ap.J., 198, 119.
- Nandy, K., Thompson, I., Jamar, C., Monfils, A., and Wilson, R., 1976, Astron. Astrophys., 51, 63.
- Nandy, K., Morgan, D.H., Willis, A.J., Wilson, R., Gondhalekar, P.M., and Houziaux, L., 1980, Nature, 283, 725.
- Peimbert, M., and Torres-Peimbert, S., 1974, Ap.J., 193, 32.
- Stasinska, G., 1980, Astron. Astrophys., 84, 320.
- Stasinska, G., 1982, Astron. Astrophys., Suppl. 48, 299.
- Van- den Bergh, S., 1981, Astron. Astrophys. Suppl. 46, 79.
- Tarrab, I., 1985, Astron. Astrophys., in press.

ON THE RELATIONSHIP BETWEEN NEUTRAL HYDROGEN MASS, LUMINOSITY AND COLOR IN IRREGULAR GALAXIES

G. Comte
Observatoire de Marseille
2, place Le Verrier
F-13248 Marseille Cedex 04
France

ABSTRACT. We present evidence for a dependence on luminosity of the neutral hydrogen mass to luminosity ratio in a large sample of irregular galaxies of moderate or low surface brightness. This is not valid for active bright irregulars of the Markaryan class. Active star-forming galaxies are segregated from quiescent ones in an (U-V) versus (log $M_H/L$) plot. This diagram can be interpreted as an evolutionary diagram if we suppose that the stellar formation in these galaxies proceeds by recurrent bursts.

1. HYDROGEN MASS AND LUMINOSITY IN LATE-TYPE GALAXIES

There has been some debate about the nature of the correlation between the gaseous mass fraction and the luminosity in galaxies (see e.g. Roberts 1972, 1974); commonly accepted views are that this relation, at a given morphological type, is linear, the HI mass being proportional to the absolute luminosity, and thus the $M_H/L$ ratio (which is distance-independent), is itself independent of the luminosity. However, Balkowski (1973) suggested that the $M_H/L$ ratio could be larger, at given type, in low-luminosity systems; Fisher and Tully (1975) brought some evidence supporting this; but Shostak (1978) found no significant luminosity dependence at given type in his sample of spiral galaxies.

Figure 1 is a plot of various literature sources concerning only Sdm, Sm, and Im galaxies, i.e. "Magellanic" objects. In abscissa is the mean logarithmic luminosity for each sample with the standard deviation as error bar, in ordinate the corresponding mean logarithmic value of $M_H/L$ with its standard deviation. All the luminosities have been adjusted to $H_o = 100$ km s$^{-1}$ Mpc$^{-1}$. The recent survey of "isolated" galaxies made at Arecibo by Haynes and Giovanelli (1984) is not included, the luminosities being based on Zwicky magnitudes without adequate corrections, but these authors give evidence for decreasing $M_H/L$ when increasing L at type Sc; the large survey of UGC galaxies by Thuan and Seitzer (1983) is used in spite of lacking precise morphological classification, but the selection criteria they used ensure that magellanic objects dominate their sample. There is a clear trend in Fig.1, but obviously, one may argue

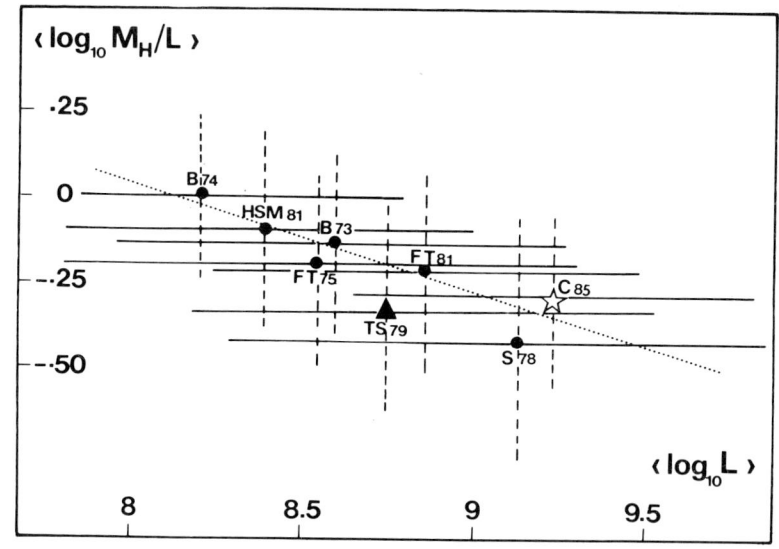

Fig.1. Mean logarithmic $M_H/L$ ratio versus mean luminosity for several samples of Sdm, Sm, Im galaxies in the literature. Sources: B73: Balkowski (1973); B74: Balkowki et al.(1974); FT75: Fisher and Tully (1975); FT81: Fisher and Tully (1981); HSM81: Huchtmeier et al.(1981), TS79: Thuan and Seitzer (1979); S78: Shostak (1978); C85: Comte (1985, in preparation), sample of 21 irregulars with detailed HI velocity fields.

against the inhomogeneity of the data.

We have undertaken to revisit this problem by using integral parameters measured in an homogeneous way for a large fraction of the galaxies from the DDO catalogue (van den Bergh, 1959). This sample contains essentially nearby late-type objects (more than 85% of Sdm's and later); these are well-observed in HI line (Fisher and Tully, 1975, 1981; Huchtmeier et al., 1981, 1983). A large photoelectric photometric survey is available from de Vaucouleurs et al.(1981, hereafter VVB1, 1983, hereafter VVB2), yielding precise asymptotic magnitudes and effective colors. The DDO sample was therefore restricted to the objects:
- with revised morphological type Sdm, Sm or Im
- with photometry in VVB1
- with at least two measurements of HI flux.

This left 119 galaxies from the original DDO list.

The luminosities and the hydrogen masses (without self-absorption corrections) were computed with the distance moduli given by VVB2 (based on revised group distances on the "short" scale equivalent to $H = 100\ km\ s^{-1}\ Mpc^{-1}$). Errors on both parameters were estimated as follows: the final uncertainty was supposed to result from the two independent errors
i) on the corrected apparent magnitude or the observed HI flux, and
ii) on the distance modulus. The errors on the magnitude can be computed

according to VVB1; the error on the distance modulus is given ("internal error") in VVB2. The error on the HI flux was estimated from the external agreement of the two (or more) published measurements and the internal uncertainty on these when given by the author; a conservative figure was kept in all cases.

Figure 2 displays the correlation between log $M_H$ and log L for the 119 objects in the restricted DDO sample. The best fit mean regression line, drawn taking into account errors on both variables, has a slope .78 ± .05, which implies a power-law of type $M_H \propto L^{.8}$. Some particular objects in Fig. 2 deserve special attention: DDO 9, 13, 17, 29, 33, 84, 86 and 154 are located well above the regression line; DDO 13 and 29 were already found to have a very important gas mass fraction (Huchtmeier et al., 1981), DDO 154 has a very extended HI envelope (Huchtmeier and Seiradakis, 1985). On the other hand, DDO 242, 216 (Pegasus dwarf), 187, 155 (GR 8) are located well under the regression line: these are either bona-fide hydrogen deficient objects (as is DDO 216), or luminosity-excess galaxies (as DDO 155 which is resolved in many clumps of blue and yellow supergiant stars: Hoessel et al., 1983) DDO 242 (NGC 5264) is an interesting case of a bright, active, dusty irregular where a fair amount

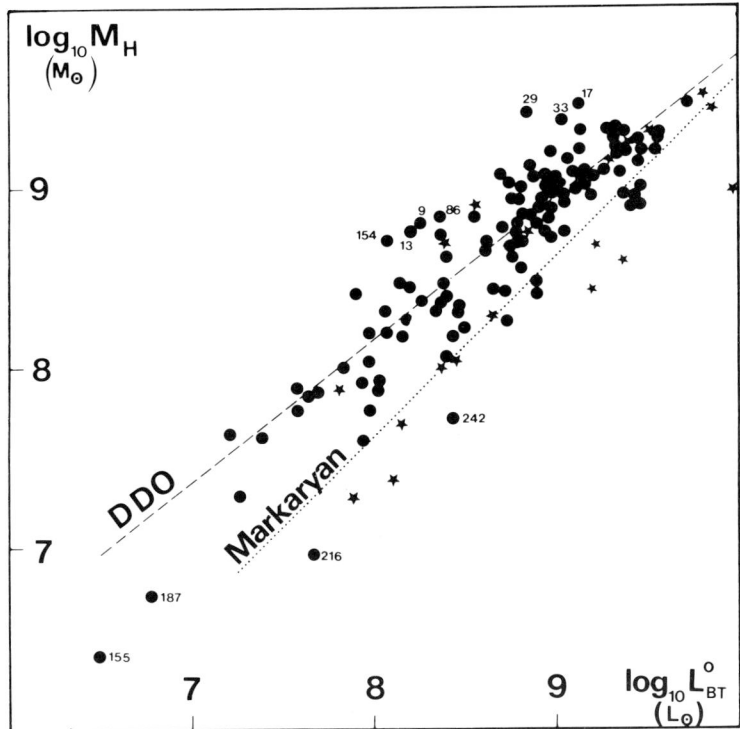

Fig.2. Hydrogen mass versus luminosity for DDO galaxies (dots) and Markaryan irregulars (stars). Lines are best-fit regression relations taking into account errors on both coordinates.

of interstellar gas might possibly be in molecular form.

A recent HI line study of dwarf irregulars inside the Virgo Cluster made by Hoffmann et al. (1985a, b) has confirmed this behaviour of $M_H$ versus L. One may argue however about the quite exotic exponent of .8 found in the $M_H$ versus L power-law: a reasonable hypothesis to explain this is to consider that it is the *total* gas mass fraction $M_g$ which is proportional to the luminosity. This includes $M_H$, the helium mass $M_{He}$ and the molecular fraction $M_{mol}$. $M_{He}$ is proportional to $(M_H + M_{mol})$ if we suppose that $M_{mol}$ is essentially made of molecular hydrogen. If, as it has been claimed by several authors, $M_{mol}$ is strongly dependent on the luminosity (Young and Scoville, 1982; Young et al., 1985), the atomic fraction in $M_g$ increases with decreasing luminosity and this causes the apparent power-law in Fig. 2. Observational tests are quite straightforward but should await powerful millimetric antennae.

To compare the DDO objects with other late-type galaxies, we have selected another subsample of objects in which the present stellar formation is very active: these are the Markaryan galaxies:

- with an apparent diameter large enough to assign a convenient morphological type Sdm, Sm or Im

- with photoelectric photometry from Huchra (1977), which was reduced to the standard $B_T$ system of de Vaucouleurs et al. (1976).

- with HI fluxes (Gordon and Gottesman, 1981; Thuan and Martin, 1981; etc...).

19 objects were selected, including Mk 209 = I Zw 36 often classified as a blue compact galaxy. Among this accordingly statistically small sample, having a span in luminosities of only two decades, we find that $M_H$ is fairly proportional to L.

## 2. $M_H/L$ RATIO AND (U-V) COLOR OF LATE-TYPE GALAXIES: A CLUE TO STARBURST EVOLUTION

Figure 3 is a plot of the corrected color index (U-V) for the 119 DDO objects (dots) and the 19 Markaryan galaxies (stars), versus the $M_H/L$ ratio. Contrary to a commonly accepted view, there appears no obvious correlation between the color (even when using a larger spectral range than the usual B-V index) and the gas content evaluated by the $M_H/L$ ratio, at a given morphological type. It is worth underlining that several objects with very high $M_H/L$ appear indeed quite red on Fig. 3. (e.g. DDO 26, 29, 33, 148, 220 etc...).

However, another point is noteworthy: there seems to exist a lower envelope for the DDO objects representative dots, and a tendancy for the Markaryan galaxies to segregate from the DDO's; that is; these latter populate preferentially the part of the diagram where there is not any DDO galaxy, as well as a narrow band near the lower envelope of the DDO dots. We may tentatively interpret Fig. 3 within the frame of multi-starburst models (Searle et al., 1973; Larson and Tinsley, 1978) as an evolutionary diagram in the following way:

1) Remember first that the DDO sample is selected on a low surface brightness criterion (van den Bergh, 1960), while the detection method of Markaryan galaxies (Markaryan, 1967: strong UV continuum on objective-

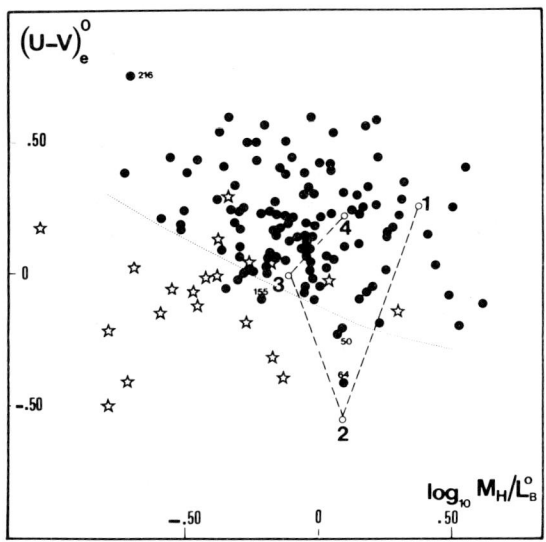

Fig.3. Effective corrected color $(U-V)_e^o$ versus $M_H/L$ ratio for DDO (dots) and Markaryan irregulars (stars). Tentative evolutionary track is explained in the text.

prism Schmidt plates) not only selects high blue surface brightness objects, but indeed forbids the detection of low blue surface brightness galaxies if they are devoid of any active star forming region: in irregular magellanic galaxies, the blue and near-UV surface brightness enhancement is linked to star-forming activity, thus in this respect the Markaryan's are all active galaxies, while the DDO's are largely quiescent ones.

2) On the lower envelope of DDO dots in Fig.3, we find some objects as DDO 155, DDO 50 (Ho II) or DDO 64, which are known either to have a high present star formation rate or to exhibit the traces of a very recent burst of star formation (e.g. Hoessel et al., 1983, 1984).

Now, if, for irregular galaxies, we accept a galactic evolution dominated by recurrent bursts of star formation, we are led to track the evolution of an object across Fig.3 along a series of loops of the kind which is sketched:

- let us start with a DDO-class object of low blue surface brightness, high $M_H$ gas fraction, low total blue luminosity dominated by some old generation of stars giving a relatively red integrated color : initial point labelled "1".

- ignite a burst of stellar formation, supposed to last an indefinitely short duration. This burst produces clusters of massive stars which radiate most of their energy in the ultraviolet, and, depending on the slope of the IMF, a low-mass accompanying population. The color of the galaxy is very sensitive to the massive star population, but the total luminosity is less sensitive at the very moment the burst takes place because of large bolometric correction. We are dealing now with a Markaryan-type galaxy : point labelled "2".

- after a few $10^6$ years, the blue massive stars have evolved into white and yellow red supergiants : the color reddens somewhat and the luminosity increases largely because of reducing bolometric correction. We reach the band where we find objects like DDO 50 or DDO 155 : point labelled "3".

- after some more $10^6$ years, supergiants begin to die, releasing part of the gas into the interstellar space whereas the luminosity decreases: the galaxy comes back to a "DDO" state (point labelled "4"), but some gas has been locked into long lived low-mass stars: these increase the total luminosity with respect to the initial observed one.

The final position of point 4 with respect to point 1 of course depends on the slope of the IMF of the burst and on the amount of gas consumed in it. Some objects may have experienced several recurrent bursts of various amplitudes to cross the whole diagram. One galaxy, DDO 216 (Pegasus dwarf) is located at the extreme upper left corner of Fig. 3: it appears extremely red and hydrogen deficient; a study of its stellar content (within reach of large telescopes or Hubble Space Telescope) would possibly probe this evolutionary scheme, since this object appears as an "exhausted galaxy".

REFERENCES

Balkowski, C. 1973, *Astron. Astrophys.* 29, 43
Balkowski, C., Bottinelli, L., Chamaraux, P., Gouguenheim, L., Heidmann, J. 1974, *Astron. Astrophys.* 34, 43
van den Bergh, S. 1959, *Publ. David Dunlap Obs.* 2, 157
Fisher, J.R., Tully, R.B. 1975, *Astron. Astrophys.* 44, 151
Fisher, J.R., Tully, R.B. 1981, Ap. J. Suppl. Ser. 47, 139
Gordon, D., Gottesman, S.T. 1981, *Astron. J.* 86, 161
Haynes, M.P., Giovanelli, R. 1984, *Astron. J.* 89, 758
Hoessel, J.G., Danielson, G.E. 1983, Ap. J. 271, 65
Hoessel, J.G., Danielson, G.E. 1984, Ap. J. 286, 159
Hoffman, G.L., Helou, G., Salpeter, E.E. 1985, Ap. J. (*Letters*) 289, L15
Huchra, J. 1977, Ap. J. Suppl. Ser. 35, 171
Huchtmeier, W.K., Seiradakis, J.H., Materne, J. 1981, *Astron. Astrophys.* 102, 134
Huchtmeier, W.K., Richter, O.G., Bohnenstengl, H.D., Hauschildt, M. 1983, *ESO preprint* No 250
Huchtmeier, W.K., Seiradakis, J.H. 1985, *Astron. Astrophys.* 143, 216
Larson, R.B., Tinsley, B.M. 1978, Ap. J. 219, 46
Roberts, M.S. 1972, *IAU Symposium 44*, 12, D.S. Evans ed., Reidel
Markaryan, B.E. 1967, *Astrofizika* 3, 55
Roberts, M.S. 1975, *in Stars and Stellar Systems*, vol. IX, p. 309, The University of Chicago Press
Searle, L., Sargent, W.L.W., Bagnuolo, W. 1973, Ap. J. 179, 427
Shostak, G.S. 1978, *Astron. Astrophys.* 68, 321
Thuan, T.X., Seitzer, P. 1979, Ap. J. 231, 327
Thuan, T.X., Martin, G.E. 1981, Ap. J. 247, 823
de Vaucouleurs, G., de Vaucouleurs, A., Corwin, H. 1976, *Second Reference Catalogue of Bright Galaxies*, The University of Texas Press (RC2)
de Vaucouleurs, G., de Vaucouleurs, A., Buta, R. 1981, *Astron. J.* 86, 1429
de Vaucouleurs, G., de Vaucouleurs, A., Buta, R. 1983, *Astron. J.* 88, 764
Young, J.S., Scoville, N.Z. 1982, Ap. J. (*Letters*) 260, L11
Young, J.S., Scoville, N.Z., Brady, E. 1985, Ap. J. 288, 487

# IZW18 AND THE SEARCH FOR VERY METAL POOR GALAXIES

D. KUNTH

Institut d'Astrophysique du CNRS
98 bis, boulevard Arago, 75014 Paris - France.

## 1. INTRODUCTION

As Lequeux has pointed out during his talk, the blue dwarf compact galaxies are experiencing strong star formation processes at present time. Nevertheless they appear to be metal deficient and gas rich. These properties led to some controversy about the very nature of these objects. One can argue that in order to maintain a low overall metallicity the averaged star formation rate must be kept quite low. A series of bursts of star formation separated by long quiescent periods would meet the requirements and the age of the galaxies can be normal. On the other hand these objects can be genuinely young namely undergo their first burst of star formation. The debate is not yet closed and I propose here to add a bit by speculating about the rarety of objects like IZW18.

## 2. WHY IS IZW18 PECULIAR ?

This object is peculiar in the sense that among the dwarf galaxies with active star formation, it is one of the bluest, faintest and the most metal deficient with a metallicity as low as $Z \simeq 1/40$ $Z_\odot$. When Sargent and myself started systematic surveys to search for more objects like IZW18, it was with the hope to find extreme metal poor cases. These objects would have been useful targets for a study of the primordial helium (see Kunth and Sargent, 1983). All searches have failed to find more cases than IZW18 which still remains the most metal deficient galaxy known which undergoes active star formation (see also McMahon et al., 1984). We further realized that this situation turns out to be quite paradoxical: indeed stars with $Z \simeq 10^{-4} Z_\odot$ do exist, as even, many dwarf ellipticals with no gas and no vigorous star formation with lower abundances ($Z \simeq 10^{-2}$ $Z_\odot$) than the so thought "unevolved" blue compact galaxies.

It is possible to convince oneself that no real observational selection effects work against finding more of IZW18 objects or with lower abundances. Idealising a galaxy undergoing a burst in primordial gas with say $Z=Z_\odot/100$, one would search for a compact galaxy (Zwicky type) a blue galaxy (Markarian) or a emission line object using an objective prism technique. Calculations show that a gas ionized by young stars would give [OIII] 5007/H$\beta \simeq 0.7$, [OII] 3727/H$\beta \simeq 0.3$ and [OIII] 4363/H$\beta \simeq 0.025$ if $Z=Z_\odot/100$ and $T_e=20000$ K. There is therefore no reasons to believe that these objects could be missed using modern observational techniques.

## 3. AN EXPLANATION FOR THE RARITY OF IZW18 AND THE LACK OF BLUE COMPACT GALAXIES AT VERY LOW METALLICITY

The explanation put forward by Kunth and Sargent (1985) goes as follows : star-forming galaxies undergoing a burst contain massive stars which enrich the gas on time-scales comparable to or shorter than the lifetime of the HII regions from which abundances are measured. We then can show that the minimum abundance value of the gas is precisely of the order (but can be larger) than the one found in IZW18.

Kunth and Sargent show that for a mass M* of stars formed the built-in heavy elements of mass yM* (where y is the yield) will enrich a gas mass Mi. The dilution factor $R=Mi/M^*$ is an unknown quantity a priori. One can speculate however that since molecular diffusion time scale is very long, and turbulent diffusion in HI gas is also very slow (for $v \simeq 10$ km sec$^{-1}$ the diffusion scale is 1 kpc in $10^{10}$ years) ; it would take too much time for the heavy elements to diffuse into the neutral gas during the lifetime of a burst. On the other hand, SN, WR stars, stellar winds ejecta produce small contaminations and free expansion shells rather small (<20 pc). We thus speculated that the mixing could occur <u>at best</u> in the HII complex itself and assumed Mi=$M_{HII}$.

We have calculated $M_{HII}$ using N the flux of Lyman continuum photons radiated per second by a star cluster and assuming that the Strömgren sphere is spherical and homogeneous. With this assumption in mind and the framework of radiative recombination theory, R can be derived but is quite sensitive to the mean density of the interstellar gas. N was taken from the computation of Lequeux <u>et al</u>., (1981) who obtain N and the cumulative oxygen production as <u>a function of time</u> of a star cluster formed in a burst. Adopting an IMF (initial mass function) with a slope x=1.5, we found R=10. The knowledge of the oxygen mass production scaled to 1 M$\odot$ of star formed during the burst leads to an oxygen abundance O/H$\geqslant 2 \cdot 10^{-5}$ by number. This value is close to the IZW18 oxygen abundance of $1.5 \cdot 10^{-5}$ and shows that one burst is enough to raise the metallicity level at such a value.

The details of these calculations are given by Kunth and Sargent who also point out further complications associated with this picture :

## 4. COMPLICATIONS AND SPECULATIONS

It can be seen from Figure 1 that under the instantaneous burst hypothesis the metals manufactured now cannot yet be seen. This comes simply because once the O stars start releasing their oxygen the Lyman continuum flux has already been reduced by 30. One could argue that metals observed now were manufactured during a previous burst. However since mixing can go on between bursts, objects more metal-poor than IZW18 should be found...We are back to the beginning.
We can still speculate that IZW18 is likely to be genuinely young but we need to relax some constraints of the above model.

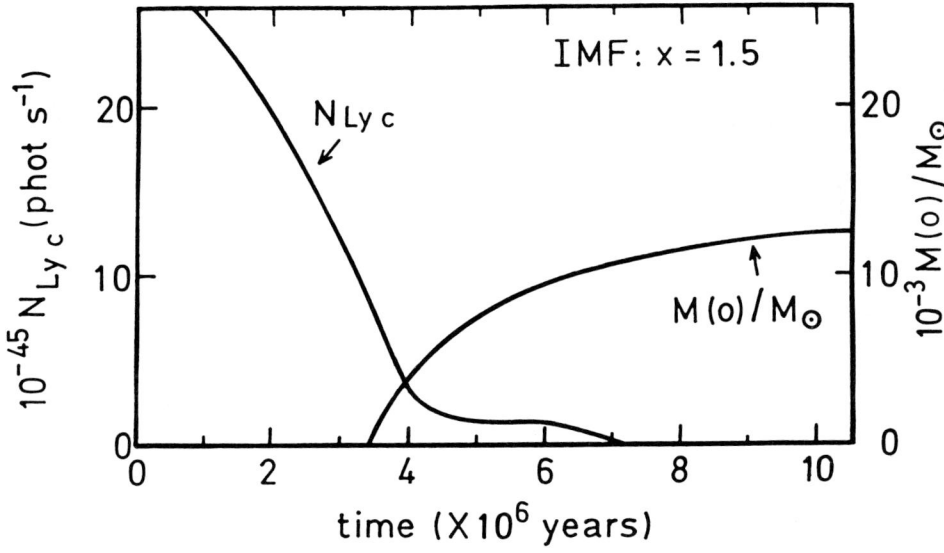

Figure 1. taken from the grid of models given by Lequeux et al (1981) for an IMF with a slope x=1.5. $N_{Lyc}$ and $M(\overline{O})/M_\odot$ (cumulative mass of oxygen produced by a single burst) are given per unit $M_\odot$ of stars formed.

One way out is to allow for a burst of longer duration. They are typically two parameters which can be used to constraint the burst strength and duration and age : F(Hβ) is a measure of the star formation rate whereas O/H constraints very much the duration of the burst since for a given star formation rate a continuous process keeps accumulating oxygen whereas as soon as the burst stops F(Hβ) rapidly decreases. Lequeux and Viallefond (1980) find that such a burst can last for $10^7$ years at a rate of 0.22 M☉/yr and account for the observed O/H abundance.

We thus reaffirm here that IZW18 is a young galaxy with an observed metallicity undiluted in the HI halo. We speculate that the HI halo of this galaxy has a much lower metallicity than the optical core. As far as metal abundance is concerned, there is no problem any more to invoke a possible relationship between dwarf ellipticals and dwarf compact galaxies as suggested by Lin an Faber (1983).

We realize in this discussion that mixing problem in giant HII regions has little been discussed in the past and this problem clearly deserves further study.

REFERENCES

- Kunth, D., and Sargent, W.L.W., 1983, Ap.J., 273, 81.
- Kunth, D., and Sargent, W.L.W., 1985, Ap.J. Letters, submitted.
- Lequeux, J., Maucherat-Joubert, M., Deharveng, J.M., and Kunth, D., 1981, Astron. Astrophys., 103, 305.
- Lequeux, J., and Viallefond, F., 1980, Astron. Astrophys., 91, 269.
- Lin, D.N.C., and Faber, S.M., 1983, Ap.J. Letters, 266, L21.
- McMahon, R., Terlevich, R., Hazard, C. and Irwin, M, 1984, in Astronomy with Schmidt-type Telescopes IAU Coll. N° 78, Ed. M. Capaccioli (Dordrecht Reidel), p.395.

# BOLOMETRIC LUMINOSITY EVOLUTION

Carol J. Lonsdale
IPAC, Caltech/JPL
CalTech 100-22
Pasadena, CA 91125
USA

ABSTRACT. The observations made by IRAS are capable of detecting evolution in the OB star formation rates of galaxies at cosmologically interesting redshifts if that evolution takes the form of star formation rates ten or more times higher than seen in nearby galaxies. Spectroscopic follow-up observations of IRAS-selected galaxies, and IRAS source counts, indicate that there may indeed exist a rare population of galaxies with such high star formation rates. The most luminous found so far approach quasars in luminosity.

## 1. IRAS SENSITIVITY TO EVOLUTION OF STAR FORMATION RATES

IRAS measured fluxes in four infrared wavelength bands, 12, 25, 60, and 100 μm. In galaxies, emission in these bands arises in large part in massive star forming complexes. The important parameter of interest for the study of galaxy evolution is the total far-infrared luminosity, L(FIR). To first order, L(FIR) for a star forming complex is the bolometric luminosity of the OB stars, and thus L(FIR) measures the star formation rate (SFR) of massive stars of a stellar population, and L(FIR)/L(blue) is a measure of the current SFR compared to the past average SFR (over the last billion years or so). The evolution of L(FIR) with time is clearly of interest for studies of galaxy evolution.

Most galaxies are detected by IRAS only in the two longest wavelength bands. Since the 60 and 100 μm bands fortunately straddle the peak of the far-infrared energy distribution in galaxies, their flux can be used to estimate L(FIR), with extrapolations to either side. However these corrections remain a source of uncertainty in deriving the bolometric luminosity.

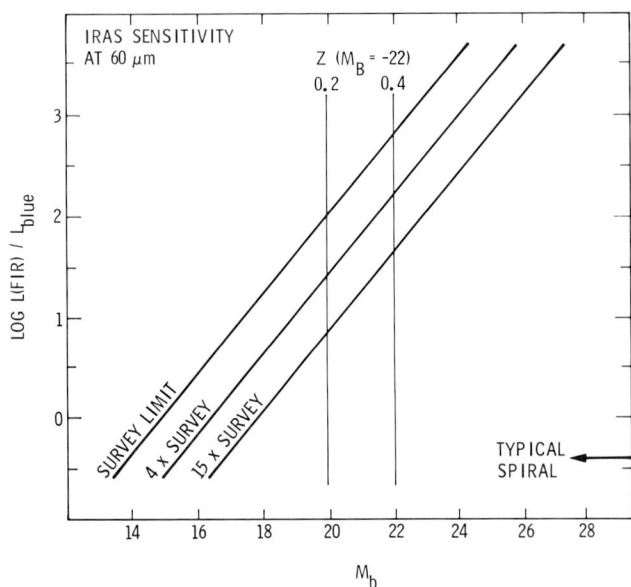

Figure 1. IRAS sensitivity to evolution of massive star formation rates, as measured by the far-infrared to blue luminosity ratio. Indicated for orientation are location of a galaxy with absolute blue magnitude -22 at redshifts of 0.2 and 0.4, and the far-infrared to blue luminosity ratio of a typical nearby giant spiral.

Figure 1 illustrates the sensitivity of IRAS to evolution in L(FIR) (calculated as described above). At a 60 μm sensitivity of about 0.5 Jy (survey mode), it may be seen that a galaxy of absolute B mag. -22 would have to have an L(FIR)/L(blue) ratio of ~100 to be detectable at z~0.2, equivalent to an SFR 100 times higher than that of the Galaxy. Clearly the study of modest SFR evolution is beyond the capability of the IRAS all-sky survey. However, IRAS also made deeper observations of selected sky areas, and sample sensitivities, comfortably above the 60 μm confusion limit, are also shown in Figure 1. At the deepest levels one may hope to probe some interesting regions of the d(SFR)/dt domain, but we are still clearly restricted to quite dramatic star formation events at redshifts of interest.

It should be noted that there are some uncertainties in the interpretation of L(FIR), besides the extrapolation to longer and shorter wavelengths. These include an observed large amount of scatter in the relationship between far-infrared and Hα fluxes in nearby disk galaxies (Lonsdale and Helou 1985), which may be due to large and uncertain extinction at Hα or dust competition for Lyman continuum photons, but which may also be due to additional dust heating by stars of later spectral type than OB. The importance of the latter depends on the reddening law, the IMF, the metallicity and HII region-molecular cloud geometry, which could be in turn a function of Hubble type, arm class, dynamical state, etc. These considerations may be insignificant in the face of luminous star forming events of the kind that IRAS is sensitive to at large distances, however it behooves us to understand nearby systems of both modest and violent SFR before interpreting distant ones.

## 2. OBSERVATIONS OF IRAS-SELECTED GALAXIES

In order to learn first hand what the properties of the galaxies that IRAS is selecting are, we have begun a large program of follow-up optical, infrared and radio observations. I will discuss the optical spectroscopy which we have done at the Palomar 5m telescope (Persson et al. 1984 - Lonsdale et al. 1985). Other data include JHK photometry, CCD images, VLA 6 and 2 cm data and Arecibo 21 cm data. The Palomar samples include the 81 galaxies detected in the IRAS minisurvey region (a small region of the survey studied in detail at the beginning of the mission - Soifer et al. 1984). We have 3400 Å - 1 μm spectra of ~50% of this sample, so far. The objects tend to be multiple, multicolored objects with high equivalent width emission lines - the median Hα equivalent width is ~30 Å. The objects tend to be small and distant, with several at $z > 0.1$ and the record holder at $z = 0.37$. The emission lines often appear to be centrally concentrated. For $H = 75$ km/s/Mpc, a few of the objects have $L(FIR) > 10^{12}$ $L_\odot$, which approaches the realm of quasars. The median luminosity for the sample is $7 \times 10^{10}$ $L_\odot$.

The excitations of the minisurvey galaxies are mostly HII-region like. On a Baldwin, Phillips and Terlevich (1981) 5007/4861 vs. 6584/6563 diagram most objects fall near the HII-region line, with a small percentage in the region occupied by power-law excited objects. The line widths are narrow: <450 km/s FWHM at Hβ for the HII-region-like spectra. None of the high excitation objects is an obvious Seyfert 1 galaxy - the most interesting is a strong-lined Seyfert 2 galaxy with spiral structure and an extended double radio source, IRAS 0421+040 (Beichman et al. 1985).

On the surface of it, it seems that the large L(FIR)s observed in the sample are caused by high SFRs, ranging up to several hundred $M_\odot$/yr in O, B and A stars. These objects would be of considerable interest to galaxy evolution studies, since they exhibit very large starbursts at cosmologically interesting redshifts. The SFRs implied in these galaxies are much higher than those discussed for the Butcher-Oemler, Dressler-Gunn blue galaxies (see paper by Dressler in this volume). Since many of the objects appear to be peculiar or multiple it may be that the star formation is interaction induced, though many of the objects are so distant that it is not possible to see direct morphological evidence for interactions.

Since such enormous SFRs are necessary to explain the high L(FIR)s observed, an alternative explanation for the far-infrared luminosity must be considered - surface star formation with an underlying far more efficient energy source, a dust-buried quasar. An optically thick dust shell of $0.5 - 1.0$ kpc radius and $6 \times 10^8$ $M_\odot$ of dust could hide a $10^{12}$ $L_\odot$ quasar and convert its energy to infrared photons effectively. To detect such a buried quasar is difficult, perhaps requiring detection of broad wings on near-infrared recombination lines.

On the other hand, there is some evidence that it is not necessary to invoke dusty quasars, except to assuage the discomfort engendered by the thought of tremendous starburst episodes. Many of the minisurvey sample follow the H$\alpha$ vs. far-infrared flux relation seen in nearby galaxies (Lonsdale and Helou 1985), in which the far-infrared flux certainly originates in star formation. This means that enough star formation is seen in the emission line spectrum to account for the infrared luminsity. Other objects do lie above the relation in the direction of higher far-infared luminosity per unit H$\alpha$ luminosity, i.e. as expected if a buried source were contributing to the far-infrared luminosity. Again it is not _necessary_ to invoke dusty quasars to explain these. In some of them, the H$\alpha$ flux may have been underestimated due to the narrow slit (2"), or the buried source(s) may be young stars. In this context it should be noted that galaxies with starburst nuclei lie above the spiral line in the same direction (high far-infrared luminosity per unit H$\alpha$ luminosity), presumably indicating higher optical depth in starburst conditions.

The luminosity function of the minisurvey galaxy sample converges with that of normal spirals at the faint end ($<10^{10}$ $L_\odot$ and with that of Seyfert galaxies at the luminous end ($10^{12}$ $L_\odot$, i.e. IRAS survey-selected galaxies range from normal, nearby spirals to a rare class of very luminous objects which can be seen to cosmologically interesting distances.

## 3. IRAS SOURCE COUNTS

I would like to close with some tantalising new results from a study of source counts in the IRAS data base. This is based on work by R. Windhorst of Mt. Wilson and Las Campanas Observatories. Compared to predicted non-evolving counts in Euclidean space, IRAS 60 μm counts in deep surveys ~4 times fainter than the survey are a factor of 5 - 10 times too high. The sources used in this comparison are known to be galaxies from spectroscopic observations at Las Campanas Observatory. A similar offset is seen in counts in some ultradeep IRAS fields, ~15 times fainter than the survey. The offset from non-evolving relativistic cosmological model counts is even larger.

It thus may be that the promise of Figure 1 will be realised - that there does exist a population of objects at cosmologically interesting redshifts which are undergoing quite significant star formation events, examples of which we are seeing in the minisurvey sample. Study of their evolution, however, should proceed with caution because of uncertainties in the internal IRAS calibration between the survey and the deeper observations, and because of uncertainties in the infrared emission mechanism and in population homogeneity. Perhaps most importantly, the estimation of the bolometric luminosity from the observed far-infrared spectral energy distribution depends more heavily on the uncertain extrapolation to short wavelengths as the redshift increases.

## 4. REFERENCES

Baldwin, J.A., Phillips, M.M., and Terlevich, R. 1981, Pub. A.S.P.,**93**,5.
Beichman, C.A., et al. 1985, Astrophys.J.,**293**,148.
Lonsdale, C.J. and Helou, G. 1985, in preparation.
Lonsdale, C.J., Persson, S.E., Soifer, B.T., Beichman, C.A., Neugebauer, G., and Houck, J. 1985, in preparation.
Persson, S.E., Lonsdale, C.J., Beichman, C.A., Soifer, B.T., Neugebauer, G., and Houck, J. 1984, Bull. Am. Astron. Soc., **16**,471.
Soifer, B.T., et al. 1984, Astrophys. J. Lett., **278**,L71.

## DISCUSSION

Renzini: What are the radio properties of the minisurvey galaxies - can they help with the quasar/starburst problem?

Lonsdale: About half of the sample were detected at 6 cm at the VLA. The radio to far-infrared flux ratios are consistent with starbursts like that in M82.

O'Connell: What infrared luminosities do the Seyfert galaxies you detected have?

Lonsdale: They have average infrared luminosities – they are certainly not responsible for all the high far-infrared luminosities in the sample.

Newbury: From your infrared fluxes for these galaxies you suggest that 100 $M_\odot$/yr of gas is converted to O and B stars. Assuming a Salpeter IMF with a lower mass cutoff at 0.1 $M_\odot$ implies a total SFR of about 20,000 $M_\odot$/yr, and a lower limit of 1 $M_\odot$ implies an SFR of about 1000 $M_\odot$/yr. Thus unless only O and B stars are made by such galaxy interactions, this implies a very high gas consumption rate.

Lonsdale: The SFRs as high as 100s of $M_\odot$/yr apply only to the most extreme objects in the sample. Your rates seem on the high side to me but I certainly agree that the infrared luminosities imply tremendous rates if a Salpeter IMF is extended to lower masses. Perhaps this means that starbursts like this simply cannot have a Salpeter-like IMF extending to low masses.

Lequeux: I would like to comment that there is a case of merger where observations strongly suggest a low-mass cutoff in the burst IMF of a few $M_\odot$. Thus the problem of gas supply to bursts may not be as severe as it looks when using, e.g., a Salpeter IMF. The object is Mkn 171 (Augarde, R., Lequeux, J., Astron Astrophys, in press).

GALAXY ENCOUNTERS AND THE HOLMBERG EFFECT

Barry F. Madore
David Dunlap Observatory
Department of Astronomy
University of Toronto
60 St. George St.
Toronto, Canada  M5S 1A1

ABSTRACT. It is pointed out that interacting double galaxies are the most common form of peculiar galaxy found in recent compilations, indicating that interactions which substantially alter a galaxy's morphology are relatively common. Continuing effects of the encounters on the star formation rate may be revealing themselves in the *Holmberg Effect*, which is a strong color correlation and coupling of apparent Hubble types found among binary galaxies.

1. Introduction.

From the initial results on extragalactic objects identified with *IRAS* sources, it has become increasingly clear that the objects most strongly emitting in the infrared are galaxies which are morphologically peculiar or disturbed through interactions, as judged by optical images. Since these encounters so profoundly effect both the structure of the galaxy and the instantaneous rate of star formation, the frequency and duration of such events should be included in any consideration of the long-term evolution of galaxies as a whole.

As a partial solution to outlining the history of galaxy evolution, as modulated by catastrophic events, Arp and Madore (1985) have recently completed a systematic search of the new *ESO/SERC* southern sky survey, for peculiar and interacting galaxies and associations. Attention was paid to recording the total number of galaxies inspected so that a relative frequency of interactions could be calculated. Out of the one hundred thousand galaxies looked at, some six percent were catalogued as peculiar. Among those six thousand peculiar systems at least thirty percent were chosen because of the noticeable effects directly ascribable to collisions, interactions and encounters. In fact, *Interacting Double Galaxies* made up the largest single category in the *Catalogue*. Table I gives the categories and the statistics for the various types of galaxies found in the *Arp-Madore Catalogue*.

2. Discussion.

Given that one galaxy in a hundred at the present time shows some sign of major disturbance due to interaction, and that these signs might be expected to last for about one free-fall time, then it is not inconceivable that in the course of one Hubble time (i.e., 100 free-fall times) every galaxy could have undergone at least one major collision or externally induced disruption. Certainly, as the mean density of galaxies increases into the past it is even more likely that interactions had a major role in the star formation histories of "individual" galaxies.

Even today the effects of tidal interaction may be more important than indicated above. Star formation may be modulated by the presence of companions, even without the major disruptions of the global structure seen in the catalogued peculiar galaxies. One striking piece of evidence for this may be the *Holmberg Effect* wherein it was noticed (Holmberg 1937) that galaxies in pairs have an unusually high incidence for being associated with galaxies of the same morphological type. This correlation has been verified by a number of authors (e.g., Karachensev and Karachenseva 1975, Page 1975, Noerdlinger 1979) for a variety of binary galaxy samples. Not only is the morphological type strongly correlated but as is shown in Figure 1 the colors of the two galaxies are strongly coupled.

Larson and Tinsley (1974) and Huchra (1977) noted that the galaxies most deviant in the colour-colour diagram, toward regions indicating strong bursts of star formation, were the closely interacting galaxies already listed in the *Peculiar Galaxy Catalog* of Arp (1964). Markarian galaxies, discovered because of their intense star formation activity giving rise to optical emission lines, are also found to have a high incidence of being found in binaries. So high is the binary frequency that at least six Markarian galaxies have companions that are also Markarian galaxies, a statistic expected to occur with a probability of less than 1:60,000 if left to chance (Karachensev and Karachenseva 1975). This latter statistic can be thought of as yet another example of the *Holmberg Effect* taken to the extreme.

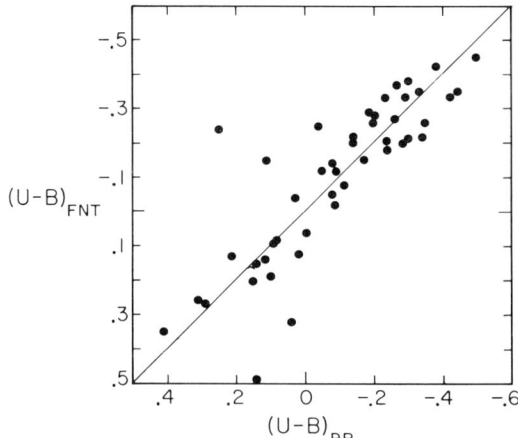

Figure 1: The correlation of the *(U-B)* colors of galaxies in binary pairs. The data are from Tomov (1978)

It is conceivable that initial conditions were sufficiently similar on scales comaparble to the binary galaxy dimensions that the resulting individual galaxies in double systems always had the same Hubble type and have simply evolved in unison. However it is of interest to note that the Hubble type of a galaxy can be derived from one or a combination of observed properties: bulge-to disk ratio, pitch angle of the arms, and/or degree of resolution into stars. Unfortunately few classifiers of galaxies explicitly state which criterion or combination of criteria define their quoted Hubble types in any specific case.

In the case of closely interacting pairs of galaxies where the star formation rate is enhanced due to the encounter the apparent Hubble type may be temporarily modified, and the criterion by which it is judged may be biased. Consider the extreme case in which star formation is extremely rampant across the face of the galaxy and its companion, being stimulated by an intense encounter which disturbs the spiral structure and results in a burst of star formation activity. During this phase the spiral structure, by definition, will not be well organized and so Hubble-typing criteria based on organized pitch angles will probably not bear too strongly on the typing. Similarly, if the burst is strong enough to affect the colours of the galaxies it should also be contributing to the luminosity in such a way that the disk-to-bulge ratio should increase. But perhaps most importantly the star formation activity will be seen as an increase in the degree of resolution of the disk. Therefore, independent of the Hubble type that the galaxy would have had in isolation an encounter which stimulates star formation will also modulate the apparent Hubble type based on blue light indicators. In fact if the interaction is in any way symmetric between the two componentsthe stimulated star formation will tend to bring the colours and Hubble types of the two galaxies into phase. After the main interaction, as the galaxies separate, the fading of the impulse should also be co-ordinated between the two galaxies and so their colors should stay correlated for some time after.

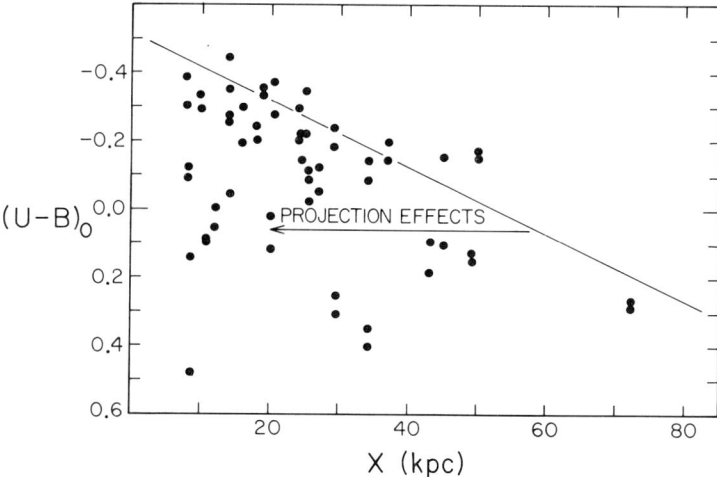

Figure 2: *(U-B)* colors of individual galaxies in binary pairs plotted as a function of their apparent separation in kpc. The upper envelope suggests that closer galaxies are bluer.

Table 1.

Categories and Relative Statistics for Peculiar and Interacting
Galaxies Found in the Arp-Madore Catalogue

| Numerical Code | Brief Description | Relative Percentage |
|---|---|---|
| 1 | Galaxies with interacting companion(s) | 5.4 |
| 2 | Interacting doubles (galaxies of comparable size) | 12.6 |
| 3 | Interacting triples | 2.0 |
| 4 | Interacting quartets | 0.4 |
| 5 | Interacting quintets | 0.1 |
| 6 | Ring galaxies or morphologically similar objects | 3.1 |
| 7 | Galaxies with (linear) jets | 2.4 |
| 8 | Galaxies with apparent companion(s) | 11.5 |
| 9 | M51-types (companion at end of spiral arm) | 1.9 |
| 10 | Galaxy with peculiar spiral arms | 4.0 |
| 11 | Three-armed spiral | 0.5 |
| 12 | Peculiar disks (major asymmetry or deformation) | 2.8 |
| 13 | Compact (very high surface brightness) galaxy | 6.4 |
| 14 | Galaxies with prominent or unusual dust absorption | 1.6 |
| 15 | Galaxies with tails, loops of material or debris | 3.4 |
| 16 | Irregular or disturbed, apparently isolated galaxy | 4.2 |
| 17 | Chains (four or more galaxies aligned) | 3.9 |
| 18 | Groups (four or more galaxies not aligned) | 4.8 |
| 19 | Clusters (only very conspicuous, rich systems) | 1.6 |
| 20 | Dwarf galaxies (low surface brightness) | 6.7 |
| 21 | Stellar objects with associated nebulosity | 0.7 |
| 22 | Miscellaneous (rare or distinctive objects) | 1.3 |
| 23 | Close pairs (not visibly interacting) | 11.3 |
| 24 | Close triples (not visibly interacting) | 5.5 |
| 25 | Planetary nebulae | 0.9 |

Some evidence for a correlation of the interaction strength with separation of the galaxy pairs is given in Figure 2. The (U-B) colors of galaxies found in binary systems (Tomov 1978) are ploted as a function of their apparent separation in kpc. Projection effects will tend to bias the data toward smaller apparent separations as compared to the true separation. With this in mind it is interesting to note the dirth of galaxies below the fiducial line drawn in for reference. The plot as given is consistent with galaxies becoming bluer as their absolute separation decreases (red galaxies at close apparent separations being due to projection).

Some tests of this hypothesis suggest themselves. If for isolated field galaxies the mass distribution (e.g., disk-to-bulge ratio) and therefore the rotation curve play an important role in regulating star formation, spiral structure, etc., then it might be hoped that these deeper seated aspects of the galaxy might not be as disturbed by encounters. Measuring the absolute size of the bulge and/or determining the types of rotation curves for galaxies in binary pairs may reveal the Hubble type that the galaxy would have if found in isolation. Similarly devising a scheme of Hubble typing based on infrared imaging would be of interest in this context too.

References:

Arp, H. C. 1966, *Atlas of Peculiar Galaxies*
  (Pasadena: Calif. Institute of Technology) = *Ap. J. Suppl*, **14**, 1.

Arp, H. C., and Madore, B. F. 1985, *A Catalogue of Southern Peculiar Galaxies and Associations*, (Toronto: David Dunlap Observatory).

Holmberg, E. E. 1937, *Ann. Lund Obs.*, No. 6.

Huchra, J. P. 1977, *Ap. J.*, **217**, 928.

Karachensev, I. D., and Karachenseva, V. E. 1975, *Sov. Astron.*, **18**, 428.

Larson, R. B., and Tinsely, B. M. 1974, *Ap. J.*, **192**, 293.

Sandage, T. 1975 in *Stars and Stellar Systems*, IX,
  *Galaxies and the Universe*, eds. A. R. Sandage, M. Sandage,
  and J. Kristian, (Chicago: Univ. Chicago Press), p. 514.

Noerdlinger, P. D. 1979, *Ap. J.*, **229**, 877.

Tomov, A. N. 1978, *Sov. Astron.*, **22**, 540.


Questions:

de Jong: You have presented very exciting material. In order to carry out a meaningful statistical analysis of this sample it is important to know how complete it is, particularly if you want to compare with "normal" galaxies, and establish the frequency of peculiarity.

Madore: Our sample includes all peculiar galaxies found down to an apparent diameter limit. Accordingly we are neither volume nor magnitude limited; but we did inspect and count all galaxies down to this same angular size limit, so in that sense we are complete. Eight percent of the sample looked at was considered peculiar enough to catalogue. It should be remembered that among the various types of peculiarities many are distance dependent at different rates.

O'Connell: Did your new survey detect any class of peculiarities which did not show up in the first Arp Atlas?

Madore: One example of a new class of object identified in the Catalogue is the so-called "bright-rimmed" spiral class. These galaxies are globally asymmetric with off-set nuclei and outer spiral structure, to one side only, compressed and enhanced along the outer perimeter.

Koo: How would you have classified the Milky Way, at a comparable distance, given the presence of the LMC and SMC (and M31) ?

Madore: Depending on whether our Milky Way shows any obvious signs of interaction with the Magellanic Clouds or not we would have classified the system as *Galaxy with Interacting Companions* or *Galaxy with Apparent Companions*. M31 would probably be too far distant to be included as a *Close Pair* by our criteria.

Eisenhardt: Is there any difference in the *Holmberg Effect* for your Code 2: *Interacting Doubles* as opposed to your Code 23: *Close Pairs* ?

Madore: The sample defining the original *Holmberg Effect* is a northern hemisphere sample, whereas our *Catalogue* is strictly drawn from the *Southern Sky Survey*. For the southern sample there is no photometery nor good Hubble types, whereas in the north the sample has not been classified on our system. However, the Larson and Tinsley (1974) study indicates that interacting doubles generally have the most deviant colors from the norm.

# THE EFFECTS OF GALAXY - GALAXY INTERACTIONS ON NUCLEAR ACTIVITY

J. M. van der Hulst
Netherlands Foundation for Radio Astronony
Postbus 2, 7990 AA Dwingeloo, the Netherlands

E. Hummel
Max Planck Institut für Radioastromie
Auf dem Hügel 68, D-5300 Bonn 1, FRG

W. C. Keel
National Optical Astronomy Observatories
P. O. Box 26732, Tucson, AZ 85726-6732, USA

R. C. Kennicutt, Jr.
Department of Astronomy
University of Minnesota
116 Church St. SE., Minneapolis, MN 55455, USA

## 1. INTRODUCTION

The notion that galaxy - galaxy interactions do affect the properties of galaxy nuclei is not new. The earliest indications come from surveys of the radio continuum emission from spiral galaxies. Sulentic (1976) and Stocke (1978) demonstrated that galaxies in close pairs show an excess of radio emission. With higher spatial resolution Hummel (1981b) showed that this increase is largely due to an increase of the emission from the central region. Also Condon et al. (1982) noted that a high fraction of spiral galaxies with a strong central radio source occur in multiple and interacting galaxies. Further evidence for enhanced nuclear activity in interacting or multiple galaxies came from studies of the IR emission (Joseph et al. 1984, Lonsdale et al. 1985), which is also enhanced like the radio emission. In addition it can be demonstrated that Seyfert galaxies have neighbours more often than field galaxies (Dahari 1984), and that a large fraction of nearby QSOs have companion galaxies (Hutchings and Campbell 1983).

In order to further investigate the above trends we have undertaken a systematic optical and radio study of a large sample of interacting spiral galaxies. We chose spiral galaxies because they have significant amounts of gas which enable us to trace the activity whether in the form of enhanced star formation or non-thermal nuclear activity. The first results from the spectroscopy of our sample have already been reported elsewhere (Kennicutt and Keel 1984, Keel et al 1985) and the purpose of this contribution is to merely summarize the main results we have obtained so far.

## 2. OBSERVATIONS AND SAMPLE SELECTION

The sample selection is described in detail in Keel et al (1985). The following samples were defined. A complete sample of interacting or multiple galaxies drawn from an unpublished catalogue of probable pairs and groups by T. S. van Albada. From this catalogue we selected all galaxies with a probability of $\geq 90\%$ that they have associated companion(s), with integrated magnitudes $B_T \leq 13$, north of $\delta = 0°$, and $8^h \leq \alpha \leq 16^h$. In addition we selected a sample of strongly interacting galaxies (very distorted galaxies and galaxies with very small projected separations) from the Arp atlas (Arp 1966) in order to study the effects of strongly disturbing encounters. For comparison we formed a control sample from the spectrophotometric survey of Keel (1983). The average distance and absolute magnitude of the complete and control samples are very similar. The Arp sample comprises galaxies which are on average four times more distant and about 0.6 magnitude brighter.

The data consists of spectrophotometry over the range 3700 - 6900 Å for the nuclei of these samples taken through a 4.7" circular aperture using the image dissector scanner (IDS) on the 1.5 m UM-UCSD telescope at Mount Lemmon. The control sample spectrophotometry was obtained with the IDS system on the 1 m Nickel reflector at Lick Observatory. A few galaxies were observed at other telescopes with similar instruments. The spectra yielded line intensities, equivalent widths and line ratios for the common emission lines $H_\alpha$, $H_\beta$, [OIII], [OII], [OI], [NII], and [SII]. The emission spectra were classified into HII region spectra, low-ionization (LINER) spectra, and high-ionization (Seyfert) spectra,

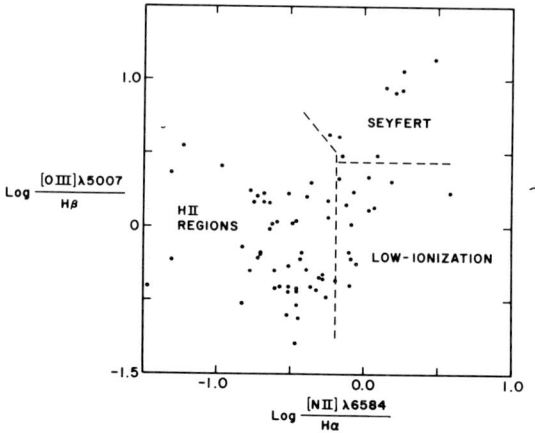

Figure 1. Line-ratio diagram [OIII]/$H_\beta$ vs. [NII]/$H_\alpha$. This was the primary tool in emission line classification. Adopted boundaries of the classes are shown as dashed lines.

essentially following Baldwin, Phillips and Terlevich (1981). The classification is graphically illustrated in Figure 1.

Further observations which are still in progress and will not be reported here include high resolution VLA observations at 21 cm wavelength of the nuclei of the galaxies in all three samples, and narrow band imaging at $H_\alpha$+[NII] and in red continuum to study the distribution of radio continuum emission and of the distribution of the emission line gas in the nuclear regions.

## 3. RESULTS

The spectrophotometry gives us various quantitative tracers of nuclear activity. We used the luminosity and equivalent width of the $H_\alpha$ line and [OIII]/$H_\beta$, [NII]/$H_\alpha$, and [SII]/$H_\alpha$ line ratios. The line ratios allow us to discriminate between the HII region type nuclei, the LINERS and the Seyfert nuclei as outlined above. The $H_\alpha$ luminosity and in particular the $H_\alpha$ equivalent width are good indicators for the present starformation rate.

In order to compare the three samples we constructed distributions of $H_\alpha$ luminosity and $H_\alpha$ equivalent width. The distributions for the control sample were corrected to reflect the absolute magnitude distributions of the complete and Arp samples in order to avoid possible luminosity biases in the comparison. This adjustment is only of significance for the comparison with the systematically brighter Arp sample. The distributions of $H_\alpha$ luminosity and equivalent width are shown in Figures 2 and 3. It is immediately obvious that both the complete and the Arp sample have a significantly increased $H_\alpha$ luminosity

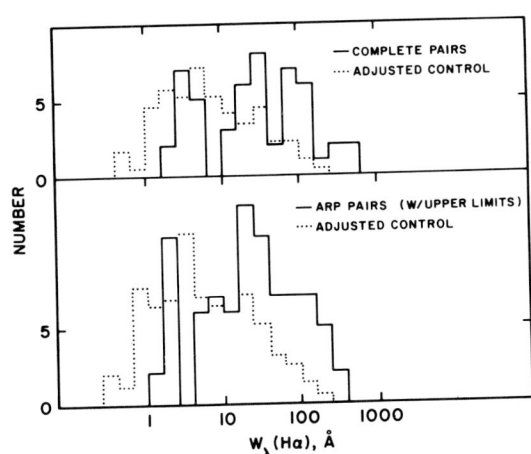

Figure 2. Distributions of $H_\alpha$ equivalent widths for interacting and field galaxy nuclei. The dashed histograms represent the field distribution, adjusted to match the Hubble type distribution of each interacting sample.

and equivalent width as compared to the control sample. These results clearly indicate that the nuclear activity, whatever its precise nature, is enhanced in galaxies which are in pairs or groups as compared to relatively isolated systems. This confirms the results found in radio continuum surveys (Hummel 1981a,b, Condon et al. 1982).

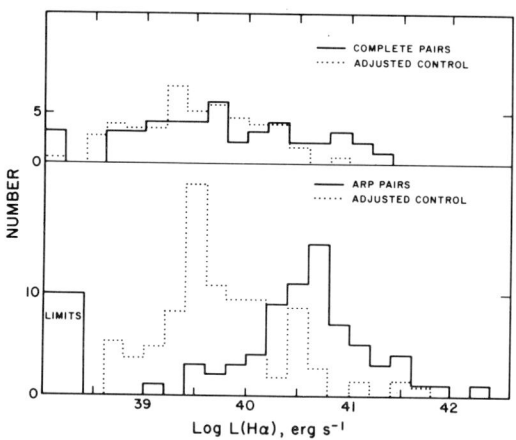

Figure 3. Distribution of $H_\alpha$ luminosity for interacting and field galaxy nuclei. The field distribution has been modified to match the $M_B$ distribution of each of the interacting samples. Upper limits are shown at the limit value; their placement does not affect the statistical significance of the comparison.

To trace the nature of the activity we used the H (HII region), L (LINER), S (Seyfert) classification and investigated the distribution of these classes with Hubble type for both the control sample and the interacting (complete + Arp) sample. The relative fractions of nuclei in these classes change strongly with Hubble type in field spiral galaxies (Stauffer 1982, Heckman 1980a, Heckman, Balick and Crane 1980) so that a comparison per Hubble type seems appropriate. The complete and Arp samples were put together because they showed similar distributions. A graphical representation of the distribution of the emission line classes with Hubble type for the control and interacting samples is given in Figure 4. The interacting sample shows more frequently HII region nuclei at early Hubble type (Sa and Sb) than does the control sample. At later type there are more Seyferts among the interacting galaxies than among the isolated galaxies. Both seem to operate at the expense of the low-ionization nuclei which become less frequent at all types in the interacting sample as if the LINER phenomenon is masked by an induced burst of starformation, or is turned into a Seyfert.

Kennicutt and Keel (1984) reported a higher incidence of Seyfert nuclei in close pairs. The present sample allows a more accurate assessment of the overall frequency of the Seyfert phenomenon. The Seyfert (mainly Seyfert 2) frequency in the interacting sample is 8.5% compared to 5.6% for the control sample and 5.1% found for a group of

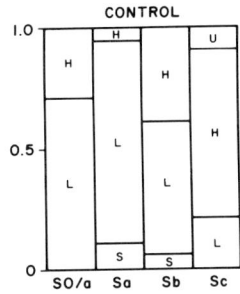 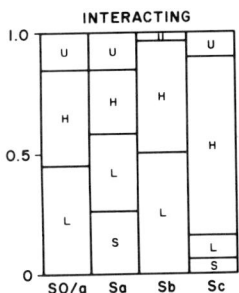

Figure 4. Distribution of emission classes with Hubble type for field and interacting (Arp + complete) nuclei. The relative fractions at each type are shown for HII, low-ionization, and Seyfert nuclei by H, L, and S; objects with very weak or undetected emission are shown by U. Their presence in the interacting samples is probably a distance effect, and shows the amount of variation possible due to very weak emission.

436 spirals by Phillips, Charles and Baldwin (1983). A large fraction of the Seyfert nuclei indeed occur in very close pairs: 6 (24%) of 25 spirals with companions within half a galaxy radius have Seyfert nuclei. It should also be noted that apart from the high incidence of Seyferts another 25% of the very close pairs have very luminous starburst nuclei. Over half of the very close pairs have unusual nuclei, and many may have Seyferts which we can not observe. On the other hand we looked for a more systematic relation between $H_\alpha$ luminosity and equivalent width, and projected separation but did not find any significant trend, although it is definitely a significant result that the galaxy - galaxy interactions do increase the nuclear $H_\alpha$ luminosities and equivalent widths.

In his radio survey Hummel found another class of objects with enhanced central radio emission: the barred spirals. The data we have allow a study of the possible effect of a bar on nuclear activity. Heckman (1980b) looked into this using fewer galaxies and found a higher incidence of nuclear HII regions in barred spirals. Here we used the somewhat different approach of constructing $H_\alpha$ luminosity functions in a fashion similar to what Hummmel (1981a,b) did for his radio survey data. We added the control and complete samples and constructed separate luminosity functions for the LINER galaxies and for the HII region-nucleus galaxies. We selected galaxies with absolute magnitudes (assuming H = 100 km/s/Mpc) between -21 and -18 . The median distances for galaxies in these two emission line classes are similar. The luminosity functions are shown in Figure 5. The HII region nuclei show the same trend as Hummel found for the radio emission: a significant enhancement of the $H_\alpha$ luminosity for galaxies with bars. For the galaxies with LINER nuclei the presence of a bar does not make any difference. These results are consistent with those reported by Heckman

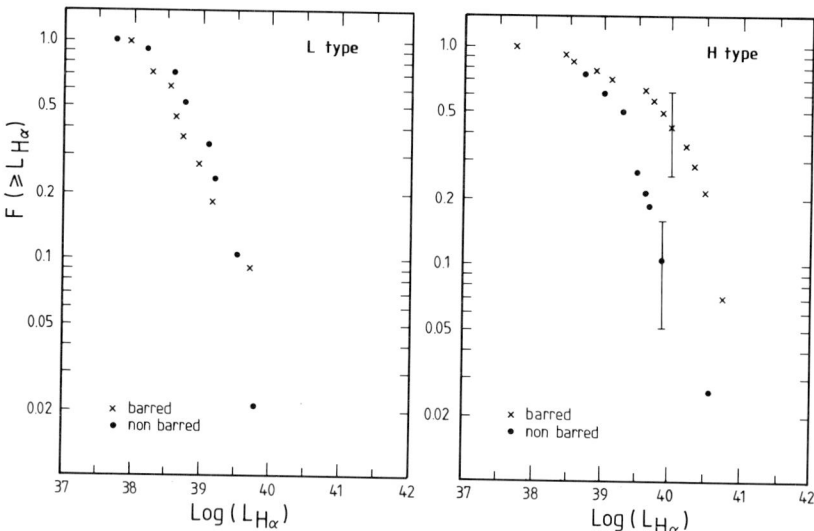

Figure 5. Fractional luminosity functions for the $H_\alpha$ emission in low-ionization (L) and HII (H) nuclei for the barred (crosses) and non-barred (dots) galaxies in the complete interacting + control sample. The L type nuclei are shown in the left panel, the H type nuclei in the right panel. Error bars are indicated in the righthand figure.

(1980b). The enhanced starformation in the nuclei in barred spiral galaxies is probably understandable in view of theories of the gas flow in and around bars as discussed by e.g. Tubbs (1982). Why, however, the galaxies with LINERs form a separate group which is as insensitive to the effect of a bar as to the effect of an encounter is at present unclear.

It is quite conceivable that the LINER phenomenon is not al all associated with nor affected by a galaxy encounter, but that the LINER phenomenon depends on details of the galaxy's own structure or is sensitive only to events very near the nucleus rather than the larger scale gas flows. It could be significant that LINERs occur predominantly in galaxies of early type which have fairly large bulge to disk ratios as already pointed out by Heckman (1980a).

## 4. CONCLUDING REMARKS

Our spectrophotometry has clearly indicated that the presence of one or more companion galaxy enhances the nuclear starformation rate in spiral galaxies of all Hubble types. The incidence of Seyfert galaxies is also significantly higher in galaxies in pairs or groups, with a preference to occur in very close pairs. The most straightforward explanation is that the interaction causes a redistribution of angular momentum in the inner disk of a galaxy which causes infall of gas in the central region. This inflow of gas may either fuel an existing nuclear engine to cause e.g. the Seyfert phenomenon or lead to a burst of starformation.

The effects of galaxy encounters on the nuclear activity of galaxies should perhaps be folded into evolutionary models since presumably many galaxies have had encounters in the past when the universe was denser.

## 5. REFERENCES

Arp, H., 1966, Atlas of Peculiar Galaxies, CIT, Pasadena and Astrophys. J. Suppl. 14, 1.
Baldwin, J., Phillips, M. and Terlevich, R., 1981, Publ. Astron. Soc. Pac. 93, 5.
Condon, J.J., Condon, M.A., Gisler, G. and Puschell, J.J., 1982, Astrophys. J. 252, 102.
Dahari, O., 1984, Astron. J. 89, 966.
Heckman, T.M., 1980a, Astron. Astrophys. 87, 152.
Heckman, T.M., 1980b, Astron. Astrophys. 88, 365.
Heckman, T.M., Balick, B. and Crane, P.C. 1980, Astron. Astrophys. Suppl. 40, 295.
Hummel, E., 1981a, Astron. Astrophys. 93, 93.
Hummel, E., 1981b, Astron. Astrophys. 96, 111.
Hutchings, J.B. and Campbell, B. 1983, Nature 303, 584.
Joseph, R.D., Meikle, W.P.S., Robertson, N.A. and Wright, G.S. 1984, Mon. Not. R. Astron. Soc. 209, 111.
Keel, W.C. 1983, Astrophys. J. Suppl. 52, 229.
Keel, W.C., Kennicutt, R.C.,Jr., Hummel, E. and van der Hulst, J.M. 1985, Astron. J. 90, 708.
Kennicutt, R.C.,Jr. and Keel, W.C. 1984, Astrophys. J. Lett. 279, L5.
Lonsdale, C.J., Persson, S.E., Mathews, K. 1984, Astrophys. J. 287, 95.
Phillips, M.M., Charles, P.A. and Baldwin, J.A. 1983, Astrophys. J. 266, 485.
Stauffer, J.R. 1982, Astrophys. J. Suppl. 50, 517.
Stocke, J.T. 1978, Astron. J. 83, 348.
Sulentic, J.W. 1976, Astrophys. J. Suppl. 32, 171.
Tubbs, A. 1982, Astrophys. J. 255, 458.

## DISCUSSION

<u>Kunth</u>: Are your galaxies at about the same redshift. This could introduce spurious effects in your $H_\alpha$ comparisons since you are steadily using the same 4.7" aperture.

<u>Van der Hulst</u>: We used 8" apertures for the control sample to ensure that the distribution of linear, projected aperture sizes for the

control and complete sample are very similar. The galaxies in te Arp sample are, however, in general a factor four more distant so that larger areas are sampled in our spectrophotometry.

Flin: How did you define your sample of isolated galaxies?

Van der Hulst: The control sample has not been subject to very stringent criteria. However, it contains no galaxies with a >80% probability of physical association with another galaxy as defined in van Albada's catalogue.

Frogel: How do you insure that the starformation rate per unit mass is really higher in the interacting galaxies, i.e. they just might be systematically more massive than galaxies in the control sample.

Van der Hulst: We use the equivalent width of the $H_\alpha$ line as an indicator of the present starformation rate and hence use a normalized quantity, i.e. a line to continuum flux ratio.

Lonsdale: A comment and two questions: I have looked at IRAS far-infrared emission of interacting versus isolated galaxies and find a similar enhancement of two to three in the far-infrared luminosity of the interacting galaxy sample. Do you really see no dependence at all on separation, even between the very wide non disturbed pairs and the close interactions? You see an enhancement of the number of Seyferts in the close pairs, do you see an enhancement of starbursts?

Van der Hulst: We expected some dependence of the starformation rate on the projected separation especially because we found so many of the very close pairs to be unusual and looked very hard. The data suggest that there is a weak dependence on separation, but it is not statistically significant. However, a moderate trend would be marked by projection and small number statistics.

GALAXIES IN THE INFRARED


Teije de Jong
Astronomical Institute Anton Pannekoek
University of Amsterdam
Roetersstraat 15
1018 WB Amsterdam
The Netherlands


ABSTRACT. In this paper we review IRAS studies of galaxies and we attempt to systematically present the new insights that have resulted from these studies.


1. INTRODUCTION

Early infrared studies of galaxies have shown that their infrared spectral energy distribution usually peaks in the far-infrared (Rieke and Lebofsky 1979; Telesco and Harper 1980). The spectra of galaxies are very similar to those of molecular cloud complexes observed in our own galaxy. This suggests that the infrared emission of galaxies is dominated by starforming regions and that the infrared radiation is emitted by dust particles heated by recently formed massive stars.

Until recently only a handful of galaxies had been studied in the far-infrared - those that were bright enough to be detectable from the NASA Airborne Kuiper Observatory. With the launch and the succesful operation of the InfraRed Astronomical Satellite (IRAS; Neugebauer et al. 1984a) this situation has dramatically improved. The IRAS Point Source Catalogue contains about 20 000 galaxies, a unique database to systematically study star formation in galaxies.

Since the far-infrared emission provides clues about the star formation rate in galaxies one would like to investigate galaxy samples at different redshift intervals to study their evolution: the theme of this Advanced School of Astronomy. It has turned out that the IRAS sensitivity is insufficient to carry out such studies on the basis of the IRAS survey data. Present estimates indicate that less than 1 % of all galaxies detected by IRAS have redshifts larger than 0.1. The question of the infrared spectral evolution of galaxies will have to be taken up by future space projects to be launched in the 1990's - ESA's Infrared Space Observatory (ISO) and NASA's Shuttle InfraRed Telescope Facility (SIRTF).

In this review paper we attempt to summarize our present understanding of the infrared behaviour of galaxies as derived from a number of recent IRAS studies. Most of these results are based on pre-catalog data. Application of the final calibration procedures and correction for hysteresis effects in the detectors has resulted in flux density revisions of at most 20 % which may affect some of the results quantitatively but does not change any of the major conclusions to be drawn here.

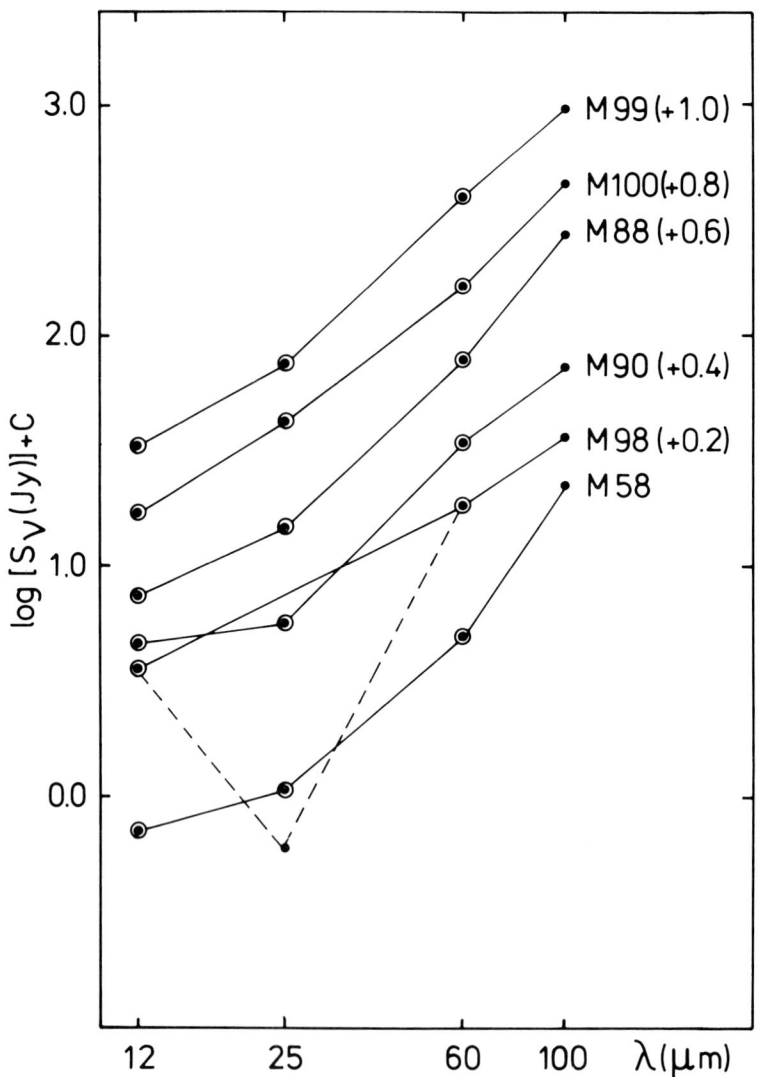

Figure 1. Infrared spectral energy distributions of Messier galaxies in the Virgo cluster core. Dots represent point source flux densties and circled dots small extended source flux densities. For clarity of illustration the energy distributions are displaced by the amounts listed in brackets.

## 2. THE INFRARED SPECTRAL ENERGY DISTRIBUTION OF GALAXIES

Typical spectral energy distributions of galaxies in the infrared are shown in fig.1 where we have plotted data for six bright spiral galaxies - those with Messier numbers - in the Virgo cluster core (de Jong 1985). All energy distributions in fig.1 are steeply rising into the far-infrared. The 60 - 100 µm color temperatures of the emission are in the range 30 - 40 K for an adopted $\lambda^{-1}$ wavelength dependence of the dust emissivity. The fact that the spectra apparently turn over beyond 100 µm indicates that an appreciable fraction of the energy has not been detected by IRAS.

Most of the Messier galaxies in the Virgo cluster are resolved by the IRAS detectors so that the point source extraction algorithm is inadequate to determine infrared flux densities. To overcome this problem the IRAS data have also been filtered using an algorithm that can handle sources with sizes up to 8 arcminutes. The database resulting from this processing forms the basis for the production of the Small Extended Source (SES) Catalog to come out in late 1985.

As indicated in fig.1 all Messier galaxies are detected by the SES processor at 12, 25 and 60 µm while at 100 µm - due to the large detector sizes - all galaxies are unresolved. One galaxy, M98, was apparently missed by the SES processing at 25 µm. The SES flux densities are typically about a factor 2 larger than the point source flux densities for galaxies at the distance of the Virgo cluster (see Table 1 in de Jong 1985). This clearly illustrates the limitations of the IRAS Point Source Catalog for galaxy work.

Due to their cluster membership all galaxies in fig.1 are at approximately the same distance so that their relative intensity is a measure of their relative intrinsic infrared luminosity. The three infrared brightest galaxies, M99 (NGC 4254), M100 (NGC 4321) and M88 (NGC 4501) are late-type spirals. Their average 40 - 120 µm luminosity is about $3 \; 10^{10}$ $L_\odot$ at a distance of 21.9 Mpc (Sandage and Tammann 1981). The infrared weaker galaxies M90 (NGC 4569), M98 (NGC 4192) and M58 (NGC 4579) are early-type spirals with infrared luminosities of order $1 \; 10^{10}$ $L_\odot$. The fact that the late-type spirals in fig.2 are brighter in the infrared than the early-type spirals is related to the star formation deficiency of early-type spirals in the Virgo cluster core to be discussed in section 4.

## 3. THE ORIGIN OF INFRARED EMISSION IN GALAXIES

Infrared emission of galaxies may originate in different galaxy components and may be produced by different radiation mechanisms. From spiral galaxies we expect to observe thermal infrared radiation emitted by cool interstellar dust particles heated by starlight from disk stars as well as emission from warm dust in molecular clouds with active star formation inside. Seyfert galaxies are known to often show additional near-infrared thermal emission from hot dust in the nucleus, probably heated by a central non-thermal source of ultraviolet radiation (Rieke and Low 1975; de Grijp et al. 1985). The infrared emission of radio-loud

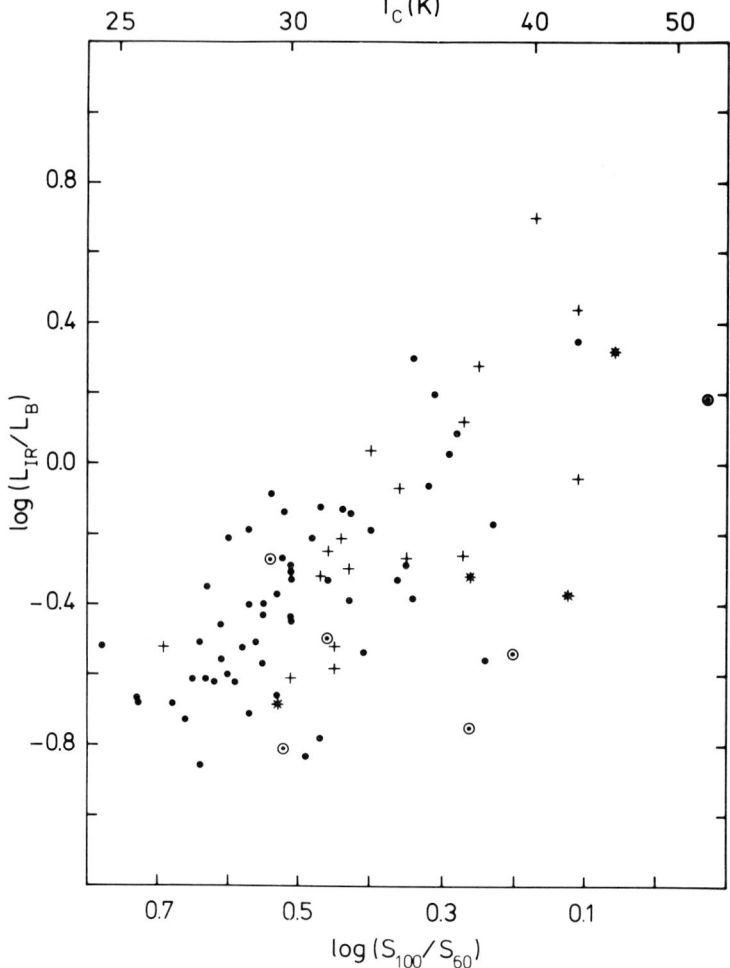

Figure 2. Far-infrared excess versus far-infrared flux density ratio for a representative sub-sample of the RSA. The upper scale gives color temperatures for a dust emissivity proportional to frequency. Separate symbols are used to indicate lenticulars (dotted circles), normal spirals (dots), barred spirals (plusses), and irregulars (stars).

quasars is probably dominated by synchrotron radiation (Neugebauer et al. 1984b).

One of the main results of the early IRAS studies of galaxies is the correlation between the far-infrared excess of a galaxy and its far-infrared colour temperature (de Jong et al 1984). These quantities are plotted in fig.2 for 89 galaxies in the Revised Shapley-Ames Catalog of Bright Galaxies (RSA, Sandage and Tammann 1981) detected during the first 100 days of the IRAS mission. The infrared excess in fig.2 is defined as $L_{IR}/L_B$ where $L_\nu = \nu S_\nu$ with $L_{IR}$ evaluated at $\nu = c/80$ μm and

$L_B$ at $\nu = c/4400$Å. Leaving lenticulars and irregulars aside, the data in fig.2 show that spiral galaxies with low color temperatures ($\sim 25$ K) emit on the average only 10% of their total luminosity in the far-infrared while infrared-warm spirals ($\sim 50$ K) emit up to 5 times more energy in the far-infrared than in the blue.

The most straightforward way to interpret this correlation is by postulating two infrared components in a galaxy: one due to interstellar dust distributed throughout the disk which reradiates a small fraction (of order the dust optical thickness of the disk) of the general visual-ultraviolet interstellar radiation field at temperatures of $\sim 25$ K; and another due to interstellar dust associated with H II regions and molecular clouds which reradiates the radiation of recently formed massive stars at a temperature of $\sim 50$ K. When the rate of star formation in a galaxy is high the warmer component increases in importance as more $\sim 50$ K radiation due to (invisible) O stars, embedded in molecular clouds, is emitted. Thus the contribution from starforming molecular clouds to the total far-infrared luminosity increases from the lower left-hand corner in fig.2 to the upper right-hand corner.

The energy emitted by starforming molecular cloud complexes is mainly produced by stars with masses $\gtrsim 10$ $M_\odot$ with main-sequence lifetimes $\lesssim 10^7$ years because less massive stars move out of the cloud during their lifetime and contribute mostly to the general interstellar infrared component as field stars. Assuming a stellar mass spectrum at birth with a slope of $-2.7$ (Garmany et al. 1982) we find for stars with $M > 10$ $M_\odot$ a conversion rate of gas into stars $\dot{M} = 3 \times 10^{-10}$ $L_{IR}$ $M_\odot$ $yr^{-1}$, where $L_{IR}$ is the far-infrared luminosity in $L_\odot$.

If we assume, guided by the data in fig.2, that typically about half the far-infrared luminosity of spiral galaxies ($\sim 1\ 10^{10}$ $L_\odot$ according to de Jong et al. 1984), is produced by starforming molecular clouds we then find that in the average spiral galaxy $\sim 3$ $M_\odot$ $yr^{-1}$ is converted into massive stars, a number comparable to what has been found for our own galaxy (Mezger et al. 1982).

## 4. A COMPARISON OF VIRGO CLUSTER AND FIELD GALAXIES

The infrared detection probability of galaxies is a strong function of morphological type. Among 165 Shapley-Ames galaxies studied by de Jong et al. (1984) no ellipticals were detected out of 26 surveyed by IRAS, while 6 out of 28 lenticulars (21%), 51 out of 62 early-type spirals (82%) and 45 out of 47 late-type spirals (96%) were detected. The detection statistics of barred and non-barred galaxies is very similar. This result is not unexpected because the amount of interstellar matter in galaxies also increases from virtually no gas and dust in ellipticals to appreciable quantities in late-type spirals.

In fig.3 we show the distribution with morphological type for two samples of galaxies: in the top panel for an optically complete sample of galaxies in the Virgo cluster core and in the bottom panel for an optically complete sub-set of the RSA, both with the same optical completeness limit. The hatched area represents the distribution of the galaxies detected by IRAS, the outer contour includes all galaxies in

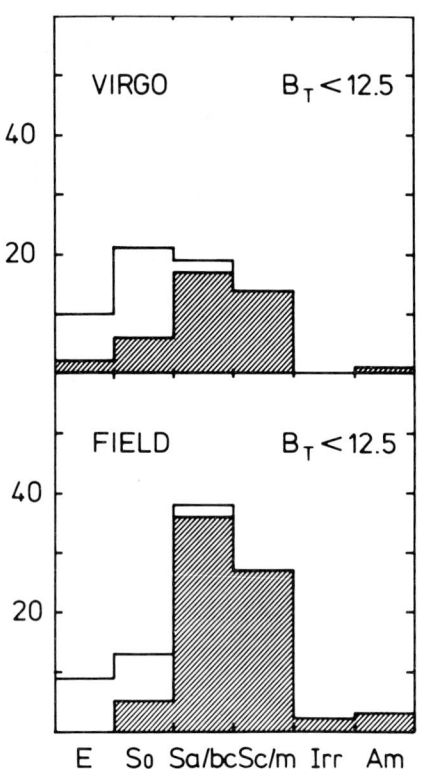

Figure 3. Infrared detection statistics of an optically complete sample of Virgo galaxies and of field galaxies as a function of morphological type. The hatched area represents the distribution of the galaxies detected by IRAS. The outer contour includes all galaxies in the sample.

the sample. The field sample consists of all RSA galaxies brighter than $B_T$ = 12.5 in an area of sky studied by de Jong and Brink (1985).

As has been known for a long time the Virgo cluster is richer in early-type galaxies, particularly in lenticulars, than the field. The fraction of galaxies detected by IRAS is similar for cluster and field galaxies. All Sc spirals and later types are detected, about 95 % of all early-type spirals are detected and about 30 % of the lenticulars. These results are overall very similar to those obtained in the earlier study of field galaxies by de Jong et al. (1984). The detection probabilities are somewhat higher than found previously because the galaxies in fig.3 are on the average more nearby due to the more strict selection criterion $B_T$ < 12.5. The only striking difference between the Virgo cluster and the field is the detection of three giant ellipticals in the Virgo cluster core, M84, M86 and M87. These galaxies will be further discussed in section 7.

To analyse in more detail the differences between Virgo cluster and field galaxies we plot in fig.4 the distribution of the infrared excesses for galaxies as a function of morphological type in the two samples. The infrared excess of a galaxy is defined as the ratio of its infrared to its blue luminosity. The infrared luminosity between 40 and 120 μm may to a very good approximation be calculated from the simple relation $L_{IR} = 1.2 (L_{60} + L_{100})$, where $L_{60}$ and $L_{100}$ are the in-band 60

GALAXIES IN THE INFRARED 117

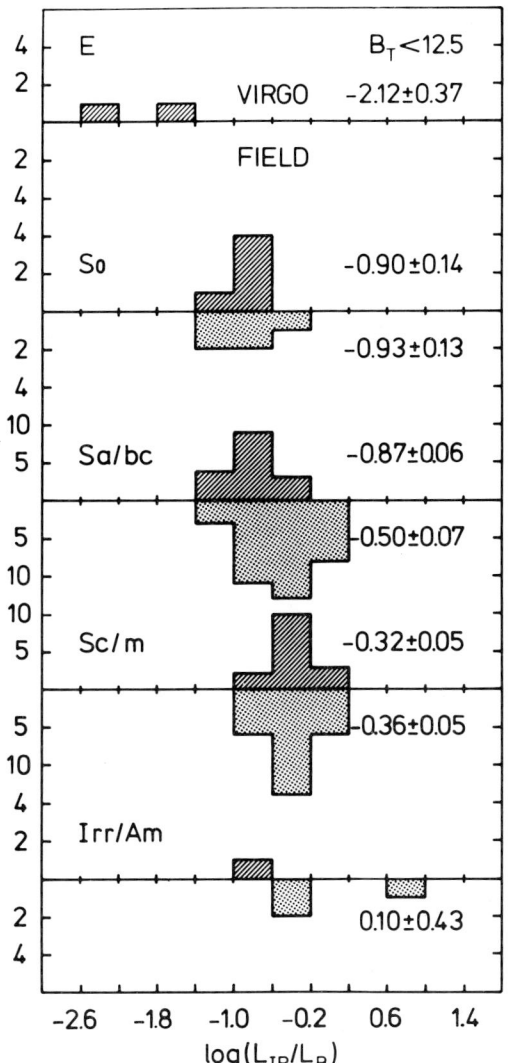

Figure 4. Infrared excess distributions of optically complete samples of Virgo and field galaxies as a function of morphological type.

and 100 μm luminosities (Joint IRAS Science Working Group 1985). The blue luminosity is obtained from $B_T^0$, the integrated blue magnitude corrected for internal absorption (galactic absorption is negligible in the direction of the Virgo cluster), by calculating the blue flux density and multiplying with the central frequency of the blue passband ($\lambda$ = 0.44 μm). The infrared excess has the advantage of being a distance-independent quantity.

The data in fig.4 show that the infrared excess of an optically complete sample both of Virgo and of field galaxies steadily increases along the Hubble sequence. Leaving the ellipticals, which will be discussed below, aside this increase can be interpreted as a gradually increasing star formation rate. Adopting the schematic two-component model of de Jong et al. (1984, see also section 3) we may explain the infrared emission of lenticulars, where no new stars are formed at present, as due to dust particles heated by the general interstellar radiation field. The magnitude of the infrared excess of ~ 0.1 is consistent with this picture because we expect in that case $L_{IR} \simeq \tau_B L_B$ with $\tau_B \simeq 0.1$ the dust optical thickness of the disk of the galaxy. Proceeding to later morphological types the contribution of infrared radiation produced by newly-formed stars still embedded in the molecular gas and dust clouds from which they formed gradually increases.

The histograms in fig.4 further show that galaxies in the Virgo cluster and in the field are overall quite similar in their infrared properties. The only clear-cut exception are the early-type spirals. While early-type spirals in the field are intermediate between lenticulars and late-type spirals, as expected, in the Virgo cluster they behave identical to lenticulars. Thus at the present epoch little star formation is going on in early-type spirals in the Virgo cluster. The data in fig.4 show that the star formation rate in early-type spirals in the Virgo cluster is more than two times smaller than in the field.

This result is similar but not as extreme as derived from a comparison of Hα data of Virgo cluster and field galaxies by Kennicutt (1983a) who found that spirals of all types in Virgo are deficient in star formation.

Ways to explain the lower star formation activity in Virgo cluster spirals have been discussed by Kennicutt (1983a). Contrary to modern trends in cluster research (Haynes et al. 1984) Kennicutt concludes that recent rapid ram-pressure stripping of interstellar matter from the galaxy by a hot cluster gas cannot be the dominant mechanism. He proposes that the most probable way to explain this effect is by slow depletion of gas from galaxies during their lifetime.

We have suggested (de Jong 1985) that this slow depletion is caused by enhanced star formation during encounters of galaxies in the cluster. While these encounters will not be as effective as those between galaxy pairs because they occur at higher relative velocities they might still induce enhanced star formation. In late-type spirals this will lead to minor fluctuations because the star formation rate is appreciable to start out with. In early-type spirals the effect may be more dramatic because the galactic wind resulting from the enhanced supernova rate during or just after an encounter may drive out interstellar matter more easily because of their low gas content.

Early-type spirals in which star formation has come to a halt altogether may turn into lenticulars. We note that a conversion of early-type spirals into lenticulars comes close to explaining the differences between morphological type distributions of Virgo cluster and field galaxies in fig.3.

## 5. STARBURST GALAXIES

According to present estimates there are about 20 000 galaxies among the roughly 250 000 sources detected by IRAS. Most of these galaxies are

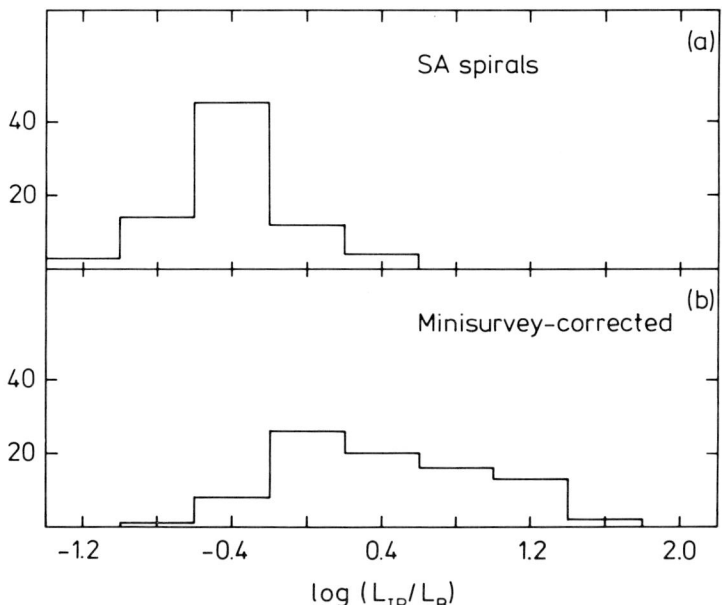

Figure 5. Infrared excess distributions of (a) an optically complete and (b) an infrared complete sample of galaxies

optically faint (B $\lesssim$ 14) and a large fraction is not even listed in any presently available catalogue. Some are so faint that they do not even show up on Palomar and/or ESO/SRC sky survey plates (B $\lesssim$ 19).

Fig. 5 shows the distributions of the infrared excesses (for a definition see section 3) in two samples of spiral galaxies, an optically complete one ($B_T$ < 12.5) and an infrared complete one ($S_{60 \mu m}$ > 0.5 Jy). The former is a sub-set of the optically complete sample of Shapley-Ames galaxies analysed by de Jong et al. (1984). The infrared sample is the IRAS mini-survey sample of Soifer et al. (1984a). Care has been taken to use optical magnitudes for galaxies in both samples determined in the same system by correcting the Zwicky magnitudes of the mini-survey galaxies listed in IRAS Circular 6 for systematic errors and for internal and galactic extinction according to recepees in the RSA (Sandage and Tammann 1981).

It can be shown that the optically complete sample is representative for the local (d $\lesssim$ 100 Mpc) population of spiral galaxies (de Jong and Brink 1985). Thus according to the data in fig.5a spirals emit on the average about 0.4 times as much energy in the far-infrared as in the visible.

If we rather arbitrarily define "starburst" galaxies as galaxies that emit more than 4 times as much energy in the far-infrared as in the visible we infer from fig.5 that, although they constitute less than 1% of all spiral galaxies, roughly 30% (~ 6000) of the galaxies detected by IRAS are starburst galaxies. The most extreme ones have recently been found to emit up to several hundred times more energy in the infrared than in the visible (Aaronson and Olzsewski 1984; Houck et al 1985).

The fraction of interacting systems among infrared galaxies is significantly higher than among local field galaxies suggesting that (distant) encounters between galaxies may play an important role in triggering bursts of star formation.

As a typical, rather nearby, example of a starburst galaxy ($L_{IR}/L_B \approx 80$) we briefly discuss the merging system Arp 220 (Soifer et al. 1984b). From the observed infrared luminosity of $2\ 10^{12}\ L_\odot$ we derive a present rate of star formation of massive stars ($M \gtrsim 10\ M_\odot$) of $\sim 600\ M_\odot\ yr^{-1}$. For an adopted gas mass of $\sim 10^{10}\ M_\odot$ this implies that all interstellar matter will be consumed by star formation in $\sim 2\ 10^7$ yrs, less than the collision time of the two merging galaxies. Arp 220 is also the most luminous OH maser source known, again indicative of an extremely high formation rate of massive stars (Baan et al. 1982).

## 6. RADIO - INFRARED CORRELATION

There are some differences of opinion in the astronomical literature about the origin of the relativistic electrons responsible for the non-thermal radio emission of galaxies. Hummel (1981) and Sancisi and van de Kruit (1981) have argued that the main sources of relativistic electrons belong to the old disk population while Klein (1982) and Kennicutt (1983b) find evidence that the non-thermal radio emission is closely associated with the young spiral arm stellar constituent.

This controversy has recently been settled by the discovery of a close correlation between the far-infrared and the radio emission of spiral galaxies by de Jong et al. (1985). This correlation is shown in fig.6 where 6.3 cm and 60 μm flux densities are plotted for about 90 galaxies. The radio data were obtained with the 100m Effelsberg telescope of the MPIfR in Bonn and the 60 μm flux densities are taken from several IRAS circulars. The line in fig.6 represents a best fit to the data and is given by the relation

$$\log S_{6cm} = (0.94 \pm 0.06) \log S_{60\mu m} - (2.34 \pm 0.06)$$

where both the radio and the the infrared flux density is given Janskys.

Since the radio emission at 6 cm is essentially non-thermal (Gioia et al. 1982) and the 60 μm emission is predominantly due to massive recently formed stars the correlation in fig.6 suggests that the non-thermal radio emission is apparently generated by relativistic electrons produced in supernova explosions of young massive stars. However some caution is required in interpreting this correlation because it might be dominated by distance effects.

To eliminate these effects de Jong et al. (1985) plotted the distribution of the logarithmic ratios of the blue, infrared and radio flux densities for two sub-sets of infrared galaxies: a representative sub-set of infrared-bright nearby spiral galaxies from the 2nd Revised Catalog of Bright Galaxies (RC2, de Vaucouleurs et al. 1976) listed in IRAS circular nr.15 and a sub-set of the several times more distant IRAS minisurvey galaxies (Soifer et al. 1984a; IRAS circular nr.6).

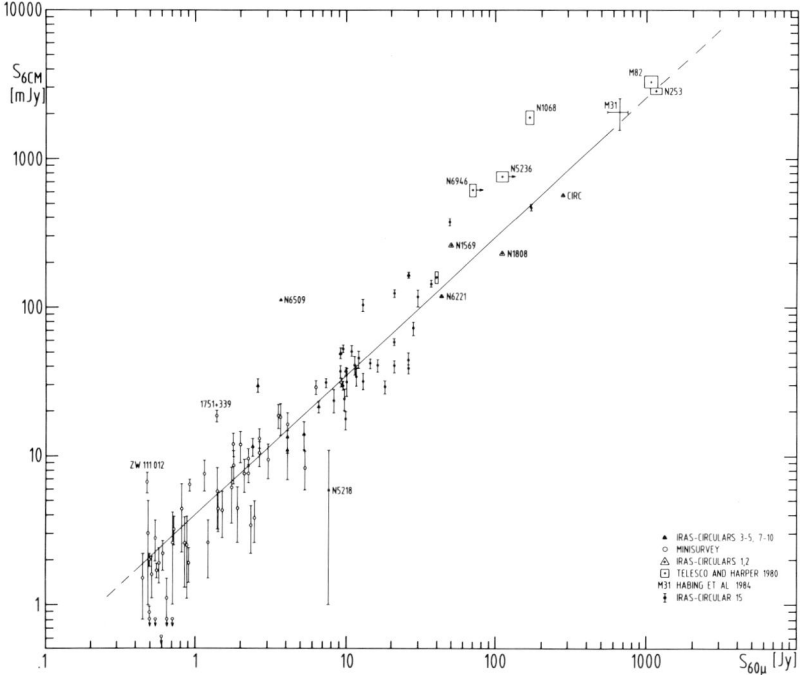

Figure 6. Plot of the 6 cm versus the 60 μm flux density of galaxies observed at Effelsberg. The radio flux density errors are indicated by error bars. The line represents a weighted best fit to the data.

The data show that the infrared-to-blue flux density ratio for the mini-survey sample is 3 times larger than for the nearby sample. Since the blue light derives from the stellar disk population - about equally from stars older and younger than $10^9$ years according to Larson and Tinsley (1978) - and since the infrared is predominantly generated by young massive stars in molecular clouds this implies that the present star formation rate (SFR) relative to the SFR integrated over the last few times $10^9$ years is apparently about 3 times larger in the mini-survey sample than in the nearby galaxy sample. This conclusion is consistent with that found by comparing the galaxy samples discussed by de Jong et al. (1984) and Soifer et al. (1984a).

The fact that the infrared-to-radio flux density ratio is the same within the errors for the two samples while the star formation activity in the mini-survey sample is about three times larger clearly establishes that the far-infrared and the radio emission originate in the same population of recently born massive stars.

## 7. ELLIPTICAL GALAXIES

As already pointed out in section 3 very few ellipticals have been detected by IRAS. Therefore it is quite remarkable that in the Virgo cluster core all three giant elliptical galaxies, M84 (NGC 4374), M86 (NGC 4406) and M87 (NGC 4486), are detected in the infrared. Together with M32 (NGC 221) and NGC 205 (Habing et al. 1984) and NGC 1052 (Neugebauer et al. 1984c) these are the only elliptical galaxies reported to have been detected by IRAS sofar. We note, however, that several studies underway at present indicate that the detection rate among dusty ellipticals (e.g. Hawarden et al. 1981) is significantly higher. Here we will briefly discuss the case of the three Virgo ellipticals, updating an earlier analysis (de Jong 1985).

The infrared spectral energy distributions of M84, M86 and M87 are remarkably similar: upper limits at 12, 25 and 100 μm and a flux density of ~ 0.5 Jy (close to the detection limit) at 60 μm. This result should, however, be considered with some caution because it is based on Point Source Catalog data. It cannot be excluded that, like for M87 at 12 μm and 60 μm (see IRAS Point Source Catalog), all three galaxies are bigger than the IRAS detectors in all wavelength bands but that they have not been detected by the SES processing in view of the low flux density levels. However, even then the flux densities are not expected to be underestimated by more than a factor 2 (see section 3 and de Jong 1985). Based on the point source flux densities we derive 40 - 120 μm luminosities of ~ 4 $10^8$ $L_\odot$ for all three ellipticals adopting a distance of 21.9 Mpc for the Virgo cluster.

M84 and M87 are strong radio continuum sources while M86 has not been detected in the radio (Kotanyi and Ekers 1983). The high radio flux of M87 ($S_{21cm}$ = 2100 Jy) allows the possibility to interpret the 60 μm and 100 μm flux densities as being due to non-thermal emission with a power law spectrum $S_\nu \propto \nu^{-1}$. The infrared and radio fluxes of M84 and M86 are however totally inconsistent with a non-thermal emission mechanism.

All three galaxies have been detected at X-ray wavelengths (0.5 - 3 keV) at widely different levels of intensity (Forman and Jones 1982). The radio and X-ray intensities are uncorrelated.

The fact that the infrared energy distributions are very similar while the X-ray and radio luminosities vary over several orders of magnitude suggests that the mechanism causing the infrared emission is of a different nature than that generating the radio or the X-ray emission.

While a non-thermal origin of the far-infrared emission of M87 cannot be excluded the most natural way to explain the infrared emission of all three ellipticals is by assuming that it is thermal emission from interstellar dust particles heated by the radiation of stars in the galaxy. Using the methods employed by Gillett et al. (1985) to analyze the IRAS data of the globular cluster 47 Tuc we derive "dirty silicate" dust masses for the three giant ellipticals in Virgo of about 2 $10^4$ $M_\odot$. This estimate is un upper limit because it is based on upper limits for the dust temperature (~ 35 K for a $\lambda^{-1}$ emissivity law of the dust). In view of the poorly known optical constants of interstellar dust at infrared wavelengths the dust mass derived should be considered as an order of magnitude estimate.

Adopting a stellar mass for the ellipticals of 3 $10^{12}$ $M_\odot$, using a stellar mass loss timescale of $10^{12}$ yrs (Gisler 1979), equal to that derived for globular clusters in our own galaxy by Faulkner and Freeman (1977), and assuming a gas/dust ratio of 1500 (for dust produced by stars with $\sim 1/6$ solar heavy element abundances) we find a rate of dust replenisment by mass loss from late-type giants of $2\ 10^{-3}\ M_\odot\ yr^{-1}$. Combining these rates with the estimated dust mass this implies that the dust in elliptical galaxies must be removed on timescales less than $10^7$ years. If the dust-producing stars were about as metal-rich as the sun this timescale decreases by about one order of magnitude.

The removal of interstellar dust in elliptical galaxies may be related to the existence of a galactic wind driven by supernova explosions (Bregman 1978). However, at typical wind velocities of about 400 km s$^{-1}$ the dust cannot be swept out of the galaxy in the time available. A plausible alternative is that the dust is destroyed locally, probably by sputtering in collisions with ions in the wind. We note that the situation in elliptical galaxies is a scaled-up version of that in the globular cluster 47 Tuc recently discussed by Gillett et al. (1985).

Our analysis of the infrared emission of Virgo cluster ellipticals suggests that their detection by IRAS is fortuitous; due to their proximity rather than to environmental or other effects related to their cluster membership.


ACKNOWLEDGEMENTS

I would like to thank Karel Brink for a critical reading of the manuscript and Cesare Chiosi for infinite patience in awaiting its arrival.


REFERENCES

Aaronson, M., and Olzsewski, E.W. 1984, Nature **309**, 414
Baan, W.E., Wood, P.A.D., Haschick, A.D. 1982, Astrophys. J.(Letters) **260**, L49
Bregman, J.N. 1978, Astrophys. J. **224**, 768
de Grijp, M.H.K., Miley, G.K., Lub, J., and de Jong, T. 1985, Nature **314**, 240
de Jong, T. 1985, Proceedings of the ESO workshop on the Virgo cluster, eds. O.-G. Richter and B. Binggeli, ESO, Garching, p.111
de Jong, T., and Brink, K. 1985, in preparation
de Jong, T., Clegg, P.E., Soifer, B.T., Rowan-Robinson, M., Habing, H.J., Houck, J.R., Aumann, H.H., and Raimond, E. 1984, Astrophys. J.(Letters) **278**, L67
de Jong, T., Klein, U., Wielebinsky, R., and Wunderlich, E. 1985, Astron. Astrophys. **147**, L6
de Vaucouleurs, G., de Vaucouleurs, A., and Corwin, H.G. 1976, Second Reference Catalog of Bright Galaxies, Univ. Texas Press, Austin (RC2)
Faulkner, D.J., and Freeman, K.C. 1977, Astrophys. J. **211**, 77
Forman, W., and Jones, C. 1982, Ann. Rev. Astron. Astrophys. **20**, 547

Garmany, C.D., Conti, P.S., and Chiosi, C. 1982, Astrophys. J. **263**, 777
Gillett, F.C., de Jong, T., Neugebauer, G., Rice, W., and Emerson, J.P 1985, in preparation
Gioia, I.M., Gregorini, L., and Klein, U. 1982, Astron. Astrophys. **116**, 164
Gisler, G.R. 1979, Astrophys. J. **228**, 385
Habing, H.J., Miley, G.K., Young, E., Baud, B., Boggess, N., Clegg, P.E., de Jong, T., Harris, S., Raimond, E., Rowan-Robinson, M., and Soifer, B.T. 1984, Astrophys. J.(Letters) **278**, L59
Hawarden, T.G., Elson, R.A.W., Longmore, A.J., Tritton, S.B., and Corwin, H.G. 1981, Mon. Not. Roy. Astron. Soc. **196**, 747
Haynes, M.P., Giovanelli, R., and Chincarini, G.L. 1984, Ann. Rev. Astron. Astrophys. **22**, 445
Houck, J.R., Schneider, D.P., Danielson, G.E., Beichmann, C.A., Lonsdale, C.J., Neugebauer, G., and Soifer, B.T. 1985, Astrophys. J.(Letters) **290**, L5
Hummel, E. 1981, Astron. Astrophys. **93**, 93
Joint IRAS Science Working Group 1985, Cataloged Galaxies and Quasars observed in the IRAS Survey, JPL D-1932
Kennicutt, R.C. 1983a, Astron. J. **88**, 483
Kennicutt, R.C. 1983b, Astron. Astrophys. **120**, 219
Klein, U. 1982, Astron. Astrophys. **116**, 175
Kotanyi, C.G., and Ekers, R.D. 1983 Astron. Astrophys. **122**, 267
Larson, R.B., and Tinsley, B.M. 1978, Astrophys. J. **219**, 46
Mezger, P.G., Mathis, J.S., and Panagia, N. 1982, Astron. Astrophys. **105**, 372
Neugebauer, G., Habing, H.J., van Duinen, R.J., Aumann, H.H., Baud, B., Beichman, C.A., Beintema, D.A., Boggess, N., Clegg, P.E., de Jong, T., Emerson, J.P., Gautier, T.N., Gillett, F.C., Harris, S., Hauser, M.G., Houck, J.R., Jennings, R.E., Low, F.J., Marsden, P.L., Miley, G.K., Olnon, F.M., Pottasch, S.R., Raimond, E., Rowan-Robinson, M., Soifer, B.T., Walker, R.G., Wesselius, P.R., and Young, E. 1984a, Astrophys. J.(Letters) **278**, L1
Neugebauer, G., Soifer, B.T., Miley, G.K., Young, E., Beichmann, C.A., Clegg, P.E., Habing, H.J., Harris, S., Low, F.J., and Rowan-Robinson, M. 1984b, Astrophys. J.(Letters) **278**, L83
Neugebauer, G., Soifer, B.T., Rice, W., Rowan-Robinson, M. 1984c, Publ. Astron. Soc. Pac. **96**, 973
Rieke, G.H., and Low, F.J. 1975, Astrophys. J. (Letters) **199**, L13
Rieke, G.H., and Lebofsky, M.J. 1979, Ann. Rev. Astron. Astrophys. **17**, 477
Sandage, A. and Tammann, G.A. 1981, A Revised Shapley-Ames Catalog of Bright Galaxies, Carnegie Institution of Washington Publication 635, Washington D.C.
Sancisi, R. and van de Kruit, P.C. 1981, IAU Symposium **94**, eds. G. Setti, G. Spada, and A.W. Wolfendale, Reidel, Dordrecht, p.209
Soifer, B.T., Rowan-Robinson, M., Houck, J.R., de Jong, T., Neugebauer, G., Aumann, H.H., Beichman, C.A., Boggess, N., Clegg, P.E., Emerson, J.P., Gillett, F.C., Habing, H.J., Hauser, M.G., Low, F.J., Miley, G.K., and Young, E. 1984a, Astrophys. J.(Letters) **278**, L71
Soifer, B.T., Helou, G., Lonsdale, C.J., Neugebauer, G., Hacking, P.,

Houck, J.R., Low, F.J., Rice, W., Rowan-Robinson, M. 1984b, Astrophys. J.(Letters) **283**, L1
Telesco, C.A., and Harper, D.A. 1980, Astrophys. J. **235**, 392

## DISCUSSION

O'CONNELL: Can you comment on the possibility that in the extreme infrared excess objects like Arp 220 the infrared luminosity is produced mainly by an active Seyfert nucleus rather than a starburst?

DE JONG: Based on the evidence that we have at the moment (mainly from optical and radio data) it seems that starbursts usually take place in the central few kiloparsecs of a galaxy while the Seyfert phenomenon is restricted to the nucleus proper. The infrared spectral energy distributions of Seyferts are usually characterized by a warm infrared component which shows up at 25 μm. Their far-infrared luminosities are more or less normal while starburst galaxies usually emit excessive amounts of energy in the far-infrared.

KOO: When claiming some evidence for early-type spirals in the Virgo cluster to possess smaller $L_{IR}/L_B$ ratios than in the field, it is important to first establish that the distribution over Sa's and Sb's is the same in both samples. Otherwise the correlation between Hubble type and $L_{IR}/L_B$ may result in the observed effect without having any real significance.

DE JONG: One of my graduate students, Karel Brink, who is here, drew my attention to this potential pitfall some time ago. It turns out that the distribution over Hubble types in the field and in the Virgo early-type spiral sample is quite similar.

DRESSLER: I wonder if your model of interaction-stimulated star formation for Virgo spirals might not produce too many gas-poor spirals in poor but relatively dense groups where the encounter velocity is lower and thus the effects more pronounced. Few (if any) poor group spirals are found to be deficient in HI.

DE JONG: First, the mechanism that I have suggested is so gentle that it only works for early-type spirals. Secondly, it does not necessarily always lead to gas-poor spirals as observed in HI because it mainly removes the gas from the central parts where most star formation takes place. Only if the gas is expelled from the galaxy altogether or if it is ionized does it lead to observable HI deficiencies.

ROCCA-VOLMERANGE: In a recent paper (Guiderdoni and Rocca-Volmerange, Proceedings of the ESO Workshop on the Virgo Cluster 1985) we have presented a compilation of spirals in the Virgo cluster and their respective H I deficiencies: early-type spirals are more likely to be very deficient than late-type ones in good agreement with the infrared excess distributions observed for Virgo spirals.

RENZINI: Since it is apparently so difficult to destroy the dust in elliptical galaxies, it may well be that evolved stars in ellipticals do not produce much dust. Of the $\sim 0.3$ $M_\odot$ lost by low-mass stars during their lifetime, some 90% is most likely lost through low mass-loss rate winds during RGB and AGB evolution and only some 10% or less (i.e. $\sim$ 0.01 - 0.03 $M_\odot$) during a final planetary nebula "ejection" phase. During the wind phases the circumstellar densities are quite low because $\dot{M}$ is low ($\lesssim 10^{-8}$ $M_\odot$ yr$^{-1}$) and therefore only very little dust may form. During PN ejection phases $\dot{M}$ is probably much higher (say $10^{-6}$ - $10^{-5}$ $M_\odot$ yr$^{-1}$) and thus also the efficiency of the nucleation process although very little mass may be ejected during these events. If so, the dust input in ellipticals may be some factors of 10 less than estimated.

DE JONG: That is an interesting suggestion. It could possibly help to solve the problem for the ellipticals but it definitely does not work in the globular cluster 47 Tuc, also supposed to consist of an old stellar population, because there we (Gillett et al. 1985) estimate the dust production rate from the observed mass loss rates of individual stars in the cluster.

IRAS FAR-INFRARED OBSERVATIONS OF INTERACTING GALAXIES

K. Brink
Astronomical Institute
University of Amsterdam
Roetersstraat 15
1018 WB Amsterdam
Netherlands

ABSTRACT. Using the large IRAS database, we investigate the effect of interaction between galaxies on their far-infrared properties by comparing the infrared-to-blue luminosity ratios of optically selected samples of interacting and field galaxies, and by briefly discussing the results of additional mapping with IRAS of a few interacting galaxy pairs.

1. INTRODUCTION

Star formation appears to be often enhanced in galaxies belonging to pairs or groups, as concluded from several studies based on optical or radio data (e.g. Sulentic 1976, Larson and Tinsley 1978). Preliminary statistical investigations of the IRAS far-infrared observational material seem to confirm this result. Soifer et al. (1984) have reported an overrepresentation of galaxies with neighbours in the infrared selected 'minisurvey' sample. Large far-infrared-to-blue luminosity ratios found for some of these galaxies suggest the occurrence of bursts of star formation due to interaction, though systematic magnitude errors and reddening uncertainties may effect the results.

2. STATISTICS

Studies of optically selected samples are useful to link galaxies with peculiar infrared/ star formation properties to those behaving 'normally'. As an illustration we compare in figure 1 the distribution of far-infrared-to-blue luminosity ratios in two samples of spiral galaxies taken from the Revised Shapley-Ames Catalog of Bright Galaxies (RSA, Sandage and Tammann 1981), which is about complete down to a magnitude $B_T = 12.5$. From each sample we have removed weak galaxies ($B_T^0 > 12.5$) and very nearby galaxies ($v_0 < 1000$ km/s). The infrared data were taken from the IRAS Point Source Catalog. The fluxes of extended IRAS sources were corrected according to a preliminary version of the

Small Extended Sources (SES) catalog.

Sample 1, the field sample, contains the RSA spiral galaxies without a close companion and not appearing in the catalogue of group galaxies compiled by Huchra and Geller (1982), which is complete for $B_T^0$ < 12.0. For consistency we have further restricted sample 1 to galaxies also having $B_T^0$ < 12.0. These requirements are met by 111 spirals, of which 110 are detected by IRAS.

Sample 2, the interacting galaxy sample, consists of those RSA spiral galaxies denoted as closely interacting by Vorontsov-Velyaminov (1959) or by Arp (1966). This is a (somewhat biased) sample selected on the basis of morphological criteria. It contains 33 spiral galaxies, of which 31 are detected by IRAS.

In computing the infrared-to-blue luminosity ratio we have used the blue magnitude corrected for galactic and internal extinction ($B_T^0$), multiplying the derived blue flux density by the central frequency of the blue passband. The far-infrared flux was calculated following the prescript given in the IRAS extragalactic catalogue (Lonsdale et al. 1985).

The two distributions shown in figure 1 differ markedly. The interacting spirals appear to have mean infrared-to-blue luminosity ratios twice as large as the field spirals. Provided they are not abnormally reddened, this indicates an enhancement of recent star formation in galaxies whose morphology is disturbed by the influence of

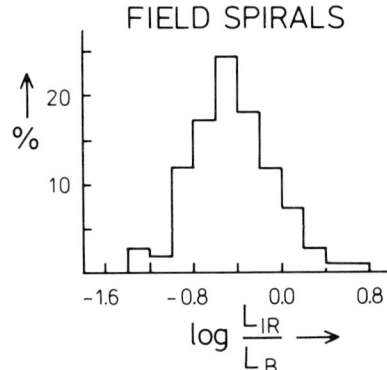

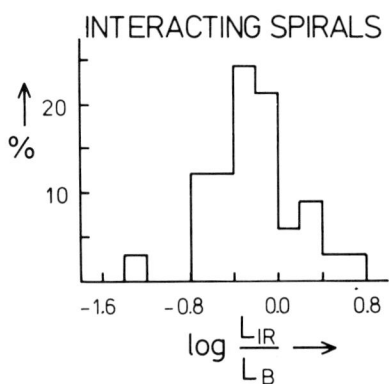

Figure 1. Normalised frequency distribution of the ratio of far-infrared to blue luminosity for field and interacting spiral galaxies.

one or more companions. (The percentage of all galaxies having close companions, that exhibit this enhancement can not be determined directly from this comparison). Since the morphological type distribution is similar for both samples the difference is not attributable to a larger fraction of late type spirals (bright in the infrared, cf. de Jong et al. 1984) in the interacting sample.

## 3. CPC FAR-INFRARED MAPS

Interaction between galaxies could stimulate star formation in several ways. Tidal forces will affect the distribution and the motion of the gas and perhaps of the stars within the galaxies (Toomre and Toomre 1972, Icke 1985), and thus cause gas accumulation or drive shocks. A collision between the gaseous disks of galaxies may produce similar effects.

Insight in the precise effects of interaction on star formation, and in the correlation of the infrared with other star formation tracers may be gained from more detailed studies of galaxy pairs resolvable by IRAS. The instrument with the best spatial resolution at far-infrared wavelengths on board IRAS was the CPC-instrument (Chopped Photometric Channel, Wesselius et al. 1985), providing a resolution of about 1.5 arcminutes.

We have inspected maps of the 50 and 100 μm surface brightness distribution in the close pairs NGC 2207/IC 2163, NGC 2992/NGC 2993, and NGC 4485/NGC 4490 and compared them with observations at other wavelengths. The data will be published in a future paper. Here we summarize the main results:
  - the infrared radiation appears to be connected with star formation: the surface brightness distribution correlates well with that of other star formation tracers, such as the radio continuum and the Hα radiation.
  - the correlation with the radio continuum is striking but does not extend to the central radio source in active galaxies.
  -the distribution of the tracers indicates the influence of interaction on star formation in these systems. The location of the major activity sites varies from pair to pair. The central regions as well as the outer parts of the galaxies can be dominating in recent star formation activity.

## ACKNOWLEDGEMENTS

Part of this work was done in collaboration with T. de Jong, W. van Oosterom and P.R. Wesselius. The author is supported by the Netherlands Foundation for Astronomical Research (ASTRON) with financial aid from the Netherlands Organisation for the Advancement of Pure Research (Z.W.O.).

REFERENCES

Arp, H.: 1966, Atlas of Peculiar Galaxies, Calif. Inst. Tech., Pasadena.
Huchra, J.F., and Geller, M.J.: 1982, Astrophys. J., 257, 423.
Icke, V.: 1985, Astron. Astrophys., 144, 115.
de Jong, T., Clegg, P.E., Soifer, B.T., Rowan-Robinson, M., Habing, H.J., Houck, J.R., Aumann, H.H., and Raimond, E.: 1984, Astrophys. J. (Letters), 278, L67.
Larson, R.B., and Tinsley, B.M.: 1978, Astrophys. J., 219, 46.
Lonsdale, C.J., Helou, G., Good, J.C., and Rice, W.: 1985, Cataloged Galaxies and Quasars Observed in the IRAS Survey, Jet Propulsion Laboratory, Pasadena.
Sandage, A., and Tammann, G.A.: 1981, A Revised Shapley-Ames Catalog of Bright Galaxies, Carnegie Inst. of Washington, Publication 635, Washington, D.C. (RSA).
Soifer, B.T., Rowan-Robinson, M., Houck, J.R., de Jong, T., Neugebauer, G., Aumann, H.H., Beichman, C.A., Boggess, N., Clegg, P.E., Emerson, J.P., Gillett, F.C., Habing, H.J., Hauser, M.G., Low, F.J., Miley, G.K., and Young, E.: 1984, Astrophys. J. (Letters), 278, L71.
Sulentic, J.W.: 1976, Astrophys. J. Suppl. Ser., 32, 171.
Toomre, A., and Toomre, J.: 1972, Astrophys. J., 178, 623.
Vorontsov-Velyaminov, B.A.: 1959, Atlas and Catalogue of Interacting Galaxies, Sternberg Institute, Moscow.
Wesselius, P.R., Beintema, D.A., de Jonge, A.R.W., Jurriens, T.A., Kester, D.J.M., van Weerden, J.E., de Vries, J., Perault, M.: 1985, IRAS-DAX Chopped Photometric Channel, Explanatory Supplement.

## II. STELLAR POPULATIONS IN THE LOCAL GROUP

J. MOULD
The stellar populations of galaxies in the local group

J. A. FROGEL
M giants in the galactic nuclear bulge

A. E. WHITFORD
Age and mass of M giants in the galactic bulge

M. AARONSON, K. H. COOK, J. NORRIS
The AGB population of nearby galaxies

W. L. FREEDMAN
M33: radial distributions and a comparison of its global luminosity function with other nearby galaxies

# THE STELLAR POPULATIONS OF GALAXIES IN THE LOCAL GROUP

Jeremy Mould
Palomar Observatory, California Institute of Technology,
and Mount Stromlo and Siding Spring Observatories

INTRODUCTION

We can study galaxy evolution in two ways. One is to use lookback time and redshift to investigate, say, the luminous output of brightest elliptical galaxies, $L_V(t)$. Aside from observational difficulties, there are some general problems obscuring this astronomical window into the past. One has to connect the present to the past, the current epoch elliptical with its precursor or precursors at redshifts of 1 or 2.

The second way to learn about galaxy evolution can be expressed in the equation:

$$L_V(t) = \int_0^t SFR(t) / m/\ell_V(t) \, dt$$

where $m/l_V(t)$ is the mass to light ratio of a standard initial mass function (IMF) stellar population of age t, and SFR(t) is the star formation rate ($M_O/yr$). Information on $m/l_V(t)$ is available from the study of star clusters and stellar evolution theory. If we can learn the star formation history of galaxies of different type from our neighbouring examples of those types, we will have a basis for understanding the luminosity evolution of galaxies.

Comprehensive reviews on the stellar populations of Local Group galaxies have been given by van den Bergh (1968, 1975). It is not possible here to cover such a large scope. Instead, I shall concentrate on subsequent developments which bear on the star formation history of these galaxies. But, to set the stage, Figure 1 shows the distribution of galaxies in the Local Group from van den Bergh (1980), as viewed from the north Supergalactic pole. This would be the picture an observer in the Hercules cluster would have (with 1 Mpc = 1/2 degree). This observer might fail to detect some of the low surface brightness dwarf spheroidals shown here, but would certainly notice two subgroups of galaxies centred on M31 and the Milky Way.

THE MAGELLANIC CLOUDS

The star formation history of nearby galaxies is reflected in their HR diagrams, luminosity functions and integrated colours. However, deduction of SFR(t) from the available data is not a straightforward exercise. Even for the Large Magellanic Cloud two quite orthogonal conclusions have recently been drawn from observations. Hardy et al (1984) infer from their monumental colour magnitude study: "it would seem that just a few billion years ago, the LMC was basically a large gaseous mass". From the integrated colours of LMC clusters Elson and Fall (1985) conclude: "our results contain no evidence for bursts in the formation of clusters, although ..... slow variations over the lifetime of the LMC cannot be ruled out."

Support for the former proposition, that the bulk of star formation in the LMC occurred a few Gyrs ago, has appeared in plenty in the last decade (e.g. Bruck and Hawkins 1981). There are five separate lines of evidence.

*1. The main sequence luminosity function.* Butcher (1977) was first to conclude from the break in the luminosity function that a sample field in the LMC was a few billion years younger than the solar neighbourhood. Completeness is the issue in this approach, and is carefully addressed by Butcher. Deeper photometry is now possible with appropriate CCD instrumentation in exceptional seeing, and the completeness can be checked.

*2. The subgiant branch.* Hardy et al (1984) conclude from the sparseness of the subgiant branch in a 6' x 12' field at the NW end of

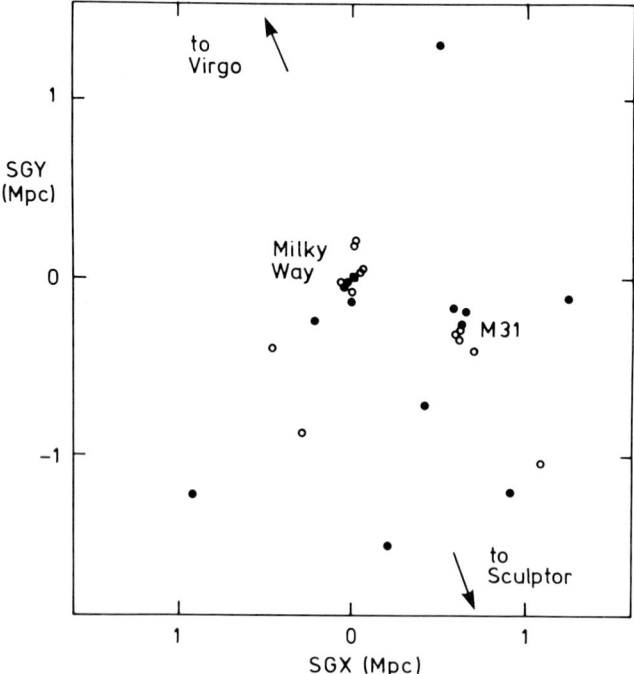

Figure 1. Distribution of Local Group galaxies in the Supergalactic Plane. Galaxies brighter than $M_V = -16$ are solid symbols; dwarfs are shown as open symbols. For orientation purposes the direction of the nearest group, the Sculptor group, is shown, together with that of the center of the Supercluster in Virgo.

the Bar that the characteristic age of the stellar population is ~ 1 Gyr, and that there is very little contribution from a population older than 3 Gyrs. These data make a strong qualitative impression, which deserves to be followed up with numerical modelling to constrain SFR(t).

3. *The horizontal branch.* A much larger field in the halo of the LMC was examined by Stryker, Butcher and Jewell (1981). They found an absence of blue horizontal branch stars, but a well developed red horizontal branch. They conclude that the LMC halo is less than 6 Gyrs old. RR Lyrae stars exist in the LMC, however. These stars are tracers of the oldest stellar population, which is not necessarily 16 Gyrs old, as evidenced by NGC 121 (Stryker, Da Costa and Mould 1985). Perhaps the LMC halo is best represented by an 12 Gyr NGC 121-like population.

4. *Carbon stars.* Richer (1981) suggested that the high surface density of carbon stars in Blanco, McCarthy and Blanco's (1980) Bar West field reflects the end product of a burst of star formation which occurred in the LMC ~ 3-5 Gyrs ago. Although it is likely that carbon stars in the LMC belong to that epoch, it is not so clear that much of a burst is required to produce them. If every star in the LMC between 1 and 2 $M_\odot$ becomes a carbon star for $10^6$ years, if there are $10^4$ of them in total (Blanco and McCarthy 1983), and if the IMF $n(m)dm$ ~ $m^{-2}dm$ between 0.1 and 100 $M_\odot$, the star formation rate* in that epoch was 0.1 $M_\odot$/yr. The current SFR in the LMC is ~ 0.67 $M_\odot$/yr according to Rocca Volmerange (1984). This is an order of magnitude calculation, but the point is: no extraordinary rate of star formation is required to produce these carbon stars.

5. *The ages of star clusters.* Aaronson and Mould (1982) pointed out that most star clusters in a magnitude limited sample in the LMC have an asymptotic giant branch (AGB) tip in the range $-5.5 < M_{bol} < -5$. This could be easily understood if most clusters were ~ 3 Gyrs old. However, the factors which govern the distribution of observable cluster properties are actually quite complex. Luminosity evolution (the fading of old stellar populations) can be expected to produce a deficit of old clusters in a magnitude limited sample, and Mould and Aaronson (1983) went on to show that the LMC AGB tip distribution was not inconsistent with a constant rate of cluster formation over 16 Gyrs. More recently, Elson and Fall (1985) have carefully parameterized the situation and reached a similar conclusion, as quoted earlier. A better age indicator than either of these is the main sequence turnoff. Hence it remains a challenge to determine accurate ages for the older clusters listed by Elson and Fall. Of the clusters with their s parameter > 40, five have RR Lyrae stars and are probably first generation clusters (NGC 1466, 1786, 2210, 2257 and Hodge 11). Two are actually 1-2 Gyr clusters (NGC 1751, 1978). Just

---

* A more direct comparison is of the formation rate of 1 to 2 $M_\odot$ stars only. This is 0.02 stars/yr 4 Gyrs ago, compared with 0.13 stars/yr currently (Rocca Volmerange 1984).

7 are candidates for the critical 2-12 Gyr period, which is in question, (NGC 1652, 1718, 1754, 1795, 1916, 2005 and SL506). None of these clusters has a measured main sequence turnoff. Ages for these clusters and others in this colour range will help to determine the star formation history of the LMC over the billion year timescales that are of concern here*.

It should be clear from the foregoing that, although there is a lot of evidence relating to the star formation history of the LMC, an unequivocal picture has yet to emerge. With Space Telescope, however, we can expect to obtain a definitive answer by applying the techniques that Miller and Scalo (1979) and Twarog (1980) have applied to the solar neighbourhood.

In closing off this section on the Clouds, I want to emphasize that the study of Magellanic Cloud star clusters can not only tell us about the star formation history of two dwarf irregulars, but also holds the key to understanding the time evolution of M/L in a stellar population. To pursue this study, we need to measure dynamical masses for the clusters. This is now a practical proposition with large telescopes in the southern hemisphere.

THE DWARF SPHEROIDAL COMPANIONS OF THE MILKY WAY

In the course of the last few years it has become apparent that, as regards their origin, the dwarf spheroidal satellites of the Galaxy have more in common with dwarf irregular galaxies than with globular clusters. Two or three billion years ago the most massive of these, the Fornax dwarf, may have been two magnitudes brighter than it is now, and have resembled IC 1613 or NGC 6822. Evidence for this late epoch of star formation is seen in the luminous AGB carbon stars, equivalent to those of NGC 1978 in the LMC, and in the blue main sequence stars observed by Buonanno et al (1985). Although similar conclusions can be supported for the Carina dwarf (Mould and Aaronson 1983), based on a 7.5 Gyr old main sequence, for the other dwarfs it would be possible to *simulate* intermediate age AGB stars by binary mass transfer within 16 Gyr old stars. Fornax is unique in requiring the transfer of more than a single 0.2 $M_\odot$ stellar envelope to achieve this. It remains to be seen, whether Leo I and Leo II, the other two dwarfs with significantly extended AGBs (Aaronson and Mould 1985), also show an intermediate age main sequence.

Further circumstantial evidence that the dwarf spheroidals were accreted by the Galactic halo exists in their distribution as a Galactic polar ring (Lynden Bell 1982). In addition, Kormendy (1985) has argued that the structural parameters of dwarf ellipticals in general resemble those of dwarf irregulars rather than normal elliptical galaxies.

*Omitted from this review is evidence cited by Frogel and Blanco (1983), Reid and Mould (1985) and Wood, Bessell and Paltaglou (1985) for a burst of star formation in the LMC very recently, i.e. within the last $10^8$ years.

## THE BULGE COMPONENTS OF LOCAL GROUP GALAXIES

In the context of galaxy evolution it is particularly important to know the size of the old population of a stellar system. A disk of constant SFR does not evolve in luminosity at all, but bulges are steadily fading. If we could look back to within 2-5 Gyrs of the initial bursts that formed galactic bulges, we would see them ~4 times brighter than they are now.

The fraction of the light that is contained in the bulge of nearby galaxies is not in general well determined, particularly where the contrast between bulge and disk is large. Table 1 gathers together what we know about key Local Group galaxies. The fraction of their V light contained in the bulge is given in magnitudes in column (3). Especially uncertain are the inferences from star counts. Bahcall, Schmidt and Soneira (1983) base their estimates on the Bahcall and Soneira (1980) model. Gilmore and Reid (1983) count many more stars beyond the 300 pc scale height disk, but the distribution of these stars (disk or flattened spheroid?) is not well determined. Column (5) of Table 1 is the value of $\Delta M_B$ adopted by Meisels and Ostriker (1984). For conversion purposes the B-V colour of a disk was assumed to be 0.55, and of a bulge, 0.90.

In the LMC the RR Lyrae stars would seem to be the best tracer of the oldest population. But how many RR Lyrae stars exist per unit mass of spheroid ? Frogel (1984) suggests 1 per $3 \times 10^4$ $M_\odot$, based on the globular cluster system; Mould (1983) takes 1 per $2 \times 10^3$ $M_\odot$, based on M3. The five RR Lyraes within 500 pc in the catalog of Woolley et al (1965) suggest 1 per $4-14 \times 10^3$ $M_\odot$ with the Bahcall, Schmidt and Soneira (1983) spheroid density. Finally, Freeman, Illingworth and Oemler (1983) claim that there is no evidence for a kinematic halo population among the globular clusters in the LMC. However, given the low velocity dispersion that one would expect for

Table 1: Bulge Fraction in Local Group Galaxies

| Galaxy | $M_V$ | $\Delta M_V$ (B-T) | Type | $\Delta M_B$ (MO) | Method | Reference |
|---|---|---|---|---|---|---|
| (1) | (2) | (3) | (4) | (5) | (6) | (7) |
| M31 | -21.1 | 1.3 | Sb | 1.5 | Surface photometry | 1 |
| Milky Way | | 3.8-1.7 | | | B and S model | 2 |
| | | 1.4-1.0 | | | Star counts | 3 |
| | -20.7 | 2.0 | Sbc | | de Vaucouleurs (1982) | |
| | | 0.7 | | | G and R counts | 4 |
| M33 | -19.2 | 4.0 | Sc | 2.6 | Surface photometry | 5 |
| LMC | -18.5 | 4.8 | Im | 4.5 | RR Lyrae cf. M3 | 6 |
| | | 2.3 | | | M(halo)=$4 \times 10^8$ $M_\odot$ | 7 |

References: 1. de Vaucouleurs (1958); 2. Bahcall, Schmidt and Soneira (1983); 3. de Vaucouleurs and Pence (1978); 4. Edmunds and Phillipps (1983); 5. Boulesteix et al (1979); 6. Mould (1983); 7. Frogel (1984).

this population (20 km/s by scaling from the Galaxy according to the ratio of disk rotational velocities), one could equally say that there is no evidence against an LMC halo.

SPHEROIDAL POPULATION OF M31 AND COMPANIONS

Stellar photometry in the outer parts of M31 and its dwarf elliptical and spheroidal companions is practicable with ground based telescopes, in the same way as we study the outer annuli of globular clusters. Indeed, the work summarized in Table 2 below has shown that the accessible stellar population in these systems resembles that of Galactic globular clusters. The absence of AGB stars in the fields examined to date implies that less than 10% of the luminous matter in the outer parts of these galaxies is in stars aged between 2 and 8 Gyrs. The dwarf spheroidal galaxy And II, however, has an extended giant branch (Aaronson et al 1985) with one or more carbon stars, just like the dwarf spheroidals of our own Galaxy. This is *prima facie* evidence of an extended history of star formation. NGC 205 has star formation continuing in the center in the present epoch.

Figure 2 shows the limits of Palomar photometry in the bulge of M31. This field contains the globular cluster G298 (see Hodge 1982). The horizontal branch might be reached in subarcsecond seeing, or, with relative ease, with Space Telescope. The horizontal branch morphology of the M31 bulge and clusters will afford a detailed comparison of the halo population in this galaxy with our own. Ground based work will permit mapping of the metallicity distribution in the M31 bulge and comparison with that of the cluster system.

The cluster system itself remains in need of a really comprehensive photometric and spectroscopic study. Searle (1982) finds evidence for a variation of the reddening law across the face of M31. He also concludes that the entire abundance distribution of the cluster system is shifted towards higher abundances than that of the Galaxy. But it remains difficult to determine whether there is a metallicity gradient in the M31 clusters, the signature of a dissipative collapse. Iye and Richter (1985) point out that it is hard to deconvolve a metallicity gradient from the reddening law, and Crampton et al (1985) indicate the difficulty posed by background contamination of the colour distribution. The solution to both these problems would seem to be a large scale medium resolution spectroscopic survey.

Table 2: Stellar Photometry in M31 and Companions

| Galaxy | Field Location | <[M/H]> | Dispersion | $(m-M)_0$ | Reference |
|---|---|---|---|---|---|
| NGC 147 | 1 kpc SE | $-1.2 \pm 0.2$ | 0.3 | 24.0 | 1 |
| NGC 205 | 2 kpc N | $-0.9 \pm 0.2$ | 0.5 | 24.3 | 2 |
| M31 bulge | 7 kpc SE | $-0.6 \pm 0.2$ | | 24.4 | 3 |
| M33 bulge | 7 kpc SE | $-2.2 \pm 0.8$ | | 24.8 | 3 |

References: 1,2. Mould, Kristian and Da Costa (1983,4); 3. Mould and Kristian (1985).

# THE STELLAR POPULATIONS OF GALAXIES IN THE LOCAL GROUP

A first attempt at a kinematic study of the M31 cluster system has been carried out by Huchra, Stauffer and Van Speybroeck (1982). They find that the cluster system rotates with a velocity of 80 ±20 km/sec compared with 230 km/sec for the disk (Roberts and Whitehurst 1975; Newton and Emerson 1977). The velocity dispersion is 130 ±16 km/sec, which is consistent with that found in the nuclear bulge by McElroy (1983). This compares with 116 ±10 km/sec for the Galactic globular cluster system (Frenk and White 1980). Freeman (1983) has shown that the metal-rich clusters in M31 lie in a rapidly rotating disk within 10 kpc of the center, while the metal-weak clusters rotate much more slowly. Galactic globular clusters have analogous kinematics according to Rodgers and Paltoglou (1984) and Zinn (1985). Amongst the goals of a complete study should be the location and measurement of the outer clusters of the M31 halo, from which the total interior mass can be determined. This halo presumably extends to 100 kpc or 10 degrees from the centre.

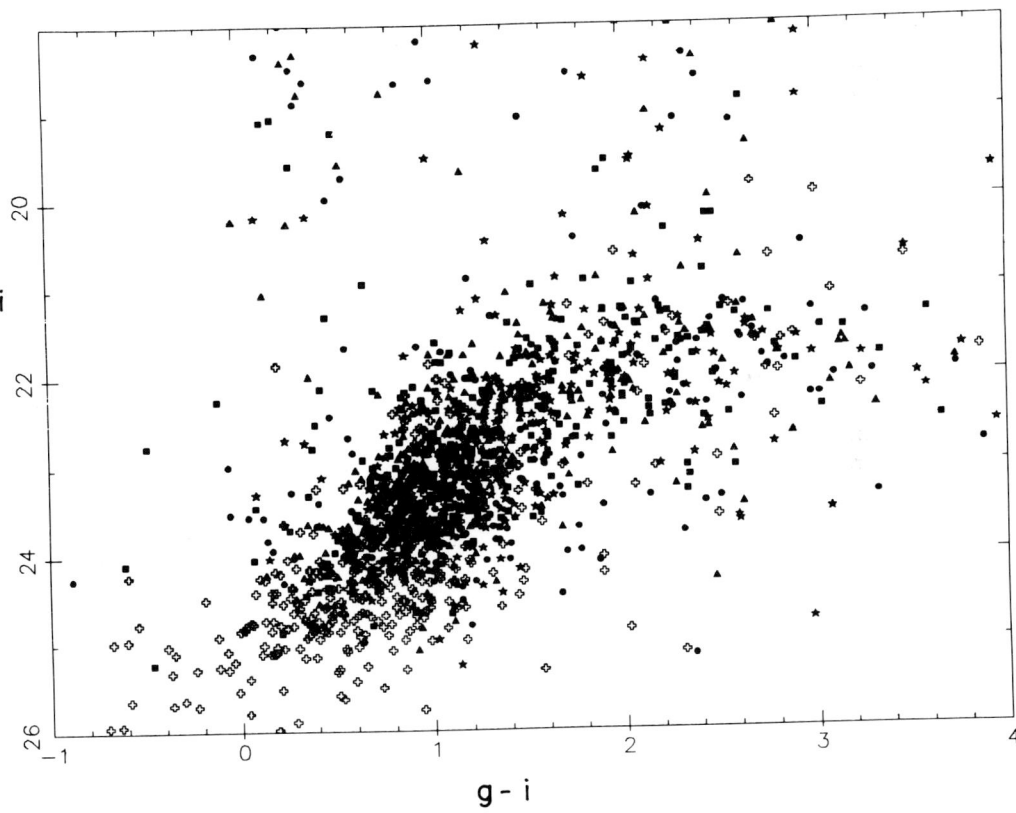

Figure 2. Preliminary colour-magnitude diagram of a field in the bulge of M31 on the Gunn photometric system obtained with 4shooter on the Hale telescope. Open symbols are stars selected on the green frame. Solid symbols are coded according to chip number.

CONCLUSION

Evolutionary models for galaxies can benefit from the study of local galaxies in two main ways. First, we can learn from their star formation history which of the parameterized evolutionary models (like those of Bruzual 1983, for example), are the closest representations of nearby galaxies of particular morphological types. We can improve the details of the parameterization to include, for example, the relation between mass and the metallicity distribution in ellipticals (and bulges ?). If late type disk systems turn out to be younger or more metal poor, we should include this in evolutionary models too.

Second, we must recognize that stellar evolution theory is the biochemistry which governs real live galaxies. This "biochemistry" must be understood, if we are going to give a realistic, rather than a fictitious, account of galaxy evolution. Stellar evolution theory has benefited most directly from confrontation with observations of star clusters. These represent template populations of different ages and metallicities. A full range of these templates is available for study in Local Group galaxies.

This work was partially supported by U.S. NSF grant AST 85-02518

REFERENCES

Aaronson, M. and Mould, J. 1982, Ap.J. Suppl. 48, 161.
---------------------- 1985, Ap.J. 290, 191.
Aaronson, M., Gordon, G., Mould, J., Olszewski, E. and Suntzeff, N. 1985, Ap.J. Letters (in press).
Bahcall, J., Schmidt, M. and Soneira, R. 1983, Ap.J. 265, 730.
Bahcall, J. and Soneira, R. 1980, Ap.J. Suppl. 44, 73.
Blanco, V. and McCarthy, M. 1983, A.J. 88, 1442.
Blanco, V., McCarthy, M. and Blanco, B. 1980, Ap.J. 242, 938.
Boulesteix, J., Colin, J., Athanassoula, E. and Monnet, G. 1979, in Photometry, Kinematics and Dynamics of Galaxies, ed. D.S. Evans, (Austin: Univ. of Texas), p.271.
Bruck, M.T. and Hawkins, M.R.S. 1981, IAU Colloquium 68, ed. A.G.D. Philip, (Schenectady: L. Davis Press), p.261.
Bruzual, G. 1983, Ap.J. 273, 105.
Buonanno, R., Corsi, C., Fusi-Pecci, F., Hardy, E. and Zinn, R. 1985, Astron. and Astrophys. (in press).
Butcher, H.R. 1977, Ap.J. 216, 372.
Crampton, D., Cowley, A.P., Schade, D. and Chayer, P. 1985, Ap.J. 288, 494.
de Vaucouleurs, G. 1958, Ap.J. 128, 465.
--------------- 1982, The Cosmic Distance Scale and The Hubble Constant, (Canberra: Australian National University).
de Vaucouleurs, G. and Pence, W. 1978, A.J. 83, 1163.
Edmunds, M. and Phillipps, S. 1983, Astr. and Astrophys. 131, 169.
Elson, R. and Fall, S.M. 1985, preprint.
Freeman, K. 1983, Internal Kinematics and Dynamics of Galaxies, ed. E. Athanassoula (Reidel: Dordrecht), p. 359.

Freeman, K., Illingworth, G. and Oemler, A. 1983, Ap.J. 272, 488.
Frenk, C. and White, S. 1980, M.N.R.A.S. 193, 295.
Frogel, J.A. 1984, Pub. A.S.P. 96, 788.
Frogel, J.A. and Blanco, V. 1983, Ap.J. Letters, 274, L57.
Gilmore, G. and Reid, I.N. 1983, M.N.R.A.S. 202, 1025.
Iye, M. and Richter, O.G. 1985, Astr. and Astrophys. 144, 471.
Huchra, J., Stauffer, J. and Van Speybroeck, L. 1982, Ap.J. Letters 259, L57.
Hardy, E., Buonnano, R., Corsi, C., Janes, K. and Schommer, R. 1984, Ap.J. 278, 592.
Hodge, P. 1982, An Atlas of the Andromeda Galaxy, (Seattle: Univ. of Washington).
Kormendy, J. 1985, preprint.
Lynden Bell, D. 1982, Observatory 102, 202.
McElroy, D. 1983, Ap.J. 220, 485.
Meisels, A. and Ostriker, J. 1984, A.J. 89, 1451.
Miller, G. and Scalo, J. 1979, Ap.J. Suppl. 41, 513.
Mould, J. 1983, Highlights of Astronomy v6
Mould, J. and Aaronson, M. 1982, Ap.J. 263, 629.
-------------------- 1983, Ap.J. 273, 530.
Mould, J., Kristian, J. and Da Costa, G. 1983, Ap.J. 270, 471.
---------------------------------- 1984, Ap.J. 278, 575.
Mould, J. and Kristian, J. 1985, Ap.J. (in press).
Newton, K. and Emerson, D. 1977, M.N.R.A.S. 181, 573.
Reid, I.N. and Mould, J. 1985, Ap.J. (in press).
Richer, H.R. 1981, Ap.J. 243, 744.
Roberts, M.S. and Whitehurst, R. 1975, Ap.J. 201, 327.
Rocca-Volmerange, B. 1984, Astr. and Astrophys. 104, 170.
Rodgers, A. and Paltoglou, G. 1984, Ap.J. Letters 283, L5.
Searle, L. 1982, Carnegie Institution of Washington Yearbook 82, (Washington: Carnegie Institution), p.622.
Stryker, L., Butcher, H. and Jewell, J. 1981, IAU Colloquium 68, ed. A.G.D. Philip, (Schenectady: L.Davis Press), p.267.
Stryker, L., Da Costa, G. and Mould, J. 1985, Ap.J. (in press).
Twarog, B. 1980, Ap.J. 242, 242.
van den Bergh, S. 1968, The Galaxy and the Local Group, David Dunlap Observatory Comm. no. 195.
-------------- 1975, Ann. Rev. Astr. Astrophys. 13, 217.
-------------- 1980, Structure and Evolution of Normal Galaxies, ed. S.M. Fall and D. Lynden Bell, Cambridge Univ. Press.
Wood, P., Bessell, M. and Paltaglou, G. 1985, Ap.J. 290, 477.
Woolley, R.v.d.R., Harding, G., Cassells, A., Saunders, J. 1965, Royal Obs. Bull. 97.
Zinn, R. 1985, Ap.J. (in press).

M GIANTS IN THE GALACTIC NUCLEAR BULGE


Jay A. Frogel
National Optical Astronomy Observatories
Cerro Tololo Inter-American Observatory
950 North Cherry Avenue
Tucson, AZ  85726


ABSTRACT. The bolometric and infrared luminosities of spheroidal systems are dominated by the light of cool stars. These stars are predominantly late type giants which are quite rare in the solar neighborhood. It is argued that the large numbers of M giants being found in the Galactic bulge are the best candidates for inclusion in stellar synthesis models for spheroidal systems. These M giants differ systematically in luminosity, color, and CO index from the old disk giants that are generally used in the models. The differences are in the direction expected for a super metal rich population. A dependence of some of the colors on galactic latitude has been detected. This could be due to a metallicity gradient. The very steep decline in numbers of the very reddest of the M giants may be connected with the metallicity gradient. When these M giants are added on to some models, the resulting integrated infrared colors reproduce the observed colors of spheroidal systems quite well.


1.  INTRODUCTION

In order to predict the spectral and luminosity evolution of galaxies via stellar synthesis models it is important to have accurate information about the cool stellar component. Cool stars, be they giants or dwarfs, dominate the bolometric luminosity of ellipticals, S0s, and the bulges of all but the latest type spirals – systems which I will collectively refer to as spheroidal. There are a number of colors and molecular band indices, particularly in the red and infrared, whose observed values are quite sensitive to stellar luminosity and to the slope of the initial mass function (IMF) of the stars. The luminosity evolution of the system will be a sensitive function of the slope of the initial mass function. In this paper I will argue that the giants in the Galactic nuclear bulge are our best chance for learning, from the ground, about the cool stellar component of spheroidal systems and describe what has been learned to date about these stars. Whitford (1985a) has provided an excellent review of this topic; he discusses a number of areas which are not mentioned here.

The stellar content of spheroidal systems of similar integrated galaxian luminosity shows little variation from one type of system to another. This fact is established by the closeness of their UV to IR energy distributions and the details of their absorption spectra (*e.g.* de Vaucouleurs 1961; Faber 1977; Aaronson 1977; Frogel, *et al*. 1978; Aaronson, Frogel, and Persson 1978). This similarity breaks down for the later type spirals - certainly for the Sc's (Turnrose 1976; O'Connell 1976; Aaronson 1977; Frogel 1985 ) - and, of course for those spirals with active nuclei.

A number of different lines of evidence demonstrate that cool stars make a significant contribution to the integrated light of spheroidal systems in the red and infrared. The data also indicate that these cool stars are giants rather than dwarfs. Examples are: the strong $1.0 \mu m$ radiation observed by Stebbins and Whitford (1948); the red $2.2-3.5 \mu m$ colors found by Johnson (1966); the strengths of TiO and Ca II absorption features measured by O'Connell (1976); the VJHK colors and CO and $H_2O$ indices (Frogel, *et al*. 1978; Aaronson, Frogel, and Persson 1978; Aaronson 1977; and the absence of the Wing Ford bands in absorption (Whitford 1977). Although the exact value differs somewhat from model to model, the cool stars appear to contribute half of the bolometric luminosity and dominate the light longward of 1.0 m but do not contribute more that 5 - 10% of the light in the visual.

A comparison of accurate infrared data for elliptical galaxies with models constructed by O'Connell (1976) and Tinsley and Gunn (1976) was presented in Frogel, *et al*. (1978) and Aaronson, Frogel, and Persson (1978). The O'Connell models fit the observed VJHK colors and CO and $H_2O$ absorption indices very well while the Tinsley and Gunn models had colors which were too blue and indices which were too weak. Differences between the model fits arose primarily because of the sensitivity of the integrated colors to the magnitudes and colors chosen for the M giants. The quality of the fit of the O'Connell models was in fact judged to be fortuitous because of the coarseness of the bins he chose to work with for the giants. In spite of these qualifications, this comparison between models and observations established that V-K provides a strong constraint on the shape of the giant branch while the CO index strongly constrains the dwarf-to-giant ratio. The $H_2O$ index sets additional restrictions on the coolest giants as its value increases rapidly for giants of spectral type M6 and later.

As the observers and the model builders showed, any significant addition of M dwarfs to the models would drive the CO index to an unacceptably low value, V-K to too red a color, and $H_2O$ to too high a value. Pickels' (1985) models, which do not use any data longward of $1 \mu m$ permit a contribution from M4-6 dwarfs to the light significantly greater that that allowed in either O'Connell's (1976) or Tinsley and Gunn's (1976) models. He relies heavily on the near infrared Na I doublet which can be difficult to measure with any precision because of contamination by terrestrial water lines. It would be interesting to see what infrared colors and indices are predicted by Pickels' models.

As Tinsley (1978) realized, the luminous late M giants which appear to dominate the observed infrared colors and bolometric luminosities of elliptical galaxies are quite rare in solar neighborhood samples of stars; thus predictions of their physical parameters such as temperature and luminosity are uncertain. What Tinsley did was to adopt a "trial and error" approach whereby she attempted to find a giant branch luminosity function that would fit the observed infrared colors while being consistent with constraints from stellar models and with the statistics for old disk stars. Her new models provided a significantly better fit to the observed infrared colors and indices. Unfortunately, because of the rarity of the M giants in the old disk population, the theoretical constraints set on them in the models were not particularly strong.

Unfortunately, colors and band indices alone appear to be inadequate in deciding between model IMFs with slope between $x = 0$ and $x = 1$, independent of the uncertainties involved in the modeling procedures. Yet it is in this domain that the rate of luminosity evolution of a stellar population changes most rapidly!

Frogel, Persson, and Cohen (1980) used models which evolved a single burst of star formation through 13 Gyr via the Ciardullo and Demarque (1977) isochrones. These basically were modified versions of the models used by Aaronson, et al. (1978; ACMM). Both Frogel, et al. and ACMM found that these models successfully reproduced broad band colors and indices for globular clusters of both high and low metallicity. However, Frogel et al. showed that the models could not reproduce colors of early type galaxies. Hence the galaxies must contain stars not present in the models and, by implication, not present in globular clusters.

Now the O'Connell, Tinsley, and Gunn models which best fit the infrared data all contained late type giants with bolometric luminosities as bright as -5.5. With the exception of a few long period variables, stars this luminous are absent from globular clusters (Frogel, Persson, and Cohen 1983) and were not included in the ACMM or Frogel, et al. (1980) models. These latter authors concluded that galaxies contain either luminous upper AGB stars with an age several Gyr younger than is typically assigned to early type galaxies, or that these systems possess extremely metal rich stars which, because of their red colors, would match the data. Implications of these results were discussed further in Gunn, Stryker, and Tinsley (1981).

The problem, then is how to determine the nature of the late type M giants which stellar synthesis models and infrared observations require to be present in elliptical galaxies. Accurate determination of the luminosity function for these stars could provide improved estimates for the luminosity evolution of the light of elliptical galaxies. From these considerations we are led to consider the stars in the nuclear bulge of the Milky Way.

## 2. BAADE'S WINDOW

### 2.1. What Is It and Why Study It?

Although optical observations of the center of our galaxy are virtually impossible because of 30 magnitudes of visual extinction, Baade (1963) recognized that essentially all of the obscuring material is closely confined to the galactic plane and quite patchy. He identified a number of "windows" through which stars of the nuclear bulge could be seen even at galactic latitudes of only a few degrees. The best studied of these windows is at $-3.9°$ and is centered on the globular cluster NGC 6522. A blue photograph of this region (Fig. 2 of Blanco, McCarthy, and Blanco 1984) shows that although there still is considerable, and variable, extinction across this field, the star density is extremely high.

The integrated light of Baade's Window is quite similar to the light of external spheroidal systems. This was demonstrated qualitatively by the photographic spectra of Morgan (1956). Whitford (1978) showed a quantitative similarity in terms of both the continuous energy distributions of the different systems and the strengths of various atomic and molecular absorption features.

A further key similarity between the Galactic nuclear bulge and elliptical galaxies is the presence in both of large numbers of late M giants. Evidence for their presence in other galaxies has been noted above. A low resolution spectroscopic survey of the Galactic center region by Nassau and Blanco (1958) revealed the presence of large numbers of M giants there. A summary of earlier work and detailed results of the most recent survey in the $-3.9°$ are given by Blanco, McCarthy, and Blanco (1984).

It is interesting to note that Blanco, *et al.* found no luminous carbon stars in Baade's Window similar to the ones which exist in such great abundance in the Magellanic Clouds. The carbon star candidates reported by Azzopardi, Lequeux, and Rebeirot (1985) appear to be relatively blue and intrinsically faint.

It is instructive to examine the relative distributions in galactic longitude of late M giants and luminous carbon stars as presented in Figures 5 and 6 of Blanco (1965). What we see is that in the general direction of the Galactic center the ratio of M to C stars is many times higher than for directions that lie outside of the solar circle. The M/C ratio is a function of age and metallicity of a stellar population (Renzini and Voli 1981; Iben and Renzini 1983). The absence of luminous C stars in Baade's Window implies a combination of high metallicity and old age.

An observation related to the absence of luminous C stars in Baade's Window is the high number of M8-9 giants relative to the M6-7 ones when compared with the Magellanic Clouds. This again can be attributed to a high metallicity for the bulge field since a high value for [Fe/H] puts the Hayashi track for a giant far to the red.

So it appears that the M giants found by Blanco and his collaborators in Baade's Window are our best hope for obtaining the information needed for the construction of accurate stellar synthesis models of elliptical galaxies and bulges of spirals. In addition they can provide the answer to many "local" problems. For example the metallicity and stellar population of the bulge itself as a function of height above the plane can be examined. If combined with data on stellar dynamics, the formation history and subsequent chemical enrichment of the spheroidal population can be investigated. The remainder of this talk will review what has been learned so far about the giants in Baade's Window and describe work in progress.

## 2.2. Earlier Research on Stars in the Nuclear Bulge

Pioneering studies of individual stars in Baade's Window were carried out by Arp (1965) and van den Bergh (1971). Two important conclusions emerged from their photographic color-magnitude diagrams were that the stars observed were old – probably as old as the oldest stars known – and metal rich – certainly as metal rich as 47 Tucanae and probably as metal rich as the sun.

A considerable effort is going into a determination of the age(s) of bulge stars via accurate c-m diagrams for main sequence stars with the CTIO 4-meter reflector. The problem is not faintness of the stars, but rather the extreme crowding. Initial results from main sequence photometry in the less crowded $-8°$ field (Rich 1984; Terndrup, Rich, and Whitford 1984; Whitford 1986) indicate that most of the stars are at least 10 Gyr old. Mould (1983) found that the velocity dispersion of a sample of Blanco, McCarthy, and Blanco's (1984) late M giants is comparable to that of the oldest stellar components of the spheroid. Wood and Bessell (1983), on the other hand, have argued that the variable stars in Baade's Window indicate an age of only a few Gyr. Elsewhere in this volume Whitford addresses the age issue of the bulge in some detail. We note that at this conference both O'Connell and Pickels have argued that their models for elliptical galaxies require substantial numbers of stars with an age less than 10 Gyr. If this turns out to be correct, there may, then, be a significant difference between the bulge stars and an elliptical galaxy population.

The first quantitative estimate for the metallicity of the bulge stars has been given by Whitford and Rich (1983) for a sample of K giants. Sixty percent of the stars in their sample have $[Fe/H]>0.0$. In addition, the total range in metallicity may be as great as 100. Clearly, if this range is typical for the stars in an elliptical galaxy, then existing models are not even coming close to reality no matter how well the broad band colors are being fit! Searle (1984) is attempting to determine the distribution over metallicity of the RR Lyrae variables in the bulge. This study, however, *a priori* excludes stars of near solar and greater metallicity since they never pass through the RR Lyrae domain in the temperature-luminosity plane.

A disquieting result has come out of the infrared observations of the K giants for which Whitford and Rich (1983) determined spectroscopic abundances. For star clusters with a range of about 100 in abundance the V-K and J-K colors of their giant branches at a given magnitude level increase monotonically with $[Fe/H]$ (Frogel, Cohen, and Persson 1983). Frogel, Whitford, and Rich (1984) have found, however, that while the more metal rich K giants in Baade's Window lie to the red of the metal poor ones in these two colors, the overall distribution of color is much too blue. Even the most metal rich stars tend to lie near or blueward of the 47 Tuc giant branch in the infrared. Following the suggestion of Frogel and Whitford (1982), it is possible that the colors of bulge giants are affected by as yet undetermined blanketing agents. As will be seen, blanketing problems seem to be present for the M stars as well.

The distribution over I magnitude for the M6-9 giants found by Blanco, McCarthy, and Blanco (1984) in the $-3.9°$ window is sharply peaked (see their Fig. 5). As discussed by Blanco, et al. "the stars presented ... constitute, therefore, a unique complete sample of the red giant population of the galactic bulge." Observation of these stars, then, should provide an unbiased data set for use in stellar synthesis models and for study of the bulge itself. I will now describe some preliminary results derived from infrared photometry of these stars plus ones in other windows. Much of this work is being carried out in collaboration with Blanco, Terndrup, and Whitford.

2.3. Infrared Photometry of Bulge M Giants at $-3.9°$

Frogel and Whitford (1982) calculated bolometric luminosities for a small subset of the BMB giants in Baade's Window and found them to be about one magnitude fainter than the giant branch used by Tinsley and Gunn (1976); nonetheless, many of the stars were luminous enough that they had to be AGB stars. As discussed by Whitford elsewhere in this volume, suggested revisions in the distance scale to the galactic center would further increase the discrepancy between the stars and the models at the same time as the problem of accounting for the most luminous stars is mitigated. Examination of our complete sample of over 200 Baade's Window stars with infrared photometry puts the empirical giant branch systematically fainter than Tinsley and Gunn's by about two mags (Frogel and Whitford 1985; Whitford 1986 , Fig. 1d). The implication for the stellar synthesis models is that the number of late M giants relative to fainter stars much be significantly increased. Whether or not the resulting luminosity function is reasonable will be decided from a careful fitting of the luminosity function for BMB's M giants onto that for the fainter K stars in the bulge window.

Globular clusters have JHK colors which are displaced from the mean line for solar neighborhood stars in the sense of having H-K colors which are too blue and J-H colors which are too red. The amount of the displacement appears to be independent of metallicity (Frogel, Persson, and Cohen 1983). Baade's Window M giants are displaced in the opposite sense from the mean field line by nearly 0.1 mags (Fig. 2 of Frogel, Blanco, and Whitford 1984). The dependence of other infrared colors on spectral type exhibited by the Baade's Window M stars is also quite different from solar neighborhood stars (cf. Fig. 1a-c of Whitford 1986 ).

It is reasonable to assume that the Baade's Window M giants are SMR. In fact, one expects the percentage of stars which are SMR to increase with later spectral types. The argument is the following: All M giants have to pass through a K giant phase, so the *minimum* percentage of SMR M stars will be given by the percentage of SMR K giants - 60% in the case of the Whitford and Rich (1983) sample. But giant and asymptotic giant branch tracks shift progressively toward cooler temperature as the metallicity is raised. Furthermore, as Whitford (1986 ) argues, an increase in the metallicity will cause an increase in TiO abundance at a fixed temperature leading to a later spectral classification since the classification is based on TiO band strengths. Suggestions as to how the SMR character of the M giants can effect their colors are dis-

cussed by Frogel, Blanco, and Whitford (1984) and, in some detail, by
Whitford (1985b). Finally, from Rich's work on the velocity dispersion
of the K giants in the -3.9° window (Rich and Whitford 1984) it appears
that they have the same dispersion as the late M giants found by BMB
(Mould 1983).

Although the CO indices of the Baade's Window M giants are rather
similar to their solar neighborhood counterparts at the same spectral
type (Whitford 1986, Fig. 1b), when plotted as a function of $(J-K)_o$ the
CO index of nearly every M1-9 bulge giant is seen to be stronger than
the mean field line (Frogel and Whitford 1985). Although the data are
few, the same is true for the K giants observed by Frogel, Whitford, and
Rich (1984).

What we have found, then, is that the colors, luminosities, and CO
indices of an unbiased sample of M giants in Baade's Window differ systematically from their solar neighborhood counterparts, *i.e.* from the
old disk giants which have been used previously in stellar synthesis
models. Since the differences are in the sense that the bulge stars are
bluer and less luminous, their relative numbers in the models may have
to be as much as two times greater than those for the presently used old
disk giants to allow the models to predict the same colors as they do
now. One of the goals of our (Frogel and Whitford 1985) infrared photometry of the stars in Baade's Window is to produce a giant branch luminosity function for use in the models. This should result in a more
accurate prediction of the luminosity evolution of the resulting stellar
system.

## 2.4. Infrared Observations of Higher Latitude Windows

In addition to the data for the stars in the -3.9° window, we have also
been studying windows at -2.9, -6, -8, -10, and -12° with identical
techniques. A few initial results have been reported by Frogel, Blanco,
and Whitford (1984, FBW) and by Blanco and Blanco (1984). The ccd spectroscopic data is still being analyzed so we do not yet have direct information on metallicity gradients. However, the colors and the number
counts by themselves have already yielded a very interesting result.

FBW divided the stars in the -3.9 and -8° windows into three groups
(cf. their Fig. 1): stars more luminous than the top of the 47 Tuc giant branch and which are redder, in $(J-K)_o$, than an extension of it;
large amplitude variables (LAVs) contained in the survey areas as identified by Lloyd Evans (1976) and Plaut (1971); everything else. *Only*
M7-9 stars and some LAVs are in the first group while the third contains
all spectral types *including* M7-9s and LAVs. For a stellar population
with a spatial density distribution which follows an $R^{1/4}$ law, the observed surface density difference between the two windows should be a
factor of 3. It was found to be 0.35 for the variables, 0.21 for
"everything else", and *0.09* for the red luminous stars. Blanco and
Blanco (1984) presented a graph showing the surface density of M5 and
later stars in the windows surveyed so far. They find that the -12°
field has about 1000 times fewer such stars than the -3.9° field whereas
for an $R^{1/4}$ law the number density should have decreased by only a factor of 10!

One possible origin of these steep declines in the reddest and latest stars would be a falloff in the number of very metal rich stars with distance from the galactic center or an overall decrease in metallicity. Since the Hayashi track for giants shifts to warmer temperatures as the metallicity decreases, the number of cool red stars would be expected to decrease. Another possibility is that these red luminous stars belong to a population with a spatial distribution different from that of the majority of the bulge stars. There is no significant difference in velocity dispersion between the red, luminous stars and the others (Mould 1983).

Preliminary analysis of the colors in the $-3.9$, $-8$, and $-10°$ windows indicates that there exist gradients in the sense to be expected if metallicity were decreasing with height above the plane. What is seen is a decrease in the difference between the observed colors for the bulge stars and the mean colors for field giants. The bulge giants appear to get bluer in $(H-K)_o$, redder in $(J-H)_o$, and weaker in CO as distance from the plane increases.

There are no apparent changes in the luminosity function of the bulge giants with latitude. More specifically, *even in the $-10°$ window there are AGB stars with luminosities significantly brighter than that at which core helium flash should occur.*

## 2.5. The Problem of Mass Loss

An important problem which can be approached with the 3.5 and 10 $\mu$m data for the bulge giants is the dependence of mass loss on metallicity. Knowledge of the mass loss rate is a key factor in predicting evolutionary lifetimes of red giants and in determining enrichment rates for the interstellar medium (cf. Iben and Renzini 1983). The lower the mass loss rate, the longer will be the lifetime of an SMR bulge giant and the higher will be the maximum luminosity it can attain on the AGB. A parameterized mass loss law such as that of Reimers (1975) does not have an explicit dependence on metallicity and is of the form $dM/dt \propto L^{3/2} T_e^{-2} M^{-1}$. Initial examination of the data for bulge giants and for globular cluster variables (Frogel 1984) does not show any significant dependence of mass loss rate on metallicity over a range in metallicity of more than a factor of 10. This is a surprising result. Perhaps much of the mass being lost from red giants is not in the form of grains that would give rise to excess emission at the observed wavelengths.

## 3. RELEVANCE OF IRAS SOURCES

Analysis of the IRAS Point Source Catalogue (1984) by Chester (1986) shows that a well defined subset of the sources - 10% of the objects in the catalogue - have a spatial distribution which closely follows the disk and bulge of the Milky Way. These "disk and bulge" sources are defined as ones that were detected in IRAS's 12 $\mu$m band and have $F_v(25/12) > 1$. CCD spectroscopy and infrared photometry of a sample of these objects by Terndrup and myself (reported in Frogel 1986) shows that they are quite similar to the reddest M giants found in the spectroscopic surveys of the bulge fields and extend the observed properties of these latter stars in a natural fashion to cooler temperatures and greater infrared excesses. It seems possible to identify all of the IRAS sources which lie in the spectroscopically surveyed windows themselves with stars that have already been found in these surveys. The number of IRAS sources in the bulge windows is about 3% of the number of M6-9 giants.

One interesting question to ask is if there is any wavelength at which the contribution of these IRAS "disk and bulge" sources will dominate the integrated light and, by implication, the integrated light from the central regions of other galaxies. Another, more complex problem is to separate the disk from the bulge sources and determine whether or not they represent the same stellar population in terms of age, metallicity, and velocity dispersion, and whether the bulge sources by themselves are from the same population as the vast number of optically more prominent late type giants found in the surveys.

## 4. THE RELATION BETWEEN THE BULGE GIANTS AND THE MODELS

An original motivation for this study was to determine if the M giants in the bulge windows are the "correct" stars to be used in stellar synthesis models for elliptical galaxies. In this context "correct" means that the proposed population is strongly constrained by observations and reproduces the integrated light of the ellipticals in terms of both the continuous energy distribution and the details of the absorption line spectrum. In so far as we believe that the giants in the bulge are an appropriate sample of stars, then the first criterion is certainly satisfied: the luminosity function, the mean colors, and the magnitudes of the these stars are well established empirically for an unbiased, complete sample.

The second of the above criterion is partially satisfied. While we have not yet calculated the effect on the optical light of a model of adding a population of bulge giants, their addition to the Aaronson, *et al.* (1977) and Frogel, Persson, and Cohen (1980) models permits the range of infrared colors and indices observed for elliptical galaxies to be reproduced by simply varying the percentage contribution of light from the bulge giants.

A number of important details need to be worked out. The photometric and spectroscopic studies of the K and M bulge giants have shown that a broad range of metal abundance exists. Stellar synthesis models have tended to consider only a mean metallicity. We also know that the luminosity sequence of elliptical galaxies is, to first order, a metallicity sequence (see review by Faber 1977). It remains to be determined, then, how to allow for variations in the mean and the dispersion of the metal abundance for a stellar population. Secondly, we have not yet determined empirical criteria for tying the M giant luminosity function to that for the less luminous K giants. This latter undertaking should be straightforward. The former will not be so. Finally, the age issue **must** be resolved.

## 5. ACKNOWLEDGEMENTS

I am grateful to my collaborators Victor Blanco, Don Terndrup, and Albert Whitford both for their collaboration on this seemingly never ending project and for their permission to discuss our unpublished results. Also, I thank the organizers of this conference for the opportunity to give this talk and to experience two weeks of really unforgettable weather. Finally, I want to express my appreciation to the people at the Microsoft Corporation for their help in answering my many inquiries about the use of their beautiful program called "Word" with which the camera ready copy of this paper was set up and printed.

## REFERENCES

Aaronson, M. 1977, Ph.D. Thesis, Harvard University.
Aaronson, M., Cohen, J. G., Mould, J., and Malkan, M. 1978, *Ap. J.*, **223**, 824.
Aaronson, M., Frogel, J. A., and Persson, S. E. 1978, *Ap. J.*, **220**, 442.
Arp, H. 1965, *Ap. J.*, **141**, 45.
Azzopardi, M., Lequeux, J., and Rebeirot, E. 1985, *Astr. Ap.*, **145**, L4.
Baade, W. 1963, in *Evolution of Stars and Galaxies*, ed. C. P. Gaposhkin (Cambridge: Harvard University Press), p. 279.
Blanco, V. M. 1965, in *Galactic Structure*, ed. A. Blaauw and M. Schmidt (Chicago: University of Chicago Press), p. 242.
Blanco, V. M., and Blanco, B. M. 1984, *Mem. Soc. Astr. Ital.*, in press.
Blanco, V. M., McCarthy, M. F., and Blanco, B. M. 1984, *Astr. J.*, **89**, 636.
Chester, T. 1986, in *First International Symposium on Results from IRAS*, ed. F. Israel (Dordrecht: Reidel), in press.
Ciardullo, R. B., and Demarque, P. 1977, *Trans. Astr. Obs. Yale Univ.*, 35.
de Vaucouleurs, G. 1961, *Ap. J. Suppl.*, 5, 233.
Faber, S. M. 1977, in *The Evolution of Galaxies and Stellar Populations*, ed. B. M. Tinsley and R. B. Larson (New Haven: Yale University Observatory), p. 157
Frenk, C. S., and White, S. D. M. 1982, *M.N.R.A.S.*, **198**, 173.
Frogel, J. A. 1984, *Mem. Soc. Astr. Ital.*, in press.
Frogel, J. A. 1985 , *Ap. J.*, in press.
Frogel, J. A. 1986 , in *First International Symposium on Results from IRAS*, ed. F. Israel (Dordrecht: Reidel), in press.
Frogel, J. A., Blanco, V. M., and Whitford, A. E. 1984, in *IAU Symp. No. 105, Observational Tests of Stellar Evolution Theory*, ed. A. Maeder and A. Renzini (Dordrecht: Reidel), p. 571.
Frogel, J. A., Cohen, J. G., and Persson, S. E. 1983, *Ap. J.*, **275**, 789.
Frogel, J. A., Persson, S. E., Aaronson, M., and Matthews, K. 1978, *Ap. J.*, **220**, 75.
Frogel, J. A., Persson, S. E., and Cohen, J. G. 1980, *Ap. J.*, **240**, 785.
Frogel, J. A., Persson, S. E., and Cohen, J. G. 1983, *Ap. J. Suppl.*, **53**, 713.
Frogel, J. A., and Whitford, A. E. 1982, *Ap. J. (Letters)*, **259**, L7.

Frogel, J. A., and Whitford, A. E. 1985, in preparation.
Frogel, J. A., Whitford, A. E., and Rich, R. M. 1984, *Astr. J.*, **89**, 1536.
Gunn, J. G., Stryker, L. L., and Tinsley, B. M. 1981, *Ap. J.*, **249**, 48.
Iben, I. Jr., and Renzini, A. 1983, *Ann. Rev. Astr. Ap.*, **21**, 271.
*IRAS Point Source Catalogue* 1985, Prepared under the direction of JISWG, U. S. Government Printing Office.
Lloyd Evans, T. 1976, *M. N. R. A. S.*, **174**, 169.
Morgan, W. W. 1956, *Pub. A. S. P.*, **68**, 509.
Mould, J. R. 1983, *Ap. J.*, **266**, 255.
Nassau, J. J., and Blanco, V. M. 1958, *Ap. J.*, **128**, 46.
O'Connell, R. W. 1976, *Ap. J.*, **206**, 370.
Pickels, A. J. 1985, *Ap. J.*, in press.
Plaut, L. 1971, *Astr. Ap. Suppl.*, **4**, 75.
Reimers, D. 1975, in *Proc. 19th International Astrophysics Symposium, Liege*, p. 369.
Renzini, A., and Voli, M. 1981, *Astr. Ap.*, **94**, 175.
Rich, R. M. 1984, *Mem. Soc. Astr. Ital.*, in press.
Rich, R. M., and Whitford, A. E. 1984, *Pub. A. S. P.*, **96**, 794.
Searle, L. 1984, *Mem. Soc. Astr. Ital.*, in press.
Stebbins, J., and Whitford, A. E. 1948, *Ap. J.*, **108**, 413.
Terndrup, D. M., Rich, R. M., and Whitford, A. E. 1984, *Pub. A. S. P.*, **96**, 796.
Tinsley, B. M. 1978, *Ap. J.*, **222**, 14.
Tinsley, B. M., and Gunn, J. G. 1976, *Ap. J.*, **203**, 52.
Turnrose, B. E. 1976, *Ap. J.*, **210**, 33.
van den Bergh, S. 1971, *Astr. J.*, **76**, 1082.
Whitford, A. E. 1977, *Ap. J.*, **211**, 527.
Whitford, A. E. 1978, *Ap. J.*, **226**, 777.
Whitford, A. E. 1985a, *Pub. A. S. P.*, **97**, 589.
Whitford, A. E. 1986, Erice conference on Spectral Evolution of Stellar Populations.
Whitford, A. E., and Rich, R. M. 1983, *Ap. J.*, **274**, 723.
Wood, P. R., and Bessell, M. S. 1983, *Ap. J.*, **265**, 748.

## DISCUSSION

**AARONSON:** What do you estimate the disk star contamination to be in your bulge samples? Could uncertainty in the field star calibration account for the offset between the bulge and field stars in the JHK plane?

**FROGEL:** The field star contamination must be negligible. Nearby giants would be bright and easily distinguishable by their near zero radial velocity. M dwarfs would be cleanly separated from the M giants. In fact in one of the higher latitude fields we have found a cluster of half a dozen M dwarfs! In answer to your second question, I would say no. The offset is too large. In any case, there is no uncertainty in the location of the several hundred globular cluster stars, nor in the several dozen fairly red giants from galactic open clusters (unpublished data) which, in the mean, lie quite close to the field line.

O'CONNELL:   Is there a resolution of the conflict between your group and Faber and French concerning possible dwarf enrichment in the core of M31?

FROGEL:   No.

DE JONG:   Will a different distance modulus for the galactic center reduce the disagreement of your data with the Tinsley and Gunn c-m diagram?

FROGEL:   No, since most revisions of the distance scale, *e.g.* that of Blanco and Blanco based on RR Lyrae variables and that of Frenk and White (1982) based on globular clusters come up with a value close to 7 kpc.   In response to SILK's question on the same topic, I would say that you do *not* want to get a distance modulus to the center from the M giants as that would assume too much about a group of stars we know too little about as yet.

NESCI:   If you use $(J-K)_o$ instead of $(V-K)_o$ for the c-m diagram, do you still get different slopes for the old disk and the bulge giants?

FROGEL:   Yes.

WHITFORD (Comment on previous questions):   The Tinsley and Gunn absolute magnitudes were derived from Eggen's moving group calibration of solar neighborhood red giants.

PICKELS:   If the offset between the observed c-m diagram for the Baade's Window M giants and the old disk (Tinsley and Gunn) c-m diagram is interpreted purely as a metallicity effect, what is the implied metallicity difference?

FROGEL:   I cannot tell since the calibration is based on globular clusters and, as you know, their metallicity barely goes up to the solar value, so that the c-m diagram for the Baade's Window M giants would involve a sizable extrapolation.   A similar answer applies to KOO's question about estimating the metallicity of the bulge stars from the CO indices - the observed values lie outside of the calibration.

PICKELS:   Presumably the derived turnoff age for NGC 6522 by Rich, *et al.* would be less than 10 Gyr if a metal rich isochrone were used?

FROGEL:   That could be true, but I do not believe that any estimates of that cluster's metallicity suggest that it is metal rich (with respect to the sun).   In response to KOO's related question, the published results of Rich, Terndrup, and Whitford have only gone so far as to exclude ages younger than 10 Gyr for stars in the $-8°$ window.   This is not to say that their data will not allow the exclusion of stars up to 15 Gyr.   I just do not think that their analysis has progressed to its conclusion as yet.

WHITFORD (Comment of the previous question): The clue to the use of 10-15 Gyr isochrones to test the age of the stars in the bulge fields may come from VandenBerg's extension of his tracks to the super-metal-rich domain. The present comparison is based on solar metallicity isochrones.

MOULD: Do you have any comment on the much higher central concentration of the Baade's Window M giants then, say, the globular cluster system as a whole and the similarity of their velocity dispersions?

FROGEL: As I have already mentioned, one possibility is that there are strong metallicity gradients affecting the M giant population. Perhaps there was a small, but sufficiently large enough difference in the collapse times of the different stellar groups to allow differential enrichment without affecting the velocity dispersion. Another thought I had is that a disk population growing exponentially in two dimensions could mimic a bulge population, or at least significantly contaminate one. However, the fact that there is no obvious difference in the velocity dispersions between the different stellar groups probably argues against this.

KOO: Do M giants of different metallicities have different functional dependencies on galactic latitude, and, (question by AARONSON) if there is a metallicity gradient in the bulge, then the number of M0+ stars in the different windows might be a better match to a de Vaucouleurs law than the M5+ stars; is this the case?

FROGEL: As I stated in the talk, Blanco, Whitford, and I have found apparent differences in latitude dependence between *photometrically* selected groups of stars. However, if the M giants are broken up into three *spectroscopic* groups, M1-4, M5-6, and M7-9, there are no significant differences in latitude dependence amongst the groups for the $-3.9$, $-8$, and $-10°$ windows. I suspect that this may reflect the rather coarse nature of low resolution spectroscopic classification schemes. In theory Aaronson's point is correct, but the metallicity differences between the windows for which the data have already been analyzed may only show in an examination of the colors.

WHITFORD (Comment on previous question): If the metallicity distribution function could be determined for the high-latitude windows, one could say whether the declining importance of the high Z tail decreases the fraction of the giants that are able to reach the M6-9 domain. It is possible that a gradient in average metallicity is a sufficient explanation for the rapid decline of M giant numbers with latitude.

FROGEL: That is correct, but I think that Marc's point is that such a gradient should show up in a change of the relative numbers of M type giants as well as in their absolute total number. Of course we do not really understand as yet the quantitative effects of metallicity on the distribution over color, luminosity, and spectral type. I hope that as we near the conclusion of this project we will get some understanding.

The CCD spectra we have been obtaining should give us a good indication of the variation in the metallicity distribution function with latitude.

KOO: Given the relatively dim $M_V$ of the Milky Way bulge compared to the $M_V$ of giant ellipticals and the "established" metallicity – $M_V$ relation for these galaxies, is it not surprising to find [Fe/H] ranging from solar to 10 times solar for 50% of the bulge giants?

FROGEL: I recollect an article recently published which claimed to show that the $M_V$ – [Fe/H] relation for bulges of spirals differs from that of ellipticals in the sense that for the former it is necessary to consider the luminosity of the entire galaxy, not just that of the bulge. Unfortunately, I cannot remember where I read this. Also, one must bear in mind that most measurements of elliptical galaxies are integrating over fairly large chunks. At the distance of the Virgo Cluster, Baade's Window would only be about 5" out from the nucleus of a galaxy.

DRESSLER (Comment on previous question): The broad line indices and colors that are used to show a luminosity-metallicity relation for E galaxies are not, as yet, firmly tied to [Fe/H]. It may well be that the range from M32 to M31 or even to NGC 4472 runs from near solar to several times solar, at least for the typical stars contributing the light of the indices. If so, the metallicity implied by these star by star measurements in the Galactic bulge may be entirely typical of an elliptical of similar luminosity, i.e. $M_B = -17$ to $-18$.

NEWBERRY: In comparison with other galactic nuclei, Whitford's spectrum of the Galactic bulge shows a large excess of blue light, especially between CaII H,K and the G band. Do you attach any significance to this?

WHITFORD: No, I would not give it much significance because of uncertainties in subtracting the foreground star light which comes from stars that are bluer. Furthermore, the areas of sky used for subtraction were about 30° away from the bulge, so I would not put much trust in the continuum slope because the corrections were so large. However, I would trust the strength of the individual features.

AGE AND MASS OF M GIANTS IN THE GALACTIC BULGE

A. E. Whitford
Lick Observatory
University of California
Santa Cruz, California 95064

ABSTRACT. Recent observations show that the M giants in the Galactic bulge are neither as luminous nor as cool as the solar-neighborhood prototypes that have generally been used in synthesizing population models; a revision in the value of $R_o$ accentuates the disparity in luminosity. Luminosities in excess of those of the long-period variables in metal-rich globular clusters are rare. Thus the argument for assigning an age $\leq$ 2 Gyr to late M bulge giants and attributing their presence to processes other than normal evolution from the dominant super-metal-rich (SMR) fraction of K giants in the bulge now finds little support. If the late giants were young an easily recognizable group of young turnoff stars should be present at the proper location on the color-magnitude diagram to be progenitors of the M7-M9 giants now seen in the window at $b = -8°$. The observations show no such turnoff group.

The properties of the late M giants in the bulge appear to be compatible with an SMR composition and an age in the range 6 to 15 Gyr. The evidence leans toward models in which the star formation rate has been negligible for a considerable part of the total evolution time.

1. INTRODUCTION

The late M giants in the nuclear bulge of the Galaxy were first identified by Nassau and Blanco (1958) on very-low-dispersion objective-prism spectra of stars in Baade's Window (BW) at $\ell = 1°.0$, $b = -3°.9$ (Baade 1951, 1963). The recent complete survey by Blanco, McCarthy, and Blanco (BMB 1984) found 300 giants of classes M6-M9 in BW within a circle 24 arcmin in diameter. When compared with giants in the solar neighborhood and resolved giants in other Local Group galaxies the bulge sample is unique in showing virtually no high-luminosity carbon stars (BMB 1984; Azzopardi, Lequeux, and Rebeirot 1984; Blanco 1985, private communication). Late M giants of the bulge population are also found in surveys of other relatively clear areas from $b = -2°$ to $-12°.6$, all near $\ell = 0°$ (V. and B. Blanco 1985). Resolution of corresponding giants in M31 and M32 has not yet been reported.

These M giants are the nearest examples of the class of very cool stars that have been found to be an essential component of the stellar population of unresolved elliptical galaxies and the bulges of spirals of Hubble type earlier than Sc. They are the source of two observed spectral characteristics of the integrated light of such systems: (1) strong IR continuum radiation [Stebbins and Whitford 1948; Johnson 1966a; Frogel, Persson, Aaronson and Matthews (FPAM 1978)], and (2) a prominent CO absorption band centered at 2.36 $\mu$m. The latter reaches appreciable strength only in late M giants (FPAM 1978). For ellipticals and spiral bulges in the upper 3 mag of the luminosity function (i.e. the galaxies dominating a magnitude-limited sample) the indices measuring these features cluster rather closely around ($V$-$K$) = 3.30 and CO = 0.16 (FPAM 1978; Aaronson 1977; Persson, Frogel and Aaronson 1979).

The M giants in the Galactic bulge should be a typical sample of this component in such populations since the spectral features of the integrated light of the patches of the BW region near NGC 6522 show close similarity to those of comparable regions of other galaxies (Whitford 1978). There had therefore seemed to be a strong possibility that the study of the numbers and the properties of these members of a resolved bulge population would resolve the difficulties that had been encountered in accounting for the cool-star component of population models. Observations of the bulge giants have, however, led to differing interpretations of their evolutionary history. The place of these most luminous members of the bulge population on the theoretical H. R. diagram should, when properly understood, be a diagnostic tool. This paper compares the modeling of the cool-star component in certain classical population syntheses with data on M giants in the Galactic bulge. It also examines the central question ''Are the late M giants in the bulge young or old?'' in the light of recent observations.

## 2. POPULATION MODELS

The problems encountered in constructing a population model for spheroidal galaxies that properly accounts for the very cool component ascribed to M giants have been reviewed elsewhere (Whitford 1985). They may be summarized as follows:

The unconstrained models of O'Connell (1976, 1980) (i.e. those not bound by rigid adherence to theoretical evolution tracks) are able to match the spectral features of the observed integrated spectra satisfactorily. The basic input comes from a library of well-observed solar-neighborhood stars, including late M giants. Star birth up to 5 Gyr ago is required to explain the UV and extreme UV features. Burstein (1985) also finds that intermediate age stars must be included.

The constraints imposed by the Tinsley (1972a, b) precepts for the evolution of galaxy populations were followed in the classical population model constructed by Tinsley and Gunn (TG 1976a). It adopts the single-burst approximation: the star formation rate SFR = 0 for galaxy ages $\geq$ 1 Gyr. The resulting ''old'' population is consistent with the lack of supergiants and observable gas and dust at a density high enough to support current star formation (e.g. Faber and Gallagher 1976). The present-day colors of giant ellipticals match the calculated $B$-$V$ color of a population of age $\sim$ 10 Gyr that has suffered progressive removal of upper main sequence stars as a result of their evolution up the giant branch (Larson and Tinsley 1978). TG also used solar-neighborhood stars as prototypes in their model. They based the calculated contribution of the late M giants for the asymptotic giant branch (AGB) not on theoretical evolution rates but on a semiempirical evaluation of the luminosities and statistics of old disk giants (Tinsley and Gunn 1976b; Eggen 1973). When the resulting model fell short of predicting the $V$-$K$ and CO indices observed by FPAM (1978), Tinsley (1978) made *ad hoc* adjustments to the statistics (or luminosity) of late giants. An updated model (Gunn, Stryker, and Tinsley 1981) used theoretical isochrones for stars of metallicity twice that of the sun, but did not change the giant-branch component of the previous model. Possible modification of the basic single-burst model to account for light in the UV and extreme UV included as one option a continuing small SFR.

Aaronson, Cohen, Mould, and Malkan (ACMM 1978) argued that solar-neighborhood stars of various ages and masses are not representative of a truly old population, and that globular cluster stars provide more realistic prototypes for a single-burst galaxy population. Their model allowed adjustment for the calculated lower temperature of galaxy giants appropriate to their higher metallicity. They did not not include any AGB stars since in general mass loss prevents globular cluster giants from rising above the luminosity of the He-core flash on their second ascent (e.g. Renzini 1977, p. 224 ff). Frogel, Persson, and Cohen (1980) contended, however, that such a globular cluster population could not provide the ''cool luminous stars'' found necessary in galaxy synthesis models. They suggested that these giants may evolve from stars younger and more massive than those in globular clusters.

## 3. CHARACTERISTICS OF BULGE M GIANTS

### 3.1 Metallicity

The metallicity of the late M giants in the bulge has not been directly evaluated. The apparent anomaly of the existence of very luminous AGB stars in an old population has raised the question of whether they may have had an origin other than simple evolution from the bulge K giants (see Section 3.2, 5.1 below). Hence the fact that the latter are dominantly super-metal-rich (SMR) (Whitford and Rich 1983) does not justify the plausible but not necessarily inevitable assumption that the M giants are also SMR. Such a conclusion is, however, supported by indirect arguments. One indication comes from the lack of carbon stars. If the initial abundance of O in these giants is well above solar, then in order to reach the crossover point where C/O > 1, second-ascent thermal pulses must dredge up an increased acount of C from the core, where it has been made by He-burning. Exhaustion or mass-loss evaporation of the hydrogen-rich envelope ends the giant evolution of such a star before it reaches the crossover point (e.g. Scalo and Miller 1981).

## 3.2 Luminosity and Mass

Bolometric luminosities derived by integration over the spectral energy distribution defined by $BVRIJHKL$ magnitudes led Frogel and Whitford (FW 1982) to conclude that in their sample of 20 BW giants those of spectral types M7-M9 are indeed AGB stars. They found $<M_{bol}> = -4.5$ for these stars, assuming a modulus $(m-M)_o = 14.8$. They argued that the bulge K giants found to be dominantly SMR by Whitford and Rich (1983) could be the progenitors of the AGB M giants because their hydrogen burning rate on the main sequence (as shown by Mengel et al. 1979) had been slower than for metal-poor stars. Thus a star with an initial mass $0.2M_\odot$ greater than that of members of classical globular clusters now becoming giants could survive 13 Gyr in the pregiant phase, and have extra mass at the start of the second ascent as fuel for carrying it up the AGB.

A calculation of the total mass loss according to a formula derived by Fusi-Pecci and Renzini (Renzini 1977, eqs. 6.17 and 6.20) showed a final luminosity $M_{bol} = -4.5$ is marginally possible, assuming $Y = 0.3$, independent of Z. Since, however, the long-period variables (LPVs) apparently reach $M_{bol} = -5.5$, a value confirmed by Wood and Bessell (1983), younger and more massive progenitors may still be required. Iben and Renzini (1983) concluded that an initial mass $M_i = 1.5M_\odot$ is necessary to account for the observed AGB stars. The corresponding pregiant age is $\sim 2$ Gyr (Mengel et al. 1979). Wood and Bessell found a pulsational mass $M_p \geq 3M_\odot$ for the longer-period LPVs in the bulge, corresponding to a pregiant lifetime of only 0.2 Gyr. For the LPVs in the bulge with P < 150 days, however, they found a luminosity and mass consistent with their being part of the metal-poor component of the bulge population to which the RR Lyrae variables belong; i.e. stars as old as the Galaxy (see Section 5.2 below).

## 4. NEWER OBSERVATIONAL DATA

A reexamination of the rather diverse evolutionary histories thus far proposed for the late M giants in the Galactic bulge (Sections 2, 3.1, 3.2) in the light of recent observational results narrows the range of possibilities.

## 4.1 Revised Distance to the Galactic Bulge

V. and B. Blanco (1985) have made a new determination of the distance $R_o$ to the Galactic center, based on a reevaluation of the distance modulus of the depth maximum in the number of RR Lyraes along the line of sight to 5 bulge windows. If Sandage's (1982) metallicity-dependent calibration of the luminosity of RR Lyraes is adopted, the Blancos find $R_o$ = 6.95 $\pm$ 0.58 kpc. Frenk and White (1982) found $R_o$ = 6.8 $\pm$ 0.8 kpc from the centroid of globular clusters having a latitude $|b| > 10°$. If other metallicity-independent calibrations are adopted the Blancos find distances ranging from $R_o$ = 7.12 to 7.95 kpc. In the following analysis the round number $R_o$ = 7.0 kpc, corresponding to $(m-M)_o$ = 14.2, is arbitrarily adopted. This is 0.6 mag less than the modulus used by FW (1982) and Wood and Bessell (1983) in calculating the luminosity of bulge stars.

## 4.2 Comparison of M Giants in Baade's Window with Local Prototypes

Fairly complete photometry in the $VJHK$ bands now exists for 209 BW giants of classes M1 to M9. For most of these CO indices were also observed (Frogel and Whitford 1985). The average colors and luminosities of the Blanco subclasses computed from these data are shown in Figure 1a to 1d. Dereddening corrections $E(V-K)$ = 1.40, $E(J-K)$ =

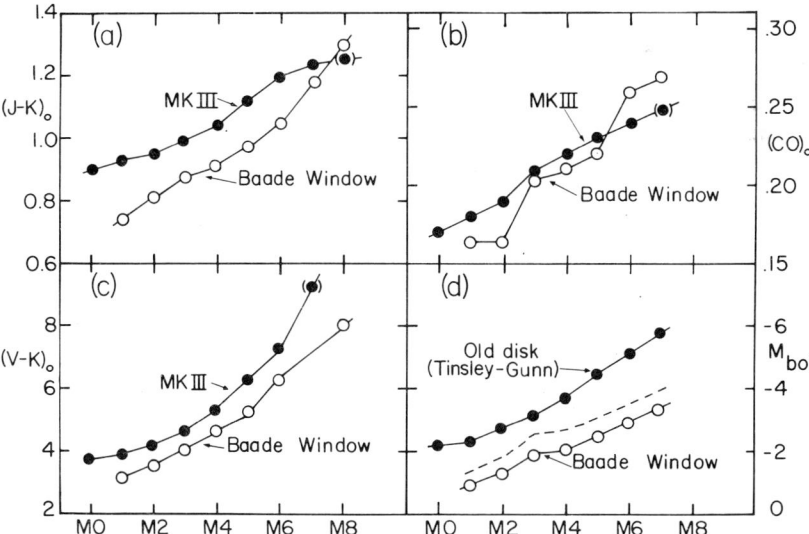

*Fig. 1. Comparison of the mean colors J-K, V-K, the CO index, and $M_{bol}$ of bulge M giants with those of solar-neighborhood giants of the same spectral classes. The dashed line shows $M_{bol}$ as calculated with the old modulus $(m-M)_o = 14.8$.*

0.26, and $E(CO)$ = -0.015 have been applied. The luminosities of the bulge giants depend on the adopted modulus $(m-M_o)$ = 14.2, $A(K)$ = 0.14, and bolometric corrections $BC_K$ taken from the $BC_K$, $(J-K)$ relation derived for the FW (1982) sample of 20 stars. Also shown in the figure are similar quantities for solar-neighbornood giants of the same M giant MKIII subclasses. The colors are taken from FPAM (1978). The luminosities are those adopted by TG (1976a); their $(R-I)_J$, $M_{bol}$ relation has been transformed to one for M giant subclasses with the aid of Johnson's (1966b) mean colors.

The possibility that the substantial difference in colors between bulge and local giants of the same M subclasses could arise from different original classification criteria is ruled out by a comparison of similarly exposed spectra of bulge giants and local MKIII standards of the same class at 10Å resolution; the near-IR molecular features used for classification on extremely compressed grism spectra (BMB 1984), when viewed at 10Å resolution, match quite closely those of MKIII standards (Whitford and Rich, unpublished). Differential blanketing between passbands by deeper molecular absorptions in SMR bulge stars is also not a satisfactory explanation, though this may occur in the $J$-band for later subclasses of the bulge giants.

A real temperature difference is quite understandable if the bulge giants are SMR. The standard equation for association-dissociation equilibrium (e.g. Aller 1963) would predict the same proportion of Ti and O associated as molecules, (corresponding, say, to given TiO feature strengths signifying a particular spectral subclass) at a higher temperature for an enhanced abundance of both Ti and O in the atmospheres of SMR stars. The higher temperature of course means a smaller $J$-$K$ and $V$-$K$ index, as seen in Figures 1a and 1d. Mould and McElroy (1978) were the first to demonstrate this effect; they showed that the temperature at which a certain TiO feature begins to appear in the spectra of cool globular cluster giants depends on the metallicity of the cluster.

The lower luminosity of bulge M giants at the same spectral class (Figure 1d) has previously been noted by Frogel (1981), FW (1982), and BMB (1984). It is likewise understandable as a consequence of the temperature-metallicity effect. A giant near the sun of MKIII class M6 would be an AGB star. A hotter SMR star in the bulge with equal TiO feature strength would be classified M6, even though in luminosity, radius, and gravity it may match (say) an MKIII giant of class M4 near the sun, a star below the RGB tip, definitely not an AGB star. The disparity in luminosity between bulge and local giants is conspicuous even with the old modulus, as is shown by the dotted line in Figure 1d.

The absence of any very significant difference in the CO index of bulge and local giants of the same subclass (Figure 1b) does not have a simple explanation since it involves the behavior of two molecules (CO and TiO) with quite different dissociation potentials; model atmosphere calculations for normal and SMR compositions are needed.

These color differences relative to local giants provide additional evidence (beyond that of the inability of bulge giants to cross the carbon-star threshold) supporting SMR abundances for the late M stars in the bulge. Both arguments are independent of the metallicity of the bulge K giants.

## 4.3 Comparison of Luminous Globular Cluster Giants with Bulge Giants

Frogel (1983) has assembled data on red variables in globular clusters. Long-period-variables (LPVs) occur only in the more metal-rich clusters and are the most luminous stars found in the globular clusters belonging to the Galaxy. Figure 2a shows the LPVs in three of these clusters plotted on a color-magnitude diagram showing $M_{bol}$ vs $(J-K_o)$. Non-variable giants and low-amplitude variables in 47 Tuc are shown for reference (Frogel, Persson, and Cohen 1981). The ridge-line for these 47 Tuc giants ends at $M_{bol}$ = -3.7, the approximate luminosity of the He-core flash in the most metal-rich clusters (Frogel, Cohen, and Persson 1983). Figure 2b shows the BW M giants relative to the same coordinates. Save for the LPVs in 47 Tuc, where means through the cycle are shown, the LPV points are from isolated observations; the range is of the order 1 mag in $M_{bol}$ and 0.10 mag in $(J-K)_o$.

The color index $(J-K)_o$ is chosen for this comparison because it

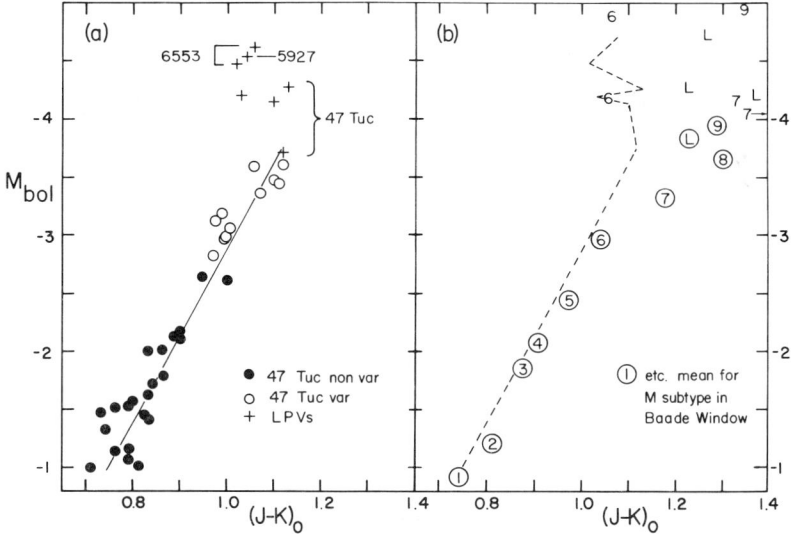

*Fig. 2a. Luminosity-color diagram of giant stars in metal-rich globular clusters.*
*Fig. 2b. Same for the mean luminosity of bulge M giants for the subclasses indicated by circled numerals. Uncircled numerals and Ls (=LPVs) denote individual AGB stars. The dashed line shows the location of the stars in Fig. 2a.*

is known for all the BW M giants in the Frogel-Whitford (1985) photometry; it is used to find $BC_K$, and has only a small range in LPVs. Frogel, Persson, and Cohen (1983) argue that blanketing of uncertain origin renders $(J-K)_o$ less reliable than $(V-K)_o$ as a temperature indicator. The suspected blanketing noted in Figure 1a as affecting the $J-K$ of late M giants appears again in Figure 2b.

The comparison shown in Figure 2a and 2b is most striking in the values of $M_{bol}$. The most luminous stars in metal-rich globular clusters are, as Frogel pointed out, AGB stars, well above the limit adopted in the ACMM (1978) galaxy population model. They occupy a domain very comparable to that of the most luminous M giants in BW, calculated with the new modulus. Two isolated non-LPV bulge giants at $M_{bol}$ = -4.9 [out of a total of 71 M6-M9 stars in the Frogel-Whitford (1985) photometry] do stand out. They could represent front-to-back deviations from the mean modulus to the order $2\sigma \approx 0.5$ mag. Or stochastic variations in the mass-loss-rate could have preserved enough mass in a star as old as the others to fuel ascent to a higher-than-average limiting luminosity. Such variations have been postulated to explain the color distribution along the horizontal branch of metal-poor globular clusters and the existence of metal-rich RR Lyrae variables (Rood 1973; Taam, Kraft, and Suntzeff 1976).

This comparison leads to the obvious inference that if giants in metal rich globular clusters can become AGB stars, even without the additional advantage of being SMR (Sec. 3.2), then equally luminous late M giants in the bulge can be old stars. How old depends on the resolution of the issue raised by Demarque (1980) who argued that metal-rich clusters are as much as 5 Gyr younger than metal-poor ones. Sandage (1982) could find no evidence for a spread in ages. Carney (1981) concluded that the uncertainties were large enough to leave the question open. VandenBerg and Bell (1985) tested age differences by fitting recent precise observations of turnoff stars in M15 (extremely metal-poor) and 47 Tuc (metal-rich) to isochrones derived from VandenBerg's (1983) evolutionary models. No significant difference in age was apparent.

## 5. AGE LIMITS FROM A COLOR-MAGNITUDE DIAGRAM

Early color-magnitude diagrams (CMDs) of the BW stars (Arp 1965; van den Bergh 1971) reached only part way down the giant branch, owing to the difficulties of iris photometry on photographs of this very crowded field. CCD photometry goes much deeper, but crowding still prevents reaching the calculated old-star turnoff at $V \approx 20.5$. On plates of a less a crowded field in the window at $b = -8°$ van den Bergh (1974) recognized the turnoff stars at $V \approx 19.75$. Photometry on CTIO CCD frames goes nearly 2 mag deeper (Terndrup, Rich and Whitford 1984; Rich 1984). Figure 3 shows the CMD of the -8° field derived from the 1983 CCD exposures. Isochrones calculated by VandenBerg (1985) for stars of solar metallicity are superposed; these incorporate the

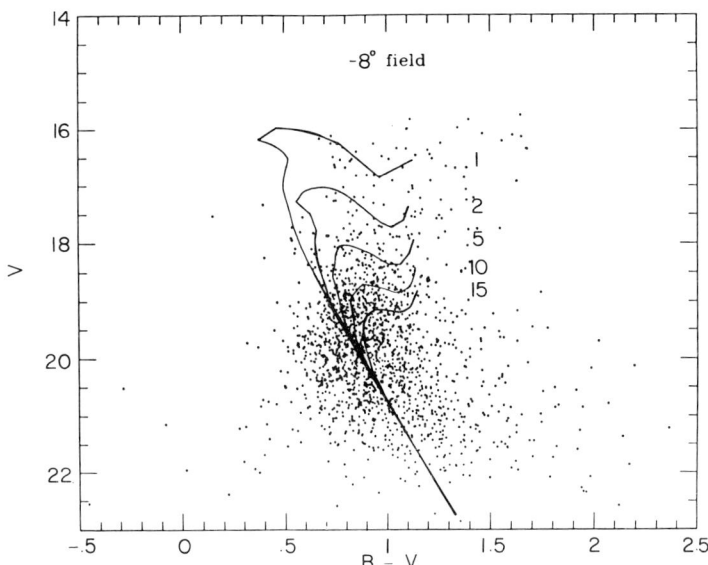

*Fig. 3. Color-magnitude diagram of stars in an area of 15 arcmin$^2$ in the bulge window at $b = -8°$. The superposed isochrones for various ages in Gyr are from VandenBerg's models. See text. Stars up to 2 mag brighter than the upper limit in this figure, measurable only on short exposures, are not shown.*

appropriate shift for reddening and for the new modulus $(m - M)_o$ = 14.2. They assume $Y$ = 0.25 and a mixing-length parameter $\alpha$ = 1.6. Isochrones for stars of SMR composition would be shifted slightly to the right and downward (VandenBerg, private communication).

Any CMD of a sample of bulge stars should show a vertical dispersion arising from a spread in the modulus along the line of sight, and a horizontal dispersion from the metallicity distribution. Whitford and Rich (1983) found a rather broad distribution among the K giants in the BW. A mixture of ages would lead to a superposition of a family of properly weighted single-age CMDs, each centered on its appropriate isochrone. The foreground population of disk stars can be statistically segregated by a modeling technique like that adopted by Arp (1965); this contamination is less serious for a line of sight at the higher latitude of the field at -8°.

Separation of these various effects must await data from additional CCD frames. Simple inspection of Figure 5, however, leads to the conclusion that the majority of the stars are at least as old as 10 Gyr.

It is possible to set an upper limit to the contribution of a younger component, particularly that presumed to exist if the late M giants in BW are indeed very luminous AGB stars. If these evolve from a minority fraction still having a significant SFR, their progenitors should show a concentration at the turnoff point for stars of the assumed mass and age; the mass suggested by Iben and Renzini (1983, Sec. 4.3.3) was $M_i = 1.5 M_\odot$ and the corresponding age is 2 Gyr (Sec. 3.2). If these AGB stars evolve from injected young stars born elsewhere (as suggested by Iben and Renzini), or from merged binaries [as suggested by Renzini (1981)], the progenitors thus acquired must still be waiting in line at the 2 Gyr turnoff to begin their relatively brief giant branch evolution.

From the times spent at various stages along a given evolution track (Mengel et al. 1979; Sweigart and Gross 1978) it is possible to estimate the number of turnoff stars corresponding to a given number of AGB stars, assuming a stable evolution pattern over the last few Gyr. The uniformity of the signature of the late M giants in elliptical galaxies and spiral bulges (Sec. 1) argues strongly against any other assumption.

If the M7-M9 giants in the bulge are AGB stars of age 2 Gyr, and if their AGB phase lasts $2 \times 10^6$ yr (Renzini 1977, eq. 4.3), then the surface density of such stars found in the grism survey of the field at $-8°$ (Frogel, Blanco, and Whitford 1984) would lead to a prediction that among the total number of stars showing on the CMD there should be about 300 stars within $\pm 0.5$ mag of the 2 Gyr turnoff. If the late M giants are not in fact all true AGB stars (as suggested by the revised luminosity of Sec. 4) then the longer evolution times for the upper 2 mag of the RGB would still require about 60 turnoff stars. It is quite clear from Figure 3 that no such number is present. The progenitors of the late M giants must then be fairly old stars. The existence of an appreciable number of stars with ages between 5 and 10 Gyr is not ruled out by the isochrone plot, however.

## 6. DISCUSSION

The newer data of Section 4 indicate an upper limit for the luminosity of bulge M giants about 1 mag above that of the He-core flash, rather than nearly 2 mag above as was the case when a luminosity $M_{bol} = -5.5$ was attributed to the LPVs. The revised upper limit is thus intermediate between that adopted in the TG (1976) galaxy population model ($M_{bol} = -5.75$) and that advocated for such models by ACMM (1978): the level of the He-core flash itself.

The mass-loss argument showing that the basic old population needs to be supplemented by higher-mass short lived progenitors to account for the late M giants (e.g. Iben and Renzini 1983) loses its force in the face of this revision. The Reimers empirical relation, as formulated by Fusi-Pecci and Renzini (1976) says that the mass-loss rate varies as $L^{3/2} T_e^{-2} M^{-1}$. Both the higher values of $T_e$ suggested by the $V$-$K$ color of bulge giants (Figure 1) and their lower $L$, relative to the previously adopted values applicable to MKIII giants, work in the direction of a reduced mass-loss rate. The luminosity effect has the greater influence; a 1 mag decrease in the luminosity during the final AGB phase when the loss rate is at its maximum would reduce the rate by a factor of 4.

The temperatures of the bulge M giants are uncertain until molecular blanketing effects in these SMR stars can be evaluated. These should be larger than those already suspected from the difference between solar-neighborhood and globular cluster stars (Frogel, Persson, and Cohen 1981). Departures from temperature-color calibrations based on broad-band magnitudes of solar-neighborhood stars (Ridgway et al. 1980) must be expected when more detailed spectral information becomes available.

Pulsation masses up to 3 $M_\odot$ for bulge LPVs derived by Wood and Bessell (1983) are subject to similar downward revision if the newer data are inserted in the pulsation equation. Since $M_p \propto L^{3/2} T_e^{-6}$ the calculated mass would be smaller if the revised luminosity is used in place of the $M_{bol}$ = -5.5 formerly adopted. If the adopted $T_e$ was too low because blanketing affected the J-K color (see Figure 1a) the sixth power dependence would sharply reduce the calculated mass. Here again more detailed spectral information is needed. A calculated mass of about 1.3 $M_\odot$ would follow from the luminosity effect alone if a correction of 0.6 mag is introduced.

The blurring effect of distance spread and metallicity spread on the CMDs of bulge star fields makes a direct age determination for the basic old population in the bulge quite uncertain (Sec. 5). A scenario which resulted in a minor but significant fraction of the total population being as young as 6 Gyr would not be ruled out by the CMD of Figure 3; adequate fuel for ascending the AGB would not be a problem for these stars. The question of whether the 6 Gyr turnoff stars are present in sufficient number to serve as progenitors of the counted late M giants in a given CCD frame can in principle be answered by the same test of the ratio of evolution times that was applied to young stars at the 2 Gyr turnoff. Statistics from additional CCD frames will be needed. The ratio increases to several thousand turnoff stars per AGB star at intermediate ages, because of a slower evolution rate. The integrated light of such a concentration on the CMD could, as noted by Renzini (1981), affect the UBV colors of elliptical galaxies; they would not follow the color evolution calculated by Larson and Tinsley (1974) for a single-burst model. Other papers at this Workshop examine the question of whether observations of distant elliptical galaxies up to a lookback time of 8 Gyr (e.g. Hamilton 1985) show colors and spectral energy distributions that indicate that such systems retained a significant SFR up to intermediate ages.

Although the early history of star formation in the bulge population and that in similar external galaxies is at present indeterminate and the single-burst approximation may later be found inadequate, there is no need to single out the progenitors of the late M giants as belonging to a special minority component: they may be presumed to be part of the basic old population of the Galactic bulge.

## 7. ACKNOWLEDGEMENTS

I am indebted to Jay Frogel, Victor and Betty Blanco, Michael Rich, and Donald Terndrup for allowing the use of observational data prior to publication. I have profited from conversations with Duane Carbon and Don VandenBerg. The work of the Santa Cruz participants in these collaborative studies of the Galactic Bulge was supported in part by Grant AST 83-12119 from the U. S. National Science Foundation.

## 8. REFERENCES

Aaronson, M. 1977, Ph.D. Thesis, Harvard University.
Aaronson, M., Cohen, J. G., Mould, J. and Malkan, M. 1978, *Ap. J.* **223**, 824. (ACMM)
Aller, L. H. 1963, *Atmospheres of the Sun and the Stars* (New York: Ronald Press), p. 131.
Arp, H. 1965, *Ap. J.* **141**, 43.
Azzopardi, M., Lequeux, J., and Rebeirot, E. 1984, *Bull. Am. Astr. Soc.* **16**, 911.
Baade, W. 1951, *Publ. Obs. Univ. Mich.* **10**, 7.
_____. 1963, in *Evolution of Stars and Galaxies*, ed. C. P. Gaposchkin (Cambridge: Harvard University Press), p. 284.
Blanco, V. M., and Blanco, B. M. 1985, *Ap. Space Sci.*, **118**, 365.
Blanco, B. M., McCarthy, M. F., and Blanco, B. M. 1984, *Astron. J.* **89**, 636 (BMB).
Burstein, D. 1985. *Pub. Astr. Soc. Pac.* **97**, 89.
Carney, B. W. 1981, in *Colloquium 68, Astrophysical Parameters for Globular Clusters*, ed A. G. D. Philip and D. S. Hayes (Schenectady: L. Davis Press), p. 477.
Demarque, P. 1980, in *Symposium 85, Star Clusters* ed. J. E. Hesser (Dordrecht: Reidel) p. 281.
Eggen, O. J. 1973, *Ap. J.* **180**, 857.
Faber, S. M., and Gallagher, J. S. 1976, *Ap. J.* **204**, 365.
Frenk, C. S., and White, S. D. M. 1982, *M.N.R.A.S.* **198**, 173.
Frogel, J. A. 1981, in *Physical Processes in Red Giants*, ed. I. Iben, Jr., and A. Renzini (Dordrecht: Reidel), p. 63.
_____. 1983, *Ap. J.* **272**, 167.
Frogel, J. A., Blanco, V. M., and Whitford, A. E. 1984, in *Symposium 105, Observational Tests of the Stellar Evolution Theory*, ed. A. Maeder and A. Renzini (Dordrecht: Reidel) p. 571.
Frogel, J. A., Persson, S. E., Aaronson, M., and Matthews, K. 1978, *Ap. J.* **220**, 75 (FPAM).
Frogel, J. A., Persson, S. E., and Cohen, J. G. 1980, *Ap. J.* **240**, 785.

————. 1981, *Ap. J.* **246**, 842.
————. 1983, *Ap. J. Suppl.* **53**, 713.
Frogel, J. A., and Whitford, A. E. 1982, *Ap. J. Lett.* **259**, L7 (FW).
————. 1985, in preparation.
Fusi-Pecci, F., and Renzini, A. 1976, *Astron. and Ap.* **46**, 447.
Gunn, J. E., Stryker, L. L., and Tinsley, B. M. 1981, *Ap. J.* **249**, 48.
Hamilton, D. 1985, preprint.
Iben, I., Jr., and Renzini, A. 1983, *Ann. Rev. Astr. Ap.* **21**, 271.
Johnson, H. L. 1966a, *Ap. J.* **143**, 187.
————. 1966b, *Ann. Rev. Astr. Ap.* **4**, 193.
Larson, R. B., and Tinsley, B. M. 1978, *Ap. J.* **219**, 46.
Mengel, J. G., Sweigart, A. V., Demarque, P., and Gross, P. G. 1979, *Ap. J. Suppl.* **40**, 733.
Mould, J. R., and McElroy, D. B. 1978, *Ap. J.* **221**, 580.
Nassau, J. J., and Blanco, V. M. 1958, *Ap. J.* **128**, 46.
O'Connell, R. W. 1976, *Ap. J.* **206**, 370.
————. 1980, *Ap. J.* **236**, 430.
Persson, S. E., Frogel, J. A., and Aaronson, M. 1979, *Ap. J. Suppl.* **39**, 61.
Renzini, A. 1977, in *Advanced Stages in Stellar Evolution*, ed. P. Bouvier and A. Maeder (Sauverny: Geneva Observatory).
————. 1981, *Ann. de Phys.* **6**, 87.
Rich, R. M. 1984, *Mem. Soc. Astr. Ital.*, in press.
Ridgway, S. T., Joyce, R. R., White, N. M., and Wing, R. F. 1980, *Ap. J.* **235**, 126.
Rood, R. T. 1973, *Ap. J.* **184**, 815.
Sandage, A. 1982, *Ap. J.* **252**, 533.
Scalo, J. M., and Miller, G. E. 1981, *Ap. J. Lett.* **248**, L65.
Stebbins, J., and Whitford, A. E. 1948, *Ap. J.* **108**, 413.
Sweigart, A. V., and Gross, P. G. 1978, *Ap. J. Suppl.* **36**, 405.
Taam, R. E., Kraft, R. P., and Suntzeff, N. 1976, *Ap. J.* **207**, 201.
Terndrup, D. M., Rich, R. M., and Whitford, A. E. 1984, *Pub. Astr. Soc. Pac.* **96**, 796 (Abstract).
Tinsley, B. M. 1972a, *Ap. J. Lett.* **173**, L93.
————. 1972b. *Ap. J.* **178**, 319.
————. 1978. *Ap. J.* **222**, 14.
Tinsley, B. M., and Gunn, J. E. 1976, *Ap. J.* **203**, 52 (TG).
VandenBerg, D. A. 1985, *Ap. J. Suppl.* in press.
————. 1983, *Ap. J. Suppl.* **51**, 29
Vandenberg, D. A., and Bell, R. A. 1985, in press.
van den Bergh, S. 1971, *Astron. J.* **76**, 1082.

———. 1974. *Ap. J. Lett.* **188**, L9.
Whitford, A. E. 1978. *Ap. J.* **226**, 777.
———. 1985, *Publ. Astr. Soc. Pac.* **97**, 205.
Whitford, A. E., and Rich, R. M. 1983, *Ap. J.* **274**, 723.
Wood, P. R., and Bessell, M. S. 1983, *Ap. J.* **265**, 748.

## 9. COMMENTS

<u>O'Connell.</u> The bulge giants seem to have a metallicity about one order of magnitude higher than 47 Tuc. If they are old (say 15 Gyr), isn't it a difficulty that they superpose so well with the 47 Tuc isochrone in your $M_{bol}$, $J$-$K$ diagram? Shouldn't they fall to the red of 47 Tuc?

<u>Whitford.</u> Yes they should, if the Yale evolution tracks are at least relatively correct for SMR giants and $J$-$K$ remains a valid temperature indicator. In the Frogel-Whitford 1982 paper this was acknowledged as a difficulty, and possible blanketing in the $J$ and $K$ broad-band magnitudes cited as a possible explanation. Mapping of actual molecular blanketing is still needed. Envelope fitting for these stars can only be approximate until SMR model atmospheres are available. Renzini's remark in a previous session of this Workshop is relevant: theoretical red giant evolution tracks are sensitive to rather small perturbations because they portray stars whose outer layers occupy a broad, shallow potential well.

<u>de Jong.</u> I think that in this connection the IRAS data on the bulge stars are very relevant. If one plots the distribution of sources with color temperatures between 200°K and 400°K on the sky the bulge shows up very clearly. This implies that very luminous AGB stars ($L \approx 10^4 L_\odot$) in a strong mass-loss phase exist in the bulge of our galaxy.

# THE AGB POPULATION OF NEARBY GALAXIES

M. Aaronson and K. H. Cook
Steward Observatory, University of Arizona
Tucson, Arizona 85721 U.S.A.

and

J. Norris
Mount Stromlo and Siding Spring Observatories
Private Bag, Woden P.O.
ACT 2606
Australia

ABSTRACT

We report the results of a preliminary survey for carbon and M stars in Local Group galaxies using a newly developed photometric technique. The surveyed objects include M31, M33, NGC 6822, IC 1613, and WLM. The C/M ratio is seen to be strongly dependent on parent galaxy absolute magnitude, a finding that has important bearing on our understanding of galaxian star formation history.

1. INTRODUCTION

In the last five years, the asymptotic giant branch has become an intensely studied region of the HR diagram. Because the stars populating the AGB are of intermediate age, they can provide a direct window to the metal enrichment and star forming rates of stellar systems. Furthermore, as several speakers have already emphasized, the AGB is an important contributor to integrated galaxian light. Hence, if we are to comprehend the evolution of very distant objects -- the central theme of this workshop -- we must have some idea of the nature and number of AGB stars to expect in different types of environments. The direct study of nearby galaxies provides the best empirical way for obtaining this information, and in this contribution we report the preliminary results of one such effort.

Until recently, detailed knowledge of the AGB has only been available for the Milky Way and its dwarf satellites. Nevertheless, some intriguing results have emerged. Blanco and McCarthy (1983) have discovered large numbers of carbon stars in the Magellanic Clouds, and derive much higher C/M star ratios in these systems than is seen in our own galaxy. Luminous carbon giants are also present in most of the halo spheroidals (Aaronson and Mould 1985), implying that star formation in these superficially Pop II-like objects has extended over a considerable period of time. A natural question arising from these various findings is to what extent the star forming history in the halo satellites is driven by tidal effects of the Milky Way.

In order to probe the AGB farther than our own backyard, beyond which conventional grism methods cannot reach, we have developed a new photometric technique utilizing CCD imaging through two narrow band filters -- one centered at $\sim 7750$Å and the other at $\sim 8100$Å. These filters monitor CN and TiO absorption, and in combination with broad band V and I photometry, allow very distant C and M stars to be easily discriminated (e.g., Aaronson, Mould, and Cook 1985). We should note that while in addition to ourselves several other groups are now pursuing similar work, the idea behind the filter method may have originated with Victor Blanco a decade ago, but it seems to have taken the development of CCD detectors before the technique could be exploited in a practical fashion.

For our preliminary survey we observed regions in five nearby galaxies, as summarized in Table 1. We will give here only a brief description of the observational details and results, as a full discussion of this work will soon be available in Cook, Aaronson, and Norris (1985).

Table 1

Surveyed Galaxies

| Name | Type | No. Fields | $I_{ap}$ ($M_I = -4.4$) |
|---|---|---|---|
| M31 | Sb | 2 | 20.0 |
| M33 | Scd | 1 | 20.1 |
| NGC 6822 | Im | 2 | 19.6 |
| IC 1613 | Im | 2 | 20.0 |
| WLM | Im | 1 | 20.3 |

## 2. OBSERVATIONS AND C/M RATIOS

During four nights in the fall of 1982, the Steward Observatory 2.3 meter telescope and an RCA CCD were used to observe a total of eight extragalactic regions (Table 1). The 0.3 arc second pixel spacing led to a final field size of 1.5' x 2.5'. The two M31 regions were roughly centered on Baade's (interarm) Field II and (arm) Field III, at radial distances of $\sim 7$ and 10 kpc, respectively. The M33 field was somewhat randomly chosen at a radial distance of $\sim 2$ kpc along the major axis in an area clear of prominent arms. A nice

# THE AGB POPULATION OF NEARBY GALAXIES

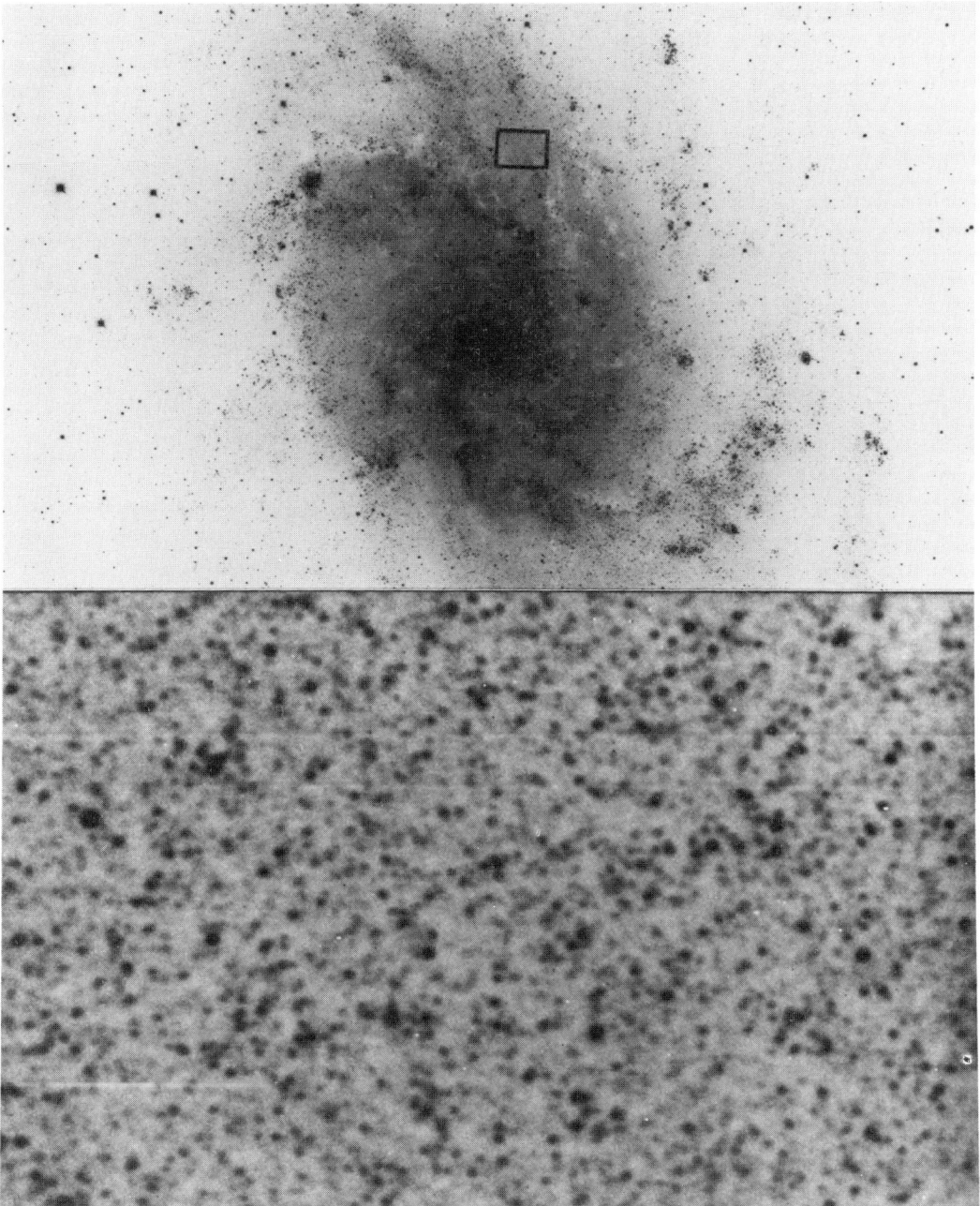

Figure 1. The upper panel shows M33, with our survey field blocked out. The lower panel is the median of three 5-minute I band CCD exposures obtained with the Steward Observatory 2.3 meter telescope. Note the good resolution of the disk into stars. North is at the top, east is to the right and the scale is given by the CCD field which is 1.5'x2.5'.

surprise was the ease with which the disks of both M31 and M33 were resolved into red giants in only moderate (~ 1.2") seeing, a point we illustrate with Figure 1. Of course, the three Magellanic irregulars we examined were also easily resolved. Note that the typical exposure times were 30 minutes in the V, "77", and "81" filters, and half this much at I.

After "cleaning up" the CCD data in the usual fashion, stellar photometry was performed using Harvey Butcher's "Mountain Photometry Code". (For reasons of efficiency, we forego in this preparatory study use of a more powerful photometry code such as RICHFELD.) Because of both variable seeing and crowding effects, the magnitude depth to which each of our fields reached was not the same. For a fair intercomparison, it was therefore necessary to chose a limiting absolute magnitude, taken as $M_I = -4.4$ mag. The apparent magnitude corresponding to this limit (which of course depends on our adopted distances and reddenings) is listed in Table 1. At $M_I = -4.4$, the photometric errors in both the V-I and "77-81" colors was < 0.1 mag.

Two-color (V-I, 77-81) diagrams were then constructed for each field, an example of which is shown for M33 in Figure 2. There we see

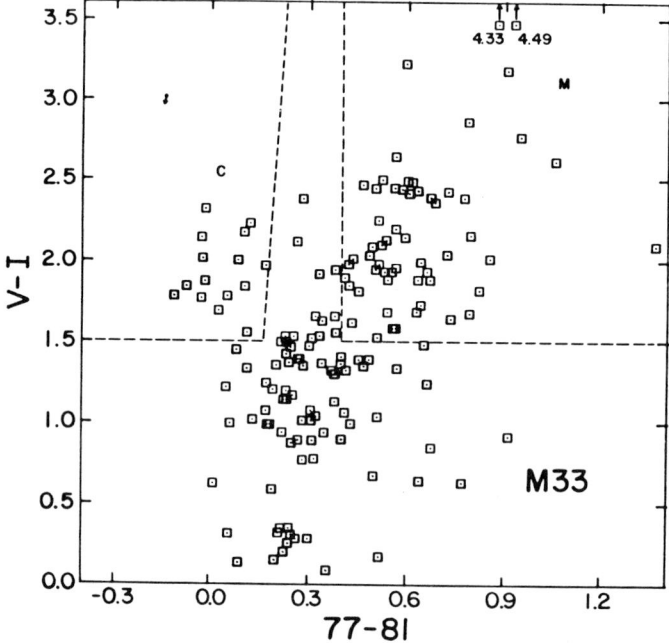

Figure 2. A (V-I, 77-81) two color diagram for our M33 field. Carbon stars lie in the blocked out area in the upper left, and M stars in the upper right. The two points labeled "C" and "M" have been spectroscopically confirmed as being these types. The small arrow in the upper left is the estimated effect of foreground reddening.

that the "giant branch" as it climbs redder splits into two forks
--the leftward one corresponding to carbon stars and the rightward one
to M stars.

How well does the method actually work at segregating C and M
types? Spectroscopic observations indicate that the answer is very
well indeed. In particular, our four band photometry of galactic
field stars shows that the 77-81 color exhibits a strong correlation
with spectral M-type and a strong inverse correlation with C-type
(Cook, Aaronson, and Norris 1985). Furthermore, we have in
collaboration with J. Mould obtained Hale 5-meter spectra of candidate
objects in seven of our surveyed fields, and the stars virtually
always turned out to be their predicted C or M type (e.g. Figure 1,
Aaronson, Mould, and Cook 1985). There is, however, a transition
region at V-I ~ 1.5 where early C, late K, early M stars may be
difficult to disentangle, though photometry more accurate than we have
obtained may be of help. "Zero band-strength" stars are also
occasionally found at redder V-I colors between the two blocked out
regions in Figure 2. This is the location at which we might expect to
come across S stars, and indeed the first such star in NGC 6822 has
been spectroscopically confirmed by Aaronson, Mould, and Cook (1985).

A survey of the C/M ratios found in all five observed galaxies is
given in Table 2, along with several other adopted galaxian properties
(see Cook, Aaronson, and Norris 1985).

Table 2

C/M Ratios

| Galaxy | $M_B^{0,i}$ | $(m-M)_0$ | C/M3+ | C/M5+ | [0] |
|---|---|---|---|---|---|
| M31 | -21.2 | 24.3 | .05 | .1 | 8.95 |
| M33 | -18.7 | 24.4 | .3 | 1.0 | 8.45 |
| LMC | -18.1 | 18.5 | .2 | 0.8 | 8.43 |
| SMC | -16.5 | 19.0 | .6 | 4.3 | 8.02 |
| NGC 6822 | -14.6 | 23.5 | 1.2 | 18.0 | 8.27 |
| IC 1613 | -14.5 | 24.4 | 2.8 | > 14 | 7.86 |
| WLM | -13.6 | 24.7 | 2.8 | > 14 | ---- |

Also included in Table 2 are C/M ratios for the Magellanic Clouds
taken from Blanco and McCarthy (1983). The new results can be
summarized as follows: In M31, we found very few C stars but
considerable numbers of M stars reaching very late spectral type. In
M33 there were a fair number of both C and M stars (Figure 2).
Finally, the three irregulars were seen to be abundant in carbon
stars, but less so in M stars, and contained very few late M stars
(none in fact later than M5 in IC 1613 and WLM -- see Table 2). For

systems where two fields were observed, no significant difference in C/M ratio was seen. In particular, the arm and interarm regions of M31 were not obviously distinguishable in this regard, suggesting that (as anticipated) we are sampling an older disk population that is little affected by the extreme Pop I component.

## 3. DISCUSSION

In Figure 3 we plot the C/M3+ ratio (i.e. for M stars of spectral type M3 and later) against parent galaxy absolute magnitude. A strong correlation is evident. It can also be seen that the Magellanic Clouds fit smoothly onto the relation defined by the other objects, which seems to answer an issue posed in the Introduction. Namely, the large C/M ratio and overall star formation history of the Clouds is probably influenced more by their internal properties rather than by their unusual location in the Milky Way's halo and any tidal effects that may have resulted therefrom.

What is the underlying cause of the trend evident in Figure 3? A clue is provided by the last column in Table 2, which lists the oxygen abundance determined from H II region emission line studies (interpolated in the case of M31 and M33 to the appropriate radii). As is well known, the gaseous metal abundance is higher in more luminous systems than in less luminous ones.

A decreasing metallicity will drive the C/M ratio up via two mechanisms: First, decreasing opacity will shift the position of the giant branch in the HR diagram blueward, and at the same time at fixed temperature the TiO bandstrength will diminish (e.g., Mould, Stutman, and McElroy 1979) -- both effects combining to yield a decrease both in the relative number of late M to early M giants and in the total number of M stars overall. (Through observations of galactic field and globular clusters giants of known abundance, we eventually hope to derive an explicit calibration between the TiO bandstrength at fixed V-I, measured by our 77-81 color, and metallicity.) Second, it becomes "easier" to make a C star because less carbon is needed to pollute the atmosphere in order to drive [C] > [O].

Abundance, however, cannot be the whole story behind Figure 3, as evidenced by the absence of luminous carbon stars in galactic globular clusters. Rather, age effects must play a role as well. In particular, we are generally sampling AGB carbon stars that have luminosities ranging between $M_{bol} \sim M_I \sim -4.5$ to $-6$ mag implying a corresponding age range from $\sim 1$ to 10 Gyr (Mould and Aaronson 1982). The stars in this sense represent a sort of "Pop 1.5" component, and their large numbers in the dwarf systems provides direct evidence that these objects have been forming stars over a major fraction of their lifetimes. The M stars we are sampling are expected to have an age distribution comparable to the C stars, though with an added younger component having initial mass $M_i > 2 M_\odot$ (e.g. Figure 7, Iben and Renzini 1983). Note, however, that the RGB and AGB contribution from very massive $M_i > 5 M_\odot$ stars is not expected to be larger than $\sim 10\%$, following Reid and Mould (1984). Some contamination from foreground

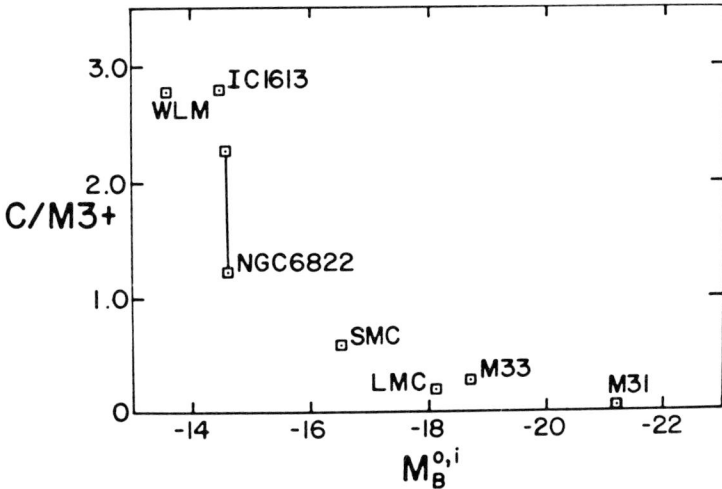

Figure 3. The ratio of carbon to M stars of spectral type M3 or later as a function of parent galaxy absolute magnitude. For NGC 6822 the lower point is the observed value and the upper one is corrected for estimated foreground contamination.

M dwarfs will also be present, but this will only materially affect the C/M ratios for NGC 6822. The latter should, in fact, be regarded as lower limits, as the Bahcall and Soneira (1980) star count model suggests that the majority of NGC 6822 M stars are interlopers.

The present discussion can be turned around somewhat and the C/M ratio used as a gauge of metal abundance at intermediate ages. The results in Table 2 then suggest that the relative gaseous abundance ranking in the sample galaxies ~ 5 Gyr ago was no different than today, and therefore that the relative enrichment rates have also been similar.

Particular attention should be called to the comparison between M33 and the LMC. Both systems have a similar absolute magnitude, and the oxygen abundance at the radial location of our M33 patch is almost identical to the mean LMC value (Table 2). Indeed, the two galaxies appear to have remarkably close C/M ratios. Yet, morphologically, these objects are quite distinct, with M33 being of de Vaucouleurs type Scd and the LMC of type Sm. This result seems to imply that disk star forming history is governed more by local conditions than by overall morphology or (again) global environment. A similar sort of conclusion has been reached by Gallagher and Hunter (1985) from their ongoing star formation studies of irregular galaxies. In this regard, it would be especially interesting to see how well the C/M ratio in M31 and M33 radially tracks the H II region abundance gradient thought to be present in these systems.

We turn now to Figure 4, which shows the observed carbon star luminosity functions. Except for the Clouds and possibly NGC 6822, their appearance is of course determined by small number statistics. Nevertheless, with this caveat in mind, the luminosity functions all appear roughly consistent. Furthermore, we have found no carbon stars brighter than $M_I \sim -6$ mag, a problem that has been an historic thorn in the side of theoretical predictions (e.g. Iben and Renzini 1983).

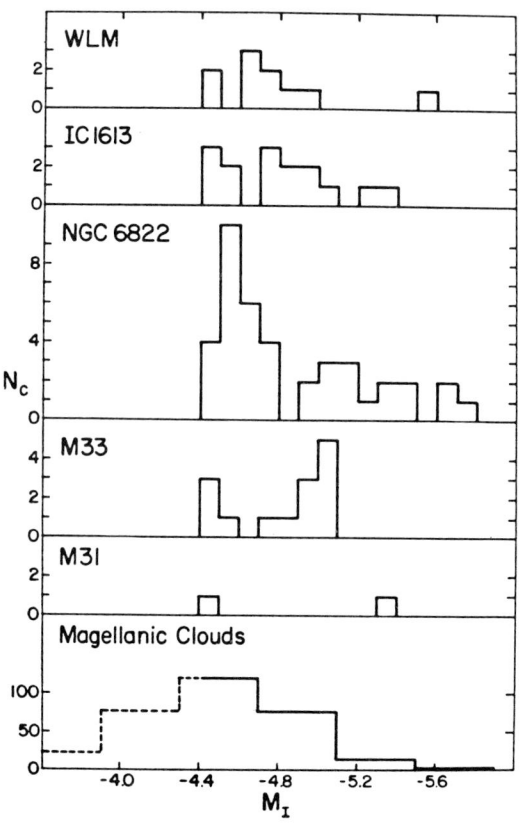

Figure 4. Observed C star luminosity functions. The Magellanic Cloud data is from Blanco, McCarthy and Blanco (1980).

This aspect of the problem can again be inverted and used to derive distance scale information from the carbon star luminosity function (which necessarily demands common IMFs and star formation rates). For instance, the results in Figure 4 can be used to argue that the relative distance moduli in Table 2 are roughly correct. In particular, our "guessed at" modulus for WLM turns out to agree quite well with Sandage's (1985) Cepheid distance. On the other hand, the results for M33 do not lend much support to Sandage's proposed increase in modulus to 25.35 mag.

Note that the total AGB luminosity function is also of considerable interest (e.g. Reid and Mould 1984), though our present sample is statistically too poor to make any discussion worthwhile. However, with a more complete survey (and assuming evolution at the AGB tip becomes better understood), it may be possible to put constraints on the star formation rate and choose, for instance, between burst models and constant or exponentially decaying models.

## 4. SUMMARY

We have developed a new and powerful method for probing the extragalactic AGB population. A quick-look survey has led to the discovery of a strong dependence between C/M ratio and galaxian absolute magnitude, which can be understood as arising from a combination of age and abundance effects. We are now in a position to completely map out the surface distribution and luminosity function of AGB stars in many nearby galaxies. In addition to possibly yielding reliable distances, such data will provide important insight into the chemical enrichment and star formation histories in varying types of systems, as well as being crucial for fully understanding the evolution of high redshift galaxies.

This work was partially supported by NSF grant AST 83-16629. M. A. thanks the Harvard-Smithsonian Center for Astrophysics, where part of the written version of this talk was prepared, for an SAO Summer Fellowship, and also for providing a warm hospitable environment.

## REFERENCES

Aaronson, M., and Mould, J. 1985, Ap. J., **290**, 191.
Aaronson, M., Mould, J., and Cook, K. H. 1985, Ap. J. (Letters), **291**, L41.
Bahcall, J. N., and Soneira, R. M. 1980, Ap. J. Suppl., **44**, 73.
Blanco, V. M., and McCarthy, M. F. 1983, A. J., **88**, 1142.
Blanco, V. M., McCarthy, M. F., and Blanco, B. M. 1980, Ap. J., **242**, 938.
Cook, K. H., Aaronson, M., and Norris, J. 1985, in preparation.
Gallagher, J. S., and Hunter, P. A. 1985, private communication.
Iben, I., and Renzini, A. 1983, Ann. Rev. Astr. Ap., **21**, 271.
Mould, J., and Aaronson, M. 1982, Ap. J., **263**, 629.
Mould, J., Stutman, D., and McElroy, D. 1979, Ap. J., **228**, 423.
Reid, N., and Mould, J. 1984, Ap. J., **284**, 98.
Sandage, A. 1985, talk presented at 165th A.A.S. meeting in Tucson, Arizona.

## DISCUSSION

J. Frogel: How do you correct your results for magnitude incompleteness, when comparing them with Blanco and McCarthy's more complete Cloud surveys?

M. Aaronson: It is true that our survey goes to only $-4.4$ mag in $M_I$, while the Blanco and McCarthy work goes a magnitude deeper. But we implicitly assume the luminosity functions are similar, so no correction is applied.

J. Mould: Isn't it possible to use the luminosity functions, when they become more complete, to separate the influence of age and metallicity on the trend in C/M ratios with type?

M. Aaronson: Assuming we have independent accurate distance, I don't see why not.

A. Dressler: You stated that a greater distance modulus for M33 would result in an anomalously bright distribution of carbon stars. Since you set the magnitude limit by <u>assuming</u> a distance modulus, are you certain that a somewhat greater distance, which would have necessitated a deeper sample to reach -4.4, wouldn't have turned up even more carbon stars, thus resulting in a typical luminosity function like that of the LMC?

M. Aaronson: Well, first, pushing M33 out a full magnitude would make some of the C stars there brighter than those found <u>anywhere</u> else. I think this would be too unusual, but we might, say, be able to push M33 out by a half magnitude. However, our M33 frame does go ~ .3 - .4 mag deeper than $M_I$ = -4.4, and we see no dramatic increase in C star number contrary to your suggestion.

D. C. Koo: What is the relative contribution of the Carbon stars to the integrated infrared light (J, H, K) of the low-metallicity-low-luminosity galaxies like the LMC?

M. Aaronson: Well for the galaxies we have looked at, we can't really estimate this because we don't yet have complete C star surface distributions. Blanco and McCarthy have made such an estimate for the Clouds, where they find the bolometric contribution of the carbon stars to be only 4%. As I mentioned yesterday, this suggests to me that the strength of any intermediate age burst was not all that large.

R. Nesci: It seems that there is a sort of threshold magnitude about -17 $M^{0,1}$: galaxies fainter than this have the ratio Carbon/M stars very sensitive to $M_B$, while the opposite happens for brighter galaxies. Would you comment on this?

M. Aaronson: This is just an artifact of the way the data is plotted. On a log C/M vs. $M_B$ diagram, the galaxies in fact define a straight line relation.

J. Frogel: It is worth pointing out that in our own galaxy the C/M star ratio is a strong function of galactic longitude -- it is quite high in the anti-center direction.

M. Aaronson: Yes. This again presumably reflects the abundance gradient that is also known to be present in our disk. In the coming season, we are hoping to obtain fields spaced along the disk in M31 and M33 to see how well the C/M ratio follows the gaseous abundance gradient.

M33: RADIAL DISTRIBUTIONS AND A COMPARISON OF ITS GLOBAL LUMINOSITY FUNCTION WITH OTHER NEARBY GALAXIES


Wendy L. Freedman
Mount Wilson and Las Campanas Observatories
813 Santa Barbara St.
Pasadena, California 91101.


ABSTRACT. The upper end of the luminosity functions for luminous blue stars is presented for a sample of 10 nearby galaxies. The slope of this luminosity function does not show significant variation from galaxy to galaxy, nor as a function of radius in M33. The blue-to-red supergiant and Wolf-Rayet to red supergiant ratios are also consistent with showing no radial trend. The radial surface distributions of neutral hydrogen, HII regions, OB stars, 6 cm radio continuum, Wolf-Rayet stars and blue surface brightness are compared. With the exception of the relatively flat neutral hydrogen distribution, all of these other components exhibit very similar scale lengths.


1. THE DATA

   Prime focus ultraviolet, blue and visual plates were obtained at the Canada-France-Hawaii telescope (CFHT) for M33, NGC 2403, and M81, and at Cerro Tololo for NGC 300. Reductions were carried out using the Automatic Plate Measuring facility in Cambridge, England which enabled the measurement of several thousand images in each galaxy. A description of the machine can be found in Kibblewhite et al. (1984). Prime focus CCD frames of M33, Holmberg IX, Sextans A, and Leo A were obtained at both the CFHT and the Kitt Peak (KPNO) 4m. The software program RICHFLD (Tody 1980) was used for the CCD reduction. These new data were supplemented with published data for the LMC (Rousseau et al. 1978), the SMC (Ardeberg and Maurice 1977), and NGC 6822 (Kayser 1967).


2. LUMINOSITY FUNCTIONS IN NEARBY GALAXIES

   The upper end of the main-sequence luminosity function is populated by hot, luminous, blue stars. Thus, apparent luminosity functions based on stars of all colors may be relatively insensitive to differences in the upper end of the mass function. Therefore, luminosity functions were constructed with a sample of only the bluest stars in each galaxy, as determined from their U-V and B-V colors. An additional advantage to restricting the sample to the bluest stars is that foreground contamination by stars in our own Galaxy is virtually eliminated. V, B and U (when available) luminosity functions were obtained. These data are described in more detail in Freedman (1985).

The V luminosity functions for the blue stars in each galaxy are displayed in Figure 1. The lines indicate the slope of the M33 luminosity function for comparison. From this figure, it can be seen that there is very little variation in the slope of the luminosity function for this sample of galaxies. Further, the slopes of the B and U luminosity functions are also similar (not shown).

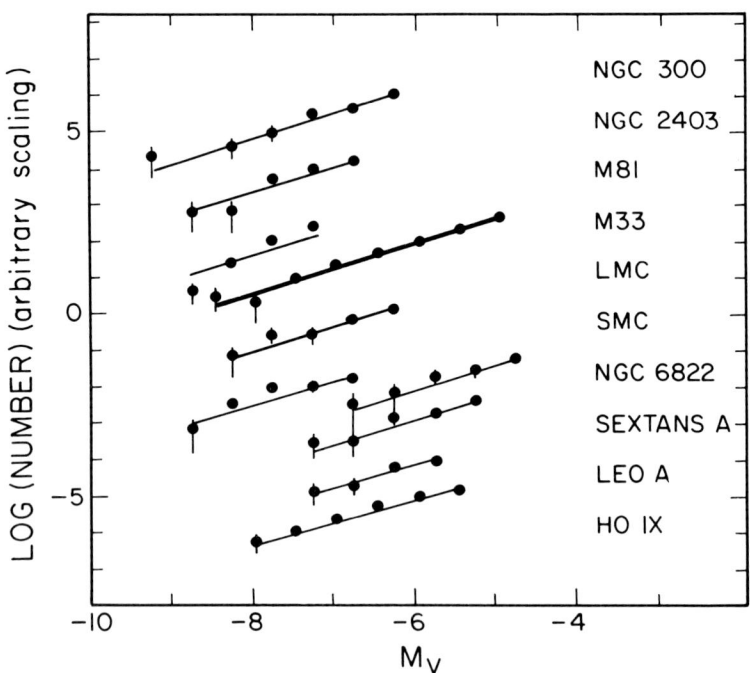

Figure 1 - The relative absolute V stellar luminosity functions for 10 nearby galaxies.

This suggests that the slope of the luminosity function is not a strong function of metallicity since the observed luminosity functions of the smaller, lower metallicity dwarf galaxies (Holmberg IX, Leo A, Sextans A and NGC 6822) do not differ in slope from the luminosity functions of the more metal rich spirals in the sample.

The number of stars in the M33 sample is sufficiently large that statistically reliable slopes can be calculated as a function of radius. This comparison is of interest since M33 has a measured abundance gradient (e.g. Blair and Kirshner 1985, and references therein). Luminosity functions for 4 regions at various radial distances in M33 are shown in Figure 2.

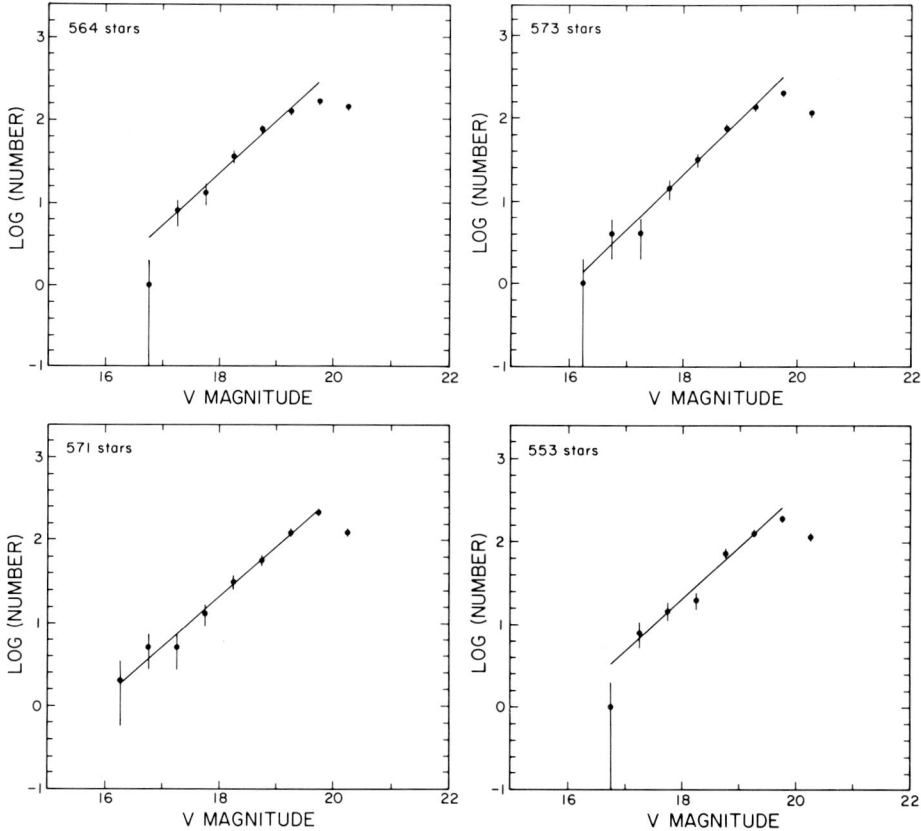

Figure 2 - The V stellar luminosity function in M33 as a function of radius, shown for 4 radial zones.

There is clearly no significant evidence for a change in the slope as a function of radius, again suggesting that the slope is relatively insensitive to metallicity. Stochastic variations due to small number statistics can be seen in the brightest magnitude bins.

For several reasons, the upper magnitude cut-offs should be viewed with caution: a) the small numbers of luminous stars make the statistics very uncertain, b) errors in the adopted distance moduli will affect the apparent cut-off magnitudes, and c) the brightest stars may not have been found due to the incompleteness of the samples, particularly in the CCD samples.

## 3. RADIAL DISTRIBUTIONS IN M33

### 3.1 Blue and Red Supergiants, and Wolf-Rayet Stars in M33

Walker (1964) first presented evidence for a radially decreasing gradient in the blue-to-red supergiant ratio in M33. Van den Bergh (1968) suggested that this ratio might be metallicity-dependent. Later, Humphreys and Sandage (1980) noted that, except for the very inner regions of M33, this ratio showed little variation. The blue-to-red ratio for the data in the present study is shown in Figure 3. This data is in good agreement with that of Humphreys and Sandage with the exception of the inner radial bins, where the numbers of stars in the Humphreys and Sandage sample is quite low. Thus within the uncertainties, there does not appear to be evidence for a very significant gradient in the blue-to-red ratio.

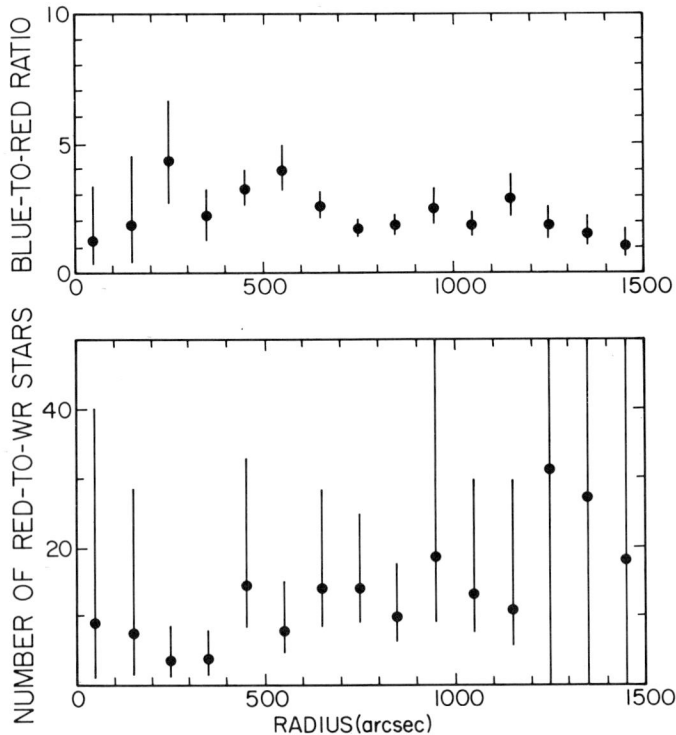

Figure 3 - The blue-to-red supergiant ratio as a function of radius in M33.

Figure 4 - The Red-to-Wolf-Rayet Star ratio as a function of radius in M33.

Maeder, Lequeux, and Azzopardi (1980) have suggested that the relative numbers of Wolf-Rayet stars and red supergiants are a more sensitive indicator of metallicity than the blue-to-red ratios. They carried out a test of this hypothesis using data for our own Galaxy, and found a factor of 90 decrease in the red-to-Wolf-Rayet ratio over a 4 kpc range in distance. A similar ratio for the data from the present M33 study is given in Figure 4, as a function of radius. The Wolf-Rayet star data are from the study of Massey and Conti (1983). These data are consistent with little or no variation as a function of radius. The uncertainties are very large due to the small numbers of Wolf-Rayet stars, particularly in the outer bins. While a shallow gradient could be supported by the present data, it is significantly smaller than that reported for our own Galaxy. Part of the difference may be reflecting the incompleteness of the galactic red supergiant catalogs at large distances from the Sun (see also Humphreys and McElroy 1985).

In Figure 5, the ratio of Wolf-Rayet stars to HII regions (as compiled by Boulesteix et al. 1974), is shown as a function of radius. Not surprisingly, the Wolf-Rayet stars and HII regions track extremely well; they are both young constituents, and each of these components fall off in surface density with galactocentric distance. Their distributions are simply a reflection of the fact that the rate of star formation is higher in the inner as compared to the outer regions of the disk.

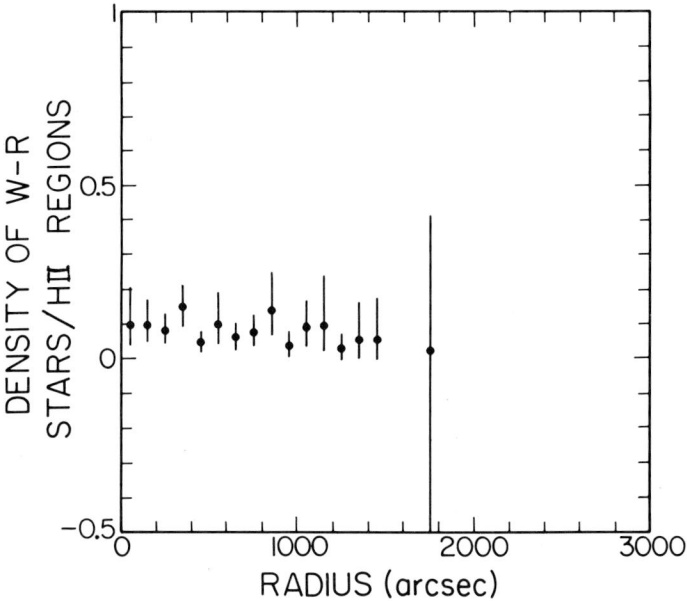

Figure 5 - The ratio of Wolf-Rayet stars to HII regions as a function of radius in M33.

## 3.2 A Comparison of the Radial Stellar and Gas Distributions

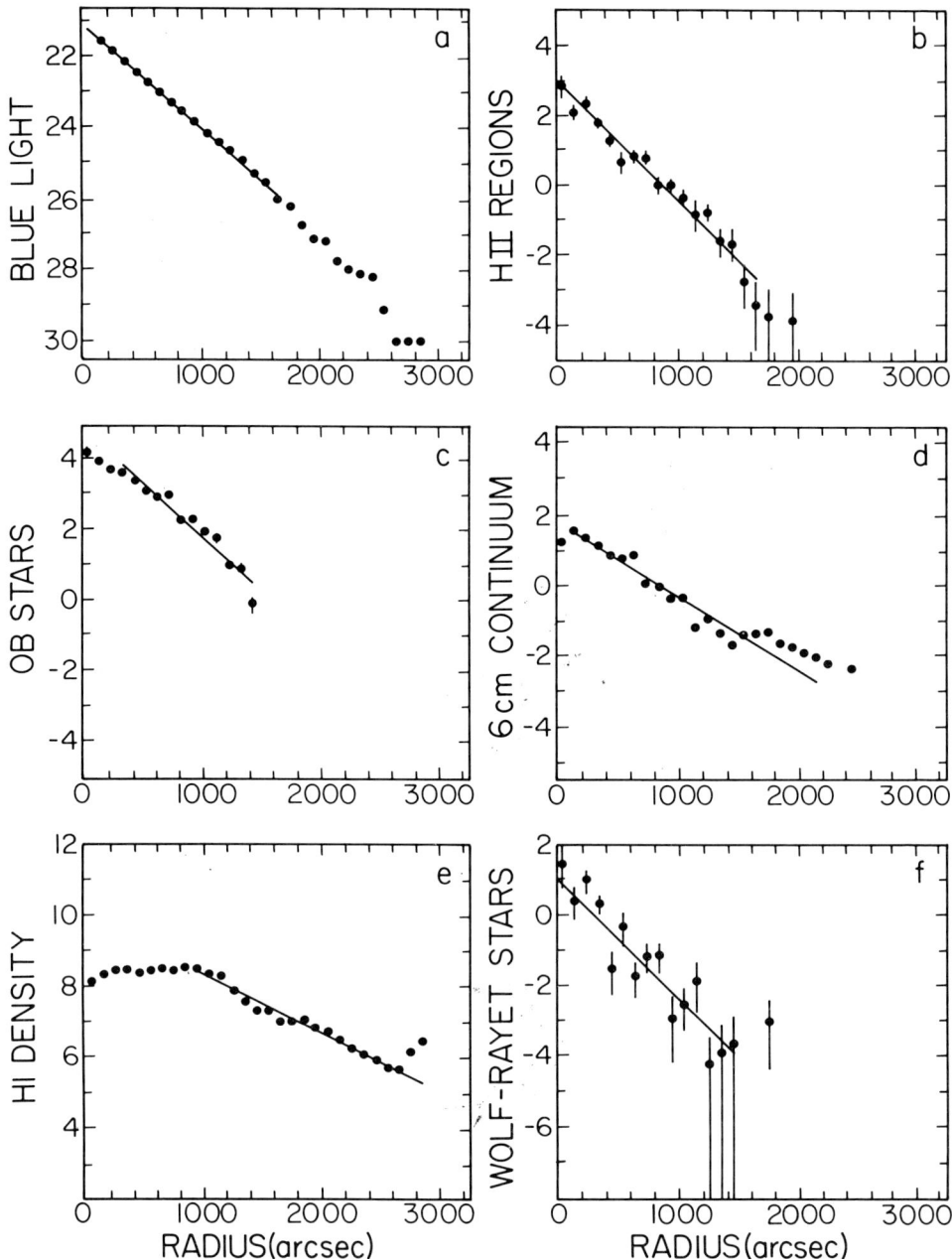

Figure 6 - The radial surface density distributions of a) blue continuum light, b) HII regions, c) OB stars, d) 6 cm radio continuum, e) neutral hydrogen, and f) Wolf-Rayet stars.

From this figure, it can be seen that the scale lengths of the HII region distribution, blue surface photometry, OB stars and Wolf-Rayet stars are all very similar. The integrated blue light distribution is reflecting the star formation over the last $10^9$ years, not only the recent star formation as delineated by the OB stars. This similar distribution suggests that the star formation history has been relatively constant in M33 over that time period. This result has been seen previously by Hodge and Kennicutt (1983) who find that the white light and HII region distributions in a sample of spiral galaxies have very similar scale lengths.

The neutral hydrogen distribution is significantly flatter than the young stars. It exhibits a flat distribution out to approximately 17 arcmin in radius, and then still decreases more slowly than the stellar components. A similar situation prevails in our own, and many other late-type galaxies. Conversely, the molecular gas component is seen to reflect the blue light distributions (Young and Scoville 1982). Thus, some mechanism re-generates the molecular gas component, and gives rise to an exponential distribution similar to the stellar distribution, while the neutral gas (from which the molecular gas must presumably originate) exhibits a flat distribution. This phenomenon is at present, an unexplained, but interesting problem in galaxy evolution.

## 4. SUMMARY

The slope of the upper end of the luminosity function is found to be constant for a sample of 10 nearby galaxies. The sample includes primarily Sc and irregular galaxies with a range of metallicities, thereby implying that the slope of the upper end of the luminosity function is independent of metallicity, and is independent of whether stars are formed in spiral arms or not. The slope of the luminosity function is also shown to be independent of radius in M33, which has a measured abundance gradient.

The ratios of blue and Wolf-Rayet stars to red supergiants are found to show very little dependence with radius in M33. The scale lengths of very recent star formation tracers are found to be similar to the scale length of the integrated blue light in M33 which suggests that the star formation history in this galaxy has been similar on a timescale of $10^9$ years.

## REFERENCES

Ardeberg, A., and Maurice, E. 1977, Astr. and Ap. Suppl., ,30, 261.

Blair, W. P., and Kirshner, R. P. 1985, Ap. J., 289, 582.

Boulesteix, J., Courtes, G., Laval, A., Monnet, G., and Petit, H. 1974, Astr. and Ap., 37, 33.

Freedman, W. L. 1985, Ap. J., in press.

Hodge, P. W., and Kennicutt, R. C. 1983, Ap. J., 267, 563.

Humphreys, R. M., and McElroy, D. B. 1984, Ap. J., 284, 565.

Humphreys, R. M., and Sandage, A. R. 1980, Ap. J. Suppl., 44, 319.

Kayser, S. 1967, A. J., 72, 134.

Kibblewhite, E. J., Bridgeland, M. T., Bunclark, P., and Irwin, M. 1984 ,in Proceedings of the Astronomical Microdensitometry Conference, NASA Conf. Publ. 2317, p. 277.

Maeder, A., Lequeux, J., and Azzopardi, M. 1980, Astr. and Ap., 90, L17.

Massey, P., and Conti, P. S. 1983, Ap. J., 273, 576.

Rousseau, J., Martin, N., Prevot, L., Rebeirot, E., Robin, A., and Brunet, J. P. 1978, Astr. and Ap. Suppl., 31, 243.

Tody, D. 1980, S.P.I.E., 264, 171.

van den Bergh, S. 1968, J. R. A. S. Canada, 62, No. 4.

Walker, M. F., 1964, A. J., 69, 744.

Young, J., and Scoville, N. 1982, Ap. J. (Letters), 260, L11.

LEQUEUX: The small variation you find in the WR/RSG and the BSG/RSG ratios may simply reflect a small metallicity gradient in this galaxy. We still know little on this gradient, and unfortunately almost all the chemical abundance determinations refer to the northern part of the galaxy.

FREEDMAN: I agree that the small variations in the ratios may be due to a shallower metallicity gradient in M33 as compared to the Galaxy. However, the existence of a gradient in M33 now seems fairly well-established (e.g., see Blair and Kirshner 1985), and it is unlikely that the metallicity gradient in the Galaxy is an order of magnitude steeper than that of M33. Therefore, I doubt that all of the difference could be attributed to metallicity alone.

KOO: What is the slope of the luminosity function of bright stars in nearby galaxies ?

FREEDMAN: The slope of the upper end of the luminosity function ($-9 < M_v < -5$ mag), based on the sample of 10 nearby galaxies discussed here is $0.7 \pm 0.1$.

MOULD: A comment. The tendency for the HI density to flatten off towards the center is common to M33, the Galaxy and M31, and seems to be more advanced with earlier Hubble type. There is a real hole in HI and $H_2$ in M31.

FREEDMAN: Yes. In late-type galaxies, there seems to be a tendency for the HI radial profiles to flatten off near the center, while in early-type galaxies (e.g., M31 and M81), a pronounced minimum is present in both HI and $H_2$. It remains an unexplained fact as to why the neutral gas exhibits a flatter distribution than the young star formation tracers (such as HII regions, Wolf-Rayet stars, etc.) in late-type galaxies. If the total gas content (HI + $H_2$) is similar to the integrated light distribution as well as the recent star formation tracers, a mechanism is needed for converting HI to $H_2$ in such a way as to produce this gradient. If there is a linear dependence of the rate of star formation on gas density, a correspondence of the total gas, young, and integrated light distributions would easily result. However, it leaves unexplained the observed abundance gradients in these galaxies. And the reason for the flatter HI component is still not clear.

A. DI FAZIO: Your IMF slope-metallicity correlation is in contradiction with that proposed by Roberto Terlevich and Melnick, in that you see little (if any) dependence of the high-mass slope on the metallicity. Maybe, this is due to the fact that Roberto and Melnick, in studying the data, varied only the slope, while keeping the minimum mass constant. Probably, if they had done their work allowing also a variation of the minimum mass of the IMF, they could have obtained a weaker dependence of the slope on metallicity. I suppose that we cannot exclude a real dependence of the slope on metallicity, but probably it is not a great one. Maybe you should have compared the slopes for various metallicity, not only making the best fits, but also statistically checking the difference or equality of those slopes. Do you agree?

FREEDMAN: The data presented here for M33 show no statistically significant difference in the slope of the luminosity function with radius (or metallicity). A limit to the dependence of the luminosity function slope on metallicity can be obtained by considering the maximum formal uncertainty in the slopes, and this gives $|\Delta(\text{slope}) / \Delta z|$ < 0.3 at the two-sigma level. This is described in greater detail in Freedman (1985). Also, as you suggest, Tervelich and Melnick varied only the slope of the mass function, and did not consider possible variations in the lower or upper mass cut-offs.

III.   TOWARDS MODELLING THE SPECTRAL EVOLUTION OF GALAXIES


A. RENZINI, A. BUZZONI
Global properties of stellar populations and the spectral evolution of galaxies

C. CHIOSI
Advancements in the stellar evolution theory: the role of convective overshooting all across the HR diagram

G. BRUZUAL A.
Spectral Evolution of galaxies

G. BARBARO, F. M. OLIVI
Spectrophotometric models of galaxies

N. ARIMOTO, Y. YOSHII
Photometric evolution of elliptical galaxy in the color – magnitude diagram

GLOBAL PROPERTIES OF STELLAR POPULATIONS
AND THE SPECTRAL EVOLUTION OF GALAXIES


Alvio Renzini and Alberto Buzzoni
Dipartimento di Astronomia,
CP 596
I-40100 Bologna


ABSTRACT. Several relevant properties of evolving stellar populations are discussed from a theoretical viewpoint. These include the absolute and specific evolutionary flux of a stellar population, the relative contributions of stars in the various evolutionary stages, the bolometric evolution, and the total and specific rates of mass return. We then introduce the concept of *phase transition* in the stellar content of a population, an event which is produced by the first appearance of stars with either C-O or He degenerate cores. The spectral energy distribution of a population is then predicted to suffer a (major) change at each of these phase transitions. Some uncertainties in both current stellar and population evolutionary models, and in population synthesis techniques are then briefly discussed, and the need is emphasized for appropriate checks and calibrations using template stellar populations. Magellanic Cloud Clusters offer the best tool for these purposes, and we then discuss in some detail current problems in dating MC clusters, as well as the origin of the blue to red transition in these clusters and its possible connection with the mentioned phase transitions. Finally, as examples of how some of these concepts could be applied to real galaxies, we discuss the possible occurrence and detection of population phase transition in high redshift elliptical galaxies, as well as the origin of the ultraviolet light of ellipticals and its predicted behaviour with look back time.


1. INTRODUCTION: GOALS AND TOOLS

The vast majority of galaxies show little or no sign of non-thermal activity, and we shall restrict our interest to these *normal* galaxies, whose light (from the UV to the near IR) comes directly from the photospheres of the constituent stars. Since these galaxies formed, their integrated spectral energy distribution (SED) has continuously evolved, just because their stellar content has been continuously changing. There are two basic mechanisms causing the stellar content of a galaxy to change. First, stars continuously die, and other, less massive stars replace them in the production of the integrated radiation. This is simply to say that the SED of a galaxy must evolve with time, because the individual stars evolve. But, second, new stars may appear in a galaxy, either because they form from its interstellar medium, or because they are incorporated into the galaxy from its environment, as a result of processes such as merging or cannibalism. The first type of evolution is purely passive, and is certainly easier to model, because the rates of star formation and accretion are assumed to be zero, except for a short period at the formation epoch of the galaxy. Conceptually, the second type of evolution does not present a much greater complexity, but requires the introduction of additional parameters describing the rates of star formation and accretion.

The immediate goal in a theoretical approach to the problem of the evolution of the SED of galaxies, is to construct model SED's as a function of time, i.e., of the age of a model galaxy. Following this preliminary step, a sequence of tantalizing applications may become possible. The comparison between the observed spectrum of a galaxy with a set of model spectra may lead to an estimate of the *age* of the galaxy, or the difference between the age of a galaxy and that of local, nearby galaxies. In principle, extending this procedure to a great number of galaxies, at greater and greater redshifts, should allow a direct estimate of the look back time as a function of redshift, independently of cosmological models. Eventually, *primordial* galaxies, at very high redshifts may be identified, and the determination of their age would set a connection between fundamental cosmological quantities such as $H_o$, $q_o$ and the redshift at which galaxies formed.

This is a very ambitious project, which may actually turn out to be unfeasible due to the complexity of evolutionary processes taking place in galaxies. But, in essence, the underlying idea is to construct a clock to measure cosmic times: an evolutionary clock. The justification for embarking in such a project, in spite of the little guarantee of success, is that astronomers can use just another cosmic clock[1]: the Hubble law, and the comparison of the two clocks is extremely meaningful for cosmology.

One can then distinguish two levels in this approach. At the first level one attempts to understand how the spectrum of galaxies evolves, while at the second level galaxies themselves are used as tools in the study of the evolution of the universe. Success at the first level is then a prerequisite to attack the second level. As already mentioned, evolutionary processes such as ongoing star formation, or star accretion, enormously complicate the picture, and likely preclude any meaningful attempt at reaching the second level. Thus spiral galaxies are not suitable cosmological probes, and we are left with ellipticals. Even then there is actually no guarantee that ellipticals are free from the complications brought about by ongoing

---

[1] A third clock, nuclear cosmochromology, can only be used for our own galaxy.

star formation and/or star accretion. We can only hope that at least a subclass of ellipticals may be well represented by a model in which a single (short) burst of star formation took place at the beginning, and in which the subsequent evolution has been purely passive. Indeed, a passively evolving model is extremely attractive, because it is relatively simple. The only clock at work is the one set by stellar evolutionary timescales, and then passively evolving galaxies (if they exist) represent the ideal tool (clock) for cosmological applications.

## 2. THE THEORETICAL FOUNDATIONS

A simple stellar population (SSP) is defined as an assembly of coeval, initially chemically homogeneous, single stars. Four main parameters are required to describe a SSP, namely its age (t), composition (Y, Z) and initial mass function (IMF). In nature, the best examples of SSP's are star clusters. Galaxies are certainly not SSP's: they contain stars of different metallicities, and of different age (perhaps with the exception of some ellipticals), and binaries are quite common objects. However, a complex population can always be *expanded* in a series of SSP's, and therefore the case of a SSP must be understood before addressing populations of more complexity. The problem of the role played by binaries however remains, and should be assessed separately. This section is devoted to a theoretical discussion of some properties of a SSP, and represents an attempt at giving a more systematic presentation to arguments and considerations which were preliminarily presented elsewhere (Renzini 1981a, Renzini and Buzzoni 1983). Some of these matters have been also reviewed by Tinsley (1980).

### 2.1 The Stellar Evolution Clock and the Evolutionary Flux

It is well known that the stellar lifetime is a function of the stellar mass, and that a star spends most of its lifetime while burning hydrogen on the main sequence. For a SSP of age t, the turnoff mass $M_{TO}$ is defined as the mass of a star which is on the verge of exhausting hydrogen at the center. Therefore, stars with initial mass $M_i < M_{TO}$ are still burning hydrogen at the center and stars with $M_i > M_{TO}$ are in more advanced evolutionary stages, or have already completed their (thermonuclear) evolution. The mass of stars lying at the main sequence turnoff (in the HR diagram of the population) is always very close to $M_{TO}$ as defined above, and this justifies the adopted morphological designation. The turnoff mass is obviously a function of the age of the SSP, and from theoretical evolutionary models (Becker and Iben 1979, Mengel et al. 1979) one derives the following analytical approximation, holding for (Y, Z) = (0.28, 0.02):

$$\log M_{TO}(t) = .0558 \log^2 t - 1.338 \log t + 7.764 \qquad (1)$$

where $M_{TO}$ is in solar masses and t in years. Figure 1 shows $M_{TO}(t)$ and its time derivative $\dot{M}_{TO}(t)$, as a function of the age of the population. The function $M_{TO}(t)$ is indeed the stellar evolution clock, i.e. the clock one would like to use to estimate the age of galaxies and to probe cosmological times.

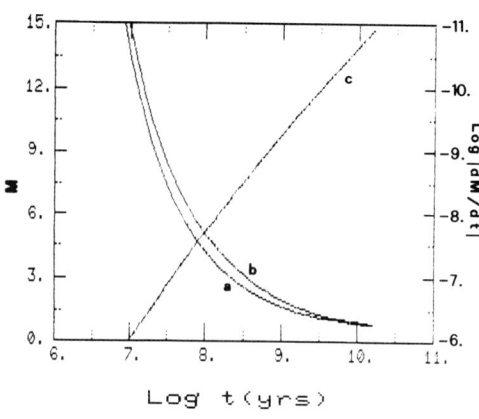

Figure 1. The turnoff mass (a), the mass of dying stars (b), and the time derivative of the turnoff mass (c) as a function of age [from Eq. (1)].

The time $t_{PSM}$ that a star of initial mass $M_i$ spends from core hydrogen exhaustion to the completion of its thermonuclear evolution is roughly given by (Renzini 1981a):

$$t_{PMS} \simeq 1.66\ 10^9\ M_i^{-2.72}\ \text{yrs}. \qquad (2)$$

Therefore, the initial mass $M_D$ of stars in the verge of completing their evolutionary cycle is approximately given by $M_D \simeq M_{TO}$ (t - $t_{PMS}$), which is also shown in Figure 1. For any given age of a SSP, stars with $M_i < M_{TO}$ are still in their main sequence stage (core hydrogen burning), stars with $M_{TO} < M_i < M_D$ are evolving in the post-main sequence (post-MS) stages, and stars with $M_i > M_D$ are *dead* remnants (white dwarfs, neutron stars, and the like). It is worth appreciating that, for any age, living post-MS stars span quite a narrow range of initial masses (cf. Figure 1). This allows us to approximate the post-MS stages with the evolutionary sequence for $M_i = M_{TO}$.

Stars continuously leave the main sequence, venture into the post-MS stages, and eventually die. The rate at which this happens is given by:

$$b(t) = \psi(M_{TO})|\dot{M}_{TO}|, \text{ (stars per year)} \qquad (3)$$

where $\psi(M_i) = AM_i^{-s}$ is the IMF of the SSP, and $\dot{M}_{TO}$ follows from Eq.

(1). The function b(t) represents the *evolutionary flux* of the population. It gives not only the rate at which stars leave the main sequence, but also the rate at which stars enter or leave any particular evolutionary phase. For instance, it gives the *death rate* of the population. To a fairly good approximation the flux is constant, past the main sequence turnoff. This comes from the narrow range of $M_i$ spanned, at any time, by post-MS stars. However, for sake of accuracy, the death rate at age t is clearly given by $b(t - t_{PMS})$, which is also shown in Figure 2 together with b(t).

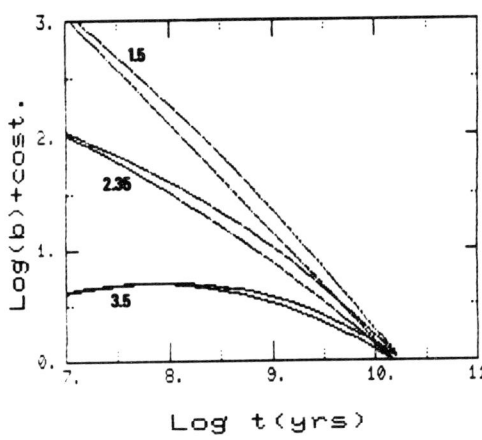

*Figure 2. The evolutionary flux b(t) from Eq. (3) for three choices of the IMF slope s, normalized at b = 1 for t = 15 Gyrs. For each value of s the upper line gives the death rate, $b(t - t_{PMS})$.*

## 2.2 The Fuel Consumption Theorem

The integrated bolometric luminosity $L_T$ of a stellar system is obviously $\Sigma_i L_i$, where the summation is extended to all stars in the system. Next, one can separate the contribution of MS stars, from that of stars in the various post-MS stages, i.e.:

$$L_T = L_{MS} + L_{PMS} = \sum_i^{MS} L_i + \sum_j n_j L_j, \qquad (4)$$

where $n_j$ is the number of stars in the "j" evolutionary stage, and $L_j$ is their average luminosity. The various post-MS stages can be classified and subdivided in whatever manner we wish to do it.

The contribution of MS stars can then be evaluated from the integral:

$$\sum_i^{MS} L_i \approx \int_{M_{inf}}^{M_{TO}} L(M) \psi(M) dM, \qquad (5)$$

where $M_{inf}$ is the lower mass cutoff of the IMF, and L(M) is the luminosity-mass relation for MS stars. To a first approximation one can adopt for L(M) the usual relation holding for zero age MS stars, which is of the form $L_{ZAMS}(M) = kM^a$, k and a being composition dependent quantities. This ignores the MS evolutionary effect, which will be largest for $M = M_{TO}$. A more accurate approach may consist in multiplying $L_{ZAMS}(M)$ by an appropriate factor $1 + c\,(M/M_{TO})^\gamma$, where $c$ is defined as $c = (L_{TO}/L_{ZAMS}|_{M=M_{TO}} - 1)$ and $\gamma = 5.5$, thus taking evolutionary effects into account.

The second term at the r.h.s. of Eq. (4) can be evaluated by noting that the number of stars in a given post-MS stage is proportional to the duration of the stage itself, and that the proportionality coefficient is just the evolutionary flux given by Eq. (3), i.e.

$$n_j = b(t)\, t_j. \tag{7}$$

On the other hand, the product $t_j L_j$ now appearing in the r.h.s. of Eq. (4) is the total energy radiated by a star during its phase j, an energy which essentially comes from nuclear burning. From these considerations we can then derive from Eqs. (4), (5) and (7):

$$L_T(t) \simeq k\psi(M_{TO})M_{TO}^{a+1}\{1/(a - s + 1) + c/(a + \gamma - s + 1)\} + \\ + 9.75\,10^{10}\,b(t)\Sigma_j\,F_j\,(M_{TO}), \tag{8}$$

where solar units are used for luminosity and mass, and where $F_j$, the *fuel consumption* during phase j, is given by:

$$F_j = m_j^H + 0.1\,m_j^{He}, \tag{9}$$

$m_j^H$ and $m_j^{He}$ being respectively the mass of H and He burned during phase j, in solar units. The coefficient 0.1 comes from the fact that the energy released by the conversion of 1 g of He into C and O is about one tenth of that released by the conversion of 1 g of H into He.

Eq. (8) is the mathematical representation of the *Fuel Consumption Theorem*, which can be formulated as follows: *The contribution of stars in any given Post-MS stage to the integrated bolometric luminosity of a SSP is directly proportional to the amount of fuel burned during that stage.*

Note that in Eq. (8) we have omitted the terms containing $M_{inf}$, and which come from the integration of Eq. (5). This is justified only insofar as $(a - s + 1) > 0$, i.e., insofar the IMF [slope] is not too steep; otherwise very low mass stars would provide a significant fraction of the total luminosity, and Eq. (8) should be accordingly modified.

For the ZAMS mass-luminosity parameters (k, a) we have used the following values, derived from the papers mentioned below and from Copeland et al. (1970) for low mass stars: (log k, a) = (0.3998, 3.4115), (0.1014, 3.8910), (-0.0336, 4.8152), and (-0.5840, 2.5540), respectively for the mass ranges $M > 4\,M_\odot$, $1.4 < M < 4\,M_\odot$, $0.57 < M < 1.4\,M_\odot$, and $0.1 < M < 0.57\,M_\odot$.

## 2.3 Fuel Consumptions

The fuel consumptions $F_j$'s are functions of the stellar initial mass

$M_i$ and composition, and are natural products of stellar evolutionary calculations. Although they are barely explicitly given in the literature, they can be (more or less) easily obtained from the data reported in papers presenting the results of evolutionary sequence calculations. Figure 3 shows the total fuel consumption, $F_T = \Sigma_j F_j$, as a function of age t and $M_i$, these quantities being related to each other through Eq. (1), i.e., $M_i = M_{TO}(t)$. Evolutionary calculations used in deriving $F_T$ include those of Iben (1965), Becker (1981), Sweigart and Gross (1976, 1978), Mengel et al. (1979), and Renzini and Voli (1981). Interpolations were used when necessary, in order to get $F_T$ for the choice (Y, Z) = (0.28, 0.02).

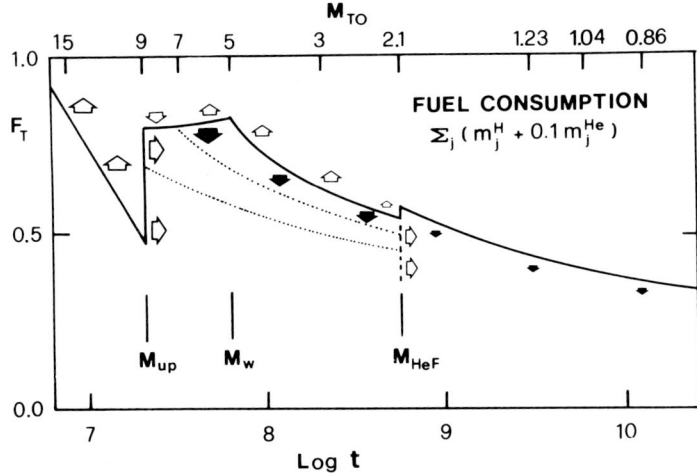

Figure 3. The total fuel consumption during the post-MS evolutionary stages (in $M_\odot$) as a function of age (lower scale) and turnoff mass (upper scale). $M_{TO}$ and t are tied together through Eq. (1). Open arrows qualitatively show the effect of convective overshooting, and filled arrows that of increasing the mass loss rate during the AGB phase. The latter effect is also pictured by the dotted lines.

Two discontinuities and one cusp are apparent in the function $F_T$ vs age (mass), respectively at $M_i$ = 9, 2.1 and 5 $M_\odot$. For general purposes, we shall designate respectively as $M_{up}$, $M_{HeF}$, and $M_w$ these three critical values of the initial mass, which in turn define four relevant mass (age) ranges. Indeed, it is worth introducing some relevant nomenclature and classification which follow from these designations (for more details, cf. Iben and Renzini 1983, 1984).

1) $M_i > M_{up}$ (Massive stars): towards the lower limit, stars in this mass range ignite carbon under semidegenerate conditions, suffer a mild carbon flash, and eventually undergo core collapse leading to a supernova event, presumably of type II (e.g. Becker and Iben 1979, Nomoto 1984). Massive stars fail to develop a highly degenerate CO core, and do not experience the *thermally pulsing* phase of the Asymptotic Giant Branch (AGB).

2) $M_{HeF} < M_i < M_{up}$ (Intermediate mass stars): stars in this mass range ignite helium nondegenerately, but following core helium exhaustion develop a highly degenerate CO core. They then experience a prolonged, thermally pulsing AGB phase terminating either with the envelope ejection and the formation of a white dwarf (WD) remnant (for $M_{HeF} < M_i < M_w$), or with carbon ignition and deflagration in a highly

degenerate CO core which has reached the Chandresekhar mass (~1.4 $M_\odot$). Following Iben and Renzini (1983), such an event is conventionally referred to as a SN I1/2 event.

3) $M_i < M_{HeF}$ (low mass stars): stars in this mass range develop a highly degenerate He core, shortly after central hydrogen exhaustion. They then experience a prolonged red giant branch phase (RGB), which is terminated by He ignition in the core leading to a mildly violent core He flash. The subsequent evolution is qualitatively similar to that of intermediate mass stars. Needless to say that the three initial masses depend somewhat on the adopted composition. This classification of stellar masses naturally leads to a classification of SSP's according to age. Indeed, we shall call *young* populations those for which $M_{TO} > M_{up}$, and then $t < t(M_{up})$, *intermediate age* those for which $M_{HeF} < M_{TO} < M_{up}$, and then $t(M_{up}) < t < t(M_{HeF})$, and, finally, *old* populations those for which $M_{TO} < M_{HeF}$, and then $t > t(M_{HeF})$. In these definitions $t(M_{TO})$ is the inverse function of $M_{TO}(t)$ given by Eq. (1). Stellar clusters, as the best examples of SSP's, can also be classified according to this scheme, and in section 4 Magellanic Cloud clusters will be discussed in this framework.

As we shall see in more detail in the following section 2.4, the distribution of stars in the HR diagram suffers nearly discontinuous readjustments as the age of a SSP grows older than the initial ages $t(M_{up})$ and $t(M_{HeF})$. Correspondingly, the SED is expected to suffer seemingly abrupt changes. For these reasons, it is worth discussing the major uncertainties affecting the theoretical determinations of the initial masses $M_{up}$, $M_{HeF}$ and $M_w$, and the effect of these uncertainties on the fuel consumptions. The value of $M_w$ is mostly sensitive to the adopted mass loss rate during the AGB phase, and Iben and Renzini (1983) give an analytical approximation of $M_w$ as a function of the mass loss parameter $\eta$. Increasing $\eta$, the fuel consumption during the AGB phase decreases, and $M_w$ increases, as qualitatively shown in Figure 3. For $\eta$ larger than a critical value, $M_w$ becomes equal to $M_{up}$ and all intermediate mass stars would leave WD remnants. The choice of the mass loss parameter does not effect $M_{up}$ and $M_{HeF}$, which, however, are mostly sensitive to another poorly known physical process: convective overshooting. The values reported in Figure 3 derive from models in which overshooting from stellar convective cores was neglected. The inclusion of this effect will decrease both $M_{HeF}$ (Barbaro and Pigatto 1984) and $M_{up}$ (Renzini et al. 1985; Chiosi, this volume) by amounts which depend on a parameter describing the extension of the overshooting region. The corresponding effect on the fuel consumption is also qualitatively shown in Figure 3.

It is worth emphasizing that we lack any satisfactory theory for both mass loss and overshooting. One is then forced to parameterize these physical processes, to obtain for a best fit between theoretical models and relevant observational constraints. In this context the most useful observations concern globular (sometimes called populous) clusters in the Magellanic Clouds. We shall further discuss this point in section 4. With this proviso, let us now return to the total fuel consumption. From Figure 3 one can appreciate that that $F_T$ is a very slow function of $M_i$ (age), i.e. it decreases by only a factor of

~2, from ~0.8 $M_\odot$ dow to ~0.35 $M_\odot$, as $M_i$ decreases from 15 to less than 1 $M_\odot$ (age increasing from ~$10^7$ yrs to over $10^{10}$ yrs). The uncertainty in $F_T$ induced by uncertainties in mass loss and overshooting can be estimated to be at most of the order of 30% for intermediate mass stars, and much less (a few percent) for low mass stars. Finally, it is worth mentioning that for stars having a convective core while on the main sequence ($M_i \gtrsim 1.2\ M_\odot$), the inclusion of overshooting will affect the $M_{TO}$-age relation by somewhat increasing $M_{TO}$ for a given age. In conclusion, we can estimate that the overall uncertainty affecting the theoretical luminosity-age relation, Eq. (8), is probably less than 30%.

## 2.4 The Evolution of Bolometric Luminosity

Figure 4 shows the evolution of the integrated bolometric luminosity of a SSP, from Eq. (8) with $F_T$ from Figure (3). Note that for $M < 0.57\ M_\odot$ the actual IMF slope we have used is the quoted s minus 2.35, so as to take roughly into account the observed flattening of the IMF at very low masses (cf. Scalo 1985, and references therein). The small discontinuity at $t = t(9\ M_\odot)$ corresponds to the sudden appearance of the AGB. A similar discontinuity at $t(M_{HeF})$ is too small to notice. Obviously enough, the flatter the IMF, the faster the rate of luminosity decline. For s = 2.35 (the Salpeter IMF) this rate is ~2.1 magnitudes per dex in age. In this framework, supposing a universe populated by identical and coeval SSP's, their intrinsic observable luminosity would then be related to the look back time $t_{LB}$ (function of redshift) through the expression:

$$L \simeq L_o (1 - t_{LB}/t_p)^{-0.85}, \qquad (9)$$

where $L_o$ is the luminosity of local, zero redshift SSP's and $t_p$ is their age.

For comparison, Figure 4 also shows the time evolution of a *complex* population (with s = 2.35), in which the star formation rate (SFR) is constant. The luminosity, $L_T^S$, of such a population is obtained from the integral:

$$L_T^S(t) = \int_0^t SFR(t - t')\ L_T(t - t')\ dt' \qquad (10)$$

where $SFR$ = const. and $L_T$ is given by Eq. (8). The region bounded by $L_T$ and $L_T^S$ will then encompass the luminosity evolution of populations in which the SFR is monotomically decreasing, e.g. SFR $\propto \exp(-t/\tau)$ with $0 \leq \tau \leq \infty$.

## 2.5 Contributions to the Total Light

In this section we illustrate the time evolution of the relative contributions of stars in the various evolutionary stages to the total

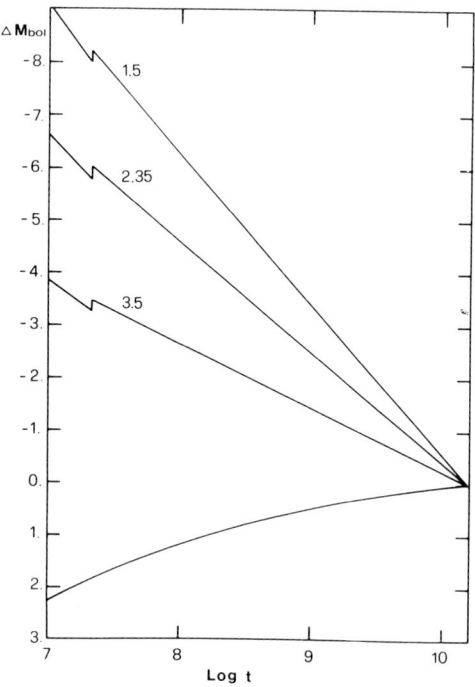

Figure 4. The luminosity evolution of SSP's for three choices of the IMF slope s, normalized at $\Delta M_{bol} = 0$ for $t = 15$ Gyrs. The lower line shows the luminosity evolution of a population with constant star formation rate, $s = 2.35$, and upper IMF cutoff at 50 $M_\odot$.

light of a SSP. From Eq. (8), these relative contributions, $L^j/L_T = L_j n_j/L_T$, are given by:

$$L^j/L_T = 9.75 \; 10^{10} \; b(t) F_j(M_{TO})/L_T \qquad (11)$$

for the post-MS stages, and by $L_{MS}/L_T$ for the MS contribution, $L_{MS}$ being the first term at the r.h.s. of Eq. (8). The various evolutionary stages are classified according to the following nomenclature:

1) Main Sequence (MS): core hydrogen burning phase.
2) Subgiant Branch (SGB): shell hydrogen burning phase, from central hydrogen exhaustion to the Hayashi line.
3) Red Giant Branch (RGB): shell hydrogen burning phase along the Hayashi line, until helium ignition.
4) Horizontal Branch (HB): core helium burning phase.
5) Asymptotic Giant Branch (AGB): hydrogen and helium shell burning phase, along the Hayashi line.
6) Post-AGB (P-AGB): final evolution from the AGB to the WD stage, through the planetary nebula phase.

This nomenclature is morphologically most appropriate for an old SSP, e.g. galactic globular clusters. For instance, the HB phase might also be called *Cepheid loop* in a young SSP containing cepheids, or *Clump Giant Branch* in an older population in which the core helium burning stars are confined to a red clump. The AGB phase is present only in SSP with $M_{TO} \lesssim M_{up}$, and the P-AGB phase only when $M_{TO} \lesssim M_w$.

Figure 5 shows $L^j/L_T$ and $L_{MS}/L_T$ vs log t, for a SSP with $(Y, Z) = (0.28, 0.02)$, $s = 2.35$ and having adopted $M_{up} = 9\ M_\odot$, $M_{HeF} = 2.1\ M_\odot$ and $M_w = 5\ M_\odot$, which corresponds to the mass loss parameter $\eta = 1/3$ (cf. Iben and Renzini 1983). The figure is largely self-explanatory, and we shall concentrate only on the "glitches" at $t = t(M_{up}) \simeq 2\ 10^7$ yrs and $t = t(M_{HeF}) \simeq 10^9$ yrs. At these epochs the stellar content of a SSP suffers two major "phase transitions". The first (or AGB) phase transition is due to the sudden appearance of AGB stars, whose relative contribution to the total light rapidly increases, reaching a maximum of around 50% at $t = t(M_w)$. Indeed, the age $t(M_{up})$ marks the transition between a young population and an intermediate age population (following the terminology introduced in section 2.3). A young population is dominated by MS stars (providing ~75% of the light), with HB stars contributing most of the residual ~25%. In an intermediate age population AGB stars gain the leadership, with MS stars stabilizing around a 30% contribution.

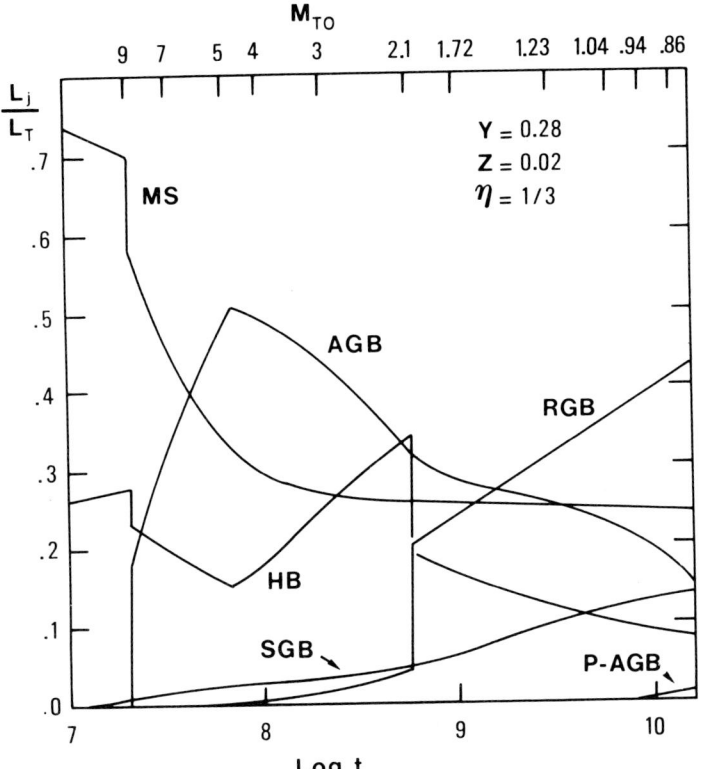

Figure 5. The relative contribution of the various evolutionary stages to the integrated bolometric luminosity of a simple stellar population as a function of age (lower scale) and turnoff mass (upper scale). The various evolutionary stages are defined in the text, composition and mass loss parameter are indicated, and s = 2.35.

The second phase transition, at $t = t(M_{HeF})$, is due to the appearance of RGB stars with degenerate He cores. The RGB luminosity extension and lifetime both dramatically increase as does the number of RGB stars. We shall then refer to this as the RGB phase transition. The AGB phase transition is virtually instantaneous, as it corresponds to a sharp bifurcation in the evolutionary behaviour of stars at $M_i = M_{up}$. In practice, the transition is completed in about one AGB lifetime ($\sim 10^6$ yrs). The RGB phase transition may not be so sharp, as the maximum core degeneracy might be a continuous, albeit very sensitive function of $M_i$, for $M_i$ decreasing across $M_{HeF}$. The reason for this uncertainty is that only a few evolutionary sequences are available, i.e. the parameter space has been insufficiently explored for $M_i \simeq M_{HeF}$. Actually, the precise value of $M_{HeF}$ and its composition dependence are still poorly known, as well as the *rapidity* of the RGB phase transition itself. In drawing Figure 5 (as well as Figure 3), we have assumed the transition to be instantaneous. This assumption should be checked (and in case corrected) by computing fine grids of evolutionary sequences for $M_i \simeq M_{HeF}$ ($\simeq 2\ M_\odot$) and for various compositions.

Anyway, the RGB phase transition marks the passage from an intermediate age to an old population (again following the terminology introduced in section 2.3). As one sees in Figure 5, RGB stars gain the leadership for $t \gtrsim 10^9$ yrs, and in a very old population they contribute up to $\sim 50\%$ of the total light. Obviously enough, the two phase transitions must be accompanied by major changes in the integrated spectrum (the SED) of a SSP. It is worth realizing that the scale in Figure 5 is logarithmic, and then changes in the various contributions are very slow with look back time. The first major readjustment one encounters is the RGB phase transition, for a look back time of $\sim 95\%$ of the age of the populations, having assumed for the latter 15 Gyrs.

It may be instructive to see in more detail the relative contributions of MS stars, and how these are affected by the adopted slope of the IMF. Figures 6a and 6b show the overall MS contribution as a function of age, for 3 values of s and for the two-slope and one-slope IMF, respectively. From these results we can infer that the SED of SSP's is not expected to much sensitive to the adopted IMF, at least when the IMF slope is less than $\sim 3 - 3.5$ for $M \lesssim 0.5\ M_\odot$. Figure 7a shows the relative contributions of MS stars of mass less than M, $L(M)/L_T$, as a function of M, for an age $t = 15$ Gyr and for three values of s, again using s - 2.35 for the IMF slope below 0.57 $M_\odot$. The function L(M) is given by:

$$L(M) = k \int_{M_{inf}}^{M} \psi(M)\ M^a\ \{1 + c(M/M_{TO})^\gamma\} dM. \qquad (12)$$

Note that with the adopted *two-slope* IMF low mass stars contribute just a few percent of the total light. On the other hand, when using a single slope IMF M dwarf may contribute a significant fraction of the total light when $a - s + 1 \lesssim 0$, as in the case s = 3.5 in Figure 7b.

Finally, Figure 8 shows the run with time of the relative contributions $L^j/L_T$ for the case of a complex stellar population, with SFR = const. Here the $L^j$ contributions are evaluated by means of integrations similar to the one in Eq. (10). Once again the Figure is self-explanatory.

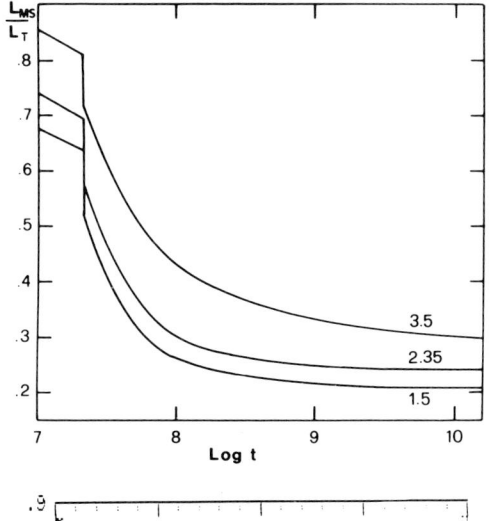

Figure 6a. The relative contribution of MS stars for three choices of the IMF slope s. For $M < 0.57\ M_\odot$ the IMF slope is $s - 2.35$.

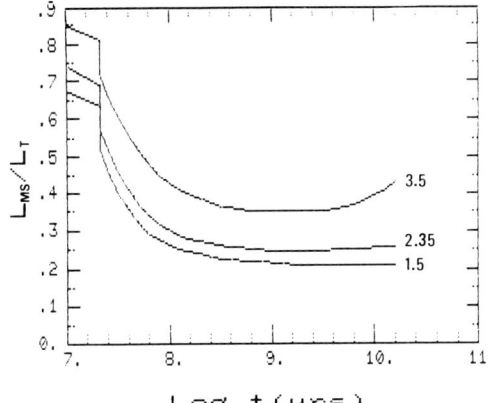

Figure 6b. The same as figure 6a but for a single-slope IMF, as indicated, and $M_{inf} = 0.1\ M_\odot$. Note the M-dwarf induced upturn for $s = 3.5$.

## 2.6 The Specific Evolutionary Flux and Its Applications

One defines the specific evolutionary flux, $B(t)$, as the ratio of the evolutionary flux to the total integrated luminosity of a SSP, i.e. $B(t) = b(t)/L_T(t)$. $B(t)$ is then the evolutionary flux per unit luminosity of the parent population, and is measured in stars per year per solar luminosity. From Eqs. (3) and (8) one has:

$$B(t) = |\dot{M}_{TO}|/\{k\ M_{TO}^a\ G(M_{TO},a,s,\gamma) + 9.75\ 10^{10}\ |\dot{M}_{TO}|\Sigma_j F_j(M_{TO})\}, \qquad (13)$$

where the function G is the expression within braces in Eq. (8). Since both $b(t)$ and $L_T$ depend linearly on $\psi(M_{TO})$, this quantity cancels in their ratio $B(t)$, and the only surviving dependence on the IMF is through the function G. One can then expect the specific flux to be nearly independent of the adopted IMF. Indeed, this can be appreciated from Figure 9, which shows $B(t)$ for three choices of the IMF slope.

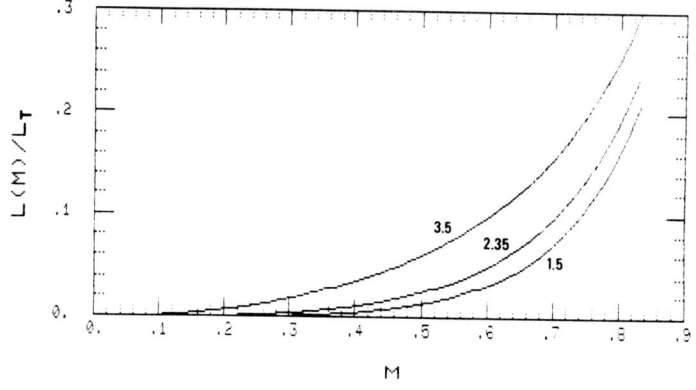

Figure 7a. The relative contribution of MS stars of mass $<M$, for three IMF slopes. Again, for $M < 0.57 \, M_\odot$ the IMF slope is $s - 2.5$.

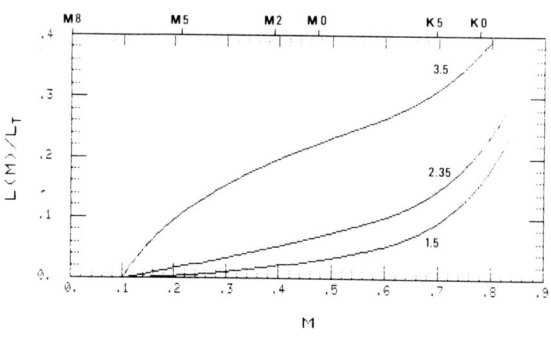

Figure 7b. The same as figure 7a, but for a single-slope IMF as indicated and $M_{inf} = 0.1 \, M_\odot$. Approximate spectral types are also given. Note the runaway M dwarf contribution for $s = 3.5$.

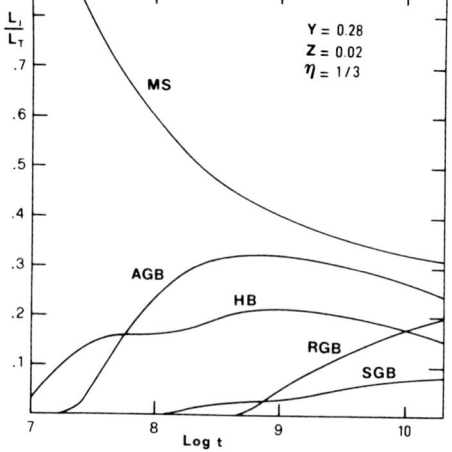

Figure 8. The same as figure 5 but for a population with constant star formation rate.

Another remarkable property of the specific flux is its near constancy while age varies by several orders of magnitude. Actually, Figure 9 shows that $B(t)$ increases from $\sim 5 \times 10^{-10}$ to $\sim 2 \times 10^{-11}$ while $t$ increases from $10^7$ to over $10^{10}$ yrs.

Since $n_j = b(t)t_j$, Eq. (7), we then have the extremely useful relation:

$$n_j = B(t)L_T t_j, \quad (14)$$

which gives the number of stars in any given post-MS stage ($n_j$), as a function the duration $t_j$ of the stage under consideration, and of the total bolometric luminosity of the sampled population. This sampled luminosity can be the total luminosity of a cluster, of an elliptical galaxy or whatever, or just the portion of this luminosity which is actually sampled by the acquisition instrument (e.g. a photometer, a CCD frame, the slit of a spectrograph, etc.). Eq. (14) allows a fairly accurate evaluation of the number of sampled stars once $L_T$ is measured, $t_j$ is derived from evolutionary models, and the age is estimated by some means. As an instructive example of how Eq. (14) can be used, let us consider a typical globular cluster with $L_T = 10^5$ $L_\odot$ and $t = 15$ Gyrs. Since we know the HB lifetime to be $\sim 10^8$ yrs, the number of HB stars in the cluster is $n_{HB} = 2 \times 10^{-11} \times 10^5 \times 10^8 = 200$ HB stars. If a CCD frame is sampling only 1000 $L_\odot$, then one has just $\sim 2$ HB stars in the frame. If the bright post-AGB lifetime is $\sim 10^5$ yrs, then Eq. (14) gives 0.2 bright post-AGB stars in the whole cluster, i.e. a 20% probability of finding one of such stars.

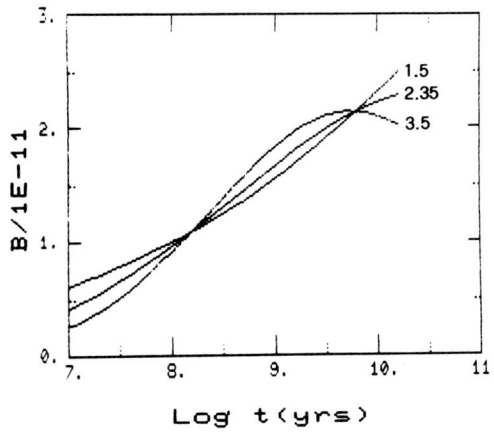

Figure 9. The specific evolutionary flux B(t) as a function of age for three IMF slopes (two-slopes IMF case).

Eq. (14) can also be used the other way around, i.e. one can estimate the lifetime of a given evolutionary stage once $n_j$ is obtained from a complete survey. For example, suppose that a complete ultraviolet survey reveals 20 bright post-AGB stars in 50 surveyed clusters, totaling $2 \times 10^7$ $L_\odot$. Then from Eq. (14) one would derive an average duration of $\sim 20/(2 \times 10^{-11} \times 2 \times 10^7) = 5 \times 10^4$ yrs for the bright post-AGB lifetime.

Therefore, possible uses of Eq. (14) include: 1) planning the search for rare stars in resolvable populations (e.g. WD's in globular clusters, Renzini 1985), 2) checking whether claimed color gradients in clusters (or galaxies) are real or mere sampling artifacts (e.g. spurious B-V color gradients in galactic globulars, Renzini 1981b), 3) estimating the lifetime of evolutionary phases (e.g. the AGB lifetime in Magellanic Cloud clusters), and many others one may invent. For several applications, one may actually prefer to use visual, rather than bolometric luminosities, which can be accomplished by integrating under the whole SED of a population, and then deriving a *bolometric correction* for the population as a whole. In this way one finds that the bolometric luminosity of galactic globular clusters is quite accurately just twice their visual luminosity, the latter being defined as $\log L_V = -0.4 (M_V - 4.7)$, where $M_V$ is the absolute integrated V magnitude of the sampled population. Therefore, in this case Eq. (14) becomes: $n_j \simeq 4 \times 10^{-11} L_V t_j$. Note, however, that the bolometric/visual conversion factor will be different for populations of different age or metallicity.

## 2.7 Total and Specific Rates of Mass Return

Before concluding their evolutionary life, stars shed a considerable fraction of their initial mass. The rate $\dot{M}$ at which a SSP returns mass to the interstellar medium is given by:

$$\dot{M}(t) = \Delta M(M_{TO}) \, b(t) = \Delta M(M_{TO}) \, \psi(M_{TO}) \, |\dot{M}_{TO}| \tag{15}$$

where $\Delta M$ is the amount of mass which is lost by a star during its whole life, i.e. $\Delta M = M_i - M_f(M_i)$, $M_f$ being the final mass, or mass of the remnant. Figure 10 shows $\dot{M}$ as a function of age, for various IMF slopes. The final mass is assumed to be $1.4 \, M_\odot$ for $M_i > M_{up}$, zero (no remnant) for $M_w < M_i < M_{up}$, and for $M_i < M_w$ it is given by the expression provided by Iben and Renzini (1983), with $\eta = 1/3$. The discontinuity at $t = t(M_w)$ is due to the transition from populations in which stars die as SN I1/2, to older populations in which stars leave WD remnants.

Finally, one can define a *specific* rate of mass return $\dot{m}$, or rate of mass return per unit luminosity, which we obtain from (15) by dividing by the total luminosity, i.e.:

$$\dot{m}(t) = \Delta M \, B(t), \tag{16}$$

which is shown on Figure 11. Note that $\dot{m}(t)$ is nearly independent of the IMF slope. This is not surprising since most of the light and all the returned mass come from evolved stars, which at any time span a very narrow range of initial masses. Note also that $\dot{m}(t)$ does not depend much on time either, and that the actual rate of mass return of a population, $\dot{M}(t)$, is simply given by $\dot{m}(t) \cdot L_T$.

## 2.8 Summary of IMF-Dependent and Independent Quantities

The IMF is such a ubiquitous ingredient in population synthesis that it is worth summarizing which quantities critically depend on the IMF, and which do not. As a general rule, *non-specific* quantities are very sensitive to the IMF slope, and specific quantities are nearly independent of it. In the first category belong the evolutionary flux $b(t)$, the total luminosity $L_T(t)$, and the rate of mass return $\dot{M}(t)$,

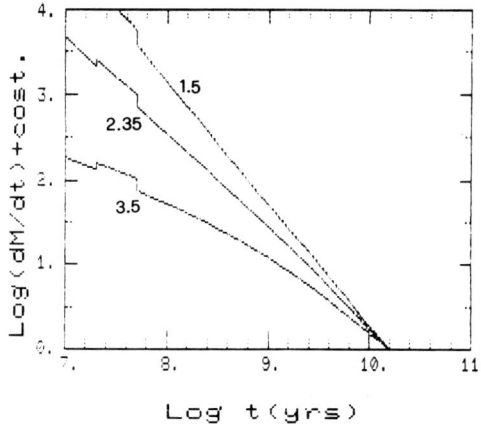

*Figure 10. The rate of mass return from dying stars, normalized at 1 $M_\odot$ per yr for t = 15 Gyrs, and for three IMF slopes.*

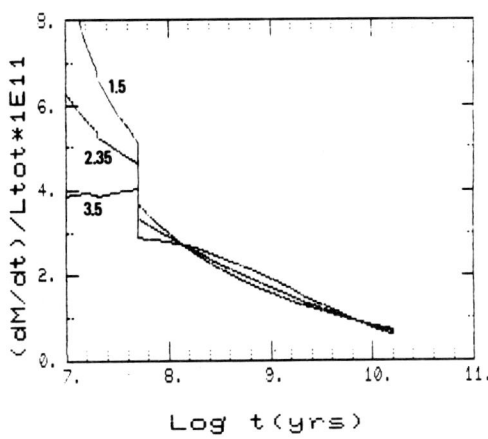

*Figure 11. The specific rate of mass return in $M_\odot$ per year and per $L_\odot$, as a function of age and for three IMF slopes.*

cf. Figures 2, 4, and 10, respectively. Examples of ~IMF-independent quantities are the specific flux B(t), the relative contributions $L_j/L_T$, and the specific rate of mass return $\dot{m}(t)$, cf. Figures 9, 6, and 11, respectively. As already mentioned, the SED of a SSP is also a specific quantity, and the spectral evolution of SSP's is nearly independent of the adopted IMF, at least insofar as the IMF is not really very steep.

## 3. FROM BOLOMETRIC TO MONOCHROMATIC

As we have seen, the bolometric luminosity of a stellar population is an important quantity, well rooted in physical principles. But observers usually wonder much more about spectra and colors, and after all it is only by interpreting this sort of data that we may hope to address the astrophysical questions briefly outlined in Section 1. However, in this section we are not going to present model SED's for stellar populations, but we shall rather discuss several methodological aspects which are relevant to population synthesis studies.

## 3.1 The "Color" of Evolutionary Stages

Although the actual calculation of synthetic SED's certainly requires elaborated calculations, it may be worth approaching the problem in a qualitative way at first. Indeed, the various evolutionary stages are typically confined to specific regions of the HR diagram, and then have characteristic temperatures (and colors). Main sequence stars are familiar to everybody, as well as the notion that the temperature and luminosity extension of the MS is a continuous function of age, with the temperature of hottest MS stars gradually decreasing as a population becomes older and older. Correspondingly, the flux at *short* wavelengths progressively decreases as the MS becomes more and more depopulated. RGB stars typically have temperatures around 4000-5000 K, the RGB location being rather sensitive to metallicity, but almost insensitive to age. Correspondingly, RGB stars will be most important in modelling the SED in the visual, red and near-IR part of the spectrum, but should have little effect at shorter wavelengths. HB stars present a much more complex problem, as the location of core He-burning models is very sensitive to both age and composition. Correspondingly, this ~10% of the total light which is provided by HB stars (cf. Figure 5) can be *spent* at quite different effective temperatures, depending on the fundamental parameters of a SSP (age, and composition). For instance, the HB will affect the visual-red part of the spectrum in the case of a stubby red HB (red giant *clump*), or the blue-UV side of the SED as in the cases of old populations with blue and extended HB's, or of young and intermediate age populations with extended *cepheid loops*. AGB stars come in two varieties: carbon and non-carbon stars. The former ones are very cool and red, and then will be important in the near-IR, but completely negligible in the optical. Non-carbon AGB stars have spectral types from late K to late M, and their spectral energy distribution is extremely sensitive to spectral type. Stars later than ~M3 are very faint in the optical (Renzini et al. 1985) and so should affect only the near-IR side of the SED. AGB stars earlier than ~M2 may affect also the optical part of the integrated spectrum. However, it is worth appreciating that the spectral type of non-carbon AGB stars can hardly be predicted by evolutionary models. Finally, the post-AGB is very hot (hotter than ~30,000 K), and will then contribute only at the UV end of the SED.

From these qualitative considerations one can anticipate the following gross features in the spectral evolution of SSP's.

1) The AGB phase transition should be accompanied by a sudden increase in the near-IR (JHK) luminosity, but should have very little effect at shorter wavelengths, with the possible exception of metal poor populations in which the AGB stars could be confined to late K-early M spectral types.

2) The RGB phase transition should considerably affect the visual to near-IR part of the SED, but again should have little effect

at blue or shorter wavelengths, particularly in the UV.

3) The SED is expected to *shrink* as a population gets older, as it becomes progressively depopulated in both its hottest* and coolest stars, i.e. as both the MS and the AGB *shorten* with age. Actually, the *width* of the SED may give a very effective tool for estimating the age from integrated spectra.

4) Finally, the latest to arrive is the contribution of hot P-AGB stars, which appears only at very late epochs (cf. Figure 5) and should produce an *upturn* in the UV spectral flux towards shorter wavelengths. Superimposed on these general trends, the contribution of HB stars will further modify the SED at wavelengths which depend on the actual temperature distribution of these stars.

## 3.2 Lights and Shades of Stellar Evolutionary Models

The theoretical calculation of synthetic SED's rests on stellar evolutionary sequences. It is therefore important to be aware of the reliability of such evolutionary sequences, at least before starting a massive production of synthetic SED's. As a rule of thumb, the luminosity of stellar models are fairly reliable, but effective temperatures are not, and unfortunately the SED of a population is most sensitive to the actual temperature distribution of the constituent stars.

The quantities entering in Eq. (8), i.e. $M_{TO}(t)$ and $F_j(M)$, are fairly reliable, the major uncertainties being related to the possible occurrence of deep extramixing processes (e.g. overshooting) which may affect both $M_{TO}(t)$ and the $F_j$'s, and to the actual rate of mass loss during the AGB phase, which affects the corresponding $F_j$. These effects must then be calibrated, and the corresponding parameters properly tuned. This can be accomplished by comparing theoretical sequences to cluster CM diagrams, an aspect that we shall further discuss in Section 4.

On the other hand, stellar temperatures (radii) are most sensitive to the treatment of the outer envelope. In the case of red giant models, the treatment of superadiabatic convection controls the location of the Hayashi track, and their effective temperature. In practice, in the framework of the mixing length theory, the red giant temperatures are sensitive to the mixing length parameter $1/H_p$, which must then be properly calibrated by fitting theoretical to observed RGB's of clusters of various compositions. This does not present major problems, while the situation is considerably more complicated in the case of models with radiative envelopes and which lie in the so-called Hertzsprung gap. In practice, this concerns stars in the core helium burning stage, whose envelope is in nearly *neutral* equilibrium (Renzini 1984). For these stars, small changes in the luminosity provided by the nuclear burning may produce very large changes in the structure of the envelope, i.e. large variations of

---

*Apart from post-AGB and hot HB stars.

radius (effective temperatures). This is a common feature of all cases in which the evolutionary track is nearly *horizontal* in the HR diagram, and indeed, if a track is horizontal this means that small changes in luminosity produce large changes in radius. Under such circumstances, any small change in the model *input physics* (particularly in opacity) may produce very large changes in the model radius. This is to say that model radii cannot be very reliable for core helium burning models, and existing tracks should be used with caution. Long-standing problems which have puzzled model makers for years, such as the blue to red supergiant ratio, the extension of the *cepheid loops*, and the HB morphology of globular clusters are to some extent a manifestation of this physical limitation of the models, i.e. when stellar radii are very sensitive to virtually *everything*, then they can hardly be accurately predicted! A more detailed discussion of these aspects will be presented elsewhere (Ferrario et al. 1985).

## 3.3 Clocks and Thermometers

The *age* information being the most interesting quantity we wish to obtain on galaxies, one has to get acquainted with the intimate structure of the stellar evolution clock. There is indeed a long way to go from the integrated spectrum of a galaxy, eventually to get its age, or the age distribution of its constituent stars. The SED is actually sensitive to the *temperature* distribution of the constituent stars, and in turn this temperature distribution is a function of age, as for instance in the case of the main sequence turnoff. Therefore, from an operational point of view, determining the age of a SSP is a two-step procedure: first one has to *measure* a temperature, and then use a theoretically calibrated relation giving temperature vs. age. Figure 12 illustrates this situation. Here the effective temperature of the main sequence turnoff is plotted against age, for two values of the metallicity (models from Mengel et al 1979). Now, suppose for a while it were possible to determine with zero error the turnoff temperature of a SSP from its integrated spectrum. Then, from Figure 12 with, say, $\log T_e^{TO} = 3.72$, one would determine an age of ~7 Gyrs if $Z = 2 Z_\odot$, or an age of ~21 Gyrs if $Z = Z_\odot/2$. That is the stellar evolution clock is extremely sensitive to composition (cf. O'Connell 1980). Chemistry plays a crucial role between thermometer and clock readings. Moreover, real galaxies are not SSP's, i.e. they are not chemically homogeneous. Elliptical galaxies in particular are likely to be constituted of stars spanning a huge range in metallicity (cf. Arimoto, this volume), and their SED is then expected to be very sensitive to the actual distribution of metallicities, *not just to the average*. In principle, one should determine the metallicity distribution *before* being in the position of dating galaxies. This may not be an easy task at all: composition and age both affect the integrated SED, and in old populations the former effect may dominate over the latter. Disentangling the two effects, and extracting the *signal* (age) from the *noise* (metallicities) may be very difficult.

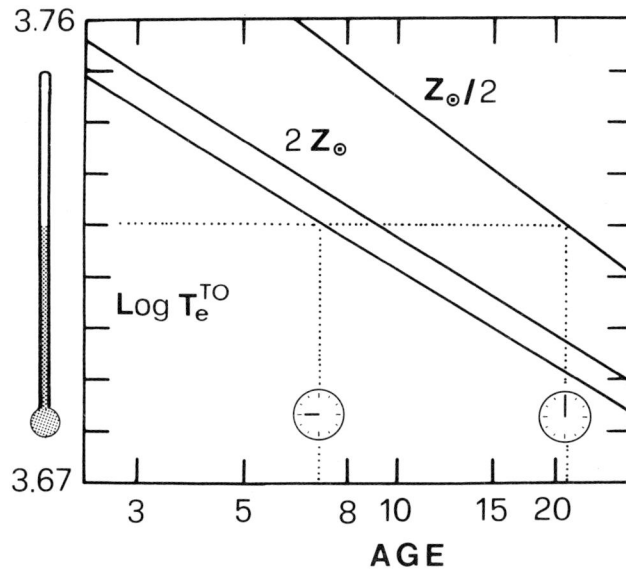

Figure 12. Thermometers and clocks. The turnoff effective temperature for two metallicities ($Z_\odot = 0.02$) and $Y = 0.2$. The middle line refers to $(Y,Z) = (0.3, 0.04)$. This illustrates how chemistry affects clock readings.

## 3.4 The Rules of the Game

From the whole previous discussion it follows that stellar evolutionary sequences and theoretical models for the spectral evolution of stellar populations should be accurately tested and calibrated before being safely applied to the study of galaxies. What one needs are suitable calibrators, i.e. template populations as close as possible to our definition of a SSP. Such template populations should satisfy several relevant requirements: (i) span the widest possible range in both age and metallicity, (ii) be populous enough so as to ensure statistical completeness for all relevant stellar evolutionary stages, in particular including those short-living stages which however may provide a significant fraction of the total light (e.g. the AGB), and (iii) be resolvable in individual stars (in order to get accurate photometry of them), and still compact enough so as to allow meaningful integrated spectroscopy.

The best approximation to a SSP that nature is able to provide us are stellar clusters. Galactic globular clusters meet quite well conditions (ii) and (iii), but span a metallicity range from ~1/100 to only ~1/5 solar, thus leaving out metal rich and super-metal rich populations. Moreover, they are all ~15 Gyr old. Globular clusters in the Magellanic Clouds are especially important, as they span the widest range in age, although in metallicity they do not go much beyond the most metal rich galactic globulars. So, unfortunately, we don't have access to super metal-rich globular clusters, which would be particularly important for the modelling of elliptical galaxies. Super metal-rich stars in the galactic bulge offer the only usable handle for the high metallicity case (cf. Whitford, Frogel, this volume). A preliminary discussion of MC clusters as template populations is given in the following Section 4.

What do we actually mean by using template populations as calibrators of theoretical models? There are indeed several aspects worth mentioning. Cluster color-magnitude diagrams offer the opportunity of testing stellar evolutionary sequences, as well as the relative contributions to the total light given by the various evolutionary stages (e.g. as shown in Figure 5). The completeness of the stellar sample for each cluster is obviously mandatory for this latter test. The quantities to be calibrated are all those *free parameters* involved in the construction of the models, e.g. mixing length, mass loss rate, overshooting, etc. Then, once accurate ages and metallicities are obtained, one gets a calibration of the integrated cluster spectra vs these two fundamental parameters. Therefore, at this stage theoretical evolutionary synthesis can be checked, as well as empirical population synthesis. Both these two techniques have been used to infer properties of galaxies, but, as far as we know, they have never been tested on template populations. Conversely, it would be extremely instructive to see what age and metallicity a population synthesis code indicates for clusters of independently known age and metallicity, when only the integrated spectrum is used. For example, given the spectrum of 47 Tucanae, which age and metallicity the synthesis code XY would indicate? One could actually push this test further, and generate a *manufactured* composite spectrum by adding the spectra of, say, M15, M3, 47 Tuc, plus those of a few younger MC clusters, in some secret proportions, and then feed the composite spectrum into the population synthesis code. What would be the answer? Knowing what we put in, it would be fascinating to see what we get out from the code. Anyway, we believe that a population synthesis code should pass this sort of test before getting a flying license to galaxies . . .

In a play on population synthesis (either empirical or theoretical) there are three main actors: the galaxy G we want to understand, the synthesis code C, and the stellar library L. Running the code means getting a number of *matrix elements*, which we may indicate as $<G|C|L>$. So the result is not just information on G, but on G *and* L. What if the library is not complete? i.e. if G contains stellar species which are not in L? Usually C is smart enough to give *an* answer anyway, and this may produce nonsense. Suppose for instance that L contains a series $L_i$ of populations with the same composition and different ages (labelled i), while G contains coeval populations of different metallicities. The code will synthesize a *best fit* galaxy $|G> = \Sigma_i a_i |L_i>$, with several non-zero eigenvalues $a_i$, and one may infer that $<G|$ has been forming stars over a prolonged period of time.

In another example, an empirical synthesis library may lack some important stellar species, e.g. carbon stars. Then C will try to get IR photons from K type stars, but in order to avoid an overproduction of yellow-red photons C may be forced to add a negative number of G type stars, and so on . . . Solutions giving a negative number of stars for some stellar species are often regarded as unphysical, and C may then be constrained to generate *non-negative* solutions. Actually, solutions with negative stars are physically more interesting than mediocre non-negative solutions! Indeed, they signal an incompleteness in L, which would otherwise remain buried in the computer when using a constrained C. So, in general, unconstrained empirical syntheses provide more informative answers, compared to those in which wise astrophysical constraints are imposed. Indeed, were L complete, C wouldn't need any constraint to reproduce G! After all, aren't constraints mere cosmetic devices to hide library incompleteness?

## 4. MAGELLANIC CLOUD CLUSTERS AS TEMPLATE STELLAR POPULATIONS

In this section we briefly touch upon the use of MC clusters as template stellar populations. There are several aspects of MC clusters which are currently investigated by quite many groups. Here we shall restrict our discussion to just a few topics, namely the problem of MC clusters dating, and the possibility of detecting the AGB and RGB phase transitions (cf. Sections 2.3 and 2.5) in the MC cluster family.

### 4.1 Dating MC Clusters

Age has been estimated for quite a number of MC clusters (cf. for instance the recent compilation by Hodge 1983). Several methods have been used to get cluster ages from stellar photometry. Namely, from the main sequence turnoff, from the luminosity of helium-burning giants, and from the brightest AGB stars. Embarrassingly enough, in several instances different investigators have offered age estimates for a given cluster which disagree by one order of magnitude. Such large discrepancies can hardly be ascribed to ordinary observational uncertainties, but rather indicate that some conceptual mistake may have been committed. Here we then focus on the potential malfunctions of the various dating methods when specifically applied to MC clusters. A few examples will illustrate the case.

For the cluster Hodge 11 (SL 868) Walker (1979) estimated an age of 0.6 Gyr, having interpreted a group of blue stars at $B - V \simeq 0.0$ and $V \simeq 19$ as the termination of the main sequence. Conversely, Stryker et al. (1984) have later shown (by deep CCD photometry) that these stars actually belong to a blue HB, and that the real turnoff must be fainter than $V \simeq 23$, thus indicating an age of $\gtrsim 10$ Gyrs. For the cluster Kron 3, Hodge (1982) estimates an age of $1 \pm 0.4$ Gyr, while Rich et al. (1984) obtain 5-8 Gyr from deep CCD photometry. In this case the lower age was derived by mistaking for cluster members bright main sequence stars which actually do not belong to the cluster, but to the SMC field in which the cluster is embedded. Deep photometry and careful statistical decontamination of field stars are then necessary prerequisites for the evaluation of reliable ages from the main sequence turnoff.

The luminosity of core helium burning giants (hereafter $L_{HeBG}$) is a function of stellar mass (and composition), and then of cluster age. The age of a cluster can then be estimated by comparing the observed $L_{HeBG}$ to a theoretical $L_{HeBG}$-age relation. This method has been recently adopted by Flower et al. (1983) and Flower (1984), and applied to several clusters in LMC (H 11, NGC 1783, 1868, 2121, 2209, and 2231). For all these clusters Flower derives an age less than 0.8

Gyr, including H11 for which Stryker et al. have established an age at least ten times larger. Indeed, this cluster dating method cannot work for all clusters, as $L_{HeBG}$ *is not* a monotonic function of age, contrary to what is implicitly assumed by Flower. For stars more massive than $M_{HeF}$ (~2 $M_\odot$) $L_{HeBG}$ decreases with decreasing mass (increasing age, cf. Flower's Figure 3). This trend, however, does not continue for stars less massive the $M_{HeF}$, which develop electron-degenerate helium cores. In fact, core degeneracy forces the helium core to grow bigger in mass before helium ignition (helium flash) terminates the RGB phase. Actually, in stars slightly more massive than $M_{HeF}$ helium ignites non-degenerately in a low mass helium core (of ~0.3 $M_\odot$, cf. Sweigart and Gross 1978), while for $M \lesssim M_{HeF}$ the core grows to 0.45-0.50 $M_\odot$ before helium ignition. Correspondingly, thanks to their more massive He cores, during the subsequent helium-burning phase low mass stars ($M < M_{HeF}$) are brighter than intermediate mass stars in the range $M_{HeF} < M \lesssim M_{HeF} + 0.5\ M_\odot$. In conclusion, $L_{HeBG}$ is a multivalued function of age, which actually flattens for decreasing M, when $M < M_{HeF}$. Therefore, Flower's method is applicable only to intermediate age clusters (following the definition in Section 2.3), and given only the cluster $L_{HeBG}$ does not allow us to distinguish whether the cluster is intermediate age, $t < t(M_{HeF})$, or old, $t > t(M_{HeF})$. This ambiguity can only be eliminated by looking at other details of the cluster HR diagram. Indeed, intermediate age clusters should contain very few RGB stars, while in old clusters the RGB should be well populated up to the flash luminosity ($\log L/L_\odot \simeq 3.2$).

An inspection of Flower's Figure 1 reveals that in all these clusters (with the exception of NGC 1868) the RGB is well populated, and actually extends up to $\log L/L_\odot \simeq 3.2$. Even more quantitatively, stellar counts show that the number of *clump* stars is comparable to that of red giants brighter than the clump itself, as expected in old clusters, but not in intermediate age clusters. Moreover, these bright red giants cannot be interpreted as early AGB stars, simply because they are too numerous. The duration of the early AGB phase is in fact ~$10^7$ yrs (cf. Iben and Renzini 1983, and Flower's Table 1), while the duration of the core helium burning phase is 1-2 $10^8$ yrs. One then expects one early AGB star for every 10-20 clump giants, while there is actually one bright red giant for every 1-2 clump giants (cf. for instance Figure 10 in Flower et al. 1983). This further confirms that bright red giants are RGB, rather than early AGB stars, as they were implicitly interpreted by Flower. Therefore, NGC 1783, 2121, 2209 and 2231 are most likely old clusters, older than $t(M_{HeF}) \simeq 1$ Gyr, as H11 which has been discussed previously. Their actual age, however, can only be obtained from main sequence photometry, since $L_{HeBG}$ becomes insensitive to age for old clusters.

The case of NGC 1868 deserves special attention. This cluster seems to have a well populated RGB, which however terminates at $\log L/L_\odot \simeq 2.4$, rather than extending up to the flash luminosity as in the other clusters in Flower's Figure 1. This suggests that NGC 1868 may be a cluster just undergoing the RGB phase transition, a possibility we shall further consider in the next section.

Finally, the brightest AGB star dating method has been extensively used and discussed by Aaronson and Mould in a series of papers (cf. Aaronson and Mould 1985, and references therein). This method is obviously applicable only to intermediate age and old populations, as young clusters don't have any AGB stars at all. It is therefore essential to decide first whether the method can be applied, i.e. whether a cluster is younger or older than $t(M_{up})$.

Our general impression is that accurate ages are available only for very few clusters, the situation being particularly confused in the age range between $\sim 10^8$ yrs to a few Gyrs. This may rapidly improve, as CCD's are currently used by many groups to gather accurate photometry of MC cluster stars. A reliable age calibration of the integrated photometric properties of MC clusters, such as the Searle et al. (1980, hereafter SWB) or the equivalent UBV Frenk and Fall (1982) classifications, should then be within reach.

## 4.2 The Blue to Red Transition in MC Clusters

It has been known for a long time that MC clusters naturally divide in two groups, blue clusters with B-V $\lesssim$ 0.4, and red clusters with B-V $\gtrsim$ 0.6, with very few clusters having intermediate colors (cf. van den Bergh 1981). Figure 13 shows the integrated B-V color of clusters (from van den Bergh) vs. the SWB type. It is apparent that all SWB I, II, and III clusters are blue, all V, VI and VII clusters are red, and that the blue to red transition takes place within the SWB type IV cluster group. (Note that the two LMC clusters shown as smaller circles are intrinsically quite faint, and then their color is dominated by statistical effects). So the transition is apparently quite sharp, i.e. the B-V color is most sensitive to age within group IV clusters. What is the origin of the blue to red transition? Various possibilities have been suggested, including: (i) an age gap (van den Bergh 1981), (ii) the AGB phase transition (Renzini 1981a), (iii) the RGB phase transition (Gascoigne 1980, Renzini and Buzzoni 1983), and (iv) the collapse to a *clump* of the *cepheid loop*. We shall now discuss in some detail these various possibilities.

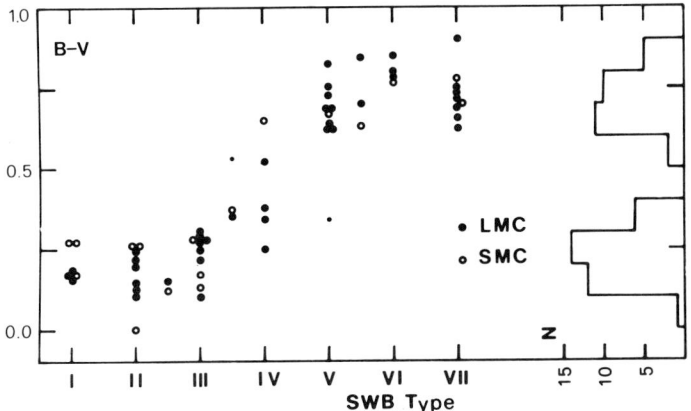

Figure 13. The integrated B-V color of LMC and SMC clusters vs the SWB type. On the right side the projected frequency histogram.

An age gap between blue and red clusters could arise either as a result of a recent burst of cluster formation, or as a result of the rapid disruption and/or fading of the young-blue clusters. Note that NGC 1866, the prototype of SWB type III clusters, is $\sim 10^8$ yrs old (Becker and Mathews 1983), and then the burst hypothesis does not look very likely.

The AGB phase transition looks more viable, NGC 1866 has an integrated luminosity of $\sim 10^6$ $L_\odot$ and should contain about 10 bright AGB stars [using an AGB lifetime of $\sim 10^6$ yrs in Eq. (14)], but it has none. This argument has been used to infer that $M_{up}$ may be as small as 5 $M_\odot$, rather than $\sim 8$ $M_\odot$ as in canonical models (cf. Renzini et al. 1985; Chiosi, this volume). There remains however the uncertainty of the effect of M-type AGB stars on the integrated B-V color (cf. Section 3.1), and therefore existing data are not conclusive in this respect. The integrated near-IR photometry of the clusters obtained by Persson et al. (1983), while confirming the importance of the AGB contribution in type V and VI clusters, don't give a clear cut answer as to whether the Blue to Red transition (in B-V) is due to the AGB phase transition. However, more clusters should be observed, since in several cases small number statistics may affect the AGB contribution.

The RGB phase transition also looks quite promising. Figure 14 shows a somewhat idealized view of what the RGB phase transition should produce in a cluster HR diagram, notably, the (sudden) development of the RGB, and the brightening of the HB clump. Note, from Figure 5, that the RGB phase transition produces an increase in the RGB contribution, which however is accompanied by a drop in the HB contribution. This latter effect is less secure, as in deriving the corresponding fuel consumptions we were forced to make *reasonable* extrapolations from existing models, in order to guess the behavior of the $F_j$'s around $M_{HeF}$. Once again, fine grids of models are required properly to map the behaviour of $F_{RGB}$ and $F_{HB}$ for stellar masses around $M_{HeF}$. This uncertainty still prevents us from making secure predictions about the size of the B-V color jump associated to the RGB phase transition, although we suspect that it should have a major effect. As we emphasized before, more models are also required in order to assess how fast the transition is.

The last possibility, i.e. the loop to clump transition, is even more difficult to assess theoretically, owing to the intimate uncertainty in the effective temperature of core helium burning models.

Ultimately, which of the above possibilities is the correct one can only be ascertained by observations. This will require good CCD color-magnitude diagrams for the clusters around the transition, i.e. SWB type III-IV-V clusters, and for all clusters with intermediate colors ($0.4 \lesssim B-V \lesssim 0.6$). Our team has already started such a project (Buonanno et al. 1985b), and at the moment reasonably good C-M diagrams have been obtained for NGC 2164 in LMC (SWB type III, B-V = 0.1) and NGC 152 in SMC (SWB type IV, B-V = 0.65), respectively the bluest type III and the reddest type IV cluster in Figure 13. Not surprisingly, the former cluster is much younger than the latter, with the main sequence extending up to $V \simeq 16$, while the turnoff is around $V \simeq 20.5$ in NGC 152. Moreover, this cluster has a well populated RGB,

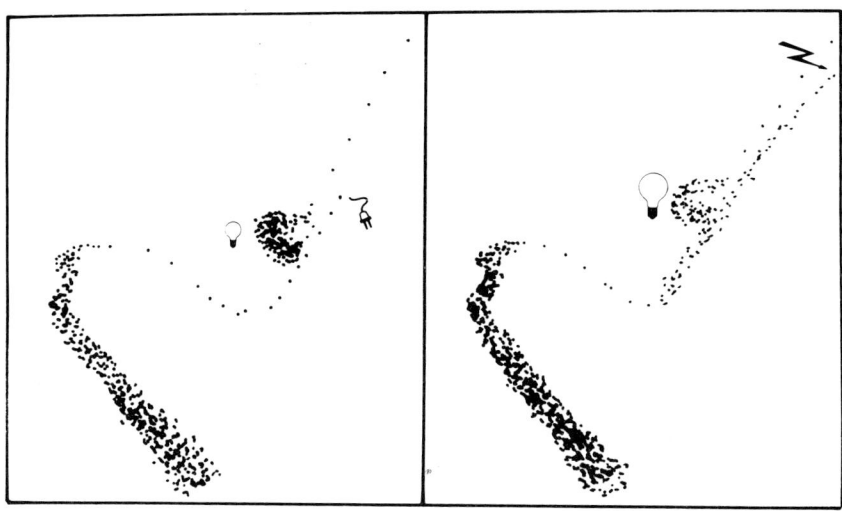

Figure 14. An artist view of the RGB phase transition. On the left panel the HR diagram of a cluster slightly younger than $t(M_{HeF})$. On the right panel the same cluster when slightly older than $t(M_{HeF})$. Note the virtual identity of the main sequences, and, by contrast, the development of the RGB. The plug indicates the point when helium ignites (non degenerately). The jagged arrow indicates the helium flash location. The growing light bulb emphasizes the brightening of the clump giants across the RGB phase transition. The AGB stays nearly the same across the transition.

while NGC 2164 does not. These data confirm that type III and IV cluster span a considerable range of ages, from $\sim 10^8$ yrs to over 1 Gyr, but are still too fragmentary to decide the issue of the blue to red transition. However, it is important to notice that existing data indicate that SWB type III clusters have not yet experienced the AGB phase transition, while the reddest type IV cluster (NGC 152) has already completed the RGB phase transition. So, apparently, both AGB and RGB phase transitions take place within the type IV group, which makes these clusters particularly interesting. A final clue comes from the case of NGC 1868 we have mentioned in the previous section. We have seen that this cluster may be just experiencing the RGB phase transition, and indeed it has an intermediate color, B-V = 0.45, which places it just in the *gap* between blue and red clusters. This may support the notion that the RGB phase transition is responsible for the bimodal color distribution of MC clusters.

## 5. POPULATION PHASE TRANSITIONS IN GALAXIES

The motivation for the rather lengthy discussion of MC clusters and phase transitions ultimately comes from a perception of potential applications to galaxies, and particularly to high redshift galaxies. Indeed, the expected *passive* spectral evolution of galaxies is very slow with look back time, and rather model dependent. On the other hand, phase transitions should occur at well defined (although not yet well determined) ages, and should then produce rather sharp changes in the SED of populations at corresponding epochs, i.e. redshifts. Obviously, the sharper such changes, the closer galaxies are to simple stellar populations, i.e. the closer to single-burst models. As already noted in the introduction, elliptical galaxies offer the best chance, if any, of detecting phase transitions.

The AGB phase transition takes place at a quite early stage in the evolution of stellar populations, i.e. at $t \simeq$ a few $10^7$ to $10^8$ yrs, and there is no chance of detecting it in high redshift ellipticals. But the RGB phase transition should take place at $t \simeq 1$ Gyr, when the bulk of star formation may be already completed, perhaps in a fraction of ellipticals. If convective overshooting is effective, $M_{HeF}$ would be somewhat decreased compared to canonical models, and then $t(M_{HeF})$ increased. In other words, convective overshooting has the effect of postponing the RGB phase transition, by perhaps as much of 1 or 2 Gyrs, and this should make it easier to detect the phase transition in high redshift ellipticals, since it should then occur at larger galaxy's age, shorter lookback time, and lower redshift. Moreover, the sharper the color jump produced by the phase transition, the shorter the star formation epoch compared to the age at which the transition occurs.

What should the RGB phase transition in ellipticals should look like, and how could it be detected? If the blue to red transition in MC clusters is really due to the RGB phase transition, we can then estimate from Figure 13 a color jump $\Delta(B-V) \simeq 0.3$, and then ellipticals may exhibit a similar change at $t(M_{HeF})$ in their rest frame color $(B-V)_{RF}$. A plot of $(B-V)_{RF}$ vs redshift should then present a break at a particular redshift $z_{PT}$, the amplitude of the break being shallower, the larger the dispersion of stellar ages within individual galaxies, and the larger the *age* dispersion of galaxies. Obviously enough, the *observational* bands in which the break may be detected will be somewhere in the near infrared, and will depend on the actual value of $z_{PT}$.

If nature was benign enough so as to have generated elliptical galaxies experiencing a measurable color jump at their RGB phase transition, then the observational determination of $z_{PT}$ will give us valuable cosmological information. Indeed, $z_{PT}$ is a function of $H_o$, $q_o$, and the redshift $z_{GF}$ at which galaxies formed, i.e. $z_{PT} = f(H_o, q_o, z_{GF})$. Therefore, determining $z_{PT}$ should provide us with a relation connecting the three other fundamental cosmological quantities.

As we have repeatedly emphasized, the detectability of the RGB phase transition in ellipticals is subject to several conditions, which may or may not be actually verified. We feel, however, that the possible existence of a sharp, model independent *feature* in the spectral evolution of galaxies is so interesting that it is worth further exploring this unique opportunity.

## 6. THE ULTRAVIOLET LIGHT OF ELLIPTICAL GALAXIES

It is now fairly well established that elliptical galaxies emit a significant fraction of their luminosity in the extreme UV (shortward of ~2000 Å), although the relative UV light ($L_{UV}/L_{TOT}$) may vary considerably from one galaxy to another (e.g. Bertola et al. 1982; Bertola, this volume). Several explanations have been put forward in order to account for the UV light of ellipticals, each appealing to a different types of hot stars, including: (i) young massive stars, (ii) hot HB stars, (iii) Post-AGB stars, and (iv) binaries of various kinds. It is worth emphasizing that a population synthesis approach can hardly distinguish among these possibilities, given the low signal-to-noise of current IUE observations. If the library stars are hot enough ($T_e \gtrsim 30{,}000$ K), each of the above stellar types fits the observations equally well (e.g. Nesci and Perola 1985). The choice among the 4 possibilities has to be made using other arguments, or has to be postponed until data of superior quality become available. We shall here briefly discuss these four possibilities, using the framework discussed in Section 2, when appropriate.

Figure 15 shows the SED of the elliptical NGC 4649, which has one of the strongest UV upturns observed so far. UV data are from Bertola et al. (1982), and near-IR photometry from Frogel et al. (1978). An integration of the SED, with some judicious extrapolation beyond 2.2 μ, gives the total bolometric luminosity ($L_{TOT}$) of this galaxy. The fluxes of three hot black bodies are also shown in the figure. Their absolute value is normalized in such a way as to give the observed flux at 1500 Å. An integration of the black body flux gives the total luminosity of the hot stellar component $L_{UV}$, and then its relative contribution $L_{UV}/L_{TOT}$, as a function of the adopted black body temperatures (see insert).

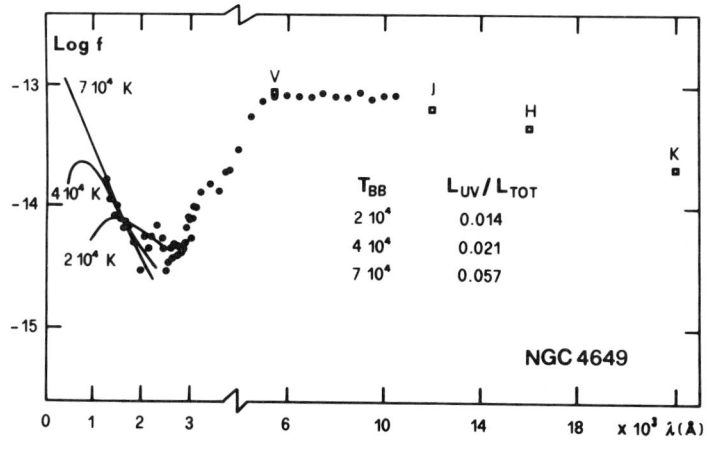

Figure 15. The spectral energy distribution (flux) of the elliptical galaxy NGC 4649. UV data from Bertola et al. (1982). IR data from Frogel et al. (1978). Three hot black bodies are made to give the observed flux at λ = 1500 Å. In the insert: the fractional luminosity emitted in the UV (including the Lyman continuum) as a function of the assumed temperature of the stars providing the observed (IUE) ultraviolet.

We can note the 20,000 K black body is too cool to fit the data shortward of 1500 Å, while the two hotter black bodies fit equally well. We can then conclude that (a) the hot stellar component must be hotter than, say, 30-40,000 K, (b) its actual temperature distribution cannot be assessed, with the available data, and (c) it contributes at least ~2% of the total bolometric luminosity of the galaxy, the precise figure depending on the unknown temperature distribution.

## 6.1 Young Stars

A recent burst of star formation can certainly account for the observed UV upturn, but this explanation has potential flaws at a closer scrutiny. Indeed, since all ellipticals observed so far exhibit a UV upturn, one is forced to conclude that some level of ongoing star formation should be present in all ellipticals, and the accumulation of lower mass stars (say, ~2 $M_\odot$) should then affect the optical and near-UV part of the spectrum. Since there is no evidence for this intermediate age component in all ellipticals, the only way of maintaining the young star hypothesis is to assume that the ongoing star formation process favours massive stars, e.g. a very flat IMF (cf. Nesci and Perola 1985). While this cannot be excluded, it nevertheless remains an *ad hoc* assumption.

## 6.2 Hot HB Stars

By analogy with galactic globular clusters, one can safely infer that ellipticals should also contain hot HB stars. The fraction of HB stars hotter than, say, 30,000 K is however expected to be a function of the actual metallicity distribution of a galaxy (once again by analogy with galactic globulars). Moreover, Nesci and Perola (1985) find a better fit of the UV upturn when using a bimodal temperature distribution for the HB stars, otherwise the UV upturn would be too shallow. This may not be completely inconceivable, as the HB location in galactic globulars appears to be non-monotonic with metallicity, and, moreover, some clusters present a gap in the HB temperature distribution (Renzini 1983, Buonanno et al. 1985a). Therefore, a continuous distribution of metallicities might produce a bimodal distribution in the temperature of HB stars, simply because evenly populated HB's (as in M3) are produced only in a restricted range of metallicities ($-1.6 \lesssim$ [Fe/H] $\lesssim -1.2$). However, the bulk of the stars in an elliptical may be much more metal rich, and an extrapolation of the trend shown by galactic globular clusters would suggest that the bulk of HB stars should be red, rather than hot. However, this extrapolation may not be legitimate, because even a modest increase (with increasing metallicity) of the RGB mass loss rate could lead to hotter and hotter HB's (cf. Renzini 1981b), as the metallicity increases from, say, that of 47 Tuc ([Fe/H] $\simeq -0.8$) up to that of most metal rich stars in ellipticals ([Fe/H] $\simeq +1.0$ ?). So, we cannot exclude this possibility, and perhaps the observation of super metal-rich HB stars in the galactic bulge can be used to check it.

Anyway, let us return to our prototype galaxy NGC 4649. Here $L_{UV}/L_{TOT}$ is larger than ~0.02, and since the HB contributes ~10% of the total light in an old population (cf. Section 2.5), this implies that more than ~20% of HB stars in this galaxy would be hotter than ~30,000 K, if hot HB stars were the only source of UV photons. Whether this is realistic or not is another question. Finally, we wish to comment on the possible correlation of the (1550-3100) UV color with the Mg-index (Faber 1983). This correlation may imply an increase of $L_{UV}/L_{TOT}$ with increasing average metallicity, a trend which again could be explained only if the HB were to get hotter with increasing metallicity.

## 6.3 Post-AGB Stars

There is no doubt that ellipticals must contain hot post-AGB stars, since dying stars necessarily go through this stage before becoming degenerate dwarfs. From the fuel consumption theorem, the relative contribution of hot post-AGB stars is proportional to the amount of fuel (mainly hydrogen) which is burned during this stage. Stars leave the AGB and venture through the Post-AGB only after having ejected most of their H-rich envelope. Hence, they retain very little fuel to burn, but they do so at such high effective temperatures that most of the released nuclear energy goes into high-value UV photons.

Figure 16 shows the mass of the residual H-rich envelope $M_N^e$ when stars reach 30,000 K, as a function of the mass of post-AGB models $M_f$. Note the extreme dependence of $M_N^e$ on $M_f$. Paczynski's (1971) models give $M_N^e \simeq 4 \cdot 10^{-5} M_f^{-7.33}$ (cf. Renzini 1981a), both masses being in solar units. Schönberner's (1983) models give slightly smaller values for $M_f \gtrsim 0.565 M_\odot$, but a much steeper slope at lower masses, i.e. $M_N^e \simeq 10^{-10} M_f^{-30.8}$ (!!), with $M_N^e$ going up by a factor of 3, as $M_f$ decreases by only ~3%! Note, however, that $M_N^e$ must flatten out for $M_f$ smaller than ~0.5 $M_\odot$. Moreover, models indicate that most of this residual envelope (~90%) is burned during the hot Post-AGB, and then the fuel consumption is ~$X^e M_N^e$, $X^e$ being the envelope hydrogen abundance ($X^e = 0.7$ in both Pazcynski and Schönberner models). Therefore, the $L_{UV}/L_{TOT}$ ratio is an extremely sensitive function of $M_f$ (note that $L_{UV} \propto X^e M_N^e$), and thus of age (cf. Figure 5), since $M_f$ is a decreasing function of age. We shall return to this point later.

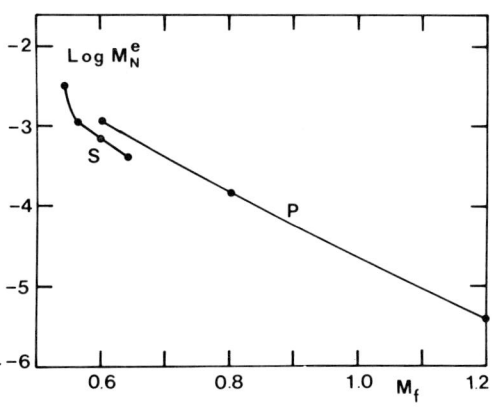

Figure 16. The mass of the residual hydrogen rich envelope, when post-AGB stars reach ~30,000 K, as a function of their actual (final) mass, both in solar units. Three models of Paczynski (1971) and four models of Schönberner (1983) are shown.

The actual flux at, say, 1500 Å will be even more sensitive to $M_f$ than $L_{UV}$ itself, since decreasing $M_f$ not only increases $M_N^e$, but the stellar effective temperature at which this fuel is burned *decreases*, i.e. the fraction of the total UV which is sampled by IUE increases. This latter effect can be appreciated by looking at the actual evolutionary tracks displayed in Figure 17. This extreme sensitivity to $M_f$ of the Post-AGB ultraviolet contribution may help to account for the great spread in the UV output among ellipticals: very small galaxy to galaxy differences in the actual distribution of $M_f$ (particularly around the smallest $M_f$ values) will indeed produce sizable differences in the Post-AGB output, particularly in the 1000-2000 Å range. Note that some small (~ few $10^{-2}$ $M_\odot$) dispersion in $M_f$ within a galaxy is expected to arise from its metallicity dispersion.

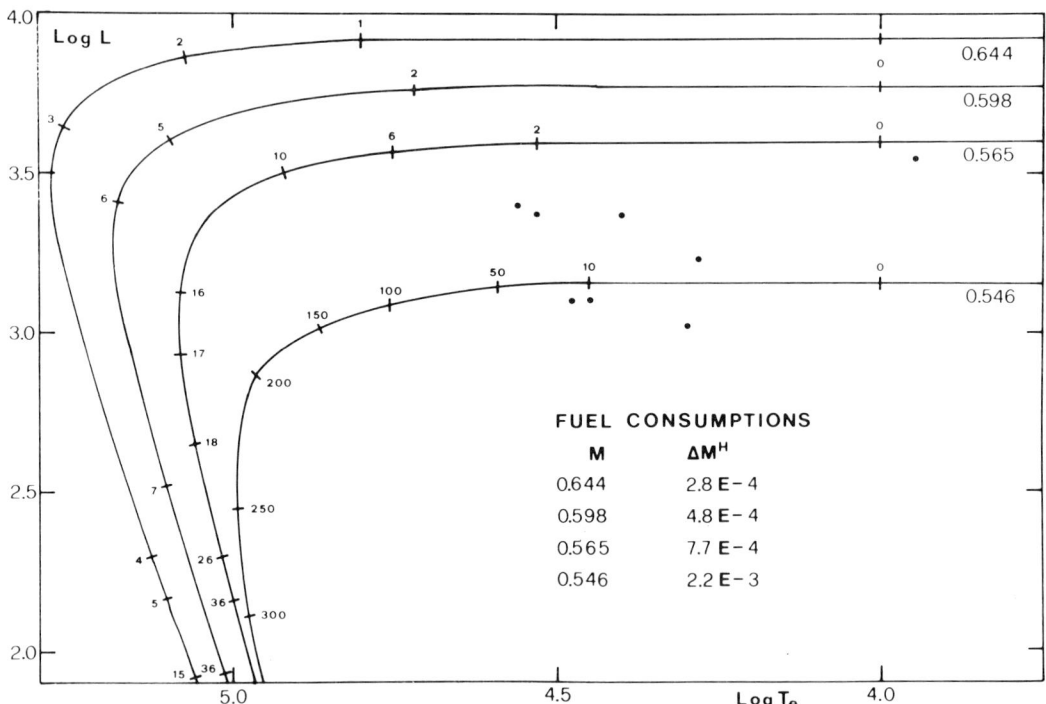

Figure 17. *The post-AGB tracks of Schönberner (1983), for the indicated post-AGB masses. Time marks (in 1000 years) give the time elapsed since models reach 10,000 K. Filled circles show the location of known post-AGB stars in galactic globular clusters (from de Boer 1985). In the insert, the corresponding hydrogen fuel consumptions in $M_\odot$ units.*

The Faber correlation between (1550-3100) and the Mg-index could be explained in this way, if $M_f$ is a decreasing function of metallicity. However, galactic globular clusters show the opposite trend, i.e. $M_f$ increasing with [Fe/H] (cf. Frogel 1983), and this trend seems to continue also for super metal-rich stars in the galactic bulge (cf. Whitford, Frogel, this volume). This is obviously derived from the observed increase in the maximum AGB luminosity (a function of $M_f$) as metallicity increases.

Then the natural question is: what's the right $M_f$? Indeed, theory can hardly predict the actual value of $M_f$ with better than ~1% accuracy as is required. Observations can help in this regard, and in Figure 17 the known Post-AGB stars in galactic globular clusters (de Boer 1985) are compared to Schönberners tracks. This comparison allows some meaningful conclusions to be drawn: (i) a value of $M_f$ around 0.53-0.55 $M_\odot$ seems appropriate for galactic globulars (in agreement with the inference from the AGB extension, cf. Renzini 1981b), and (ii) *known* Post-AGB stars should represent only the low-temperature end of the actual distribution. Indeed, these stars were discovered from optical observations, which undoubtedly would have missed equally bright (in bolometric) but much hotter stars, with very large bolometric corrections. Note that the hottest stars in Figure 17 have temperatures around 35,000 K, while tracks extend to ~100,000 K, and a much longer time is actually spent at high temperatures. Therefore, galactic globulars should contain many still undetected hot Post-AGB stars, and Eq. (14) helps in making some predictions. The 0.546 $M_\odot$ model fades to 100 $L_\odot$ in ~0.5 million yrs (cf. Figure 17), and then a typical cluster ($10^5$ $L_\odot$) should contain, on average (!), one Post-AGB star brighter than 100 $L_\odot$. Actually, it is better to say that 100 clusters (~$10^7$ $L_\odot$) should contain ~100 such stars. The "Ultraviolet Imaging Telescope" (UIT) will easily observe all these stars (see, for instance, Bohlin et al. 1983), allowing a stringent comparison of theoretical Post-AGB tracks with observations, and a direct measurement of the $L_{UV}$ (Post-AGB)/$L_{TOT}$ ratio for an old stellar system such as the system of globular clusters. Similarly, UIT and/or ST ultraviolet imaging of the galactic and M31 bulges should allow the same tests for old, metal rich populations.

Finally, we conclude this section by noting that with $M_f \simeq 0.53 - 0.55$ $M_\odot$, the Post-AGB energy output is of the order of ~1% of the total luminosity, quite consistent with the observed $L_{UV}/L_{TOT}$ ratio in ellipticals.

## 6.4 Binaries

When dealing with population synthesis problems, one should never forget that some 50% of all stars are probably members of binary systems, and that a good fraction of binary systems form close enough so as to allow Roche-lobe overflow(s) in the course of their evolution. Binaries form such a huge zoo, that we are forced to restrict our attention to some of the situations in which one of the two components become hot enough so as to produce UV photons. We shall just list a few cases: (i) a primary component fills its Roche-lobe while climbing the RGB. After shedding most of the envelope the primary becomes a Helium WD, evolving along a *Post-RGB* track (similar to Post-AGB tracks) and burning (at hot effective temperatures) a residual envelope which can be considerably more massive than in the case of single Post-AGB stars. Therefore, one such star may give as many UV photons as do several single stars of the same initial mass during their whole Post-AGB. (ii) In a system with a somewhat wider initial separation, the core helium flash may precede the completion of envelope removal from the primary, which will then become a very hot HB star (e.g. Renzini et al. 1977). (iii) Finally, when the turn arrives for secondary components to fill their Roche-lobe, accretion onto the primary (now either a He or a CO white dwarf) may occur at such a rate as to ensure quasi-static burning of the accreted material (e.g. Iben 1982). Just one of these objects may then produce (over its lifetime) as many UV photons as several hundred single Post-AGB stars. These systems may be SNI progenitors; if so their UV contribution should correlate with the SNI rate (Greggio and Renzini 1983), although it is worth mentioning that the currently more fashionable SNI progenitors (i.e. double-degenerate systems, cf. Iben and Tutukov 1984) would deliver many fewer UV photons.

In conclusion, binaries offer a variety of possibilities, but the current uncertainties in the distribution functions of the initial binary parameters, and in the evolution of such systems, prevent any stringent prediction about their actual contribution to the UV light of ellipticals.

## 6.5 The Predicted Evolution of the UV Light With Look Back Time

In the previous sections we have discussed pros and cons of the various candidate UV producers. Here we show that perhaps the most elegant way of solving the problem is to look at the variation of $L_{UV}/L_{TOT}$ with look back time (with ST, of course).

Indeed, the contribution of each of the four candidates is expected to change in quite different ways with look back time (LBT). More specifically:

(i) Young Stars. The $L_{UV}/L_{TOT}$ contribution from young stars should go as the rate of star formation. In a simple-minded model, the SFR may be proportional to the rate of mass return $\dot{M}(t)$, cf. Section 2.7. Therefore $L_{UV}/L_{TOT}$ [$\propto \dot{M}(t)$], should *increase* with LBT as does $\dot{M}(t)$.

(ii) HB Stars. It is well known that the HB location moves to the red when age is decreased (cf. Rood 1973, Renzini 1977). A few billion yrs ago the (now) blue HB of a cluster such as M13 was as red as that of 47 Tuc! This is to say that the development of blue HB's is quite a recent event, which started just a few billion years ago. Therefore, the HB contribution to the UV is predicted to *disappear* within a LBT of 2-3 Gyrs, i.e. at redshifts as small as ~0.2. Moreover, since the HB moves to the red with increasing LBT, the rest frame UV spectrum should *soften* with increasing redshift.

(iii) Post-AGB Stars. As shown in Figure 5, and discussed in Section 6.3, $L_{UV}/L_{TOT}$ is an extremely sensitive function of $M_f$. Since $M_f$ increases with LBT (as does $M_{TO}$), $L_{UV}/L_{TOT}$ should drop quickly, and in a few Gyrs back the UV excess of ellipticals should virtually disappear (again at redshifts as small as ~0.2-0.3). Contrary to case (ii), sampled Post-AGB stars become hotter with LBT (cf. Figure 17), and then the UV spectrum should fade but get *harder* with increasing redshift.

(iv) Binaries. This case is a bit more difficult. The UV light from binaries should be proportional to the death rate of binary components, and then roughly proportional to the evolutionary flux of the population b(t), as defined in Section 2.1. Then, since $L_{UV}/L_{TOT} \propto$ b(t), this UV contribution should increase with galaxy's redshift (cf. also Greggio and Renzini 1983).

In summary: In cases (i) and (iv) the UV output increases with redshift. In cases (ii) and (iii) it decreases, getting either softer (ii) or harder (iii). The only remaining ambiguity is between cases (i) and (iv), and other tests can be envisaged to distinguish between them. In conclusion, ST observations of ellipticals at low redshifts ($z \lesssim 0.3$) will not only solve the problem of the UV producers, but are also likely unambiguously to detect the only sizable evolutionary effects at low redshifts. Indeed, in cases (ii) and (iii) the UV is the most rapidly changing part of the SED of a low redshift elliptical, while in other spectral ranges the evolution is very slow (cf. Bruzual, this volume).

## 7. CONCLUSIONS: THE NEXT FEW STEPS

In this section we summarize what we think are the most urgent theoretical and observational programs, which may sharpen our understanding of the spectral evolution of galaxies, and improve the performances of the stellar evolution clock as a tool for measuring cosmological times.

The role and characteristics of the AGB and RGB phase transitions need to be clarified, in particular pinpointing the precise age at which they occur, and assessing the size of the associated changes in the SED. To do so, CCD and near-IR observations of MC clusters are particularly appropriate, especially for SWB type IV clusters. Integrated UV spectroscopy (with IUE) for as many as possible of these clusters would also be useful. On the theoretical side, many new stellar evolutionary tracks around $M_{HeF}$ are required, as well as the synthetic SED's which could be constructed using these tracks.

Concerning the issues discussed in Section 6, a survey of hot Post-AGB stars in the galactic globulars and in the galactic bulge (with UIT) would be particularly valuable. Again, many Post-AGB evolutionary tracks should be computed, particularly for small $M_f$ values ($M_f \simeq 0.52 - 0.55$).

In modelling SED's one should pay particular attention to the wings of the SED, i.e. to the near-UV and near-IR. Indeed, as we emphasized, the wings are much more sensitive to age, than the optical body of the SED.

Acknowledgments. Over the past few years we have interacted about these matters with so many astronomers that it would be really difficult to recollect them all. Anyway, we want to particularly mention here Marc Aaronson, Roberto Buonanno, Art Code, Carlo Corsi, John Danziger, Sandra Faber, Jay Frogel, Flavio Fusi Pecci, Jay Gallagher, Laura Greggio, Icko Iben, David Koo, Richard Kron, Jeremy Mould, Colin Norman, Bob O'Connell, Luisa Pigatto, Michael Rich, Bob Rood, Allan Sandage, Leonard Searle, Jo Silk, Allen Sweigart, Massimo Tarenghi, and Albert Whitford. We wish also to remember with particular devotion Beatrice Tinsley, who pioneered population synthesis studies and first conceived several of the concepts we have developed in this paper. One of us (A. R.) is grateful to the Space Telescope Science Institute for its hospitality during the period in which most of this paper was written and set up. Finally we are particularly indebted to Michael Fall for his critical reading of the manuscript and to Ms. Dorothy Schlogel for her invaluable editorial cooperation.

REFERENCES

Aaronson, M., Mould, J. R. 1985, Ap. J., **288**, 551.
Barbaro, G., Pigatto, L. 1984, Astron. Astrophys., **136**, 365.
Becker, S. A. 1981, Ap. J. Suppl., **45**, 475.
Becker, S. A., Iben, I. Jr. 1979, Ap. J., **232**, 831.
Becker, S. A., Mathews, G. J. 1983, Ap. J., **270**, 155.
Bertola, F., Capaccioli, M., Oke, J. B. 1982, Ap. J., **254**, 494.
Bohlin, R. C., Cornett, R. H., Hill, J. K., Smith, A. M., Stecher, T. P., Sweigart, A. V. 1983, Ap. J. Lett., **267**, L89.
Buonanno, R., Corsi, C. E., Fusi Pecci, F. 1985a, Astron. Astrophys., **145**, 97.
Buonanno, R., Corsi, C. E., Fusi Pecci, F., Renzini, A. 1985b (in preparation).
Copeland, H., Jensen, J. O., Jørgensen, H. E. 1970, Astron. Astrophys., **5**, 12.
de Boer, K. S. 1985, Astron. Astrophys., **142**, 321.
Faber, S. M. 1983, Highlights of Astron., **6**, 165.
Ferrario, L., Greggio, L., Renzini, A. 1985 (in preparation).
Flower, P. J. 1984, Ap. J., **278**, 582.
Flower, P. J., Geisler, D., Hodge, P., Olszewski, E., Schommer, R. 1983, Ap. J., **265**, 15.
Frenk, C. S., Fall, S. M. 1982, M.N.R.A.S., **199**, 565.
Frogel, J. A. 1983, Ap. J., **272**, 167.
Frogel, J. A., Persson, S. E., Aaronson, M., Matthews, K. 1978, Ap. J., **220**, 75.
Gascoigne, S. C. B. 1980, Star Clusters, ed. J. E. Hesser (Dordrecht: Reidel), p. 305.
Greggio, L., Renzini, A. 1983, Astron. Astrophys., **118**, 217.
Hodge, P. 1982, Ap. J., **256**, 447.
Hodge, P. 1983, Ap. J., **264**, 470.
Iben, I. Jr. 1966, Ap. J., **143**, 516.
Iben, I. Jr. 1982, Ap. J., **253**, 244.
Iben, I. Jr., Renzini, A. 1983, Ann. Rev. Astron. Ap., **21**, 271.
Iben, I. Jr., Renzini, A. 1984, Phys. Reports, **105**, 329.
Iben, I. Jr., Tutukov, A. V. 1984, Ap. J. Suppl., **54**, 335.

Mengel, J. G., Sweigart, A. V., Demarque, P., Gross, P. G. 1979, Ap. J. Suppl., **40**, 733.
Nesci, R., Perola, G. C. 1985, Astron. Astrophys., **145**, 236.
Nomoto, K. 1984, Ap. J., **236**, 430.
Paczynski, B. 1971, Acta Astronomica, **21**, 417.
Persson, S. E., Aaronson, M., Cohen, J. G., Frogel, J. A., Matthews, K. 1983, Ap. J., **266**, 105.
Renzini, A. 1977, Advanced Stages of Stellar Evolution, ed. P. Bouvier and Maeder (Geneva), p. 149.
Renzini, A. 1981a, Ann. Phys. Fr., **6**, 87.
Renzini, A. 1981b, Effects of Mass Loss on Stellar Evolution, ed. C. Chiosi and R. Stalio (Dordrecht: Reidel), p. 319.
Renzini, A. 1981c, Astrophysical Parameters for Globular Clusters, ed. A. G. D. Philip and D. S. Hayes (Schenectady: L. Davis), p. 72.
Renzini, A. 1983, Mem. Soc. Astron. It., **54**, 335.
Renzini, A. 1984, Observational Tests of Stellar Evolution Theory, ed. A. Maeder and A. Renzini (Dordrecht: Reidel), p. 21.
Renzini, A. 1984, Astron. Express, (in press).
Renzini, A., Bernazzani, M., Buonanno, R., Corsi, C. E. 1985, Ap. J. Lett., (in press).
Renzini, A., Buzzoni, A. 1983, Mem. Soc. Astron. It., **54**, 739.
Renzini, A., Mengel, J. G., Sweigart, A. V. 1977, Astron. Astrophys., **56**, 369.
Renzini, A., Voli, M. 1981, Astron. Astrophys., **94**, 175.
Rich, R. M., Da Costa, G. S., Mould, J. R. 1984, Ap. J., **286**, 518.
Rood, R. T. 1973, Ap. J., **184**, 815.
Scalo, J. 1985, Fundamentals Cosmic Phys., (in press).
Schönberner, D. 1983, Ap. J., **272**, 708.
Searle, L., Wilkinson, A., Bagnuolo, W. G. 1980, Ap. J., **239**, 803.
Stryker, L. L., Nemec, J. M., Hesser, J. E., McClure, R. D. 1984, Structure and Evolution of the Magellanic Clouds, ed. S. van den Bergh and K. S. de Boer (Dordrecht: Reidel), p. 43.
Sweigart, A. V., Gross, P. G. 1976, Ap. J. Suppl., **32**, 367.
Sweigart, A. V., Gross, P. G. 1978, Ap. J. Suppl., **36**, 405.
Tinsley, B. M. 1980, Found. Cosmic Phys., **5**, 287.
van den Bergh, S., 1981, Astron. Astrophys. Suppl., **46**, 79.
Walker, M. F. 1979, M.N.R.A.S., **186**, 767.

## DISCUSSION

FROGEL. I disagree with your statement that the integrated spectrum of a SSP model is independent of the IMF. In models calculated in papers by Aaronson, Cohen, Mould and myself, the broad band colors such as U-V, V-K change very strongly as you go from a Salpeter IMF to a dwarf dominated IMF.

MOULD. To support Jay's remarks, V-K color is sensitive to IMF slope in the models of Aaronson et al. (1978). With $n \simeq M^{-4}$, M dwarfs dominate the light at 2 μ.

RENZINI. Perhaps I wasn't too clear. Certainly, if M dwarfs dominate, the spectrum must be sensitive to the IMF. What matters here is how the slope of the ZAMS mass-luminosity relation (a) compares to the IMF slope (s), at the low mass end. If $a - s + 1$ is positive, then the spectrum is independent of s. But as $a - s + 1$ approaches zero, and then becomes positive, one has a sort of runaway with M dwarfs dominating the light. So I think we agree. However, I think your CO index observations and the Wing-Ford band strengths indicate that in real ellipticals M dwarfs don't dominate, and then $a - s + 1$ is indeed positive in these galaxies.

O'CONNELL. Where in your % light diagram are the red supergiants which appear in clusters like h and χ Persei at ages of a few $10^7$ yrs?

RENZINI. They are presumably core helium burning stars, so they are labelled as HB. However, part of this stage may also be spent as a blue supergiant.

O'CONNELL. You emphasized the theoretical uncertainties in the evolutionary tracks. It's important that the empirical synthesis approach on clusters can help provide information on the deficiencies in the theory and perhaps determine the free parameters in the theory to better than the 30% error you mentioned.

SILK. If you want to use an evolutionary synthesis code to help search for evidence of young galaxies, it would be important to have a calibration against systems that are young and much more metal-poor than LMC, for example such as IZw18.

AARONSON. This point will come up again I'm sure, but one type of cluster you don't have to check against the evolutionary codes are metal rich clusters of varying ages. This sort of population may be very important for synthesizing integrated galaxy light.

RENZINI. I fully agree. This is why metal rich stars in the galactic bulge are so important.

DI FAZIO. Can you comment on the possibility of synthesizing a "post-protogalaxy", once a global evolutionary hydrodynamical model gives you the calculated IMF and the metallicity distribution obtained at virialization?

RENZINI. That would be fancy! Ultimately this is what one would like to do: an encyclopedic approach. But I'm not sure the state of the art is ripe yet.

AARONSON. Is there a "canonical" mass that is the major contributor to the 10 billion years old light in your constant star formation model?

BUZZONI. If we define this "canonical" mass $M_{can}$ (at a given total age T) as

$$M_{can} = T^{-1} \int_{M_{inf}}^{M_{max}} |dt/dM_{TO}| \phi(M_{TO}) dM_{TO}, \text{ with}$$

$$\phi(M_{TO}) = \int_{M_{inf}}^{M_{TO}} L(M) \psi M dM / \int_{M_{inf}}^{M_{TO}} L(M) \psi dM,$$

then one finds that in general $M_{can}$ is of the order of the turnoff mass of the oldest simple population, or $M_{can} \simeq 1 M_\odot$ for T = 10 Gyrs.

RENZINI. Yes, these stars are not so bright, but have been accumulated over a long period of time.

CAPUZZO DOLCETTA. I would like to emphasize two points. There are many papers dealing with population synthesis, and they usually succeed in reproducing colors in various wavelength ranges. However, only very seldom they give fluxes in the whole spectral range, and do a complete comparison with the observations, which I think is necessary to say something reliable about the synthesized stellar systems (see e.g. Altamore et al., <u>A. & A.</u> 1983, 1984). Yet, another fundamental check is the correspondence of the fractional contributions of the various stellar types to the total emitted power, as these contributions can easily be estimated by theory. With regard to the synthesis of MC clusters I agree with you on the importance of using them as calibrators of stellar population programs, but I know (as I've been working on this project for some time) that this is quite difficult because of the lack of complete stellar evolutionary tracks for the relevant masses and compositions.

BRUZUAL. You look very pessimistic about applying population synthesis techniques to external galaxies, basically because we still do not understand nearby stellar systems. However, despite the little we know, some "satisfactory" results have been obtained in applying these techniques to galaxies. Do you think this is pure chance? Could you comment on this?

RENZINI. Maybe I gave a wrong impression, I don't want to be pessimistic! Concerning your specific question, I have the feeling that it may actually be "too" easy to get some sort of agreement insofar only the optical part of the spectrum is synthesized. Putting together a main sequence and a standard red giant branch should not be that wrong after all. But it is in the wings of SED that the deficiencies of current models will most likely become apparent. And, as I tried to emphasize, the UV and near-IR wings are expected to be much more sensitive to relevant parameters (e.g. age) than the optical core of the SED.

FROGEL. What was the IMF for the spiral (constant star formation rate) models? How sensitive is the luminosity evolution of spirals to the IMF slope?

RENZINI. It was the Salpeter IMF, s = 2.35. We did not compute models with other slopes, but I suspect that the luminosity evolution of spirals should be <u>less</u> sensitive to IMF, compared to single burst models.

AARONSON. Have Christian and Schommer found a B-V jump in their M33 cluster sample, where they also report a wide range of cluster ages?

RENZINI. No, they didn't. They have blue and red clusters, as well as clusters with intermediate colors. If this lack of bimodality is due to inaccurate photometry I don't know.

FROGEL. If you subtract <u>all</u> the AGB stars from the integrated color of the MC clusters, you see the jump even in the JHK colors.

RENZINI. Oh, very interesting. This would favour the RGB phase transition as producing the jump.

SILK. What is the metallicity range between blue and red clusters? In order to apply the color information to synthesized color evolutionary models, will it not be necessary to disentangle age from metallicity?

MOULD. Isn't true to say that the B-V discontinuity with age is partly an artifact, or at least is artificially enhanced, by the low metallicity of the older MC clusters? When you add the ultraviolet dimension, you see that B-V turns around at SWB class VI.

RENZINI. Oh yes, that's the bending end of the SWB fishing hook! So metallicity really drops in SWB types VI and VII, but the color transition takes place within type IV, and then these latter clusters should not exhibit a very wide range of metallicities. The effect of the drop in metallicity is probably that of making the B-V range of red clusters narrower than it would have been, were all the clusters at the same metallicity. However, I agree with Jo's comment.

KOO. The recent (preprint 1985) work R. Elson and M. Fall shows that the bimodal distribution in B-V of MC clusters is purely an artifact of the degeneracy in B-V of the UBV two color plot. Of more importance than the above effect, to any claim for abrupt color changes with small changes in age, the SWB type IV group may well include a very wide range in age, thus also artificially enhancing any apparent color change with age.

RENZINI. I agree with your second point completely. Existing CM diagrams for type IV clusters are still too few, and then one cannot exclude that blue type IV clusters are $\sim 10^8$ yrs old, and red ones $\sim 10^9$ yrs old. Concerning your first point, I've not seen the paper you are referring to, but it seems to me that if there's a degeneracy, this is in the SWB Q's. In other words, I think that if one cluster has B-V = 0.3 and another B-V = 0.6, then there must be a significant difference in their stellar content, even if their Q's are the same. What I tried to emphasize, is that it would be very interesting to figure out what this difference is.

FROGEL. In IR colors you also see big ranges or gaps which don't exist in optical colors or SWB's indices. These gaps, and lacks of them in other colors, provide, as was said, important diagnostics.

DRESSLER. Armando Manduca has done just what you suggested in trying to fit Searle's observations of MC clusters. He used modified Yale evolutionary tracks and was able to reproduce visual colors, Balmer and metal line strengths for clusters with a wide range of ages.

RENZINI. I'm not familiar with Manduca's work; however, Yale tracks are not so finely spaced (in mass) as to properly map the RGB phase transition. Also, they don't include the AGB.

CAPUZZO DOLCETTA. Yes, I computed synthetic colors myself, using the Yale tracks (Mem. S. A. It. 1984, in press). There is indeed a change in slope around B-V = 0.5 in the cluster evolutionary track in the integrated V-(B-V) diagram, and which is due to the appearance of a well populated red giant branch. This may be relevant to the B-V gap, but no firm conclusions can be drawn until we have a detailed grid of stellar models around the critical mass for degenerate He cores. This will then allow to estimate the actual rate of color change around B-V = 0.5.

DRESSLER. Let me change the subject. It would appear that direct UV imaging by ST of ellipticals as far away as Virgo will immediately discriminate, for example, between hot, young stars and a hot horizontal branch, merely by resolving the former.

RENZINI. Oh yes, and there's more that ST can do, for example UV galaxy photometry at moderate redshifts (z from zero up to 0.3-0.4). This should allow disentangling the other (old stars) possibilities. As I said, the UV will likely show evolutionary effects in low redshift ellipticals.

# ADVANCEMENTS IN THE STELLAR EVOLUTION THEORY: THE ROLE OF CONVECTIVE OVERSHOOTING ALL ACROSS THE HR DIAGRAM

Cesare Chiosi
Istituto di Astronomia
Universita' di Padova
Vicolo Osservatorio 5,
35122, Padova, Italy

ABSTRACT

In this paper we present recent calculations of stellar structure in which the effects of convective overshooting are taken into account during the core H- and He-burning phases of stars in the mass range 1.3 $M_\odot$ to 100 $M_\odot$. In addition to this, we briefly review the effects of mass loss in luminous stars of all spectral types, and in red giant and asymptotic giant branch stars. Finally, the effects of the novel cross section for the $^{12}C(\alpha,\gamma)^{16}O$ nuclear reaction are also illustrated.
The main purpose of this review resides however in lending convincing support to the idea that convective cores of real stars are greater than commonly supposed in classic models. To this aim, several observational facts that could not be explained by classic models are reanalyzed in the light of the new ones. Since a much better agreement between theory and observations is now possible, we are inclined to conclude that convective overshooting may be of paramount importance in theories of stellar structure and that convective cores in real stars ought to be larger by approximately one pressure scale height than predicted by classic models. The implications of these new stellar models in theoretical models of photometric evolution of galaxies are numerous and easy to foresee.

INTRODUCTION

Normal galaxies (with little or no evidence of non thermal emission) are mostly made of stars of different mass, chemical composition and age, which contribute in various proportions to the building up of the integrated spectral energy distribution. The stellar content of a galaxy however is not constant with time, but it continuously varies, firstly because stars evolve and die and others, less massive, take over in the production of the integrated energy emission, secondly because new stars may be generated. These facts make the integrated light emitted by a galaxy at any age a result of the history of star formation. A necessary step to understand and to model the spectral evolution of a galaxy is, among others, the detailed knowledge of physical properties of stars all over their life. The more we refine this information, the more successful the theoretical models of spectrophotometric evolution of galaxies will be.

The question then arises whether current theories of stellar evolution are in full agreement with observational data, in particular with those provided by the advent of highly sensitive instrumentation.

Purposes of this paper are the critical discussion of several observational facts, whose interpretation by means of classic stellar models is difficult, and the presentation of recent advancements in the theory of stellar evolution aimed at improving upon those disagreements. It goes without saying that, if our goal is achieved, only these new stellar models ought to be used in the construction of spectro photometric models of galaxies.

The amount of observational data available today on star clusters of different ages and chemical abundances makes it possible to study stellar evolution from an empirical point of view. Even if the general properties of stellar evolution are known from long time, still it has become evident over the recent years that many points of contradiction between classic theory and current observation exist implying a deep revision of stellar models. In this paper we will limit ourselves to consider only those observational facts that led us to revise the current models of massive and intermediate mass stars. Populous young clusters of the Magellanic Clouds (LMC and SMC) are particularly suited to this purpose as they are sufficiently populated throughout the various evolutionary stages, thus allowing us to compare them safely with theoretical predictions, even for the very short lived evolutionary phases. On the contrary, this is not easily feasible with galactic clusters as they are known to contain much fewer stars. Therefore, the comparison can be made only collecting data for many individual clusters, and constructing composite HR diagrams. Since many reviews exist describing the properties of massive (Chiosi 1982a,b; Maeder 1984; Chiosi and Maeder 1985) and intermediate mass stars (Iben 1974; Iben and Renzini 1983, 1984), we will not go into any detail relative to classic models. On the contrary, we will focus on four main lines of work which, if they will receive general consensus, may deeply change the classic scenario. In fact, over the past years, our appreciation of convective overshooting, of mass loss by stellar wind, of the rate of several important nuclear reactions, and of stellar opacity has been changed either by improved observational information, laboratory experiments an/or theoretical considerations. It goes without saying that not all of the above physical processes are known with the same degree of confidence. In fact, while the occurrence of mass loss is indicated by observations, despite the uncertainty still affecting the determination of mass loss rates, only very indirect arguments can be put forward to lend support to the existence of convective overshooting. Furthermore, if the rates of nuclear reactions are supported by laboratory experiments, the estimate of the true stellar opacity, particularly in the range of temperature and density correspondent to the ionization of CNO and heavier elements is still controversial and all modifications to current opacities advanced insofar have received neither general consensus nor clearcut confirmation or disproval by theoretical calculations. Accordingly, the evolutionary scenarios, in which one or more of the above physical processes have been incorporated, are reflective of the underlying uncertainty. In fact, while the effects of mass loss have been thoroughly investigated, those of convective overshooting all over the life of a star now begin to be assessed, those of opacities different from those currently in use are still in an exploratory stage.

In the following, we will mainly concentrate on convective overshooting, firstly describing the far reaching consequences of this deeply seated phenomenon, secondly searching as many as possible observational facts, which, discrepant with classic models, may now be interpreted by the new ones. The great advantage shown by the new models over the old ones will be taken as strongly supporting the existence of convective overshooting in real stars. Furthermore, since many important details of this phenomenon are still poorly known, this way of proceeding will implicitly tell us more about the basic physical processes in which convective overshooting roots. The effects of mass loss on stellar evolution will be recalled wherever appropriate, while those of opacities somewhat different from the classic ones will be only shortly mentioned without any particular emphasis.

The plan of this paper is as follows. In section I, we briefly report on several observational facts that inspired the revision of classic models of massive and intermediate mass stars. In section II, we concisely discuss the fundaments of the four physical ingredients of model construction we have been referring to insofar. In section III, we present the main result for models with convective overshooting, new $^{12}C(\alpha,\gamma)^{16}O$ reaction rate and mass loss by stellar wind (when necessary), all over the mass range in which convective overshooting may be effective. Section IV applies the results to massive stars and points out the uncertainties that still exist with the new models. Section V deals with several important consequences of the new models in the domain of intermediate mass stars. Finally some concluding remarks are drawn in section VI.

## 1. SHADES OF THE OBSERVATIONAL SCENARIO

From the large body of literature dealing with observations of star clusters in our own galaxy and nearby galaxies (LMC and SMC), we have selected the following points relative to massive and intermediate mass stars:

### 1.1 Massive Stars

The catalogue of all known supergiants, O type stars and less luminous B type stars in our galaxy with MK spectral types and luminosity classes compiled by Humphreys and McElroy (1984) provides the most extended source of data for galactic luminous stars. Similar, though less complete, catalogues for a few nearby galaxies (LMC, SMC and others) have been compiled by Humphreys (1982,1984). Since WR stars are commonly understood as the descendents of bright O-type stars via the mechanism of mass loss by stellar wind (cfr. Chiosi 1982a,b for recent reviews of the subject), before comparing stellar counts with theoretical predictions, the Humphreys and McElroy catalogue has to be complemented by the list of known galactic WR stars (van der Hucht et al 1981). The HR diagram of luminous stars constructed with those catalogues reveals several features that have to be matched by theoretical models. Since they have been amply discussed by Humphreys (1982), Chiosi (1982b) and Chiosi and Maeder (1985), further description here is perhaps of little interest. I shall rather begin with a few comments and then concentrate on results of stellar counts which have driven most of the recent theoretical work done in this context. First of all, this major source of data suffers from a certain degree of incompleteness which is difficult to assess. This may be particularly severe for the earliest O-type stars for which the bolometric corrections are the greatest. The catalogue can be considered as reasonably complete only down to $M_{bol} = -8$ and within 2.5 Kpc of the Sun. Secondly, the question arises whether the Teff resulting from hydrostatic evolutionary computations is comparable with the Teff given by observations. Strong stellar wind may in fact produce a pseudo photosphere in the flow itself, thus indicating a Teff which may be significantly cooler than the one derived from hydrostatic atmospheres (de Loore et al 1982; Bertelli et al 1984). Finally, stellar counts per spectral type indicate that the star frequency distributions seem not to mimic the distribution of relative lifetimes one would expect from theoretical models. It appears as there is a deficiency of bright O-type stars near the zero age main sequence. This seems to occur among O-type stars brighter than $M_{bol} = -8$ or equivalently more massive than about 30 to 40 $M_\odot$. Is this indicating that stellar models are in error or that the majority of O-type stars are already evolved ? We will touch upon this point later on. Furthermore, Meylan and Maeder (1982) and Bertelli et al (1984) suggested that a deficiency of core H-burning stars (main sequence band) with respect to the evolved ones is likely to exist also in other luminosity intervals. Bertelli et al (1984) examined the luminosity range $- 7 > M_{bol} > -9$ limited to the distance of 2.5 Kpc from the Sun, in order to minimize effects of incompleteness of any type. The star counts are given in Table 1. Out of these data, it seems that some 40 % of the stars falls outside the region of core H-burning, in that contrary to the theoretical expectation of about 10 to 20 %. The different grouping of spectral types given in the first column of Table 1 shows the maximum extension of the main sequence band allowed by constant mass models and in presence of mass loss by stellar wind. The discrepancy is well evident. It appears as either star counts are still severely biased by incompleteness and/or selection effects or the theoretical main sequence band ought to extend at least up to the spectral type A0. Conversely, if incompleteness is the cause, it can be easily seen that in order to reconcile classic theory with observations the number of stars in the spectral range O to B0.5 (considering the most favourable case of models with mass loss) should be increased by more than a factor of 2. The argument of a lower Teff caused by the mass outflow, which would alter the correspondence between hydrostatic models and real stars, cannot be safely used in this case as the mass loss rates pertinent to the luminosity range in question are fairly modest. If we apply the argument that the observed O type stars are already evolved from the main sequence (cfr. Garmany et al 1982 for a similar discussion), it would imply that more than 50 % of the core H-burning lifetime of a star with initial mass of about 20 to 30 $M_\odot$ is spent while the star cannot be observed because still embedded in its parent cloud.

Although the timescale necessary to evaporate the parent cloud left over

Table 1
Stars of the Humphreys and McElroy (1984) sample with luminosity in the $-7 > M_{bol} > -9$ and within 2.5 Kpc of the Sun. The total number of stars is 743. The WR stars of van der Hucht et al (1981) are also included.

Constant mass evolution: the main sequence band extends up to Sp: O9.5

| Spectral Range | O | B | A | F | G | K | M | WR |
|---|---|---|---|---|---|---|---|---|
| Star Numbers | 270 | 380 | 14 | 7 | 5 | 2 | 45 | 20 |
| Percentages | 37.4 | 51.1 | 1.9 | 0.9 | 0.7 | 0.3 | 6.1 | 2.7 |

Evolution with Mass Loss: the main sequence band extends up to Sp: B0.5

| Spectral Range | O-B0.5 | B1-B9 | A | F | G | K | M | WR |
|---|---|---|---|---|---|---|---|---|
| Star Numbers | 503 | 147 | 14 | 7 | 5 | 2 | 45 | 20 |
| Percentages | 67.7 | 19.8 | 1.9 | 0.9 | 0.7 | 0.3 | 6.1 | 2.7 |

Evolution with mass loss and convective overshooting: the main sequence band extends up to Sp: B1

| Spectral Range | O-B1 | B1.5-B9 | A | F | G | K | M | WR |
|---|---|---|---|---|---|---|---|---|
| Star Numbers | 572 | 78 | 14 | 7 | 5 | 2 | 45 | 20 |
| Percentages | 77.0 | 10.5 | 1.9 | 0.9 | 0.7 | 0.3 | 6.1 | 2.7 |

by a star forming event is not very well known (Appenzeller 1980), hardly it can be as high as a few millions years. It may well be that a significant fraction of massive stars potentially are still in this phase and therefore not included in the above stellar counts (infrared emitters ?), but in percentage likely less than 50 %. A plausible guess may be about 10 to 20 % (Metzger 1976; Garmany et al 1982). It has been also suggested that in OB associations very massive stars form after less massive stars (Doom et al 1985). Although the conclusion by Doom et al (1985) is based on the assumption that the Humphreys and McElroy (1984) catalogue, from which the data for individual associations are derived, is complete down to $M_{bol} =- 5$ and definetely it is not, while it can be applied to single stellar associations, certainly it cannot be applied to such a composite HR diagram, which samples stars of any age. A possible way out of the dilemma was indicated by Bressan et al (1981), and Bertelli et al (1984), who considered the effects of convective overshooting during the core H- and He-burning phases in addition to those of mass loss by stellar wind. The main sequence band of the models with convective overshooting was found to extend up to the spectral type B1 (cfr. the last case of Table 1) and more important the lifetime spent by the models in the spectral range O9.5 to B1 turned out to be of the order of 20 % of the total core H-burning lifetime. This result greatly alleviated the discrepancy indicated by the stellar counts. However since a residual 10 % could not be accounted for also by models with convective overshooting, Bertelli et al (1984) suggested that only a suitable combination of mass loss, convective overshooting, atmospheric effects on Teff and a gentle increase in the opacity of middle temperature region ($5 \times 10^5$ to $5 \times 10^6$ °K) could entirely remove the discrepancy. In fact both concur to spread the main sequence band toward lower effective temperatures. However, as massive stars turned out to be a poor laboratory for testing the real occurrence of the above physical implementations, the attention was addressed to stars of lower mass, which also were known to present many controversial problems.

## 1.2 Intermediate Mass Stars

The main source of data for galactic stars in this mass range is the catalogue of Mermilliod (1981). As for LMC clusters many different sources have been considered that will be quoted wherever appropriate. Out of the available material and previous studies on the same subject, we have selected the following points of controversy. They are briefly summarized below in order of increasing complexity.

i) Widening of the Main Sequence Band. Similarly to massive stars, also the main sequence of galactic open clusters in the mass range of Pleiades to Hyades appear to be wider than expected (Maeder and Mermilliod 1981). The same property is also shown by LMC clusters, like NGC 1866 and others of similar type (Becker and Mathews 1983).

ii) Morphology of HR Diagrams. Furthermore, the overall morphology of the above clusters seems to possess features that cannot be easily explained. In fact, looking at the composite HR diagram of Mermilliod (1981) for galactic open clusters it turns up that very few stars are located in the middle of the Hertzsprung gap with the exception of a few Cepheids and /or composite spectroscopic binaries. In addition to this, one also notices that red giants are rather grouped showing two distinct trends in the pattern of red giant concentration. Red giants of relatively aged clusters (from Hyades to NGC 6475) occur at about the same colour ($(B-V)o = 1.0$), while the absolute magnitude becomes higher. On the contrary, red giants of younger clusters become redder and redder as the absolute magnitudes get higher and the luminosity class gradually changes. It is clear that the relative number of stars in different areas of the HR diagram (main sequence, blue and red giants, asymptotic giant branch stars) are related to the lifetime of the underlying evolutionary phase. Even if the correspondent evolutionary phases, namely core H-burning, core He-burning (in the loop and along the Hayashi line) and double shell stages beyond central He-burning are well assessed, the precise relative duration of these stages has never been carefully tested on observational basis. The only study we are aware of is by Lindoff (1969), who analyzing 108 galactic open clusters found that giant star (blue and red) lifetime versus main sequence lifetime indicated by observations was much shorter (by about a factor of 2 to 3) than predicted by those days evolutionary models. Amazingly enough, the disagreement is still there even with more recent models (Becker 1981). In addition to this, the HR diagram seems also to indicate that in galactic environment (chemical abundances) extended loops during the core He-burning phase are not likely to occur or that the time spent in the "blue loop" is much shorter than the time spent along the Hayashi line. On the contrary, the existence of blue loops in intermediate age clusters of LMC is well documented by their HR diagrams (NGC 1866 as a prototype). However, even in this case, the ratio of blue to red giants is not entirely compatible with classic models. Observations indicate 1, while theory yields 0.3 (Becker and Mathews 1983). Variations in the chemical composition parameters seem insufficient to remove the disagreement.

iii) Lack of Bright AGB Stars. The remarkable absence in NGC 1866 and other similar clusters of LMC of very bright AGB stars, which on the contrary are expected to occur in a cluster with age of about $86 \times 10^6$ yr and a turnoff mass of 5 $M_\odot$ (Becker and Mathews 1983). It has been suggested that a rate of mass loss (either in stationary or superwind mode) much greater than customarily assumed for stars in this phase (cfr. Iben and Renzini 1983, 1984) may result in an early termination of the AGB phase, thus accounting for the observed lack of very luminous AGB stars. As it will be discussed later on, this is not a viable explanation, or at least inadequacy in the mass loss rates is not the sole cause of disagreement.

iv) Quasi Old Clusters. The peculiar morphology of galactic as well as LMC clusters with age in the range 1 to $2 \times 10^9$ is another puzzling problem. In fact, Barbaro and Pigatto (1984) found that, while clusters older than 2 to $3 \times 10^9$ yr generally agree with theoretical predictions for their red giant star luminosity function, theory fails in interpreting the red giant distribution in clusters of slightly lower age. In fact, their behaviour is typical of even younger clusters, in that a well developed red giant branch is not observed. On the theoretical side, this can be explained supposing that those red giants, which on the basis of the cluster turn-off mass ($< 2$ $M_\odot$) are expected to be in shell H-burning phase, to develop a highly degenerate He core and therefore to eventually undergo core He-flash, actually evolve as more massive stars. Since the minimum core mass for non degenerate He ignition is 0.33 $M_\odot$, to which an initial mass of 2.2 to 2.3 $M_\odot$ is customarily assigned, everything occurs as if stars of initial mass as low as 1.3 $M_\odot$ or thereabouts were able to build He cores more massive (or as massive as) the above limit without passing through a phase of degeneracy in the core. The explanation of this dilemma, suggested by Barbaro and Pigatto (1984) and confirmed by Bertelli and Bressan (1985a), was attributed to convective overshooting during core H-burning phase.

v) Age Discrepancy. It has been pointed out (Hodge 1983) that ages of LMC clusters derived from the terminal AGB luminosity (Mould and Aaronson 1982) are in disagreement with ages derived from the main sequence turn-off (or termination) luminosity (Hodge 1983) and/or red giant clump luminosity (Flower 1984). While the ages derived from the last methods are in satisfactory agreement, they are too low when compared to Mould's and Aaronson's estimates for most of the clusters in common. Furthermore, the discrepancy is the highest for young clusters and it gets negligible for the oldest ones. To get rid of the difficulty both Mould (1983) and Hodge (1983) suggested that more mass has to be lost during the ascent of the AGB and/or the planetary nebula ejection phase. However, the same arguments against more substantial mass loss advocated for the AGB termination problem, hold even in this case and other causes of disagreement are likely to exist.

vi) AGB Star Luminosity Function. Another facet of the lack of very luminous AGB stars resides in the disagreement between theoretical and observational luminosity functions for field stars of LMC studied by Reid and Mould (1984). In brief, the observed luminosity function not only shows very few stars brighter than $M_{bol} = -6$, but also it decreases with increasing luminosity steeper than predicted by classic models of AGB stars under any plausible assumption for the star formation rate and initial mass function. To overcome the difficulty Reid and Mould (1984) suggested that more mass has to be lost by AGB stars. As discussed by Renzini (1984a), Bertelli et al (1985) and Chiosi et al (1985), this is not a viable solution.

## 2. PHYSICAL FUNDAMENTS OF EVOLUTIONARY MODELS

In this section, we briefly summarize the main ideas relative to a few points of stellar structure which are at the base of the most recent computations, even if they have not yet received general consensus.

i) Convective Overshooting

In classic models of stars, the boundary of the convective core is defined by condition $\nabla_R = \nabla_A$, which is equivalent to say that the core may extend up to the layer where the buoyancy acceleration of convective elements vanishes. The velocity however does not get zero at this layer, implying that convective elements may penetrate (overshoot) into the formally stable radiative zones, thus increasing the mass of the convective core. Due to well known uncertainties in the physics of convection and mixing processes, contrasting conclusions have been reached by different authors, which go from considering convective overshooting negligibly small to claiming that it is of paramount importance. Among others, we recall Maeder (1975, 1976), Cogan (1975), Roxburgh (1978), Cloutman and Whitaker (1980), Maeder and Mermilliod (1981), Bressan et al (1981), Matraka et al (1982), Doom 1982a,b, 1985), Bertelli et al (1985a). In particular, Bressan et al (1981) have shown the importance of convective overshooting in massive stars, while Bertelli et al (1985a) have investigated its far reaching effects in the domain of intermediate mas stars. These authors describe convective overshooting by means of the mixing length theory of convection and propose a formalism containing the mixing length scale of motion ($\ell = \lambda$ Hp) as an adjustable parameter ($\lambda$) to be eventually fixed by comparing model results with observations. In this formalism, when $\lambda = 0$ the classic condition is recovered.

ii) Mass Loss by Stellar Wind
Luminous early type stars and late type giants and supergiants are known

to lose mass from the surface at rates that may significantly affect their evolution. The subject has been reviewed so many times that a detailed presentation of current mass loss rates and physical processes powering the wind of different types of star is superfluous. An extended review of the topic can be found in Chiosi and Maeder (1985), to whom we refer for more information. It suffices to recall here the mass loss rate parametrizations that have been used in model computations we are going to describe. a) Massive stars have been evolved taking into account mass loss according to the following prescriptions for the rates: Chiosi and Olson (1984) for O-B type stars, Barlow and Cohen (1977) for A to M supergiants, the rates however scaled to the results of Jura and Morris (1981), and Barlow et al (1981) for the so-called WR stage. This is assumed to begin when the following conditions are satisfied: surface abundance of hydrogen less than 0.1 (in mass) and Teff of the models greater than 20000 °K. b) A few exploratory sequences for intermediate mass stars have been computed using either Waldron's (1984) parametrization, however rescaled as described in Bertelli et al (1985a) to obey the constraint imposed by globular clusters (Fusi-Pecci and Renzini 1976), or Reimers' (1975) formula with the parameter $\eta$ comprised in the range 0.3 to 1. Using Bertelli's et al (1985a) notation, Waldron's rates multiplied by the factor 0.2 to 0.3 are equivalent to Reimers' rates for $\eta = 1$.

iii) Nuclear Reaction and Neutrino Emission Rates

The major novelty in this context is the recent determination of the cross section for the $^{12}C(\alpha,\gamma)^{16}O$ reaction by Kettner et al (1982) together with the theoretical reanalysis by Langanke and Koonin (1982). The new rate runs 3 to 5 times faster than the classic rate of Fowler et al (1975). Here the new rate is taken to be 3 times the old value. All other reaction rates pertinent to the evolutionary phases under consideration are as in Fowler et al (1975). When appropriate, neutrino energy losses have been included. These are from Beaudet et al (1967), using their interpolation formula which is accurate within 20%.

iv) Radiative Opacities

Opacity of the stellar material was derived by interpolation (in temperature, density and chemical composition) of the opacity tables of Cox and Stewart (1970) for all the computed models. However, several independent arguments have been advanced (Simon 1982; Iben and Renzini 1984; Bertelli et al 1984) suggesting that current opacity calculations may actually underestimate the true opacity in the middle temperature region ($5 \times 10^5$ to $5 \times 10^6$ °K) where the main source of opacity are the bound-bound and bound-free transitions involving occupied levels in highly ionized species of elements from carbon to iron. Starting from the modification proposed by Bertelli et al (1984) and a suggestion advanced by Renzini (1984b), Bertelli and Bressan (1985b) adopted the following relation:

$$\kappa = \kappa_{XY} + A \Delta Z (1 + \chi f(\rho,T) \exp(-4\alpha \log(T/T_0))) , \qquad (1)$$

where $\kappa_{XY}$ is the opacity of a metal free mixture of H and He. A $\Delta Z$ gives the contribution of heavy elements to the classical opacity. This is evaluated subtracting at each value of $\rho$ and T the opacity of a metal free mixture with given X and Y from the opacity of a mixture having the same X and Y but metal abundance Z. Furthermore, $f(\rho,T)$ is a suitable function defining in the $\rho$-T plane a band along which the opacity enhancement given by the exponential term is allowed to occur. $T_0$ is the central value of the temperature interval ($10^6$ °K). The request of matching the opacity increase proposed by Simon (1982) and Bertelli et al (1984) allows us to determine $\alpha$, $\chi$ and $f(\rho,T)$. Since all this is highly speculative, the models computed with the above modification to the opacity will not be described here. The results of those preliminary computations have been briefly summarized by Chiosi (1986).

## 3. EFFECTS OF CONVECTIVE OVERSHOOTING AND NEW $^{12}C(\alpha,\gamma)^{16}O$ RATE

To fully appreciate the differences with respect to standard theories of stars given by convective overshooting it is worth introducing five critical values of the initial mass, which in turn define six mass ranges in which stellar evolution proceeds differently.

1) $M_i > M_{mas}$ (massive stars): above this limit stars proceed through a series of nuclear burnings in non degenerate conditions towards the construction of an iron core, subsequent photodissociation instability with core collapse and supernova explosion (cfr. Woosley et al 1984). The current value of $M_{mas}$ is 12 $M_\odot$.

2) $M_{up} < M_i < M_{mas}$ (quasi massive stars): stars in this mass range ignite carbon under mildly degenerate conditions, suffer a mild carbon flash but burn carbon non violently. Their subsequent evolution is fairly complicate, but eventually terminated by core collapse leading to a supernova explosion (Nomoto 1984). These stars and those of the previous range fail to develop a highly degenerate CO core, and do not experience the thermally pulsing AGB phase. The current value of $M_{up}$ is about 9 $M_\odot$ for Pop I and lower than this (about 7 $M_\odot$) for Pop II chemical composition (Becker and Iben 1979).

3) $M_{HeF} < M_i < M_{up}$ (intermediate mass stars): stars in this mass range ignite helium non degenerately but following He-exhaustion develop a highly degenerate CO core. They undergo a long thermally pulsing AGB phase, terminated either by envelope ejection and formation of a white dwarf (for $M_{HeF} < M_i < M_w$) or carbon ignition and deflagration in a highly degenerate core which has grown to the Chandraseckar limit of 1.4 $M_\odot$. This event is usually referred to as a supernova of type I1/2 (Iben and Renzini 1983). The critical mass $M_w$ above which this may occur is determined by the efficiency of mass loss by stationary wind or by so-called "superwind". Current estimates set $M_w$ around 5 $M_\odot$ (Iben and Renzini 1983, 1984).

4) $M_i < M_{HeF}$ (low mass stars): stars in this range of mass develop a highly degenerate He core after central hydrogen exhaustion. They develop the red giant branch along which they suffer significant mass loss by stellar wind. The red giant branch is terminated by the violent

ignition of He-burning in the core (He-flash), when the core mass has grown to 0.4 to 0.5 $M_\odot$ depending on the chemical composition. Their subsequent evolution is quite similar to that of intermediate mass stars. The classic value of $M_{HeF}$ is 2.2 to 2.3 $M_\odot$ for Pop I and about 1.8 $M_\odot$ for Pop II chemical composition. Within this mass range we also define another mass, $M_{LC}$, above which stars possess convective cores on the main sequence. This mass represent the minimum value for stars being affected by convective overshooting during their core H-burning phase. A provisional estimate sets $M_{LC}$ at about 1.2 $M_\odot$ for Pop I composition (Bertelli et al 1985b).

i) The Core H-burning Phase
Since the major effects of overshooting on core H-burning models are already known (Maeder 1976; Maeder and Mermilliod 1981; Bressan et al 1981; Matraka et al 1982; Bertelli et al 1985a), the discussion will be kept very short. Models with overshooting possess more massive convective cores, run at higher luminosities and live longer than classical models. They also extend the may sequence band over a wider range of Teff. The above effects depend on $\lambda$ and on the stellar mass. The relative increase in the mass of the convective core is $\Delta(Mr/M)$ = 0.16 when $\lambda$ varies from 0 to 1, while at given $\lambda$ the increase in core mass is greater at lower masses. The increase in the lifetime mimics the

Table 2
(Lifetimes of models with overshooting, $\lambda$ = 1, X = 0.700 and Z = 0.020)
The models are from Bertelli et al (1985b)

| $M/M_\odot$ | t | $t_H$ | $t_{He}$ | $t_{Heb}$ | $t_{Her}$ | $t_{He}/t_H$ |
|---|---|---|---|---|---|---|
| 1.2 * | 5.32(9) | 4.72(9) | 1.47(8) | ------ | 1.47(8) | 0.03 |
| 1.4 * | 3.55(9) | 3.25(9) | 1.41(8) | ------ | 1.41(8) | 0.04 |
| 1.5 * | 3.03(9) | 2.73(9) | 1.85(8) | ------ | 1.85(8) | 0.07 |
| 1.6 | 2.59(9) | 2.30(9) | 2.76(8) | No Loop | 2.76(8) | 0.12 |
| 1.7 | 2.18(9) | 1.95(9) | 2.30(8) | No Loop | 2.30(8) | 0.12 |
| 3.0 | 4.60(8) | 4.27(8) | 3.26(7) | No Loop | 3.26(7) | 0.08 |
| 5.0 | 1.25(8) | 1.18(8) | 7.06(6) | 3.59(6) | 3.20(6) | 0.06 |
| 6.0 | 8.34(7) | 7.88(7) | 4.62(6) | 2.37(6) | 2.07(6) | 0.06 |
| 7.0 | 8.04(7) | 7.70(7) | 3.35(6) | 1.98(6) | 1.37(6) | 0.05 |
| 9.0 | 3.55(7) | 3.34(7) | 2.10(6) | 1.20(6) | 9.00(5) | 0.06 |
| 20.0 | 1.13(7) | 1.03(7) | 5.82(5) | 4.13(5) | 1.70(5) | 0.05 |
| 60.0 | 4.36(6) | 4.03(6) | 3.29(5) | 3.29(5) | No Red | 0.08 |
| 100.0 | 3.55(6) | 3.26(6) | 2.80(5) | 2.80(5) | No Red | 0.08 |

Note: Ages are in years. The new rate of $^{12}C(\alpha,\gamma)^{16}O$ nuclear reaction is used. Masses with an asterisk undergo core He-flash. Massive stars (20, 60 and 100 $M_\odot$) are calculated with mass loss by stellar wind (see text)

dependence of the increase in the core mass on the star mass. The same is true for the increase in the stellar luminosity. The variation in the range of Teff's covered by core H-burning models, namely the extension of the main sequence band, increases with the stellar mass. Massive stars ($M_i > 40\ M_\odot$) would spread all across the HR diagram, it were not for the contrasting effect of mass loss (cfr. Chiosi and Maeder 1985).

ii) The Core He-burning Phase
The overluminosity caused by overshooting during the core H-burning phase still remains during the shell H- and core He-burning phases. The mass of the H-exhausted core, $M_{He}$, and the mass of the CO rich He-burning convective core are increased by $\Delta(Mr/M) = 0.13$ when $\lambda$ increases from 0 to 1. As a consequence of the higher luminosity, the lifetime of the He-burning phase ($t_{He}$) gets shorter in spite of the increase in the core mass by overshooting. This, combined with the longer H-burning lifetime $t_H$, makes the ratio $t_{He}/t_H$ fairly low (from 0.12 to 0.06 when the stellar mass varies from 1.6 $M_\odot$ to 9 $M_\odot$. In Table 2 we summarize the lifetimes for the set of models with $X = 0.700$ and $Z = 0.020$. The location in the HR diagram of core He-burning models can be schematically summarized as follows. In massive stars, where mass loss may occur even during the main sequence phase, it is almost entirely dominated by this phenomenon and it may take place either partly in the red and partly in blue, or entirely in the red, or entirely in the blue (cfr. the presentation of current scenarios by Chiosi and Maeder 1985). In the range of quasi massive and intermediate mass stars, it is well known that extended blue loop may develop. Their extension however depends on chemical composition and details of model structure (see the discussion of Renzini 1984c on this subject), and furthermore on overshooting, rate of the $^{12}C(\alpha,\gamma)^{16}O$ nuclear reaction, and finally on mass loss during the red giant phase. In brief, convective overshooting strongly decreases the loop extension, mass loss along the Hayashi line makes the loops even redder, while the increse in the nuclear rate of the above reaction acts in the opposite sense. Finally, the loop extension depends also on the stellar mass: in general they begin and get bluer at increasing mass.

iii) The critical Masses $M_{mas}$, $M_{up}$ and $Me_{HeF}$
In virtue of the larger helium core and carbon-oxygen core left over at the end of core H and He-burning respectively, the relation between the initial mass and $M_{He}$ and $M_{CO}$, which defines the above critical masses, is different with the new models. The new critical masses are given in Table 3 for the two sets of chemical composition. The most important result is that both $M_{up}$ and $Me_{HeF}$ are significantly lower in models with overshooting. This means that no AGB phase is now expected above the new $M_{up}$, while no prolonged RGB phase occurs for stars as low as 1.6 $M_\odot$. The impact of this finding on observational front is straightforward and of paramount importance. Lower values of $M_{up}$ have been also suggested by Renzini et al (1985), Castellani et al (1985), while a suggestion for a lower value of $M_{HeF}$ is by Barbaro and Pigatto (1984). Remarkably, in models with overshooting, $M_{HeF}$ has been found not to depend on Z (Bertelli et al 1985b), while its possible dependence on X has not yet been tested. Finally, we recall that a different $M(M_{He})$ relation now

holds all over the mass range in which convective overshooting is in operation. This is particularly relevant in conjunction with the problem of chemical enrichment per stellar generation (cfr. Chiosi and Matteucci 1984). Furthermore, the drastic lowering of $M_{up}$ towards $M_w$ makes the occurrence of type I1/2 supernovae very unlikely.

Table 3
(Critical Masses $M_{mas}$, $M_{up}$ and $M_{HeF}$: in $M_\odot$)

| X | Z | $M_{mas}$ | $M_{up}$ | $M_{HeF}$ |
|---|---|---|---|---|
| 0.700 | 0.020 | 9 | 6 | 1.6 |
| 0.700 | 0.001 | 8 | 5 | 1.6 |

iv) Effects of the $^{12}C(\alpha,\gamma)^{16}O$ rate on the CO core

With the old value for the rate of this nuclear reaction almost equal abundances of $^{12}C$ and $^{16}O$ are expected in the CO core at the end of core He-phase, the abundance of $^{12}C$ being moreover a decreasing function of the stellar mass (Arnett 1972). With the new rate very little carbon is left in the core, the final abundance of it being lower than 10%. This result is also partly caused by the prolongement of the He-phase due to the replenishment of novel fuel by overshooting. Furthermore, if future measurements of the cross section will confirm the high value estimated today, the drastic underabundance of $^{12}C$ may have strong implications for the subsequent evolution of massive as well as intermediate mass stars. We will touch upon this point later on.

## 4. OBSERVATIONAL IMPLICATIONS FOR MASSIVE STARS

Under the combined effects of overshooting and mass loss, the band of core H-burners may now extend up to the spectral B1 in the luminosity range correspondent to stars of initial mass from 20 to 60 $M_\odot$. At higher masses the core H-burning band shrinks towards the zero age main sequence as it happened with classic models losing mass at substantial rate. The advantage here is that the same result is obtained without enormous losses of mass, owing to the larger convective cores, which favour the appearance of CNO processed material at the stellar surface with consequent shrinkage of the radius. The results for the mass range 20 to 60 $M_\odot$ are particularly interesting because, due to the longer lifetime of the H-phase spent in the spectral range B0-B1 (20% of the total), which approximately amounts to three times the core He-burning lifetime, the excess of stars outside the classic main sequence band (cfr. section 2.1) can be almost entirely accounted for. The core He-burning phase takes place partly in the red and partly in the blue for initial masses in the range 20 to 40 $M_\odot$ or thereabouts. Approximately 30% of the He-burning lifetime is spent by a typical 20 $M_\odot$ star as a red supergiant. Stars more massive than 40 $M_\odot$, even though they may reach the red supergiant region, spend there very short time and run back soon to the blue side of the HR diagram (Wolf Rayet star progenitors ?). The behaviour of stars in the mass range 10 to 20 $M_\odot$ is not entirely clarified. Likely they will spend the whole core He-phase as red supergiants. Stars of 10 to 12 $M_\odot$ likely are the most massive objects in which extended loops may develop with Pop I composition. The results of stellar counts performed by Bertelli et al (1984) for models with overshooting and mass loss by stellar wind are shown in the last case of Table 1. Even if the main properties of the evolution of massive stars in occurrence of mass loss are reasonably well understood, still they are hampered by the uncertainties affecting the determination of the mass loss rates (cfr. Chiosi and Maeder 1985 for more details).

Perhaps the most interesting advancement in the theory of massive stars is related to the new cross section for the $^{12}C(\alpha,\gamma)^{16}O$ reaction. In fact, as a consequence of the very little carbon left in the core following He-exhaustion, massive stars may be able to skip the C-burning phase. As found by Wilson et al (1985) this has profound consequences in subsequent evolution. Omitting all details, their 25 $M_\odot$ at the end of evolution builds an iron core of about 2.2 $M_\odot$ instead of the classic 1.4 $M_\odot$. A black hole is likely to form in this event. It is clear that the particular value of the initial mass above which this may occur, depends on the amount of carbon present in the core at the end of the He-phase, which in turn depends on the reaction rate and prolongement of He-burning by overshooting.

## 5. OBSERVATIONAL IMPLICATIONS FOR INTERMEDIATE MASS STARS

In this section we will touch upon the use of models with overshooting in the interpretation of the HR diagrams of intermediate age and quasi old clusters, in the problem of clusters dating and age discrepancy, and finally in the problem of the AGB star luminosity function. Although the adopted chemical compositions may not be fully suited, we shall consider only LMC clusters as they are the best laboratory where such a comparison can be successful.

### i) Intermediate Age Clusters (NGC 1866)

NGC 1866 is particularly suited to this purpose, because it is well populated and it has been studied recently by Becker and Mathews (1983) on the basis of classic models. Those authors assigned NGC 1866 a turn-off mass of about 5 $M_\odot$, a chemical composition Y = 0.273 and Z = 0.016, and an age of $86 + 5 \times 10^6$ yr. However they failed in reproducing the relative stellar frequencies (main sequence versus red giant stars, blue versus red giants) and the lack of bright AGB stars. Table 4 contains the star counts derived from the photometric study by Robertson (1974) plus the list of superluminous stars by Flower (1981). Since the main sequence stars may be affected by incompleteness in that the majority of them fall below the survey limit (18 mag), we prefer to use the giant star population to compare theory with observations. Omitting all the details of the discussion which can be found in Bertelli et al (1985a), we propose the following scenario: turn-off mass of about 4 $M_\odot$, age of about $210 \times 10^6$ yr, an equal percentage of blue and red stars (cfr. the lifetimes of Table 4). With the above choice, the number of main sequence stars in the observational sample would amount only to 10% of the total pertinent to a turn-off mass of 4 $M_\odot$. The rest falls below the survey limit. The lack (paucity in general) of very bright AGB stars can be understood by the much lower $M_{up}$ indicated by models with overshooting. The few superluminous stars are compatible (in number) with being core C-burners (cfr. Flower et al 1980). Mass loss by stellar wind may help to remove them away from the AGB area. If all this is correct, it would imply that $M_{up}$ for the chemical composition of NGC 1866 is down to about 4 $M_\odot$. Models of ours with X = 0.700 and Z = 0.001 yield $M_{up}$ = 5 $M_\odot$. Other chemical compositions haave not yet been explored. Direct measurements of the chemical abundances in stars of NGC 1866 would be highly useful.

Table 4
(Stellar counts in NGC 1866)

| Note | MS | BG | RG | AGB | SLS |
|---|---|---|---|---|---|
| (1) | 206 +14 | 46 + 10<br>    - 7 | 47 + 29<br>    - 7 | 14 + 13<br>    - 5 | 6 |
| (2) | 116 | 60 | 137 | 40 | |
| (3) | 21.1 (7) | 0.79 (7) | 0.79 (7) | << 5 (6) | |

(1) Star counts from Becker and Mathews plus the superluminous stars (SLS) of Flower; (2) Star numbers predicted by Becker and Mathews; (3) Theoretical lifetimes for a 4 $M_\odot$ star with convective overshooting ($\lambda$ = 1) and chemical composition X = 0.700 and Z = 0.020

ii) Quasi Old Clusters (NGC 2190 and NGC 2162)

The HR diagram of clusters like NGC 2190 and 2162, recently obtained by Schommer et al (1984) with CCD photometry, are reflective of the lower $M_{HeF}$ given by models with overshooting, similarly to those clusters analyzed by Barbaro and Pigatto (1984). In fact, in the study by Chiosi and Pigatto (1984) an age of about $1 \times 10^9$ yr has been assigned and more important it has been shown that they possess a single peak luminosity function for red giants. The correspondent turn-off mass is about 1.8 $M_\odot$. This well to models with overshooting which predict $M_{HeF}$ as low as 1.6 $M_\odot$ and core He-burning to occur along the Hayashi line, to which a clump of red stars corresponds on observational side. With classic models of the same age and chemical composition, there would be a prolonged red giant branch followed by core He-burning in the horizontal branch (red in this case). A double peak luminosity function for red stars is therefore expected, contrary to what observed.

iii) Clusters Dating and Age Discrepancy

As already anticipated in section 1, three methods exist to date clusters, namely the luminosity of the main sequence turn-off (and/or termination), the luminosity of the He-burning red giants, and the luminosity of the brightest AGB stars. Each rests upon suitable luminosity versus age relationships derived from theoretical models. Among other things, the theoretical relations depend on the chemical composition, which has to be specified a priori. Unfortunately, good abundance determinations are available only for a few clusters. Furthermore, the calibrating relationships suffer from all uncertainties affecting stellar models. In spite of it, age compilations have been derived for quite a number of clusters (Hodge 1983; Flower 1984; Mould and Aaronson 1982). The results are fairly discouraging as ages derived by different methods may disagree by large factors. It goes without saying that each method competes with intrinsic observational as well as theoretical difficulties that often invalidate the whole results.

a) The turn-off method is hampered by the lack of very good photometry down to magnitudes faint enough to delineate the unevolved portion of the main sequence. This is particularly severe for LMC clusters older than a few $10^8$ yr or thereabouts.

b) The red giant luminosity method requires that the luminosity of core He-burning models is a monotonic function of the age. This is true for stars more massive than $M_{HeF}$ (cfr. Renzini and Buzzoni 1985, this conference).

c) The AGB luminosity method can be applied only to clusters with turn-off mass below $M_{up}$. However, the greatest uncertainty with this method resides in the low number of AGB stars usually present in a cluster, which makes the identification of the maximum AGB luminosity quite uncertain. Furthermore, the uncertainty in the mass loss rates for the red giant and asymptotic giant phases strongly affects the theoretical age calibrating relationships.

Chiosi et al (1985) revisited the whole problem of clusters dating in the light of models with overshooting. The calibration relations for the turn-off (termination), red giant and AGB star methods are shown in Fig 1,2 and 3 respectively. All details relative to the construction of these relations can be found in Chiosi et al (1985).

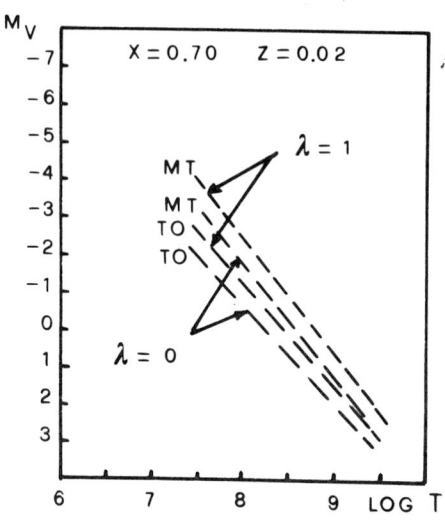

Fig. 1 Turn-off (TO) and main sequence termination (MT) luminosity versus age relationships for classic models ($\lambda = 0$) and models with overshooting ($\lambda = 1$)

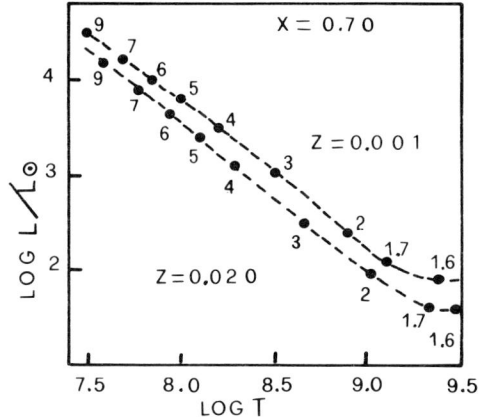

Fig. 2 Red giant stars (core He-burners) versus age relations for vaious metallicities. $\lambda = 0$ indicates classic models, while $\lambda = 1$ shows models with overshooting

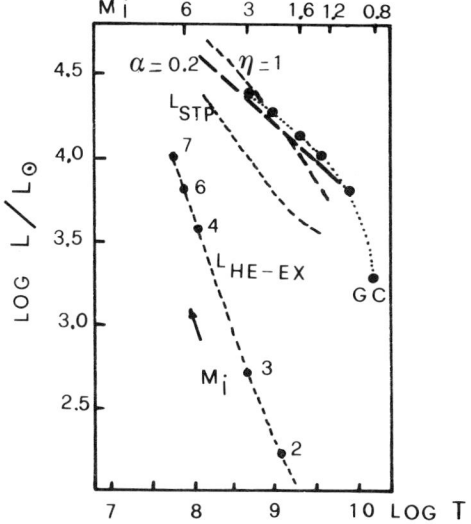

Fig. 3 Maximum AGB luminosity versus age relation for $X = 0.700$, $Z = 0.02$ and mass loss rates as indicated. The luminosity at the core He exhaustion and beginning of thermally pulsing AGB phase are also indicated. The full dots show the relation used by Mould and Aaronson (1982)

The results of the above study can be summarized as follows:
1) Due to the higher luminosity and longer lifetime of the core H-burning phase of models with overshooting, we expect that for any given turn-off (termination) magnitude of clusters whose turn-off mass has a convective core, the correspondent age is greater than classic estimates. Similar increase in the age is also expected from the red giant method for all clusters which have a turn-off mass greater than $M_{HeF}$. The relations below may be used to convert ages derived from classical models into the new ones:

$$\text{Log } T = 0.78 \text{ Log } T_{old} + 2.337 \qquad \text{(turn-off method)}$$

$$\text{Log } T = 1.05 \text{ Log } T_{old} + 0.092 \qquad \text{(red giant method)}$$

where the age T is given in years.

2) Following the procedure outlined by Iben and Renzini (1983, 1984) however adapted to the new models, the relation of Fig. 3 between the maximum AGB star luminosity and age is derived. The rate of mass loss during the RGB and AGB phases is from Reimers (1975) with $\eta = 1$ or equivalently from Waldron (1984) scaled by the factor $\alpha = 0.2$. Ages derived from the novel relation with overshooting are only modestly different from the classic ones. This surprising result can be understood as due to the fact that AGB evolution is mainly driven by the CO core mass and it depends little on the past history. What is actually changed is the correspondence between the initial mass and total lifetime.

3) The three methods applied to a few clusters, for which all necessary data were available, give ages that are in much better agreement. Even if difficulties may arise with the interpretation of those clusters in which the red giant method has been applied, that is to know a priori whether or not their turn-off mass is above $M_{HeF}$ (cfr. Renzini and Buzzoni 1986, this conference), the net result remains that ages based on the main sequence turn-off and red giant luminosity get closer to those based on AGB luminosity. Even with all reservations caused by the uncertainties discussed insofar, the so-called "age discrepancy" if not completely ruled out is greatly alleviated when overshooting is taken into account.

iv) The AGB Luminosity Function for Field Stars of LMC

Here we discuss the AGB luminosity function obtained by Reid and Mould (1984) for a selected area of LMC. The luminosity function (number of stars per magnitude bin) is shown in Fig. 4 and compared with the theoretical prediction based on the standard theory of thermally pulsing AGB stars. All details relative to the procedure and assumptions for the initial mass function and star formation rate are given in Reid and Mould (1984) and therefore omitted here. It suffices to recall that the particular case shown in Fig. 4 refers to a constant star formation rate and the Salpeter initial mass function. Chiosi et al (1985) have repeated the whole analysis using models with overshooting in the previous evolutionary phases. The mass loss rates were the same as

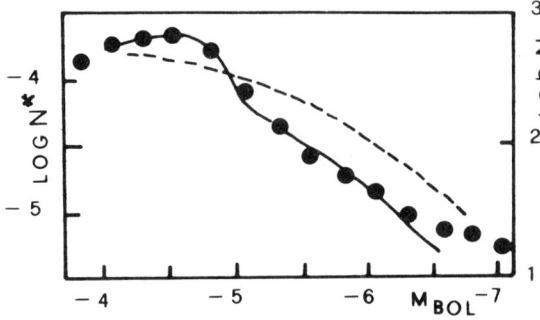

Fig. 4  The luminosity function for AGB stars. N and N$^*$ are the observed (full dots) and predicted (continuous and dashed lines) number of stars per magnitude bin. They differ by a normalization factor. The dashed line visualizes the standard case by Reid and Mould, while the solid line shows the result for models with overshooting

reported in the previous paragraph. All other assumptions are as in Reid and Mould (1984). Two major novelties are however important. First, the contribution of early AGB stars neglected in Reid and Mould (1984) was taken into account. Second, it has been found that also normal red giants (core He-burners) with initial mass in the range 4.5 to 7 $M_\odot$, being overluminous because of overshooting, may contaminate the lower magnitude bins. No such a contamination was expected to occur with classical models because the stars in question were either too faint or too short lived to appreciably alter the statistics. After normalization of the theoretical stellar numbers per magnitude interval at the second 0.25 mag bin, we find the relation shown, in Fig. 4 by the continuous line. The normalization is possible as the constant in the initial mass function is not specified. The agreement with observations is remarkably good. The net improvement is mostly due to the novel features of models with overshooting, and only marginally to the mass loss rates we have adopted. The same result would not have been possible by varying the mass loss rates alone.

## 6. CONCLUDING REMARKS

Before concluding this paper, we like to touch upon a few more points that may be affected by the evolutionary models we have been presenting insofar.

i) The N(log P) distribution of Cepheid Stars

In a recent paper, to which the reader should refer for a better understanding of what follows, Serrano (1983) analyzed the number frequency - period distribution (N(LogP)) of Cepheids in the light of classic models. The distribution is specified by three parameters: the short period cutoff $P_0$, the period $P_1$ of the maximum of the distribution and the rate at which Cepheids with $P > P_1$ decay in number with respect to the period. These two periods, weighted on the initial mass function, correspond to the minimum masses whose He-burning loops enter the instability strip at the red side and spend the whole He-phase in the loop within the instability strip, respectively. Equivalently, the periods $P_0$ and $P_1$ depend on the location and inclination in the HR diagram of the blue loop band with respect to the Cepheid instability strip. The He-band is known to depend on the chemical composition. Limiting the discussion to the sole solar vicinity to minimize effects of variation of chemical abundance with galactic site, the observed difference $LogP_0 - LogP_1$ is in the range 0.3 to 0.6, while the theory predicts 0.06 (Becker et al 1977). Furthermore, there is an excess of long period Cepheids which can be explained only invoking a two component birth rate, more efficient for massive stars (Becker et al 1977). It is an easy matter to show that Becker's (1981) models yield too small a $LogP_0 - LogP_1$ difference for any combination of chemical abundances. To overcome the problem we may either make wider the instability strip (Pel and Lub 1978) or increase the slope of the blue loop band. It turns evident that models with overshooting possess the latter requisite and, as shown by Bertelli et al (1985a), not only the period difference is matched, but also the excess of long period Cepheids is accounted for.

ii) The Distance Modulus of LMC

The most common approach to the determination of the distance to a star cluster is the main sequence fit method. The recent development in CCD technique makes it feasible also for clusters of LMC, which were heretofore hampered by the lack of sufficiently deep photometry down to the unevolved portion of the main sequence. Using this technique Schommer et al (1984) derived a true distance modulus of $(m-M)o = 18.2\pm0.2$ mag. However, since the main sequence and/or isochrone fitting turn out not to be as objective as one may wish, this being even more crucial when a few tenths of magnitude are in question, Chiosi and Pigatto (1985) preferred to use the method of the luminosity function for both main sequence and red giant stars, whose advantages over other procedures have been illustrated by Paczynski (1984). The clusters used by Schommer et al (1984) to estimate the distance modulus of LMC are NGC 2190 and NGC 2162, whose observational properties already described in section 5, were found to be in conflict with classic theory but in agreement with those for models with convective overshooting. When the new models are adopted to derive the luminosity functions of main sequence and red giant stars and the luminosity function method is used, the distance modulus of LMC is back to the canonical value of 18.6 mag.

iii) Overshooting in Low Mass Stars

Stars with initial mass in the range $M_* < M_i < M_{HeF}$, where $M_*$ is the minimum mass for a star whose He core is able to reach the critical mass for He-ignition in degenerate material (0.4 to 0.5 $M_\odot$ depending on chemical composition), undergo He-flash followed by convective core He-burning and therefore may be affected by convective overshooting during this phase. Furthermore, stars comprised between 1.2 $M_\odot$ and $M_{HeF}$ (now 1.6 $M_\odot$) develop a convective core during the core H-burning phase and therefore may suffer the effects of convective overshooting during both phases. Models of ours with initial mass 1.2, 1.3, 1.4, 1.5 $M_\odot$ belong to this group and have been followed until the core He-exhaustion stage (cfr. Table 2). However they have been evolved at constant mass during the red giant branch phase, along which mass loss is on the contrary known to occur with decrease in total mass by about 20%. Although the evolution of the central core is expected not to be affected by mass loss in the previous stages, the location of these models in the HR diagram may depend on the amount of mass lost in RGB phase. Therefore they are only a particular case of a larger network of possible models. Stars of lower mass with convective overshooting only during the core He-burning phase have not yet been calculated. In concluding this section we like to recall that another line of work has been developed over the past years dealing with the so-called "semiconvection" and "breathing pulses" of convection in the lates stage of core He-burning of horizontal branch stars (Castellani et al 1985, and references therein). Although the two phenomena in question are not formally equivalent to convective overshooting, they lead to similar results in that larger masses of the CO rich convective core are obtained.

In this paper we reported on studies of the effects of convective overshooting on stars of all masses and evolutionary phases in which this phenomenon may be effective. The aim was not only to discuss convective overshooting from the viewpoint of general interest toward this particular physical process, but also to find astrophysical tests of its occurrence in real stars and hopefully to indirectly assess the actual extension of convective cores. Looking at the results we have presented insofar, convective overshooting turns up to be a very promising tool for removing or at least alleviating some of the discrepancies that were known to exist between current theories of stellar structure and crucial observational facts. Whether or not the arguments presented in this paper have been convincing and the goal achieved is difficult to say. Certainly this line of work deserves more careful studies.

This work has been supported by the National Group of Astronomy (GNA) and the Italian Space Research Program (PSN) of the National Council of Research of Italy (CNR) under contracts n. 8302422-02 , n. 83-018 and n.84-040.

REFERENCES

Appenzeller, I. 1980. in Star Formation, 10th Advanced Course of the Swiss Society of Astronomy and Astrophysics, Saas-Fee, p.3
Arnett, W. D. 1972, Astrophys. J. 176, 681
Barbaro, G., Pigatto, L. 1984. Astron. Astrophys. 136, 355
Barlow, M. J., Cohen, M. 1977. Astrophys. J. 213, 737
Barlow, M. J., Smith, L. J., Willis, A. J. 1981. M. N. R. A. S. 196, 101
Beaudet, G., Petrosian, V., Salpeter, E. E. 1967. Astrophys. J. 150, 979
Becker, S. A. 1981. Astrophys. J. Suppl. 45, 475
Becker, S. A., Iben, I. Jr. 1979. Astrophys. J. 238, 831
Becker, S. A., Iben, I. Jr., Tuggle, R. S. 1977. Astrophys. J. 218, 633
Becker, S. A., Mathews, G. J. 1983. Astrophys. J. 270, 155
Bertelli, G., Bressan, A. 1985a. Preprint
Bertelli, G., Bressan, A. 1985b. In preparation
Bertelli, G., Bressan, A., Chiosi, C. 1984. Astron. Astrophys. 130, 279
Bertelli, G., Bressan, A., Chiosi, C. 1985a. Astron. Astrophys. 150, 33
Bertelli, G., Bressan, A., Chiosi, C., Angerer, K. 1985b. Preprint
Bressan, A., Bertelli, G., Chiosi, C. 1981. Astron. Astrophys. 102, 25
Castellani, V., Chieffi, A., Pulone, L., Tornambe', A. 1985. Astrophys. J. In press
Chiosi, C., 1982a. in Wolf Rayet Stars: Observations, Physics, Evolution, ed. C. de Loore, A. Willis, p. 323, Dordrecht: Reidel
Chiosi, C. 1982b. in The Most Massive Stars, ESO Workshop, ed. S. D'Odorico, D. Baade, K. Kjar, p. 27
Chiosi, C. 1986. In Luminous Stars and Stellar Associations, ed. P. S. Conti, C. de Loore, E. Kontizas, Dordrecht: Reidel. In press
Chiosi, C., Bertelli, G., Bressan, A., Nasi, E. 1985. Astron. Astrophys. submitted
Chiosi, C., Maeder, A. 1985. Ann. Rev. Astron. Astrophys. (1986)
Chiosi, C., Matteucci, F. 1984. in Stellar Nucleosynthesis, ed. C. Chiosi, A. Renzini, p. 359, Dordrecht: Reidel

Chiosi, C., Olson, G. L. 1984. Unpublished
Chiosi, C., Pigatto, L. 1985. Astrophys. J. Submitted
Cloutman, L., Whitaker, R. 1980. Astrophys. J. 237, 900
Cogan, B. C. 1975. Astrophys. J. 201, 637
Cox, A. N., Stewart, J. N. 1970. Astrophys. J. Suppl. 19, 243
Doom, C. 1982a,b. Astron. Astrophys. 116, 303, 308
Doom, C. 1985. Astron. Astrophys. 142, 143
Doom, C., De Greve, J. P., de Loore, C. 1985. Astrophys. J. 290, 185
Flower, P. J. 1981. Astrophys. J. 249, L11
Flower, P. J. 1984. Astrophys. J. 278, 582
Flower, P. J., Geisler, D., Hodge, P. Olszewski, E. W. 1980. Astrophys. J. 235, 769
Fowler, W. A., Caughlan, G. R., Zimmermann, B. A. 1975. Ann. Rev. Astron. Astrophys. 13, 69
Fusi-Pecci, F., Renzini, A., 1976. Astron. Astrophys. 46, 447
Garmany, C. D., Conti, P. S., Chiosi, C. 1982. Astrophys. J. 263, 777
Hodge, P. W. 1983. Astrophys. J. 264, 470
van der Hucht, K., Conti, P. S., Lundstrom, I., Stenholm, B. 1981. Space Sci. Rev. 28, 227
Humphreys, R. M. 1982. in The Most Massive Stars, ESO Workshop, ed. S. D'Odorico, D. Baade, K. Kjar, p. 5
Humphreys, R. M. 1984. in Observational Tests of the Stellar Evolution Theory, ed. A. Maeder, A. Renzini, p. 279, Dordrecht: Reidel
Humphreys, R. M., McElroy, D. B. 1984. Astrophys. J. 284, 565
Iben, I. Jr. 1974. Ann. Rev. Astron. Astrophys. 12, 215
Iben, I. Jr., Renzini, A. 1983. Ann. Rev. Astron. Astrophys. 21, 271
Iben, I. Jr., Renzini, A. 1984. Physics Reports 105, n. 6, 329
Jura, M., Morris, M. 1981. Astrophys. J. 251, 181
Kettner, K. U., Becker, H. W., Buchman, L., Gorres, J., Kravinkel, H., Rolfs, C., Schmalbrok, P., Trauttvetter, H. P., Vlieks, A. 1982. Z. Phys. 308, 73
Langanke, K., Koonin, S. E. 1982. Nuclear Physics A410, 334
Lindoff, U. 1969. in Mass Loos from Stars, ed. M. Hack, p. 106, Dordrecht: Reidel
de Loore, C., Hellings, P., Lamers, H. 1982. in Wolf Rayet Stars: Observations, Physics, Evolution, ed. C. de Loore, A. Willis, p. 53, Dordrecht: Reidel
Maeder, A. 1975. Astron. Astrophys. 40, 303
Maeder, A. 1976. Astron. Astrophys. 47, 384
Maeder, A. 1984. in Observational Tests of the Stellar Evolution Theory, ed. A. Maeder, A. Renzini, p. 299, Dordrecht: Reidel
Maeder, A., Mermilliod, J. C. 1981. Astron. Astrophys. 93, 136
Matraka, B., Wassermann, C., Weigert, A. 1982. Astron. Astrophys. 107, 283
Mermilliod, J. C. 1981. Astron. Astrophys. 97, 235
Metzger, P. 1976. Proceedings of 3rd Eur. Astr. Meeting, ed. E. K. Kharadze, p. 369, (Tblisi)
Meylan, G., Maeder, A. 1982. Astron. Astrophys. 108, 148
Mould, J. 1983. in Structure and Evolution of the Magellanic Clouds, ed. S. van den Bergh, K. S. de Boer, p. 195, Dordrecht: Reidel
Mould, J., Aaronson, M. 1982. Astrophys. J. 263, 629

Nomoto, K. 1984. in Stellar Nucleosynthesis, ed. C. Chiosi, A. Renzini, p. 239, Dordrecht: Reidel
Paczynski, B. 1984. Astrophys. J. 284, 670
Pel, J. W., Lub, J. 1978. in The HR Diagram, ed. A. G. Philip, D. S. Hayes, p. 229, Dordrecht: Reidel
Reid, N., Mould, J. 1984. Astrophys. J. 284, 98
Reimers, D. 1975. Mem. Soc. Roy. Sci. Liege 8, 369
Renzini, A. 1984a. in Stellar Nucleosynthesis, ed. C. Chiosi, A. Renzini, p. 99, Dordrecht: Reidel
Renzini, A. 1984b. Private communication
Renzini, A. 1984c. in Observational Tests of the Stellar Evolution Theory, ed. A. Maeder, A. Renzini, p. 21, Dordrecht: Reidel
Renzini, A., Bernazzani, M., Buonanno, R., Corsi, C. E. 1985. Astrophys. J. Lett. In press
Robertson, J. W. 1974. Astron. Astrophys. Suppl. 15, 261
Roxburgh, I. 1978. Astron. Astrophys. 65, 281
Schommer, R. A., Olszewski, E. W., Aaronson, M. 1984. Astrophys. J. 285, L53
Serrano, A. 1983. Rev. Mexicana Astron. Astrophys. 8, 131
Simon, N. R. 1982. Astrophys. J. 260, L87
Waldron, W. L. 1984. Preprint
Wilson, J. R. Mayle, T., Woosley, S., Weaver, T. 1985. Preprint
Woosley, S., Axelrod, T. S., Weaver, T. 1984. in Stellar Nucleosynthesis, ed. C. Chiosi, A. Renzini, p. 263, Dordrecht: Reidel

DISCUSSION

MOULD: How uncertain is $\lambda$ ?
CHIOSI: The derivation of $\lambda$ from first principles is not yet feasible, due to all uncertainties and complicacies of theories of convection and mixing processes. The way we addressed the problem and the comparison of model results with observations allow us to estimate $\lambda = 1$.

CAPUZZO DOLCETTA: As far as I know convective overshooting was usually treated by the Schwarzschild criterion. What is the physics of your new criterion ?
CHIOSI: I would pose the question in a different way. The Schwarzschild criterion assumes the boundary of convective cores (regions) to occur at the layer where the buoyancy acceleration is zero. Convective overshooting, on the contrary, seeks to determine the layer where the velocity of convective elements gets zero. Therefore, it is a more complex description of convective motions, which may or may not rest on the mixing length theory of convection, depending on the assumptions made and formalism used. The major problem with convective overshooting is whether or not it is an efficient phenomenon, and as you know contrasting conclusions have been reached. We are inclined to think that convective overshooting is indeed an important physical process.

CAPUZZO DOLCETTA: Do you know whether someone has considered the effects of viscosity on convective elements ?

CHIOSI: Yes. Maeder (1975, 1976; Astron. Astrophys. 40, 303; 47, 384).

MADORE: Could you comment on the application of "turbulent diffusion" to the problems discussed earlier in the context of convective overshooting.
CHIOSI: Turbulent diffusion is parametrized by the scale length of diffusion (cfr. Cloutman and Whitaker 1980, Ap. J. 237, 900). Therefore the effects of turbulent diffusion, which are similar to those of convective overshooting, depend on this parameter and will be also uncertain as long as a satisfactory theory of turbulence is missing.

FROGEL: Is the lifetime increased just because the core is allowed to grow larger even though luminosity is greater ?
CHIOSI: Yes. We must however distinguish between core H- and He- burning phases. The core H-burning lifetime is significantly increased by the growth of the convective core, while the consequent increase in luminosity as the evolution proceeds has no visible effect. On the contrary, the core He-burning lifetime gets shorter because of the higher luminosity and insufficient core feeding by overshooting. The total lifetime is however always longer than in classic models.

MOULD: How do you decide the age of NGC 1866 to see any discrepancy in the main sequence turn-off luminosity ?
CHIOSI: Since the main sequence stars may be more severely affected by problems of observational incompleteness, we use the red giant stars to derive the cluster age. To do so, we impose that evolutionary tracks (or isochrones) match the mean luminosity of the red and/or blue clump stars. This has been done with models with overshooting. The difficulties encountered with classic models are described by Becker and Mathews (1983). In brief, not only the stellar counts (main sequence versus giant stars, blue versus red giants) were in disagreement with theoretical predictions, but also the 5 $M_O$ track (turn-off mass) had the core H-burning phase that could cover only a little portion (in luminosity) of the luminosity interval interested by main sequence stars. Remember that evolutionary tracks run almost vertically in the V - (B-V) plane during this phase. This simply means that the observed main sequence band is wider than predicted by the classic models. The difficulty is removed with the new ones, but in this case most of the main sequence stars pertinent to the novel turn-off mass we assign to the cluster should be still below the survey limit of the material used to perform the analysis.

FROGEL: For NCG 1866 how much width do you expect to the main sequence because of overshooting ?
CHIOSI: Likely by as much as indicated by the composite HR diagram for galactic clusters of Mermilliod (1981).

FREEDMAN: Have you determined a distance modulus using the luminosity function method and standard models without convective overshooting ? In other words how much of the difference in distance modulus between

Schommer et al versus your new distance modulus is due to the difference between the two methods (main sequence fitting versus luminosity function method) as opposed to the different stellar evolutionary models used ?

CHIOSI: As I said, the two clusters (NGC 2162, NGC 2190) used to discuss the distance modulus of LMC appear to belong to that class of clusters (Barbaro and Pigatto 1984) which, expected to undergo He flash in virtue of their turn off mass ($< 2\ M_0$), seem to ignite He-burning non violently. This advised us not to use standard models in the final discussion. In my opinion, the difference in distance modulus with respect to Schommer's et al estimate is due to the combined effects of the different method and properties of the new models (more luminous).

MOULD: How did you decide the ages of the two Schommer-Olszewski-Aaronson clusters in a distance independent way ?

CHIOSI: In brief, we proceed as follows: isochrones and luminosity functions (main sequence and red giant phases) are derived from evolutionary tracks. The same is also obtained for classic models. Then we select that particular isochrone (or narrow range of) that fits both the main sequence and red giant population at the same time. For the reasons discussed above, this can be achieved reasonably well only for models with overshooting. This procedure allows us to derive a first estimate of the age and distance at the same time. This latter is then refined by the detailed comparison of the luminosity functions (theory versus observations). All this of course depends on the chemical composition we have adopted.

AARONSON: How did you obtain the luminosity functions you used to derive the LMC distance, and did you correct for incompleteness ?

CHIOSI: From the published HR diagrams, and no correction for incompleteness has been applied. The result is therefore preliminary, even though the observed luminosity functions mimic so closely the expected ones that the problem of incompleteness ought to be marginal.

COMTE: Can you fix the precision you may obtain on the LMC distance modulus by your method ?

CHIOSI: Of the order of $\pm\ 0.25$ mag.

FROGEL: What effect does a change in age have on your fit to Reid and Mould luminosity function ?

CHIOSI: We derived AGB star luminosity functions for various ages of the underlying stellar population at given initial mass function (Salpeter) and star formation decay rate. The results were almost insensitive to changes in age by a factor of two. They were, on the contrary, much more sensitive to the star formation rate and initial mass function. These are however constrained by other observational data and the only parameter left over is the AGB star evolution.

AARONSON: What happens to globular cluster ages with convective overshooting ?

CHIOSI: Since globular clusters have turn-off masses below 1.2 $M_0$, the minimum mass for stars being able to develop convective cores during the main sequence phase (with our abundances), there is no effect of convective overshooting. Core He-burning, on the contrary, may be affected by this phenomenon. However, evolutionary sequences in this mass range have not yet been calculated.

RENZINI: In my talk I pointed out that Flower's (1984) clusters have well populated red giant branches, and then must be older than the age at which RGB stars with degenerate helium cores appear. This implies that the luminosity of the clump stars cannot be used to derive the cluster age. Did you take this into account, when deriving new cluster ages with your overshooting models ?

CHIOSI: The calibrations I have presented cover the whole range of mass where convective overshooting is effective (cfr. Fig 2). If the clusters in question have turn-off masses greater than $M_{HeF}$, the ages we have estimated are correct (within the observational uncertainties). If the turn-off mass is below this limit, then the ages may be underestimated. Only a good photometry down to the unevolved portion of the main sequence will eventually discriminate the type of cluster. In fact, Flower's HR diagrams do not give a clearcut answer to the question whether or not all those clusters have well developed RGB's. With the exception of SL 868 and NGC 1783, almost all bright red stars in the remaining clusters are declared as "probable field stars". It seems to me that the situation is more uncertain than we may wish. In any case, the net result remains that models with overshooting give much longer ages (turn-off and/or red giant stars), thus improving upon the age discrepancy problem, which in my opinion is not entirely of observational nature.

TOSI: What are the effects of overshooting on stellar nucleosynthesis, apart from the obvious consequences of reducing $M_{up}$ ?

CHIOSI: We have not yet explored the nucleosynthetic aspect of the new models in more detail than briefly reported in my talk.

# SPECTRAL EVOLUTION OF GALAXIES

Gustavo Bruzual A.
Centro de Investigaciones de Astronomía (CIDA)
Apartado Postal 264
Mérida 5101-A
Venezuela

ABSTRACT. A summary of the current situation in the subject of spectral evolution of galaxies is presented. First the existing uncertainties in the interpretation of the stellar populations in nearby galaxies are analyzed and the model dependency of the results is discussed. Then a few samples of faint galaxies with measured redshift and photometric properties are used to derive the degree of evolution observable in these samples with respect to nearby galaxies. Finally, recent results that claim that 3CR radio galaxies show evidence of dramatic evolution are discussed. It is concluded that for the vast majority of galaxies spectral evolution is a slow process which may escape detection. In a given volume element galaxies of very different populations coexist, producing a wide spread in the observable properties at any redshift.

## INTRODUCTION

This is the first paper on the subject of the school, namely Spectral Evolution of Galaxies. I will introduce what I think are the most important aspects of this area. This will not be a review talk, but only an introductory one. Despite everything that have been said by the people working in stellar structure and evolution during the last few days (see Renzini, this conference) against the idea of doing population and evolutionary synthesis of stellar systems other than the nearest stellar clusters, I still think that by applying these techniques to a wide variety of systems we can learn many things. We can derive information about these systems and also about the techniques that we are using to study them.

In order to study spectral evolution of galaxies we have to be able to understand the population of nearby galaxies. Even though this may sound as a trivial statement, we are far from having a complete understanding of the type of stars that are present in most galaxies near us. In some cases different techniques will give very different results.

In the following section I will discuss some of the current problems in understanding the dominant stellar populations in nearby early and late type galaxies.

## MODELS FOR THE SPECTRAL EVOLUTION OF GALAXIES

The models for the spectral evolution of galaxies constructed by the author (Bruzual 1981, 1983a,b,c,d; Koo 1985) have been used by several authors as a guide in the interpretation of the photometric data of distant galaxies. In the rest of this paper these models will be used with the same purpose. For reasons of space no details are given here about the models. The reader should consult the references mentioned above.

## NEARBY GALAXIES: EARLY-TYPES

Since many participants in this conference will address the question of stellar populations in elliptical galaxies, I will only include a few comments on the current situation. There are two main problems that remain to be solved. One is the source of the UV light in elliptical galaxies. The other problem is the age of the dominant population in these systems.

Since the first IUE observations of elliptical galaxies became available it was clear that the flux level in the region from 2000 to 3300 Å is higher than one would expect from a typical mix of a metal rich population of stars (Bruzual 1981,1983). The source of this light is not well established. There are several possible sources that range from an underlying population of metal poor horizontal branch stars to a large number of metal rich white dwarfs or recently formed massive main sequence stars.

Most of the evolutionary synthesis models build so far (Tinsley 1972, Tinsley and Gunn 1976, Bruzual 1981, 1983) have ignored important phases in the evolution of post main sequence stars. In particular, the horizontal branch evolution was not treated properly by these authors. The main reason for this was the very uncertain evolutionary tracks available at the time. This models cannot say anything about the contribution of the horizontal branch stars to the light of a galaxy. The situation should improve in the near future as more and more evolutionary tracks that incorporate mass loss and different metal content are becoming available (see Barbaro, this conference).

Another source of uncertainties in the understanding of the population of nearby elliptical galaxies is the use of the same non-evolving giant branch (such as Tinsley and Gunn 1976) for stars less massive than the sun. In the Tinsley and Gunn luminosity function of low mass red giants observed in the vicinity of the sun not all the evolutionary phases are represented.

The metallicity of the giant stars from which the Tinsley and Gunn luminosity function was derived is typically lower than the metallicity of stars in the centers of elliptical galaxies. Thus as a whole the Tinsley and Gunn luminosity function has a mean color that may be too blue compared with the mean color of the giants in the center of an elliptical galaxy. This brings as a consequence in an evolutionary synthesis that very old ages (greater than 12 Gyr) are needed to reproduce the observed colors of these galaxies. With a luminosity function that reflected the higher than solar metallicity observed in the centers of ellipticals, one could reproduce the observed properties with younger models.

Wyse (1985) has studied the effects of including stars in the horizontal branch and post red giant phases in the evolutionary synthesis approach. As a result she can produce elliptical galaxies that become red faster (due to post red giant evolution of relatively massive stars) and argues for a lower formation redshift for these galaxies.

O'Connell (1980 and this conference) derives much younger ages (around 6 Gyr) for the stars in the main sequence turnoff of these galaxies. He uses a population synthesis approach and thus the methods are not directly comparable. However, the differences implied by the different programs are worrisome.

Besides the theoretical progress that is being made in this subject it is likely that the problem of the age of the dominant population in elliptical galaxies, and hence of the source of the UV light, remains unanswered until high resolution observations of elliptical galaxies become available from Space Telescope and other instruments.

## NEARBY GALAXIES: LATE-TYPES

The interpretation of the spectrum of spiral and irregular galaxies is more complicated than that of elliptical galaxies. The UV SEDs show large differences from galaxy to galaxy and from region to region in a given galaxy. The inclusion in the aperture of an active region of star formation shows up as a very steep gradient shortward of 2000 Å. Statistical fluctuations in the number of stars included in the detector can produce noticeable differences in the shape of the SED. The average surface brightness over the entire galaxy is lower in irregulars and spirals than in ellipticals. Thus bright spots are preferentially selected for observation. The effect of dust in the light coming out of these regions is very uncertain and in general is not accounted for. Because of a combination of all of these factors the average spectrum of late type galaxies of a given morphology is not so well defined as for ellipticals.

Since late-type galaxies are recognized as regions of active star formation it is reasonable to assume that young, massive main sequence stars are the source of most of the UV emission and of a large fraction of the flux at longer wavelength.

The theoretical interpretation of the spectrum of late-type galaxies is uncertain essentially because of lack of information about three important quantities: (a) the slope of the IMF, (b) the age of the dominant population, and (c) the amount of light absorbed by dust as a function of wavelength. Other factors, such as the star formation rate (as long as it is large enough as for young stars to dominate the spectrum), the upper mass limit for star formation or the chemical abundance, do not seem to be as critical as the ones mentioned above.

Figure 1 shows the spectrum of the irregular galaxy NGC 4449 from the UV to the IR in a log-log scale normalized to 1 at $\lambda = 5500$ Å. The spectrum shown was provided by R. S. Ellis (private communication). The fluxes reported by Huchra and Geller (1982) for NGC 4214 and NGC 4670 are shown in the same scale. These data provide a good opportunity to

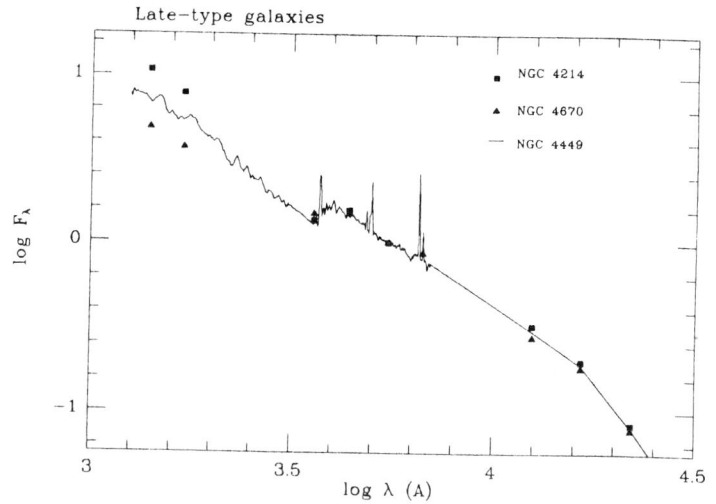

Figure 1.- Observed spectral energy distribution of NGC 4449 (Ellis, private communication). For comparison broad band fluxes are shown for NGC 4214 and NGC 4670 normalized at 5500 Å (Huchra and Geller 1982).

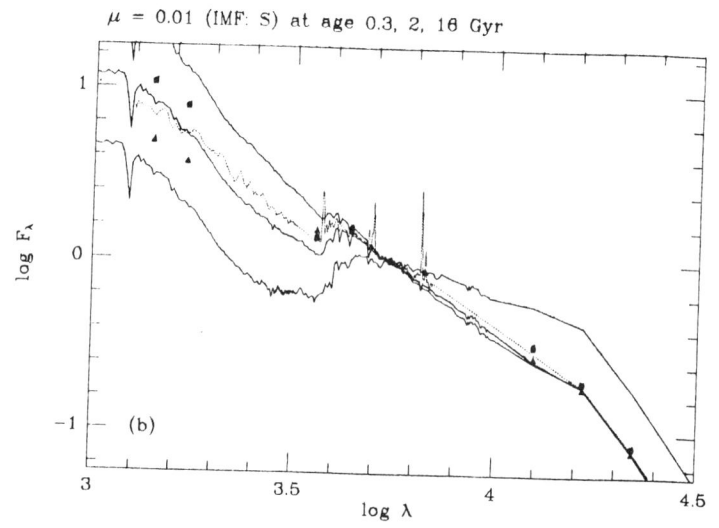

Figure 2.- SEDs for a $\mu = 0.01$ model at tg = 0.3, 2, and 16 GYR (top to bottom in the UV; reverse in the IR) assuming the Salpeter IMF. The obseved SED shown in Figure 1 is repeated as a dotted line.

# SPECTRAL EVOLUTION OF GALAXIES

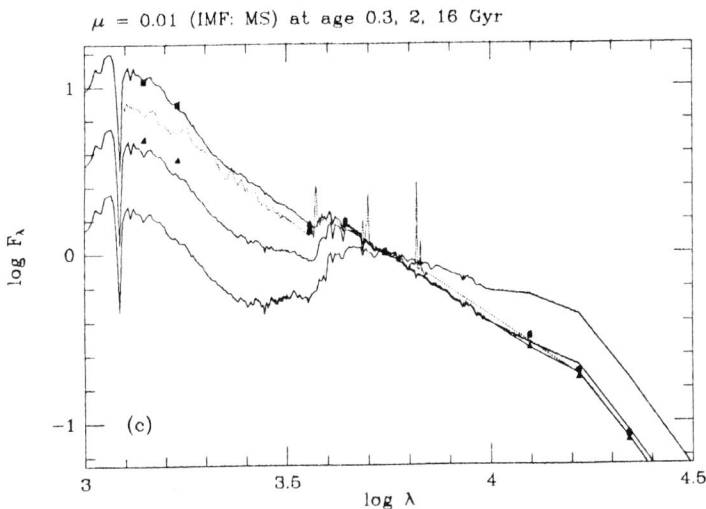

Figure 3.- SEDs for a $\mu = 0.01$ model at tg = 0.3, 2, and 16 GYR (top to bottom in the UV; reverse in the IR) assuming the Miller and Scalo (1979) IMF. The obseved SED of Figure 1 is repeated as a dotted line.

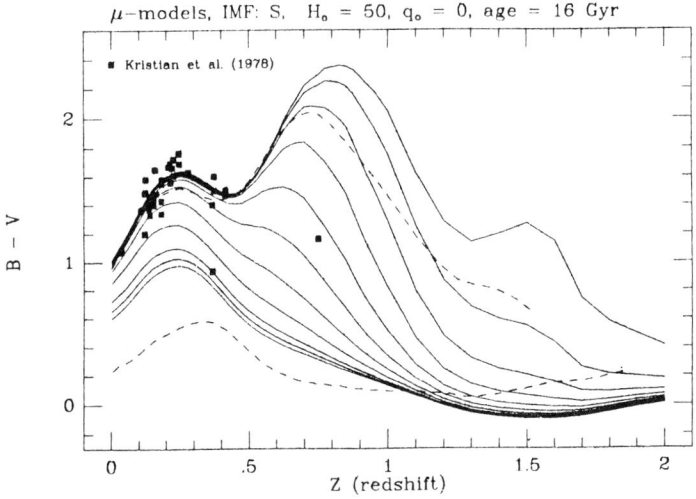

Figure 4.- Behavior of B-V with redshift. Models with the following values of $\mu$: 0.01 (bluest), 0.05, 0.1, 0.2, 0.3, 0.4, 0.5, 0.6, 0.7, 0.8 and 0.9 (reddest) at age 16 Gyr for the Salpeter IMF.

study the effect of some of the factors mentioned above in a galaxy SED. In the rest of this paper the evolutionary models of Bruzual (1981, 1983) will be used frequently. The notation used is the same as in these papers.

Figure 2 shows the spectrum for a $\mu = 0.01$ model at ages 0.3, 2, and 16 Gyr (top to bottom in the UV; reverse in the IR). The data of Figure 1 are repeated for comparison. The model was built with the Salpeter (1955) IMF in the range from 0.08 to 75 $m_\odot$. Figure 3 shows the respective spectra for the same model but using Miller and Scalo (1979) IMF in the range from 0.08 to 25 $m_\odot$.

These figures show the interdependence of galaxy age and stellar IMF mentioned above: for a given age the UV spectrum is very sensitive to the IMF; for a given IMF the UV spectrum is very sensitive to galaxy age. In both cases shown, ages as youg as 2 Gyr seem to be ruled out by the observations. Ages < 1 Gyr seem most likely for the dominant population in these galaxies. The spectra of late-type galaxies of earlier type than NGC 4449 do not require such extreme young ages.

## FAINT GALAXY SAMPLES: COLOR EVOLUTION

From the behavior of the faint galaxy colors as a function of redshift one expects to extract useful information about spectral evolution of galaxies. However, as it will become apparent below, this interpretation is subject to the same uncertainties discussed in the case of late-type galaxies. In the rest of this section a brief discussion will be given of the interpretation of the photometric data of various galaxy samples.

In a general sense the existing faint galaxy samples can be grouped in three different categories according to the criteria under which they have been selected.

## CLUSTER GALAXY SAMPLES

These samples are chosen by systematically studying several galaxies belonging to a cluster. The membership of the galaxies must be established. Cosmological information is derived by choosing clusters at different redshifts. In many instances only the brightest galaxy in a cluster is observable.

In Figures 4, 5, and 6 the B-V colors for first ranked cluster galaxies from Kristian et al. (1978) are compared with several model predictions. In all of these diagrams the top dashed line represents the K-correction corresponding to an observed elliptical galaxy spectrum shifted to the same z. The dashed line toward the bottom part of the diagrams shows the K-correction corresponding to NGC 4449. These two lines have been included to show the expected range of behavior of B-V vs z in the absence of evolution.

Figure 4 shows the behavior of B-V vs z for $\mu$-models when the Salpeter (1955) IMF is used. The following values of $\mu$ have been used: 0.01 (bluest), 0.05, 0.1, 0.2, 0.3, 0.4, 0.5, 0.6, 0.7, 0.8, and 0.9 (reddest). To compute these colors Ho was taken = 50, qo = 0, and the galaxy age was taken as 16 Gyr. Figure 5 is identical to Figure 4 except

# SPECTRAL EVOLUTION OF GALAXIES

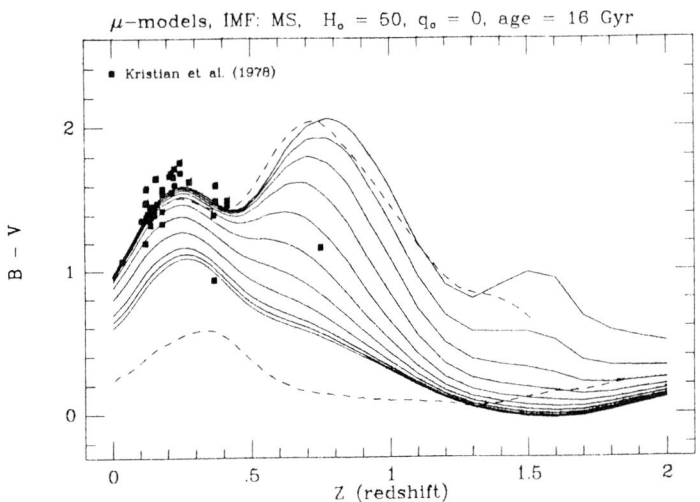

Figure 5.- Behavior of B-V with redshift. Models with the following values of μ: 0.01 (bluest), 0.05, 0.1, 0.2, 0.3, 0.4, 0.5, 0.6, 0.7, 0.8 and 0.9 (reddest) at age 16 Gyr for the Miller and Scalo (1979) IMF.

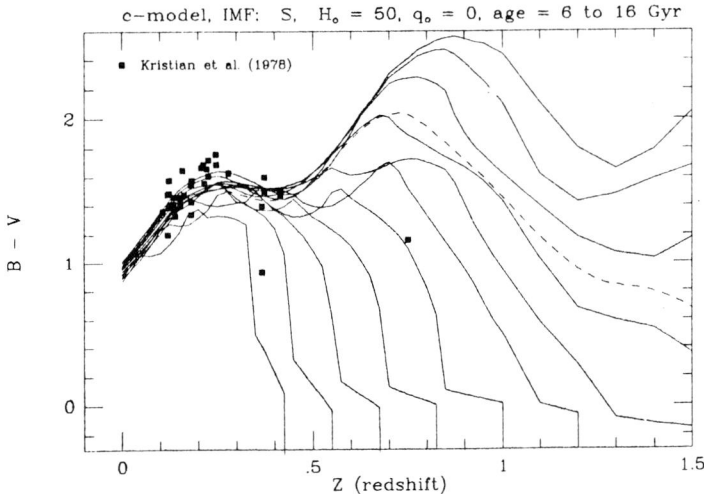

Figure 6.- Behavior of B-V with redshift. c-model ($\tau = 1$) at tg = 6 to 16 Gyr in 1 Gyr steps for the Salpeter IMF.

that in this case the Miller and Scalo (1979) IMF was used to compute the μ-models.

Figure 6 shows the B-V vs z dependence for a c-model ($\tau = 1$) when the galaxy age tg is assumed to vary from 6 (bluest) to 16 Gyr (reddest) in 1 Gyr steps. The Salpeter IMF was used in this case. The vertical segments in the color lines at the bottom of the figure point to the value of the "formation redshift" for the same cosmology as before and the corresponding galaxy age.

The following conclusions can be derived from inspection of figures 4 to 6.

- The first ranked cluster galaxies are distributed in the color-redshift plane around the lines corresponding to μ-models representing elliptical galaxies ($\mu > 0.4$). This is expected since these galaxies are typical examples of elliptical systems.

- In the range from $z = 0.1$ to $0.3$ the B-V color of these galaxies is redder than the color corresponding to a redshifted nearby elliptical galaxy (K-correction). This effect, pointed out by Ciardullo and Demarque (1978), is not understood at present.

- Figures 4 and 5 show that star formation taking place in various amounts in different galaxies may be the cause of the color spread seen in the color-z plane. The models represented in these two figures undergo star formation in a continuous fashion throughout all their lifetime.

- Figure 6 shows, however, that a "spread in galaxy age" is another possible cause of the bluer colors of distant galaxies. This mechanism works even for an initial burst population. If the burst did not take place too long ago, the main sequence stars present in the galaxy are responsible of the blueing of the galaxy colors with z.

BLIND DEEP SAMPLES

These are magnitude limited faint galaxy samples chosen without any information about the nature of the objects included. Kron, Koo, and Windhorst (1985) and Koo (1981,1985) have done the most careful work with a complete magnitude limited sample. Figures 7, 9, and 11 show the behavior of U-J, J-F, and F-N with z, respectively, for μ-models. The data points are from Windhorst (1984). These galaxies have been detected in radio. Figures 8 and 10 show the behavior of U-J and J-F with z, respectively, for μ-models. The data points are from Koo (1985). The dashed lines have the same meaning indicated above. The cosmology is the same that was used in the previous figures.

From their data and a comparison with the author's models the following conclusions emerge:

- Spectral evolution of galaxies is a slow process. At least to $z = 0.7$, a large fraction of galaxies is consistent with no or very slow spectral evolution. The observed colors of galaxies fall in the range expected for slowly evolving stellar populations.

- To the limit of their sample, the early stages of galaxy formation have not been detected. If the opposite were the case, an absence of red galaxies should be apparent at their highest z values. Red galaxies, instead, are well represented at all z's.

- Up to their limit of $z = 0.7$, galaxies of all color classes seem

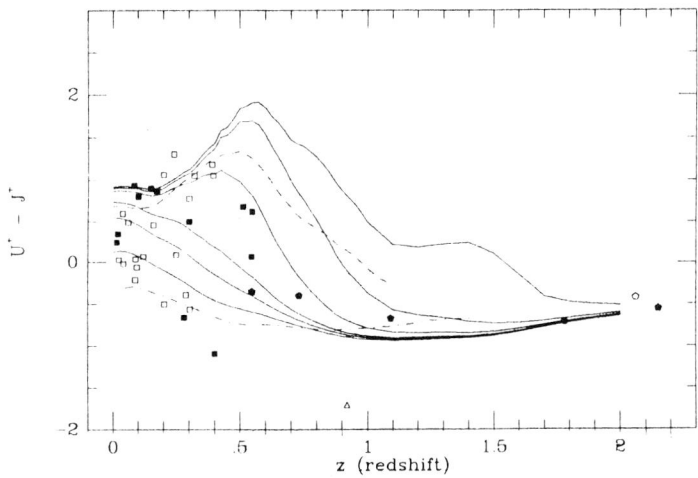

Figure 7.- Behavior of U-J with redshift. Models with the following values of μ: 0.01 (bluest), 0.05, 0.3, 0.5, 0.7, and 0.95 (reddest) at age 16 Gyr for the Salpeter IMF. The data points are from Windhorst (1984).

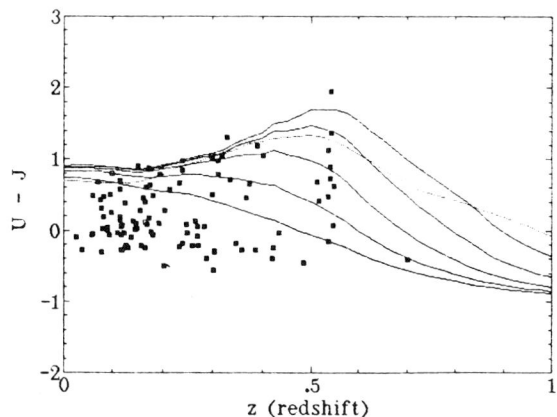

Figure 8.- Behavior of U-J with redshift. Models with the following values of μ: 0.3 (bluest), 0.4, 0.5, 0.6, and 0.7 (reddest) at age 16 Gyr for the Salpeter IMF. The data points are from Koo (1985).

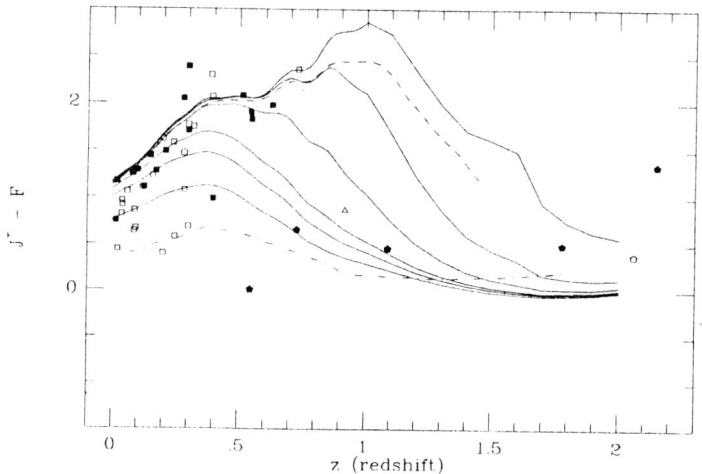

Figure 9.- Behavior of J-F with redshift. Models with the following values of µ: 0.01 (bluest), 0.05, 0.3, 0.5, 0.7, and 0.95 (reddest) at age 16 Gyr for the Salpeter IMF. The data points are from Windhorst (1984).

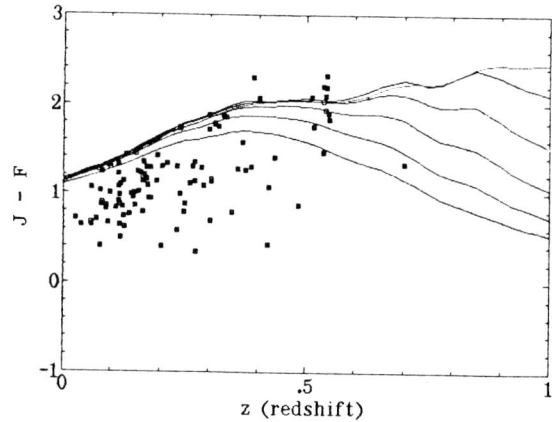

Figure 10.- Behavior of J-F with redshift. Models with the following values of µ: 0.3 (bluest), 0.4, 0.5, 0.6, and 0.7 (reddest) at age 16 Gyr for the Salpeter IMF. The data points are from Koo (1985).

# SPECTRAL EVOLUTION OF GALAXIES

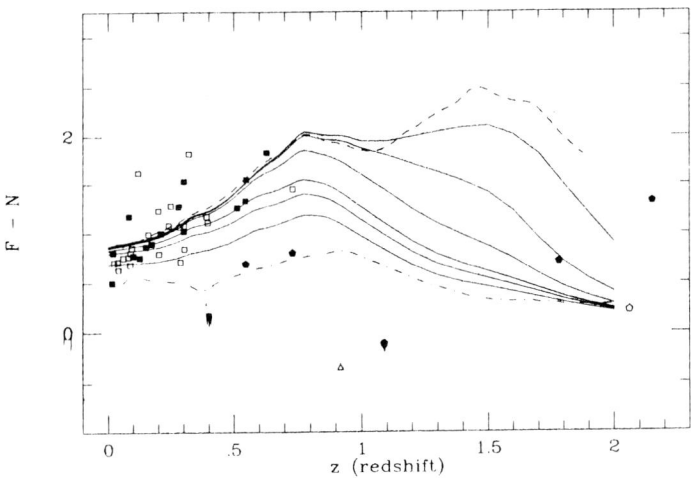

Figure 11.- Behavior of F-N with redshift. Models with the following values of μ: 0.01 (bluest), 0.05, 0.3, 0.5, 0.7, and 0.95 (reddest) at age 16 Gyr for the Salpeter IMF. The data points are from Windhorst (1984).

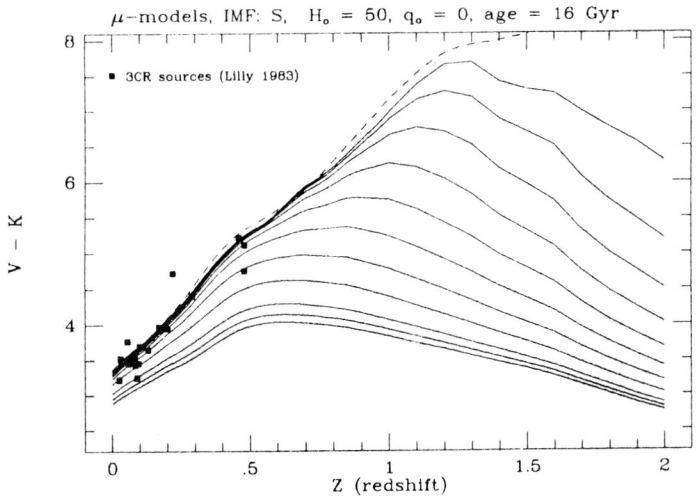

Figure 12.- Behavior of V-K with redshift. Models with the following values of μ: 0.01 (bluest), 0.05, 0.1, 0.2, 0.3, 0.4, 0.5, 0.6, 0.7, 0.8, and 0.9 (reddest) at age 16 Gyr for the Salpeter IMF. The data points are from Lilly (1983).

to coexist at different epochs. The most natural interpretation of their color vs. z diagrams in terms of Bruzual's models is that the reddest galaxies at any z will evolve into galaxies similar to present day elliptical galaxies. The bluest galaxies at any z seem to be the progenitors of present day spiral and irregular galaxies. There is no evidence in their data to support the view that these blue systems will become as red as nearby ellipticals. These bluer galaxies are especially prominent in the sample of Koo (1985) shown in figures 8 and 10.

## PECULIAR GALAXY SAMPLES

These samples contain galaxies that are chosen according to a criterium that singles them out. For example, the optical counterparts of 3C radiosources galaxies. There is some danger in applying the conclusions derived from peculiar galaxies to the general population of galaxies.

It has been claimed by several authors (e.g. Djorgovski and Spinrad, 1985; Lilly 1983) that they can detect spectral evolution in the 3C sources beyond z of 1 and up to the limit of their current data (z of about 2). Their data (not reproduced here) show indeed that the distant galaxies are markedly brighter in the ultraviolet than nearby elliptical galaxies. However, these galaxies are known to have a non-thermal continuum. Before these colors are taken as evidence of spectral evolution it should be proved that the observed colors reflect the colors of the stellar population and are not affected by this emission. As an illustration Figures 12 and 13 show the behavior of V-K and R-K with z for $\mu$-models, respectively. The data points are from Lilly (1983). Figure 14 shows the B-V colors for the four more distant galaxies in Djorgovski and Spinrad (1985) sample. In many of their plots the evidence in favor of evolution is more striking than in this diagram.

One point in favor of their interpretation is the fact that the observed points cluster remarkably well around the model prediction for a softly evolving stellar population. If the non-thermal emission were the dominant effect in these galaxies, there is no a *priori* reason to expect that this emission will evolve in time as to mimic the emission from an evolving stellar population, and not in a more or less random fashion.

Thuan et al. (1984) find that the strong 3CR radio galaxies and the optically selected radio-quiet field galaxies cannot be distinguished by their infrared and optical-infrared colors. They conclude that the colors are probably due to a stellar population and also claim that they see in their sample the effects of a mildly evolving population.

Figures 15 shows the time evolution that is needed in the spectrum of an elliptical galaxy in order to reproduce the magnitude and color evolution detected by Djorgovski and Spinrad (1985).

A less dramatic evolution (2 burst galaxy instead of exponential decay) is shown in Figure 16. If the galaxy is observed close to the second burst, it will look bluer and brighter (in the UV) than the pre-burst galaxy.

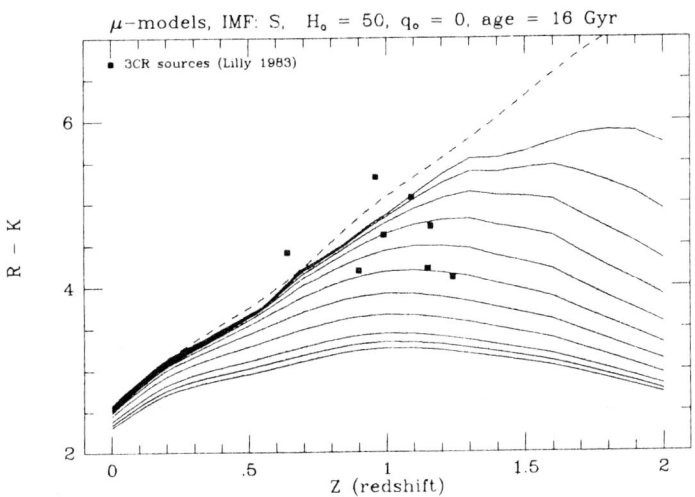

Figure 13.- Behavior of R-K with redshift. Models with the following values of $\mu$: 0.01 (bluest), 0.05, 0.1, 0.2, 0.3, 0.4, 0.5, 0.6, 0.7, 0.8, and 0.9 (reddest) at age 16 Gyr for the Salpeter IMF. The data points are from Lilly (1983).

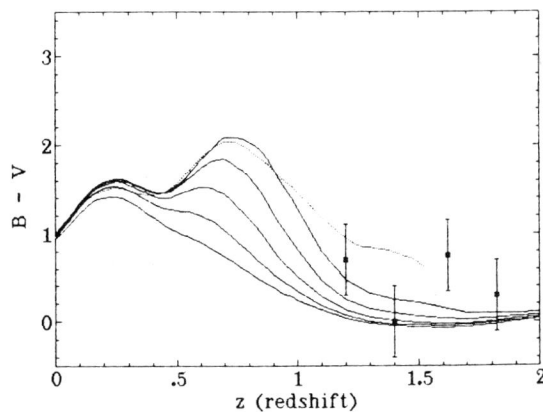

Figure 14.- Behavior of B-V with redshift. Models with the following values of $\mu$: 0.3 (bluest), 0.4, 0.5, 0.6, and 0.7 (reddest) at age 16 Gyr for the Salpeter IMF. The data points are from Djorgovski and Spinrad (1985).

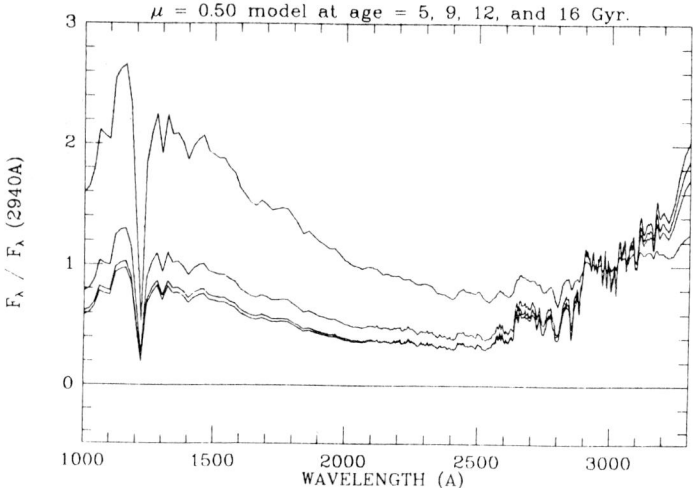

Figure 15.- SEDs for a $\mu = 0.50$ model at tg = 5, 9, 12, and 16 Gyr (top to bottom).

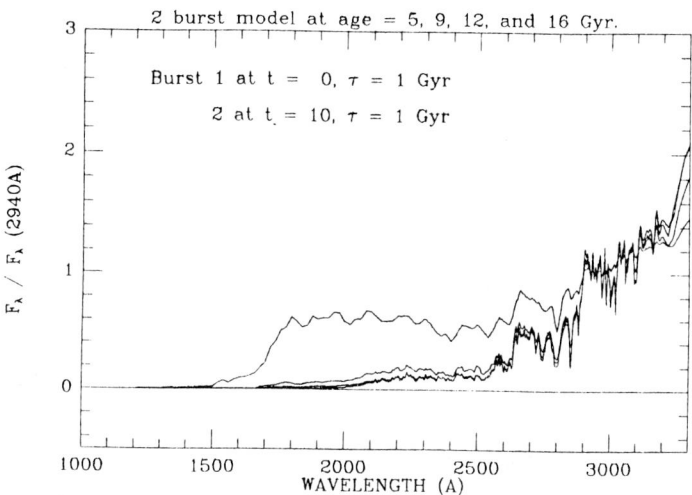

Figure 16.- SEDs for a c-model ($\tau = 1$) in which a second burst of star formation occurs at tg = 10 Gyr. The SEDs are shown at tg = 5, 9, 12, and 16 Gyr. The top line corresponds to 12 Gyr, the other lines are as in Figure 15.

## CONCLUSIONS

The quality of the data on faint galaxy photometry has improved by a large factor in the recent past. However, the detection of spectral evolution remains still uncertain. At most, the data provide some information on the relevance of spectral evolution for the problems of interest.

Much more data are needed before the effects of evolution can be accounted for properly and the cosmological problem approached in a safe ground.

Even though the 3C radio galaxies seem to show a detectable amount of spectral evolution, the selection effects and peculiarities of this sample have not been fully evaluated. Thus it seems dangerous to extend this result to the general population of galaxies.

The deep galaxy samples have shown clearly that different galaxy types coexist at a given redshift, and that evolution, if it occurs, is much slower than previously thought. The early stages of galaxy evolution occur long before the epoch that is being sampled.

The particular cosmological model used in the interpretation of the observations is of minor importance for the conclusions. The evolutionary effects are much greater than the differences introduced by different cosmologies.

## REFERENCES

Bruzual A., G. 1981, Ph. D. thesis, University of California, Berkeley.
Bruzual A., G. 1983a, Ap. J., 273, 105.
Bruzual A., G. 1983b, Rev. Mexicana Astron. Astrofis., 8, 29.
Bruzual A., G. 1983c, Ap. J. Suppl., 53, 497.
Bruzual A., G. 1983d, Rev. Mexicana Astron. Astrofis., 8, 63.
Ciardullo, B., and Demàrque, P. 1978, in IAU Symposium 80, The HR Diagram, eds A. G. D. Philip and D. S. Hayes (Dordrecht:Reidel), p. 345.
Djorgovski, S. and Spinrad, H. 1985, Ap. J. (in press).
Huchra, J., and Geller, M. 1982, in Four Years of IUE Research, eds. Y. Kondo, J. M. Mead, and R. D. Chapman, NASA:CP-2238.
Koo, D. C. 1981, Ph. D. thesis, University of California, Berkeley.
Koo, D. C. 1985, A. J., 90(3),418.
Kristian, J., Sandage, A., and Westphal, J. A. 1978, Ap. J., 221,383.
Kron, R. G., Koo, D. - C., and Windhorst, R. A. 1985, Astr. Ap. (in press).
Lilly, S. J. 1983, Ph. D. thesis, University of Edinburgh, Scotland.
Miller, G. E., and Scalo, J. M. 1979, Ap. J. Suppl., 41, 513.
O'Connell, R. W. 1980, Ap. J., 236, 430.
Salpeter, E. e. 1955, Ap. J., 121, 161.
Tinsley, B. M. 1972, Astr. Ap., 20, 383.
Tinsley, B. M., and Gunn, J. E. 1976, Ap. J., 203, 52.
Thuan, T. X., Windhorst, R. A., Puschell, J. J., Isaacman, R. B., and Owen, F. N. 1984, Ap. J., (submitted).
Windhorst, R. A. 1984, Ph. D. Thesis, Sterrewacht Leiden (Holland).
Wyse, R. F. G. 1985, Ap. J. (in press)

# DISCUSSION

O'CONNELL: Your arguments that little or no spectral evolution has been detected at z around 1 contradicts some strong recent claims by Lilly and Longair, Djorgovski and Spinrad, and Schild. Where have they gone wrong? I think Lilly and Longair ruled out contamination by nonthermal radiation in their 3CR sample. All groups also claimed a systematic color change consistent with a $\mu = 0.5$ model, which perhaps suggests they are observing a uniform sample of galaxies (ellipticals?) at different redshifts.

BRUZUAL: Despite all that have been said about spectral evolution of galaxies, very few attempts have been made to define in a unique way what would be considered evidence for spectral evolution. There is no doubt that we expect passive evolution to occur in galaxies, produced by the evolution of their stellar content. This is the kind of evolution that I have intended to model in the work I have just described. The interaction with the environment, which can change completely the passive evolution scheme, has been neglected.

How to proceed in order to detect passive evolution is a different matter. The color vs. z lines, of which I showed several examples, indicate the expected behavior in this plane of a sample of galaxies undergoing passive evolution at different rates. These lines are, in general, bracketted by the lines corresponding to the K-corrected color derived from an observed spectrum. In order to detect passive evolution we need to study a complete sample of galaxies, up to a value of z which is high enough to allow us to discriminate between the pure K-corrected color and the evolutionary prediction, and see if galaxies avoid the region of the color-z plane that corresponds to non-evolving E-galaxy spectra. These galaxies would then be seen at earlier stages of their evolution, when upper main sequence stars make an important contribution to their light, which makes the corresponding colors bluer than the non-evolving K-corrected color at the same z.

As it is apparent from the color vs. z diagrams shown, several star formation histories will lead to galaxies with very similar colors at z $< 0.5$. This is the case for the $\mu$-models with $\mu > 0.5$. When we study a complete sample, as the one mentioned above, we should be able to tell which star formation history is followed by all or most galaxies, i.e. determine the most appropriate value of $\mu$.

All the authors you mentioned in your question have detected galaxies which at their measured z appear bluer than nearby ellipticals would at the same z. However, even if non-thermal radiation is ruled out as the source of the excess blue light, still we do not know what kind of galaxies these systems are. Assuming that these authors are observing the brightest galaxies inside the volume element at the given z, all we can say is that these galaxies are definitely bluer (within the photometric errors) than the brightest galaxies in nearby clusters would appear when K-corrected to the same z. It could happen that these high z galaxies correspond to systems that will evolve into galaxies that are bluer than nearby ellipticals, or to systems that will "evolve" along the K-correction line of some population mixture. In both of these cases the observations mentioned will be related more to the evolution of the distribution of galaxies in color class as a function of redshift, than to the spectral evolution of galaxies properly.

The idea of spectral evolution, in its most simple minded version, implies that past a given value of z we should not see galaxies that are consistent with the non-evolving K-corrected colors of an old population. What I meant on my talk is that up to z of 0.7, Koo, Kron, and Windhorst sample contains galaxies at all z's which have colors that are consistent with a very slowly or non-evolving spectrum of an old population. In their sample all color classes are well represented and coexist at a given time (see Figures 8 and 10). The bluer galaxies in their sample could be interpreted as non-evolving younger populations, and then one could argue that this sample is consistent with no spectral evolution at all, even though this is against our intuitive notion of passive evolution.

Claims that spectral evolution has been detected on the basis of a single data point in the color vs. z plane should be taken with caution. For those galaxies which have several colors measured, the degree of evolution implied should be the same independently of the color used. This is not always the case. Unfortunately the complete samples that are available now do not go deep enough in redshift, whereas the distant galaxies with good photometry are isolated cases, without any information about the nature of neighboring galaxies. This situation should improve in the near future.

RENZINI: Among hot stars in ellipticals you have mentioned young massive stars and hot horizontal branch stars. These types of stars may or may not be present in ellipticals, but there is another kind which certainly must be present, and these are post AGB stars.

BRUZUAL: Yes, I am aware of this fact. Most of the population synthesis models that have been constructed, including mine, lack some important stellar groups, such as the ones you mentioned, and even more important, a more accurate giant branch. For the evolutionary synthesis approach we need detailed evolutionary tracks for the missing groups. In some cases the tracks are available, but the spectra or some representative magnitudes for these stars along their evolutionary tracks are not. Approximate values for the bolometric correction and colors of some galactic stars that are thought to be in the post-AGB phase can be found in the literature. However, we do not know how well the local stars represent the stars in the center of giant ellipticals. If there are differences due to metallicity, age, mass, etc., the use of the local values may introduce unknown systematic errors in the galaxy models which will make the interpretation of observations equally or more uncertain that it is now (without the post-AGB). Until a complete and self-consistent set of observational properties becomes available for these stellar types, I think that it is better not to include them in population synthesis models (for a different approach see the paper by Wyse mentioned in the references).

There is no question that current population models, such as the ones I have just described, are only a first order approximation to the actual stellar populations present in galaxies. I think that they cannot be completely wrong, because there is quite good agreement with very different types of observations. I think that the models are still useful to understand, to first order, distant galaxy properties, as well as to make predictions of various properties of faint galaxy samples. The kind of models I have presented are not accurate to describe the stellar population in nearby stellar clusters, such as the ones you need to build the population template you mentioned in your talk. This was never the purpose of my models. Still I think that some understanding about galaxy evolution or lack of evolution has been obtained from simple minded, not very accurate models such as Tinsley's or mine.

PICKLES: What value of $\mu$ fits present elliptical colors if star formation started 16 Gyr ago? How good is the fit over the whole spectrum?

BRUZUAL: You need values of $\mu > 0.5$. In the visible and IR the fit is very similar for all of these models and is reasonably good. In the UV the lower $\mu$ models fit the data best because the upper main sequence stars present in larger amounts in these models produce the required UV flux observed in "normal" ellipticals. To reproduce hotter ellipticals with $\mu$ models you need either values of $\mu$ lower than 0.5, or younger ages. Even though the source of UV light in ellipticals may not be the main sequence stars, I think that the models mentioned give you an indication of the amount of energy involved in the process. This information can be used to derive the number of other possible candidates.

PICKLES: (1) Would a set of metal-rich models (at some $\mu$) enclose the observed galaxy colours which are redder than the present model predictions?, and (2) If so, and if these observed colours are not photometric errors, would all galaxy colours then fit with variable $\mu$ or variable amounts of recurrent star formation?

BRUZUAL: I have not built the models that you mention but I think that it is possible to fit nearby ellipticals with a metal richer stellar population. More important than the value of $\mu$ will be the galaxy age. Metal rich stars evolve more slowly and the galaxy as a whole will become red at an older age. However, at any age, stars will be redder and these two effects may cancel each other. The post-AGB stars should also be included. These stars will also contribute to make the models redder at a younger age (Wyse 1985).

As you are aware, to build these models we need a complete stellar library, which includes metal-poor, metal-rich and solar metallicity stars. From my experience with the present models I think that we need accurate giant branches for all these metallicities because, to a first approximation, the spectrum of an old population is very similar to the combined spectrum of the giant branch. These giant branches should also include differences due to different stellar masses and, hopefully, all the evolutionary stages mentioned by Renzini. The amount of information needed is quite large and I think that we are far from solving the problem using population synthesis. It is likely that analytical work, such as Arimoto's, will help us understanding this problem.

ROCCA-VOLMERANGE: Are your far UV spectral data from Kurucz's models or from the IUE Spectral Atlas?

BRUZUAL: I use Kurucz's models only for stars hotter than 30 or 40 thousand K. For cooler stars I use IUE data obtained by myself and by other authors, but not from the Atlas compiled by C. C. Wu. I am now in the process of revising my library of stellar spectral energy distributions in order to incorporate this Atlas.

DE YONG: Could you explain in some detail the models that you used to compute the spectra of dusty galaxies shown in your talk?

BRUZUAL: This work is still in progress in colaboration with G. Magris at CIDA.

We solve the equation of radiative transfer for a mixture of stars and dust grains distributed homogenously in an infinite plane parallel configuration. The effects of absorption and multiple scattering of light due to dust grains with the same optical properties as the interstellar dust in the Galaxy are included.

The values of the albedo and the asymmetry factor published in the literature allow us to solve the problem in the range from 1000 to 10000 A. The results are computed in the form of correction factors as a function of wavelength, inclination angle, and optical depth, $\tau$, of the system. These factors can be used to correct the spectrum of a mixture of stars and dust to obtain the spectrum of the stellar component alone.

It is found that for $\tau < 1$ the correction factors depend critically on the inclination angle. Since the surface brightness profiles of spiral galaxies do not show this dependence, we assume that for these galaxies $\tau > 1$. For values of $\tau > 1$ the difference between our correction factors and the standard reddening law is large, especially in the ultraviolet region of the spectrum. We think that our correction factors should be prefered over the standard reddening law to unredden galaxy spectra.

From the corrected spectra we compute the properties of disk galaxies in several photographic and photoelectric bands. Thus we compute magnitude and color vs. redshift for galaxies seen at different inclinations angles and with different values of the optical depth through the plane of the galaxy.

DE YONG: Now that you have apparently gone through the trouble of solving the full radiative transfer problem of a dusty disk with embedded stellar sources you might just as well go one step further by taking into account that the dust layer and the source layer have different scale heights. This problem has been solved before by people interpreting observations of the diffuse galactic light in order to predict what an observer in the disk would observe, rather than one outside the disk that you are more interested in.

DE YONG: I also think that it would be important to apply your calculations to try to improve the (too) simple recipes that are generally used for calculating corrections for internal reddening in galaxies as a function of inclination angle and effective wavelength of the band used.

BRUZUAL: Thank you very much for your suggestions. We will incorporate them into our project.

# SPECTROPHOTOMETRIC MODELS OF GALAXIES

G. Barbaro[*‡], F. M. Olivi[‡]
[*]Institute of Astronomy, University of Padua
vicolo dell' Osservatorio 5
35100 Padova
Italy
[‡]International School for Advanced Studies
strada costiera 11
34014 Trieste
Italy

ABSTRACT. This paper analyses the basic assumptions for the computation, by the evolutionary synthesis, of galaxy spectral energy distributions. In particular it is considered how the models are affected by the unsatisfactory knowledge of some evolutionary phases (e.g. core He-burning of low mass stars), the omission of some phenomena in the calculation of stellar evolutionary tracks (e.g. overshooting from convective core) or the neglect of stars in some advanced phases, whose contribution usually is thought unimportant (e.g. AGB and p-AGB stars). The input from the theory of stellar atmospheres is also analysed. As intermediate step in the process of galaxy synthesis, models are presented for single generations of stars of different ages; for old generations two chemical compositions are used: $Z=10^{-4}$ and $Z=10^{-2}$. Such models are compared with the observed spectra of clusters in the LMC and of globular clusters of our galaxy. Galaxy models are then constructed by changing the values of some parameters and allowing for the presence of two populations.

## 1. INTRODUCTION

With the exception of the nearest ones, the study of the galaxies is based on the information included in their integrated light, since the totality or the majority of their population cannot be resolved into single stars. Therefore methods have been developed which allow the analysis of the integrated light to ascertain the nature of the hidden objects responsible for the emission. Although the problem has been faced since nearly half a century ago (Wipple 1935), only in the last two decades have two efficient tecniques been developed to cast light on the populations of unresolved galaxies: the population synthesis and the evolutionary synthesis. This paper deals only with the second method by which the integrated light of a galaxy can be modelled by making suitable assumptions on the stellar birth rate and following the evolution of its population; in this way it is possible not only to derive the population and the spectral energy distribution (SED) at the present time but also their past evolution can be reconstructed. At least in principle one cannot enucleate the problem of the population and spectral evolution from the general evolutionary frame and particularly one cannot neglect dynamic and chemical evolution. The present knowledge of the star formation theory allows one to foresee how important is the dynamic behaviour of the interstellar medium for the astrogenetic processes and suggests an even more important role during the collapse of the protogalaxies. On the other hand the importance of the chemical evolution is evident as it defines one of the basic parameters characterizing the stellar populations: the metal content.

Although Larson and Tinsley (1974) followed simultaneously the dynamic, chemical and photometric evolution of a galaxy model, still the inclusion of the dynamic factors in the integrated light synthesis at present requires considerable effort which is not rewarded by proportionate advantages, since the present status of the theory of star formation is unable to correlate the star formation rate to the state of the interstellar medium. The chemical evolution, on the contrary, can be computed without particular difficulties, at least in an approximate form, to derive the time dependence of the metal content of the interstellar gas and therefore the initial chemical composition of each generation of stars which enters in the model synthesis. The distiction between populations of different chemical composition seems important in models whose light is dominated by old stars since, from the theory of stellar evolution it is well known that the metal content strongly affects the lifetime and effective temperature during some evolutionary phases.

The synthetic approach is generally based on the following simplifications:
- the non-thermal emission from the interstellar gas is negligible; this automatically rules out the possibility of modelling active galaxies, whose activity cannot be ascribed to the stars;
- the thermal emission from the interstellar gas is negligible; Huchra (1977) in fact has shown that the continuum emission from the interstellar gas is negligible compared to that of stars;
- the absorption of light from gas and dust can be neglected. Although this approximation is fairly correct for the models of early type galaxies, it does not seeem justified in the case of late-type galaxies, where gas and dust are present in considerable amount; Bruzual in this workshop has presented a first attempt to evaluate the intrinsic absorption of light.

Therefore, on these grounds, the integrated light of a galaxy is the superimposition of the light emitted from all the stars that populate the system.

Let $f_\lambda(m,\tau,Z)$ and $C(m,t)$ denote respctively the monochromatic flux emitted by a star of mass m, age $\tau$ and metal content Z and the stellar birth-rate, then the integrated monochromatic flux of the galaxy is given by:

$$F_\lambda(T) = \int_0^T \int_{m_L}^{m_U} C(m,t) \, f_\lambda(m,\tau,Z) \, dt \, dm \qquad (1)$$

where T is the age of galaxy and $(m_L, m_U)$ the stellar mass interval.

In principle the evaluation of $F_\lambda$ requires the contribution of several theories:

a) The star formation theory which should supply us with the stellar birth rate $C(m,t)$; however, as already remarked, the present status of development of this theory does not allow the derivation of reliable estimates for the birth rate in terms of physical conditions of the interstellar gas.

b) Theory of chemical evolution which yields the function $Z(t)$ describing the evolution of the metal content.

c) Theory of stellar evolution by which the luminosity and effective temperature of the stars are evaluated as functions of the mass, chemical composition and age. Some difficulties can arise due to the uncertainties connected with the treatment of advanced phases, whose contribution cannot be neglected without affecting the model and to the lack of a sufficient number of evolutionary computations.

d) Theory of stellar atmosphere from which we derive the SED $f_\lambda$ in terms of effective temperature, surface gravity and chemical composition.

In what follows of this paper we analyse in detail how the uncertainties of the theories of star interiors and stellar atmospheres can affect the integrated SED and what consequences are ecountered when neglecting some particular evolutionary phases.

Model spectra are derived for single generations and then compared with the observed spectra of same cluster of the LMC and of globular cluster of our galaxy.

SEDs for galaxies are computed for suitable stars formation rates and allowing for two populations of different metal content.

## 2. INPUT FROM THE THEORY OF STELLAR STRUCTURE

### 2.1. Central H-burning

It has been shown (Maeder 1975, 1976; Bressan et al. 1981) that the inclusion of overshooting from the convective core affects both the age and the extension of the Main Sequence band: therefore tracks, in which core convection during the Main Sequence phase is treated by accounting for the overshoot, could modify in some way the ultraviolet and optical portion of the spectrum. More important probably are the uncertainties in the evolution of massive stars ($m \gtrsim 20 \, M_\odot$). In this case the temperature distribution of early type stars, as deduced from the theory, is in strong disagreement whith the observed distribution (Humphreys 1984) in the sense that, according to the former there should be a large gap in correspondence to the early B types (Bertelli et al. 1984) which, on the contrary, does not appear in the observed HR diagram. The consequences of this disagreement should influence the UV and blue part of the spectrum, not however in a dramatic way.

2.2. Central He-burning of intermediate mass stars

The overshooting from convective core during both core H-burning and He-burning affects the optical and infrared portion of the SED, because both the loop extension and the corresponding lifetime are significantly modified (Chiosi, this workshop).

2.3. Central He-burning of low mass stars

Here we refer to stars undergoing He-flash. Their mass is less than 2.2 M⊙ according to Iben (1967a) and Sweigart and Gross (1978) for Pop.I stars and is less then than 1.6 if overshooting is taken into account during core H-burning (Barbaro and Pigatto 1984, Bertelli et al. 1985). Their evolution during core He-burning is affected by considerable uncertainties and this in turn influences the evaluation of the UV flux. The analysis of the HR diagram of globular clusters has revealed that the morphology of their horizontal branches (HB) cannot be interpreted, whithin the classical evolutionary theory, in terms of change of only one parameter: the metal content. Evidence is growing, on the contrary, that a further parameter is required but its identification is, however, still doutbful. Faced by these difficulties galaxy modelmakers have adopted different solutions: some have simply ignored the He-burning for such stars, others have adopted a semiempirical approach by distributing a suitable number of stars along the zero age horizontal branch according to a temperature distribution which depends on one or more parameters and which can be constrained by the observations. Others have adopted the theoretical approach including the effects of mass loss during the shell H-burning which determines the correct mass of stars burning helium. Rood (1973) suggested that the temperature distribution of the HB stars cannot be interpreted unless the amount of mass lost during the red giant branch changes from star to star giving rise, on the HB, to the following distribution

$$P(m,t) = g(t) R(m) \qquad (2)$$

$$R(m) = (m - m_c)(m_i - m) \exp\left[-\left(\frac{m - m_{fl}}{\sigma}\right)^2\right] \qquad (3)$$

where $g(t)=const=g$ and $m_{fl}$ is the most probable value of the mass at the moment of the flash, which can be evaluated from the formulae of Reimers (1975) or of Fusi-Pecci and Renzini (1975), $m_i$ and $m_c$ are respectively the original mass and the core mass of the stars undergoing the flash; the constant $\sigma$ has the value 0.025, determined by the fitting with the observed distributions. While the evolutionary approach prevents the modelling of anomalous globular clusters, i.e. clusters whose HB differs from what is expected by the theory for a given metal abundance, it provides the way of computing the HB evolution of a generation of stars, and this indeed is otherwise impossible.

Moreover one can guess at the inaccuracy of the results, if HB stars are omitted in the synthesis of a metal-poor generation. Sil'chenko (1983) has analyzed the HR diagrams of the globular clusters collected in the Dudley Catalogue and computed for each cluster the integrated colours $(B - V)_o$ by summing up the contributions of all the stars placed in the HR diagram. To test if the stars considered in the HR diagram are representative of the whole population of the cluster he has evaluated the difference $\delta$ between the integrated $(B - V)o$ colour as determined from the integrated light and the same colour as determined from the stars considered in the HR diagram. Retaining only those clusters for which $\delta$ is small, Sil'chenko has found that, while for metal-rich clusters ( [Fe/H])-1.1) the contribution to the integrated colour $(B - V)o$ of HB stars is small, for metal-poor clusters the corrections cover a large range, the maximum value being of about 0.2 magnitudes.

2.4. Double shell phase (AGB)

This phase has usually been neglected partly because it was believed (or hoped) that its contribution to the total light of a generation was small and partly due to the lack of the suitable theoretical background necessary to evaluate this contribution. Renzini and Buzzoni(1983) and Renzini (this workshop) have stressed the importance of such stars in the synthesis of the integrated light of a generation. According to them the largest percentage contribution from AGB stars to the total bolometric luminosity is reached for a generation of intermediate age $(t=10^8$ yr); it then decreases as the age increases, but still for $t \gtrsim 10^{10}$ yr is not negligible.

For lower ages the luminosity of AGB stars goes to zero through a discontinuity at an age of $2 \times 10^7$ (evaluated in an evolutionary scheme which does not consider the overshooting from the convective core) namely in correspondence to a mass of 9 M☉: for larger masses the C+O core is no longer degenerate. The models of single generations in which the contribution of AGB stars is added according to Renzini's prescription can show how the spectral energy

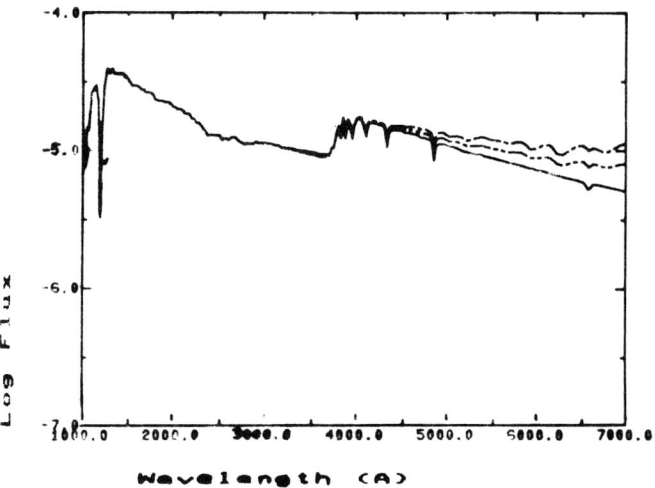

Fig. 1. Contribution from AGB stars.

distribution is modified. Fig. 1 shows (lower curve) the spectrum of a generation $10^8$ yr old with metal content $Z=10^{-2}$, computed with the method and under the assumptions described in Sect. 4; while in this case the AGB contribution is not included, in the upper curve the light of these stars is considered by supposing that their effective temperature is 3400°K and their total bolometric luminosity is 50% of the luminosity of the whole generation. The intermediate curve is instead derived by the assumption that AGB stars contribute with 30% of the total luminosity. As it appears from the comparison of the spectra, AGB stars do not affect the spectral distribution for $\lambda < 4000 A$; but in the red region (and even more in the near infrared) they sensibly influence the flux.

Table I shows, for generations of different age, how the UBV colours are modified by the inclusion of AGB stars when their total luminosity is evaluated according to Renzini.

TABLE I

| $t(10^6 yr)$ | $L_{AGB}/L_{TOT}$ | $\Delta (U-B)_o$ | $\Delta (B-V)_o$ |
|---|---|---|---|
| 22.5 | 0.20 | 0.04 | 0.14 |
| 31.7 | 0.30 | 0.05 | 0.18 |
| 100. | 0.50 | 0.08 | 0.22 |
| 250. | 0.44 | 0.06 | 0.15 |
| 1600. | 0.15 | 0.02 | 0.03 |

One can ask in what direction and to what extent Renzini's prescription is modified by use of the new theoretical background in which overshooting is taken into account. Due to increased mass of the C+O core during the thermal pulses AGB phase, the luminosity increases, being a linear increasing function of the core mass (Paczynski 1970). On the other hand correspondingly the mass of the envelope is reduced and this implies a shorter lifetime. It is impossible to ascertain, without a detailed evolutionary computation, whether the first effect or the second prevails; possibly the change will be relevant. Moreover the age at which in a generation the AGB will develop is modified, as Chiosi (this workshop) has shown that the corresponding mass will be lowered up to about 6 M$\odot$.

2.5. Post-AGB phase

Also for this phase, according to Renzini and Buzzoni (1983), although the corresponding lifetime is very small, the contribution to the total light is not negligible in generations older than 10 billion years. Due to the high temperatures reached during this phase the influence is expected mainly in the far UV. In Sect. 6 a galaxy model will be presented in which the contribution of post-AGB stars is roughly estimated.

3. INPUT FROM THE THEORY OF STELLAR ATMOSPHERES

The most adopted and useful models are those of Kurucz (1979) because:
- they cover a wide spectral region: from 229 A to 20000 A;
- they have been computed for several values of the metal content ($Z=1/2$ Z$\odot$; $Z=1/20$ Z$\odot$; $Z=1/200$ Z$\odot$);
- the range in effective temperature and surface gravity is wide; for the temperature the models are computed in the following intervals

$Z = 1/2$ Z$\odot$     5500°K $\leq T_e \leq$ 50000°K
$Z = 1/20$ Z$\odot$; $Z = 1/200$ Z$\odot$     5500°K $\leq T_e \leq$ 10000°K.

The interval of effective temperatures and surface gravities covered by the models of Kurucz is however narrower than the whole temperature and gravity range described by the evolutionary tracks and therefore one must make up for this lack. Most resonable solutions are:
- for low temperatures (T < 5500°K) the observed UV and optical spectra of cool stars of the late type can be used;
- for extremly high temperatures (T > 50000°K) a good approximation is the black body spectral energy curve because the blanketing effect becomes negligible, all the elements, even the heavier ones, being almost completely ionized;
- for high temperatures (10000°K < T < 50000°K) in the case of a composition more metal-poor than the solar one Kurucz's models for Z=1/2 Z☉ can be used; Capuzzo-Dolcetta et al. (1981) in fact have shown, by comparing model with Z=1/2 Z☉ and models with lower metal abundance in the temperature range 10000°K-25000°K that the differences between models with different metal content but same effective temperature, tend to decrease and become meaningless as the temperature increases.

Kurucz's models of stellar atmospheres have been reckoned under the following assumptions: 1) parallel-plane geometry, 2) hydrostatic equilibrium, 3) local thermodynamic equilibrium, 4) radiative or convective transfer of energy, 5) the molecular component is omitted in the equation of state. These approximations are generally fairly correct; in some situation however they can lose their validity, for instance hypotesis 5) is not realistic for the atmospheres of cool stars while condition 3) fails for systems whith low surface gravity.

Kurucz's models are in good agreement with the observed spectra of A and F stars (Malagnini et al. 1982) and B stars (Malagnini et al. 1983) at least for what concerns the continuum in both the UV and the optical regions.

The most recent and refined evolutionary tracks for very massive stars (cf. review paper of Chiosi and Maeder, 1985) imply effective temperatures as high as 150000°K (however such high temperatures are reached only for a very small fraction of the total lifetime, of the order of $10^5$ yr). For such models as far reproduction of the continuum is concerned one can use Kurucz's models when $T_e$ < 50000°K. This approximation can be applied also to those models which, owing to the loss of a large amount of mass, have a surface composition deficient in H, as it turns out that atmospheric models of pure He, at high temperature, are not very different from the models with normal composition (Underhill 1980). For $T_e$ > 50000°K the black body emission curves can be a reasonable approximation since they fit very well both Kurucz's models for $T_e \approx$ 50000°K and the observed spectra of O type stars.

Alternatively to the models of stellar atmospheres one can use a library of observed stellar spectra which covers the required range of effective temperature, surface gravity and whose wavelength interval is sufficiently large. To build up a library of observed spectra the best material is offered by the spectrophotometric observations which analyse the energy distribution in several points and within rather narrow bands so that the corresponding flux can be considered almost monochromatic. The observations most useful for an evolutionary synthesis are those at low dispersion, since the goal, at present, is to reproduce the behaviour of continuum without taking into account the detailed spectral features. Straizys and Sviderskiene (1972) have derived standard energy distributions in the optical region for several stars of different spectral types and luminosity classes by combining observations of several authors with atmospheric models. Spectrophotometric UV observations are presented by:

a) the Bright Stars Catalogue which gives monochromatic fluxes in 61 bands for 1971 stars observed by the TD1 satellite (Jamar et al. 1976, Macau-Hercot et al. 1978);

b) the UV Stellar Spectral Atlas of the OAO-2 satellite (Code and Meade 1979, Meade and Code 1980) with the fluxes of 330 bright stars corresponding to 120 bands in the wavelength range 1200 - 3580 A;

c) the IUE Stellar Spectral Atlas which covers the spectral region between 1150 and 3200 A.

The derivation of a stellar flux library from the collections of observations of different sources encounters several difficulties due to the uncertainty of the relations between the observed data and the theoretical quantities Te and g. The method generally adopted to compute these relations essentially consists in deriving from the atmosphere models some parameters and comparing them with the same observed quantities relative to several spectral and luminosity classes until the fit is obtained; in this way the required correspondence is obtained.

## 4. THE SPECTRA OF YOUNG STELLAR GENERATIONS

Since there is no definite indication whether and how the initial mass function $\varphi(m)$ of the stellar generations changes during the history of a system like a galaxy, usually the conservative hypothesis is made that $\varphi(m)$ does not evolve. This is equivalent to assume that the birth-rate $C(m,t)$ is the product of the function $\varphi(m)$ of stellar mass and the of a function $B(t)$ of the time. If $\varphi(m)$ is normalized to 1 in the stellar mass interval $(m_L, m_U)$ $B(t)$ gives the number of stars, of any mass, which are born per unit time.

Relation (1) can then be written in the following form:

$$F_\lambda(T) = \int_0^T B(t)\, f_\lambda^*(\tau)\, dt \qquad (4)$$

where

$$f_\lambda^*(\tau) = \int_{m_L}^{m_U} \varphi(m)\, f_\lambda(m,\tau,z(t))\, dm \qquad (5)$$

is the integrated monochromatic flux of a stellar generation of age $\tau$ and metal abundance $Z(t)$. Therefore the computation of a galaxy model can be performed in two steps: a) calibration of the integrated SED of the stellar generations of different age; b) computation of the galaxy integrated spectrum as superposition of the contributions of all the generations, each of them weighted by the function B(t). This fact suggests that the first test to verify the agreement of the models with the observations should be the comparison of the integrated spectra of single generations with the observed spectra of star clusters. This two-step procedure is not essential, still the spectra of the single generations can be derived by choosing a birth rate defined by an impulsive function.

In the following a series of population I models with $t<10^9$ yr will be described. They have been derived under the following assumptions:
a) for stars with $20 \leq m \leq 100$ (M☉) the evolutionary tracks of Bressan et al. (1981) and Bertelli et al. (1981, unpublished) have been adopted, in which overshooting from the convective core and mass loss have been taken into account; the evolution has been followed up to the end of the phase of core He-burning which, due to the large amount of mass lost, takes place almost completely in the region of high temperatures, close to the Main Sequence and to its left;
b) in the mass range $2.2 \leq m \leq 15$ (M☉) the evolutionary tracks of Iben (1965, 1966, 1977b) have been used;
c) although the core He-burning is included in the theoretical background, no attempts have been made to consider the AGB and the post-AGB phases. The consequences of this omission have been analysed in Sect. 2;
d) the metal abundance is $Z=0.01$;
e) the initial mass function is the Salpeter function with $\alpha=2.35$.

The models have been grouped in order of increasing age.
1) $1\times10^6 \leq t \leq 3\times10^6$. All the stars are supposed to belong to the Main Sequence; this is true for the more massive ones since their MS lifetime is larger than $3\times10^6$ yr. Stars of low mass which are still in the contraction phase prior to the H ignition, have been treated as ZAMS stars; their contribution is not relevant to the total light in any wavelength region considered. The integrated spectrum is very similar to the spectrum of a hot star: the monochromatic flux $f_\lambda^*$ is a monotonically decreasing function of $\lambda$.
2) $3\times10^6 \leq t \leq 8\times10^6$ (the evolved stars have masses in the range $29 \leq m \leq 100$ M☉). $f_\lambda^*$ is still monotonic; its slope however is smaller both owing to the disappearance of the most massive stars from the Main Sequence and because there is now a number of evolved stars in the region of low and intermediate temperatures.
3) $8\times10^6 \leq t \leq 7\times10^7$ (mass of evolved stars: $5.6 \leq m \leq 29$ M☉). The Balmer discontinuity and some H lines appear on the spectrum and gradually become more marked. $f_\lambda^*$ shows a maximum in the UV region which moves at high wavelengths (still remaining in the far UV), a minimum below the Balmer discontinuity at $\lambda \approx 3650$ A and moreover a further relative maximum at $\lambda \approx 3850$ A. According to the evolutionary scheme adopted, which does not take into account the overshooting from the convective core during core H-burning, in this age interval the contribution of AGB stars begins to appear, which however has not been included; therefore the SED in the red region should have a milder slope (cf. the spectra of Fig.1).
4) $7\times10^7 \leq t \leq 3\times10^8$ (mass of evolved stars: $3.1 \leq m \leq 5.6$ M☉). In the spectrum the Balmer discontinuity and the lines are clearly evident. The

two maxima at first tend to be level and then the maximum in the optical region at $\lambda \simeq 3850$ A becomes prominent. The UV flux progressively decreases, as the hottest stars gradually evolve from the Main Sequence becoming colder.

5) $3 \times 10^8 \lesssim t \lesssim 8 \times 10^8$ (mass of evolved stars: $2.2 \lesssim m \lesssim 3.1$ M⊙). The Balmer lines and the discontinuity still appear, other features however are present: the lines of the ionized metals FeII, SiII ect. The maximum in the optical region becomes more prominent and moves at higher wavelengths; the other maximum in the UV region becomes fainter.

In Fig. 2 some spectra are presented with the purpose of showing how they evolve when the generation ages. In a more synthetic way two fundamental types of spectra can be distinguished: a) spectra of very young generation with a monotonically decreasing monochromatic flux, these spectra are very much like the spectra of an O star and clearly the integrated light is dominated by such stars; b) spectra of intermediate generation in which the SED exibits two maxima and the Balmer discontinuity is always present; the spectra of this generation is dominated by A and F stars, whose spectra also show these features.

Models have been computed also by changing the exponent of the initial mass function. Adopting $\alpha = 3$ the new models, in respect to the previous set, exhibit the following behaviour:

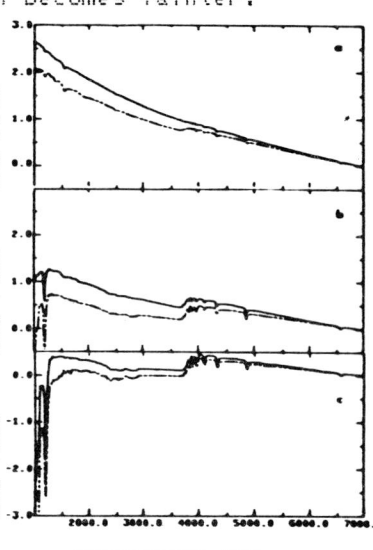

Fig. 2. Generation models for several ages (million years): a) t=1 and 8, b) t=40 and 150, c) t=300 and 500.

a) the difference between two models with different $\alpha$ and the same age is more marked when $t < 10^7$ yrs. due the different relative number of massive stars; this difference tends to disappear as the generation grows older;

b) owing to the lower relative number of massive stars in the case $\alpha = 3$ which allows the emergence of the contributions of less massive stars, the Balmer discontinuity appears in the integrated spectrum earlier than in the other case.

The best candidates for a test of these models are the globular-like clusters of the Magellanic Clouds. The HR diagram for many of these clusters is available, in this way we can derive independent estimates of the age; moreover several determinations of their chemical composition are also published. Searle et al. (1973, SWB) have developed a classification of the clusters of the LMC based on two reddening-free parameters derived from Gunn's photometric system (Thuan and Gunn, 1976); according to this classification the clusters are arranged in a sequence along which both age and metal content are changing. The correlation between the types defined by SWB and the age is still uncertain and different age intervals are defined for the classes from different authors (e.g. Cohen 1982, Hodge 1983).

Cohen et al.(1984) have derived UV spectroscopic observations with the IUE for 17 globular clusters of the LMC which cover the whole range of the SWB classes. From this sample we have excluded those clusters which do not belong to the age range here considered, for the remaining 14 the colour C1=(1345-3150) and C2=(1750-2670) have been derived from their Table I and have been corrected for reddening in the same way as described by the authors, namely by separating the absorption due to our Galaxy from that correspondent to the LMC.

Fig. 3 shows the two colour diagram of the clusters; the theoretical calibration line, which has been derived from the synthetic spectra previously analysed, is also plotted (full line). From the figure it results that, neglecting the subdivision in SWB classes, the observed colours, after correction for reddening, are in reasonable agreement with the theoretical predictions and the dispersion is small. However taking into account the SWB types, we note that there is a non negligible mixing of objects of different types. This fact may have several causes: a) the evaluation of the reddening is not correct, a better estimate of this correction could move the objets in the right position since the reddening line is almost parallel to the theoretical locus; b) there are errors in the observations; c) the criteria according to which clusters are put into classes are such that only to a first approximation do the SWB types correlate with age. In particular the cluster NGC2100 [C1=-1.55, C2=-0.71] deviates significantly from the average positions of the sample clusters belonging to the same type. By making use of the theoretical colour-age relations average values of the age have been derived for the first four SWB classes.

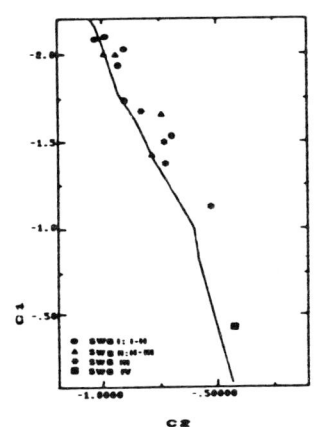

Fig. 3. UV two-colour diagram for LMC clusters.

TABLE II

| SWB | C1 | C2 | t($10^7$yrs) | Δt($10^7$yrs) |
|-----|-------|-------|--------------|---------------|
| I   | -2.11 | -1.00 | 3.5          | 1.4           |
| II  | -1.54 | -0.79 | 10.4         | 5.0           |
| III | -1.52 | -0.75 | 10.6         | 7.5           |
| IV  | -0.94 | -0.65 | 19.2         | ---           |

The dispersion in age, shown in column 5, is computed from the dispersion of the observed colours and does not take into account the possible uncertainties connected with the theoretical calibrations; moreover, due to the small number of clusters considered, it is only indicative. The whole sample has been compared by Cohen et al (1984) with theoretical models but, since their models have ages ranging from $2\times10^8$ to $1.2\times10^{10}$, their analysis mainly refers to the last SWB types and there is only a small overlap with the models here considered.

Cohen et al (1984) have also shown the UV spectra of four clusters. These spectra are presented in Fig.4 with model spectra chosen in order to give the best fit with the observed spectra. Although no attempt has been made to correct the cluster spectra for reddening, the agreement is rather good and the continuum is very well reproduced with the exception of the very far UV ($\lambda < 1200$ A).

The spectrum of the cluster NGC2004 is fitted by a synthetic spectrum with age intermediate between $4 \times 10^7$ and $7 \times 10^7$ yr (from the UV colours we derived $t = 3 \times 10^7$ yr). Hodge (1983) derives from the HR diagram an age of $8 \times 10^6$ yr and Cohen (1982) estimates an average age of $10^7$ yr for SWB type I clusters. The cluster NGC1711, according to the fitting with the present models is $7 \times 10^7$ years old (and from UV colours is $1.4 \times 10^7$ years old). According to Hodge (1983) SWB II clusters have $1.3 \times 10^7 < t < 4 \times 10^7$ while according to Cohen (1982) $t = 5 \times 10^7$ yr.

For the cluster NGC1866 we derive from the fitting of the spectrum an age intermediate between $1.5 \times 10^8$ and $3 \times 10^8$ yr, from the UV colours $2 \times 10^8$ yr; the age, estimated from the HR diagram (Hodge 1983) is $8 \times 10^7$ and Cohen for SWB III clusters gives $t = 5 \times 10^8$ yr. The cluster NGC1856 is fitted by a synthetic spectrum of a generation $5 \times 10^8$ years old; from the HR diagram an age is derived of $1.2 \times 10^8$ yr, while Cohen (1982) derives for SWB IV clusters an age of $10^9$ yr.

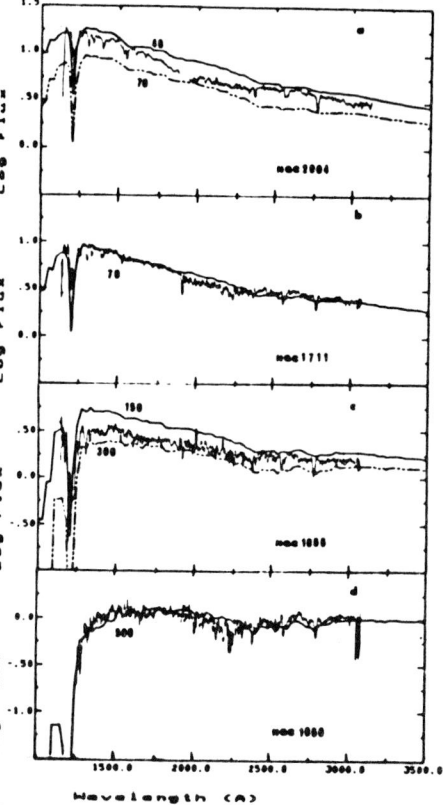

Fig. 4. Spectra of some young clusters (from Cohen et al. 1984) compared with models. Figures give the model age in $10^6$ yr.

## 5. THE SPECTRA OF OLD STELLAR GENERATIONS.

### 5.1 Introduction

The models of old stellar generations ($t > 10^9$ yr) have been computed for two chemical compositions: a) $Y = 0.30$ $Z = 10^{-2}$, b) $Y = 0.30$ $Z = 10^{-4}$. In both cases the portion of the isochronous lines before the He flash has been taken from Ciardullo and Demarque (1977). The temperature of their red giant branches however has been corrected to modify the effect of the mixing length parameter $\alpha = 1/H_p$ which is equal to 1 in the stellar models used by Ciardullo and Demarque. The same authors suggested in fact that the best fit with the observed red giant branches of old open clusters is reached with $\alpha = 1.5$. The correction is given by

$$\log T_e = 0.176 [0.02 \log L/L_\odot + 0.143] \qquad (5)$$

(Ciardullo e Demarque 1979). The same correction has been applied also to the portions of Iben's evolutionary tracks near to the Hayashi line, which have been used in deriving the models of the younger generations.

The mass loss during shell H burning has not been considered in the isochrones of Ciardullo and Demarque (1977): this omission has negligible consequences on the evolution along the red giant branch, since in this phase the evolution is essentially independent of the mass. However the loss of mass along the red giant branch is important in determining the evolutionary behaviour after He flash, as has already been remarked in Sect. 3. In this spirit we have adopted for the stars in the phase of core He burning the distribution function (2); using this relation the contribution of these stars is given by

$$F_\lambda = \int \int g\, R(m)\, f_\lambda(m,\tau,Z(t))\, dm\, d\tau \qquad (6)$$

where $\tau$ is the age starting from the He flash.

Expression (6) can be put in the following form:

$$F_\lambda = g \int_{m_c}^{m_i} R(m)\, W_\lambda(m)\, dm \qquad (7)$$

with

$$W_\lambda(m) = \int_0^{t_{He}(m)} f_\lambda(m,\tau,Z(t))\, d\tau \qquad (8)$$

where $t_{He}(m)$ is the lifetime during core He burning of a star of mass m. According to these relations, to derive $F_\lambda$ one has to integrate along each evolutionary track to get the contribution W(m) of all the stars of given actual mass m during the different phases of core He burning and then the final result is obtained by the integration over the masses in the interval $(m_c, m_i)$.

The evolutionary tracks of Sweigart and Gross (1976) have been used. The function R(m) weights the function $W_\lambda(m)$ in such a way that contributions to the integral (7) from masses far from $m_{fl}$ become negligible; therefore it will be sufficent to compute the integral over a smaller mass interval [than that defined by (7)]. This circumstance prove useful since the evolutionary tracks of Sweigart and Gross do not cover always the required mass range.

For the stellar spectra Kurucz's models (1979) have been used and, where necessary, they have been supplemented in the way described in section 3.

5.2 Integrated spectra with $Z=10^{-2}$.

These have been collected in three groups according to their age:
1). $1 \times 10^9 \leq t \leq 4 \times 10^9$. The monochromatic flux in the UV region gradually decreases. The maximum in this region moves beyond the discontinuity at

$\lambda = 2800$ A due to MgII; moreover there are present the absorption features of FeII, SiII, etc. In the optical region the Balmer lines are weaker and weaker and tend to disappear.

This group of spectra correspond to generations with evolved stars in the mass range 1.3-2.1 M⊙; according to the adopted evolutionary background such stars, different from the evolved stars of the younger generations considered in Sect. 4, develop during the shell H burning a degenerate core and this fact is responsible for the sudden appearance of the red giant branch.

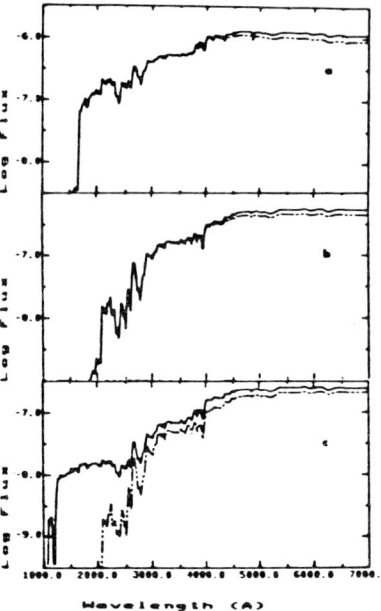

Fig. 5a shows the spectrum of the generation $2.\times 10^9$ years old as representative of the group. The same figure also presents the spectrum from which the contribution of the He burning stars has been subtracted. As is well known, in the case of Pop. I, these stars occupy in the HR diagram the "clump" (Cannon 1970). The two spectra are not very different, however the clump stars increase the whole flux and reduce the slope of the spectrum in the red region.

2) $4 \times 10^9 \leqslant t \leqslant 1.2 \times 10^{10}$. The UV flux still decreases, being produced by the emission of the hottest main sequence stars. The absorption lines of FeII and SiII are still evident; in the optical region the flux has a maximum which, as the age increases, moves towards the red region.

Fig. 5b shows the spectrum of the generation with $t = 6 \times 10^9$ yr with and without the contribution of clump stars. In this case also the inclusion of clump stars tends to flatten the monochromatic flux in the red region. The difference between the two spectra is small compared with the previous case: this derives from the reduced ratio of clump stars to the total

Fig. 5. Spectra of old generations with $Z = 10^{-2}$. The lower curve of each panel gives the SED computed without clump stars.

number of red stars (stars in phase of shell H burning and core He burning) (Barbaro and Pigatto 1984).

3) $1.2 \times 10^{10} \leqslant t \leqslant 1.6 \times 10^{10}$. In the far UV region the monochromatic flux reverses the trend till now observed and begins to increase with age. The reason for this behaviour is that the clump now changes into a proper HB, gradually populated by hot stars. The flux from these stars opposes the gradual decrease of the UV flux from the stars of the MS turnoff point as the

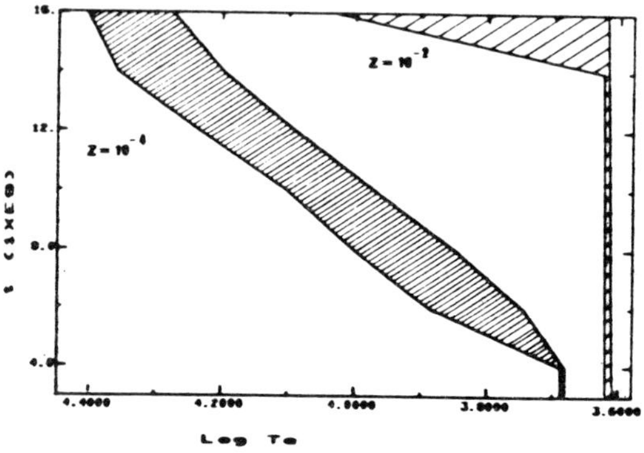

Fig. 6. Extension of the HB as function of age.

generation grows older. The temperature of the stars during the core He burning, for the two considered composition, are shown as function of the age in Fig. 6. The extension of the HB is defined in such a way as to include 90% of the stars, leaving outside 5% from both sides.

In the optical region the H and K lines of CaII are present. In Fig. 5c the spectrum of the generation $1.6 \times 10^{10}$ years old is shown; the lower curve is the spectrum when He burning stars are included; the difference between the two curves clearly indicates that HB stars are responsible for the flux in the region $1000 < \lambda < 2500 \text{Å}$.

## 5.3 Integrated spectra with $Z=10^{-4}$

1) $1 \times 10^9 \leq t \leq 5 \times 10^9$. In the far UV the monochromatic flux quickly increases with $\lambda$, beyond the Ly$\alpha$ the continuum flattens. The Balmer discontinuity and some lines of the same series are present.

In Fig. 7a the spectra of the generation with $t=2 \times 10^9$ and $t=4 \times 10^9$ are shown; from the comparison of the two curves it turns out that the UV flux decreases as the age increase: this flux is produced by the stars of the MS turnoff point while the HB stars, having low effective temperatures, do not emit appreciably in the UV (cf. Fig. 6).

2) $5 \times 10^9 \leq t \leq 9 \times 10^9$. At first the decrease of the UV flux continues but at $t \simeq 8 \times 10^9$ the HB begins to extend towards high temperatures and its stars radiate significantly in the UV, thus incresing the flux.

Fig. 7b presents the spectral distribution for the generation $t=8 \times 10^9$ yr.

3) $9 \times 10^9 \leq t \leq 1.2 \times 10^{10}$. Compared to the optical the UV flux increases and a maximum appears at $\lambda=1250$ Å; just to the right of this maximum there is a large plateau extending up to 3500 Å. There is also a second maximum at about 4000 Å. The flux increases as the age increases.

4) $1.2 \times 10^{10} \leq t \leq 1.6^{10}$. Qualitatively these spectra are similar to those of the previous group; however the intensity of the Balmer lines and the discontinuity decreases, while the maximum in the UV grows larger and moves at the left of the Ly$\alpha$ line. Fig. 7c shows the spectra with $t=1.2 \times 10^{10}$ and $t=1.6 \times 10^{10}$ yr.

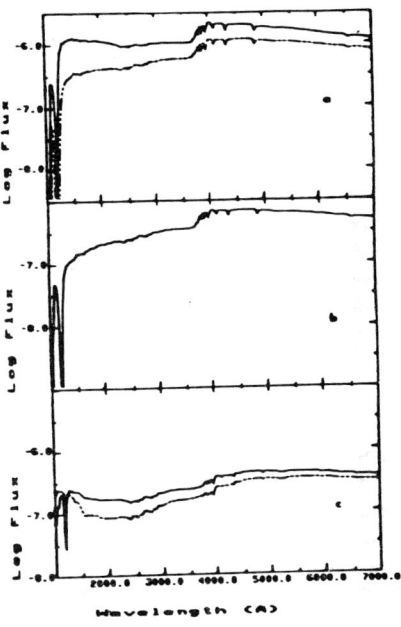

Fig. 7. Spectra of old generations with $Z=10^{-4}$: a) t=2 and 4; b) t=8; c) t= 12 and 16 (ages are in $10^6$ yr).

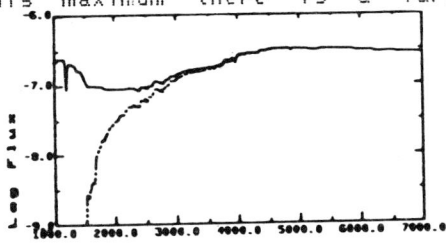

Fig. 8. SED of the generation $Z=10^{-4}$ $t=1.6 \times 10^{10}$ (upper curve). Lower curve gives SED when HB stars are neglected.

In Fig. 8 the spectrum with $1.6 \times 10^{10}$ yr (upper curve) is compared with the spectrum obtained when HB stars are omitted. The HB, in this case, is populated by hot stars with a strong emission in the UV. For $\lambda > 3000$ A the two spectra are similar, this showing that HB stars are too hot to emit significantly in this region.

It may be of interest to compare the spectra of two generations of the same age but different metal content. In the case of $t=2 \times 10^9$ yr (Fig. 9a) the monochromatic flux in the far UV ($\lambda < 1500$ A) is much larger in the metal-poor spectrum and the difference decreases as $\lambda$ increases and vanishes for $\lambda \approx 5000$ A. The different behaviour in the UV region is not ascribable to the HB which is red in both cases (cf. Fig. 6) but to the different temperature of the brightest MS stars. Fig. 9b shows the spectra with $t=8 \times 10^9$ the relative behaviour is qualitatively similar to that of the previous case: the monochromatic flux of the model with $Z=10^{-2}$ begins to decrease, as the wavelength decreases, earlier than in the model with $t=2 \times 10^9$ yr and this depends on the different temperature of MS turnoff stars in the two cases. Moreover in the UV the difference between the two spectra is still larger than in the previous case since now the UV flux is contributed also by the HBs, which have different extensions: HB stars of the model with $Z=10^{-2}$ being red while those of the model with $Z=10^{-4}$ are blue. The Fig. 9c refers to the generations

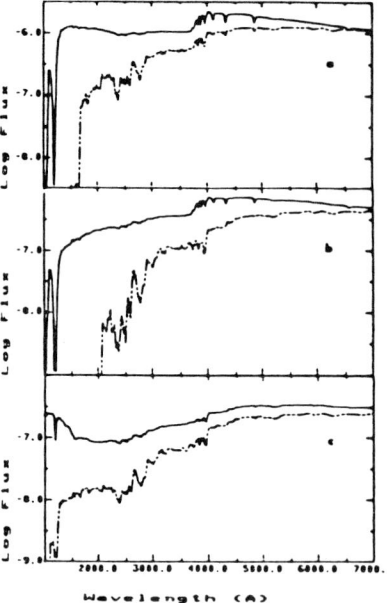

Fig. 9. Comparison of generations with same age and different Z: a) $t=2 \times 10^9$; b) $t=8 \times 10^9$; c) $t=1.6 \times 10^{10}$.

with $t=1.6 \times 10^{10}$ yr. There is still a difference in the UV portion of the spectrum: this is partly due to the MS stars and partly to the HB which extends, although to a different extent, up to the blue. While the monochromatic flux of the case $Z=10^{-4}$ has a minimum for $\lambda = 2000$ A and then increases as $\lambda$ decreases, the spectrum with $Z=10^{-2}$ continuously decreases as $\lambda$ decreases: this difference in the far UV spectrum arises from the behaviour of the bluest HB stars.

The energy distributions of old generations have been tested by comparing them with the spectral distribution of globular clusters. These have been obtained from the ANS observations of de Boer (1985) who gives the average fluxes corresponding to 13 wavelengths (in several cases this number is smaller) covering a wavelength interval from 1550 A to 5500 A for many globular clusters. For some of then de Boer's data have been supplemented with Faber's data (1973): in this case the energy distribution extends up to 7400 A. Moreover for several of the considered clusters Nesci (1978 and private communication) has derived IUE ultraviolet spectra.

Table III gives for each of the analysed clusters the metal abundance taken from Harris and Racine (1979) and HB types from Dickens (1972). In Fig. 10 the SEDs derived from the de Boer's observations, corrected from reddening according to Seaton (1979), and the theoretical curves which give for each case the best fit are shown.

TABLE III

| CLUSTER |         | [Fe/H] | HB type |
|---------|---------|--------|---------|
| 104     | (47 Tuc)| -0.44  | 7       |
| 362     |         | -1.20  | 5       |
| 1904    | (M79)   | -1.58  | -       |
| 5272    | (M3)    | -1.57  | 4       |
| 5904    | (M5)    | -1.25  | 3       |
| 6093    | (M80)   | -1.54  | -       |
| 6205    | (M13)   | -1.42  | 1       |
| 6341    | (M92)   | -2.12  | 2       |
| 6356    |         | -0.37  | 7       |
| 6441    |         | -0.24  | -       |
| 6624    |         | -0.34  | -       |
| 6637    | (M69)   | -0.47  | 7       |
| 6656    | (M22)   | -1.69  | 2       |
| 6752    |         | -1.62  | 1       |
| 7099    | (M30)   | -2.03  | 1       |

In the following we analyse in some detail, the clusters and the possibility of fitting their distributions with our models. With regard to this it must be remarked that the impossibility of getting the fit in some cases can be due to the small number of the explored chemical compositions.

NGC104: there is some disagreement between the UV spectrum of Nesci and de Boer's curve. The best fit is obtained with the synthetic spectrum with $Z=10^{-2}$ and $t=(1.5 - 1.6) \times 10^{10}$ yr (in this age range the UV spectrum changes very quickly due to the rapid evolution of the HB).

NGC362: Nesci's spectrum and de Boer's UV portion of the energy distribution agree fairly well. The whole distribution curve is intermediate between the two synthetic spectra respectively with $Z=10^{-4}$ and $t=1.6 \times 10^{10}$ yr and $Z=10^{-2}$ and $t=1.6 \times 10^{10}$ yr. Since Harris and Racine (1979) give [Fe/H]=-1.20, it seems that the best fit could be obtained with a model with an intermediate metal abundance.

NGC1904 (and NGC6656): the fit with the model $Z=10^{-4}$ and $t=1.5 \times 10^{10}$ yr is very good.

NGC5272: this cluster has a spectrum which is intermediate between those with $Z=10^{-4}$ $t=1.6 \times 10^{10}$ yr and $Z=10^{-2}$ and $t=1.6 \times 10^{10}$ yr.

NGC5904 (and NGC6093): in the UV the curves derived from de Boer (1985) are in agreement with the corresponding spectra of Nesci. On the whole wavelength range in both cases the spectral distribution is fitted by the model with $Z=10^{-4}$ and $t=1.6 \times 10^{10}$ yr, although the metal abundance, according to Harris and Racine (1979), is intermediate.

NGC6205: the observed spectrum gives a good fit with the synthetic curve with $Z=10^{-4}$ and $t=1.6 \times 10^{10}$ yr. However in this case also the metal abundance, derived from Harris and Racine, is not so extreme.

NGC6341: the two observed sets of data (de Boer and Nesci) agree in their common portion. de Boer's curve is fitted by the model with $Z=10^{-4}$ and $t=1.6 \times 10^{10}$ yr although same disagreement is found in the far UV and in the interval (3000 - 4500A).

<u>NGC6441</u>: the fit with the curve $Z=10^{-2}$ and $t=1.6\times10^{10}$ yr is rather uncertain. Probably this could depend on the bad quality of the UV data.

<u>NGC6624</u>: de Boer's curve is fitted by the model with $Z=10^{-2}$ and $t=1.5\times10^{10}$ yr. Nesci's spectrum, compared with this model, has a larger monochromatic flux for $\lambda<2000$ A. Perhaps a better fit in this case can be obtained by increasing to $1.55\times10^{10}$ yr the age of the model.

<u>NGC6637</u>: the SED of this cluster is similar to that of NGC 6356 which shares also the same metal abundance and the same HB type. The fit is obtained with the model $Z=10^{-2}$ and $t=1.5\times10^{10}$ yr.

<u>NGC6752</u>: its energy distribution is fitted by the model $Z=10^{-4}$ and $t=1.6\times10^{10}$ yr.

<u>NGC7099</u>: as Harris and Racine (1979) give a very low metal content, the fit should be reached with $Z=10^{-4}$ but neither the model with $t=1.6\times10^{10}$ yr nor the model with $t=1.5\times10^{10}$ yr give a good result.

As has already been mentioned, the models for the group of globular clusters have been derived by supposing that the only parameter is the metal content: metal-poor clusters display a blue HB while metal-rich ones have red HBs; however the necessity of a second parameter, whose nature is still uncertain, has been stressed by several authors. If the globular clusters constitute a bi-parametric family one expects that there are clusters which can be interpreted in terms of the present calibrations and others which cannot be interpreted.

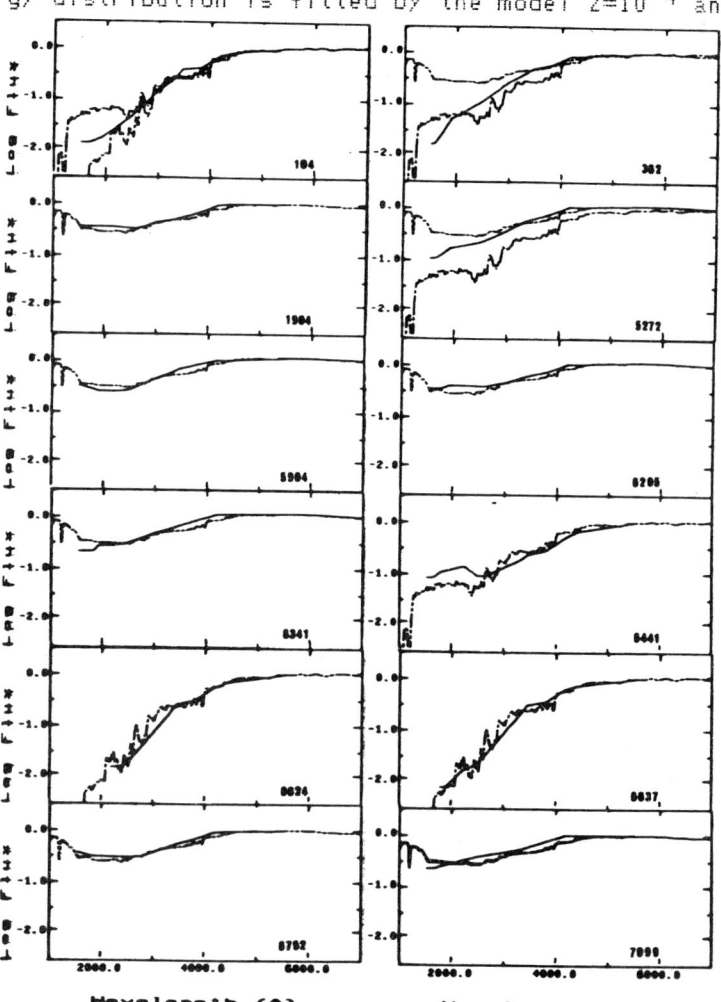

Fig. 10. SED of 12 globular clusters from data of de Boer (1985) and Faber (1973) (solid line of each panel). Superimposed are models whose identification parameters are quoted in the text.

However the difference between SEDs of normal and anomalous clusters for the same type of HB is not relevant, at least if one neglects the consideration of the spectral features. This is the case, for instance, of the couples of clusters: NGC6205, NGC7099 and NGC6341, NGC6656. Both members of each couple have the same type of HB and different metal content (cf. table III); their SEDs are not much different. It is interesting also to compare clusters with the same metal content. Clusters NGC6752 and NGC5272 have approximately [Fe/H]~-1.6: the former however has, according to Dickens (1972), a type 1 HB, while the latter a type 4. The comparison of the two spectra (Fig. 11a) shows that UV fluxes are very different, the lower flux corresponding to the redder HB.

The same situation is offered also by NGC6205 and NGC362 which have [Fe/H]~-1.4 and whose HBs are respectiwely of type 1 and 5 (Fig. 11b). Provided that the estimates of [Fe/H] are sufficently correct these cases represent a confirmation of the existence of a second parameter.

## 6. GALACTIC SYNTHETIC SPECTRAL ENERGY DISTRIBUTIONS

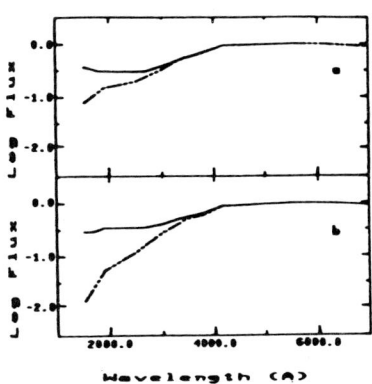

Fig. 11. a) SED of NGC 6752 (upper curve) and NGC 5272; b) SED of NGC 6205 (upper curve) and NGC 362.

Galaxy models have been derived under the following assumptions:
- the stellar birth rate is a decreasing exponential function of time: $B(t) = A e^{-\beta t}$ with a time scale $\tau = 1/\beta$ where $\beta$ has been varied in the interval (0,10), the time being measured in tens of billion years. A more detailed anlysis involving birth rates of different form, and particularly including stars bursts, will be discussed in a further paper.
- the initial mass function is the Salpeter function with $\alpha = 2.35$ and the calibrations of the stellar generations adopted are those described in the previous sections.
- the age of the galaxies is $1.6 \times 10^{10}$ yr.
- Account is taken, although in an approximate way, of the chemical evolution of the interstellar medium. The evolution of the metal abundance is schematized by means of a one-step function:

$$Z = 10^{-4} \qquad t < t^*$$
$$Z = 10^{-2} \qquad t > t^*$$

this rough approximation of the function $Z(t)$ which can be derived from the simple model of Talbot and Arnett (1971), is required because only the calibrations with $Z=10^{-4}$ and $Z=10^{-2}$ are available. On the other hand, since the simple model for the chemical evolution of a closed system is also rather inaccurate, $t^*$ should be treated as a parameter.

Fig 12 shows same synthetic spectra computed neglecting the metal-poor population ($t^* = 0$) and for different values of the star formation timescale corresponding to $\beta = 0, 2, 5, 10$. The spectral distributions are normalised at $\lambda = 7000$ A.

Each curve can be schematically interpreted as the superposition of two fundamental spectra whose relative importance depends on the value of $\beta$: the former corresponds to a very young generation (and as already remarked, is very similar to the spectrum of an early type star) the latter to an old generation. When the star formation rate is constant ($\beta = 0$) the spectrum of the young generations dominates and the resulting spectral distribution is characterized by a large UV flux. As $\beta$ increases the decreasing star formation rate depresses the relative contribution of the young generations and the far UV flux becomes negligible. It must be remarked however that the models ignore the metal-poor generations which, for advanced ages, can

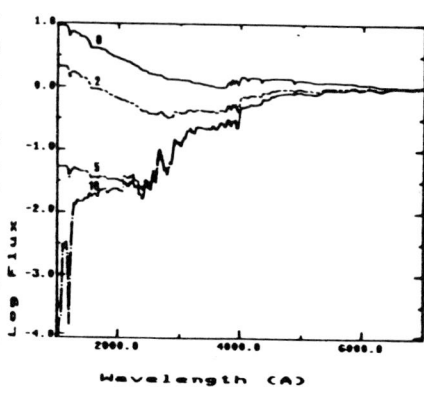

Fig. 12. Galaxy models with $t=1.6 \times 10^{10}$, $t^*=0$. Figures give the value of $\beta$.

contribute to the UV flux with their hot HB stars and moreover p-AGB stars have also been neglected. Fig. 13 shows how the spectra are modified when the metal-poor population is included contributing to the total flux with a fraction which depends on the parameter $t^*$. Obviously the inclusion of the metal-poor generations does not sensibly modify the spectrum when $\beta$ is small and the galaxy spectrum is dominated by the light of the young massive stars; as $\beta$ increases, however, this population affects mostly the UV region. Although the amount of the correction due to the metal-poor population depends on the value of $t^*$, for large values of $\beta$, the inclusion of Pop. II blue HB stars can markedly affect the far UV flux. We can speculate how the modification of the exponent of the IMF from 2.35 to 3 can alter the previous results. A steeper IMF, by decreasing the number of massive stars, produces the same effect that is obtained by raising the value of $\beta$. At first sight it seems therefore impossible to distinguish between the effects of $\alpha$ and $\beta$ on the SED; however in the far UV the maximum which appears in the spectra of young generations is strongly dependent on the slope of the continuum in this spectral region: a useful parameter for this correlation could be the ratio of the maximum flux to the minimum flux both in the UV region.

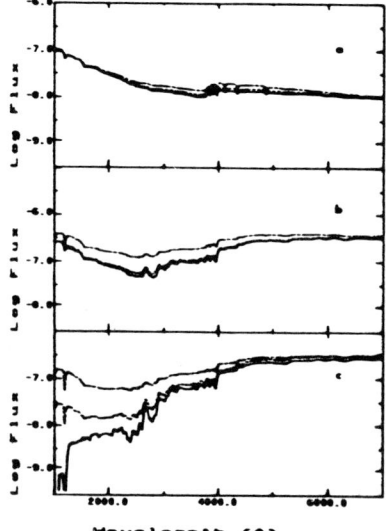

Fig. 13. Galaxy models. From top to bottom within each panel: a) $\beta=10$ $t^*=0$, 8, 14; b) $\beta=3$ $t^*=0$, 0.4, 4; c) $\beta=10$ $t^*=0$, 0.12, 1 ($t^*$ in $10^9$ yr).

The comparison of the synthetic spectra with the observed ones is rather difficult owing to the lack of observations in the UV: in fact while UV observations of early type galaxies are relatively abundant, the same type of observations for late type galaxies is scarse since these objects, due to their low surface brightness, cannot be easily observed with the present instruments.

The most useful data are:
- standard SEDs for the different types of galaxies, obtained by collecting observations of several authors: they represent average distributions for a given morphological type and are usually derived with the purpuse of determining the K-correction. Distributions of this type have been published by Pence (1976) and by Coleman et al. (1980). The UV spectral distributions of Pence are based on the wide band photometric system of the OAO-2 satellite.
- IUE measures of the UV spectrum of single elliptical galaxies (Bertola et al. 1980, Oke et al. 1981, Bertola et al. 1982): these data are particularly interesting since they have revealed that there is an increase of monochromatic flux shortwards below $\lambda = 2000$ Å.

While a detailed analysis of the observational data will form the subject of a further paper, at present we restrict ourselves to compare our models with the standard curves of Pence.

Fig. 14 shows Pence's curves for the different types on which some models are overimposed. The figures on the diagrams refer to the parameters of models. The following conclusions arise: the models which neglect the metal-poor population ($t^* = 0$) are not able to fit the observed curves (see for instance Fig. 14a for Im - Sdm galaxies); a better agreement is obtained with models which account for the presence of this population. However in this case also some disagreement remains since in the far UV Pence's curves are steeper than the models.

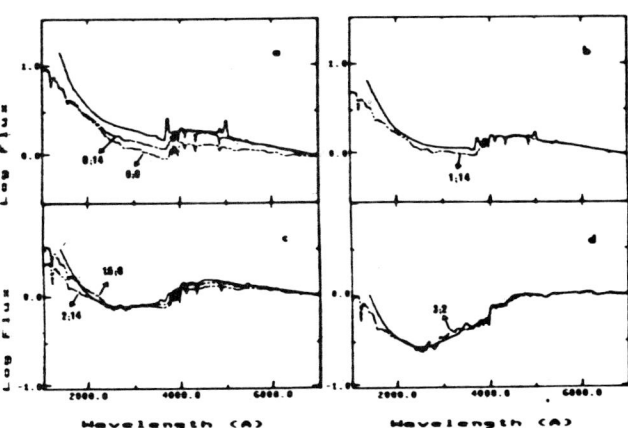

Fig. 14. Comparision of Pence's curves (solid line) with the models: a) Im-Sdm; b) Scd; c) Sbc; d) Sab. The quoted numbers are respectively the values of $\beta$ and $t^*$ in billion years.

Provided that the observed curves are correct, such disagreement can be removed if a larger number of massive stars is included in the models: this result can be obtained either with an IMF flatter than that used ($\alpha = 2.35$) or by adding a burst of star formation. With reference to Fig. 14a if we add to the model with $\beta = 0$ and $t^* = 14$ a burst of star formation of the length of $10^6$ yr and whose intensity is 10 times as large as the constant value of the birth-rate of the underlying model, the agreement with the observed curve is farly good.

Elliptical galaxies have been the subject of exhaustive analyses in the last years (cf. Wu et al. 1980) and during this workshop (Bertola,

Bruzual, Nesci); the conclusion was that, at present, it is impossible to single out univocally a model able to reproduce the optical and UV spectrum. Models with a suitable percentage of hot stars can explain the upturn of the monochromatic flux in the far UV; the nature of such hot stars cannot however be determined in an unique way: in fact it has been suggested that they are young massive stars resulting from a burst of star formation, or blue Pop. II HB stars or active white dwarfs (Nesci and Perola, 1983), or nuclei of planetary nebulae, O and B subdwarfs, etc.

We only want to add a further model which accounts for both Pop. II HB stars and p-AGB stars. The model is characterized by the following parameters: age of $1.6 \times 10^{10}$ yr, $\beta = 10$, $\alpha = 2.35$, $t^* = 1.2 \times 10^8$. Furthermore, according to the prescription of Renzini and Buzzoni (1983), it has been assumed that all the generations older than $1 \times 10^{10}$ yr do have p-AGB stars and their contribution to the bolometric light is 10% of the total luminosity of each generation. The mean temperature adopted for such stars is $20000°K$. The model is shown in Fig. 15, with the spectrum of NGC4649 (Bertola et al., 1982); from the comparison it turns out that the upturn in the far UV is reproduced fairly well; there is yet a discrepancy in the interval 2500-4000 A. It must be remarked, however, that Pence's standard curve for ellipticals, which is also shown in Fig. 15, fits the model at these wavelengths. Within this model, the different upturn, observed in different galaxies, could be produced by fluctuations of the temperatures of p-AGB stars.

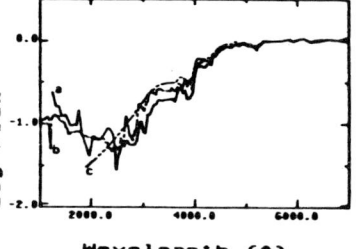

Fig. 15. Comparison of the spectra of NGC 4649 curve a) with the model including the contribution of the post-AGB (curve b). Pence's curve for ellipticals is also displayed (curve c).

AKNOWLEDGMENTS.

We are indebted to C. Chiosi, G. Bertelli and A.G. Bressan for having supplied us with their results on stellar evolution before publication and to C. Chiosi for several enlightening discussions. Moreover we thank R. Nesci who kindly made available his IUE spectra of globular clusters. This work has been financially supported by funds of the C.N.R. and of the M.P.I.

REFERENCES.

Barbaro, G.; Pigatto, L.; 1984, Astron. Astrophys. 136, 281.
Bertelli, G.; Bressan, A.G.; 1985, preprint.
Bertelli, G.; Bressan, A.G.; Chiosi, C.; 1984 Astron. Astrophys. 130, 279.
Bertelli, G.; Bressan, A.G.; Chiosi, C.; 1985, Astron. Astrophys. in press.

Bertelli, G.; Bressan, A.G.; Chiosi, C.; 1981, unpublished.
Bertola, F.; Capaccioli, M.; Holm, A.V.; Oke, J.B.; 1980, Astrophys. J. 237, L65.
Bertola, F.; Capaccioli, M.; Oke, J.B.; 1982, Astrophys. J. 254, 494.
Bressan, A.G.; Bertelli, G.; Chiosi, C.; 1981, Astron. Astrophys. 102, 25.
Cannon, R.D.; 1970, M.N.R.A.S. 150, 111.
Capuzzo-Dolcetta, R.; Kurucz, R.; Rossi, R.; 1983, Mem. S.A.It. 54, 829.
Chiosi, C.; Maeder, A.; 1985, Ann. Rev. Astron. Astrophys., in press.
Ciardullo, R.B.; Demarque, P.; 1977, Yale Trans. 33.
Ciardullo, R.B.; Demarque, P.; 1979, Dudley Obs. Rpt. 14, 317.
Code, A.D.; Meade, M.R.; 1979, Astrophys. J. Suppl. 39, 195.
Cohen, J.G.; 1982, Astrophys. J. 258, 153.
Cohen, J.G.; Rich, R.M.; Persson, S.E.; 1984, Astrophys. J. 285, 595.
Coleman, G.D.; Wu, C.C.; Weedman, D.W.; 1980, Astrophys. J. Suppl. 43, 393.
de Boer, K.S.; 1985, Astron. Astrophys. 142, 321.
Dickens, R.J.; 1972, M.N.R.A.S. 157, 281.
Faber, S.M.; 1973, Astrophys. J. 179, 731.
Fusi-Pecci, F.; Renzini, A.; 1975, Astron. Astrophys. 39, 413.
Harris, W.E.; Racine, R.; 1979, Ann. Rev. Astron. Astrophys. 17, 241.
Hodge, P.W.; 1983, Astrophys. J. 264, 470.
Huchra, J.P.; 1977, Astrophys. J. 217, 928.
Humphreys, R.M.; 1984, Observational tests of the stellar evolution theory; IAU Symp. 105, eds. Maeder A., Renzini A.; 279.
Iben, I.; 1965, Astrophys. J. 142, 1447.
Iben, I.; 1966, Astrophys. J. 143, 483, 505, 516.
Iben, I.; 1967a, Ann. Rev. Astron. Astrophys. 5, 571.
Iben, I.; 1967b, Astrophys. J. 147, 624, 650.
Jamar, C.; Macau-Hercot, D.; Monfils, A.; Thompson, G.; Houziaux, L.; Wilson, R.; 1976, Ultraviolet Bright Stars Spectrophotometric Catalogue, ESA SR-27.
Kurucz, R.; 1979, Astrophys. J. Suppl. 40, 1.
Larson, R.B.; Tinsley, B.M.; 1974, Astrophys. J. 192, 293.
Macau-Hercot, D.; Jamar, C.; Monfils, A.; Thompson, G.; Houziaux, L.; Wilson, R.; 1978, Suppl. to the Ultraviolet Bright Stars Spectrophotometric Catalogue, ESA SR-28.
Maeder, A.; 1975, Astron. Astrophys. 43, 61.
Maeder, A.; 1976, Astron. Astrophys. 47, 389.
Malagnini, M.L.; Faraggiana, R.; Morossi, C.; 1983, Astron. Astrophys. 128, 375.
Malagnini, M.L.; Faraggiana, R.; Morossi, C.; Crivellari, L.; 1982, Astron. Astrophys. 114, 170.
Meade, M.R.; Code, A.D.; 1980, Astrophys. J. Suppl. 42, 283.
Nesci, R.; 1981, Astron. Astrophys. 99, 121.
Nesci, R.; Perola, G.C.; 1983, Mem. S.A.It. 54, 851.
Nussbaumer, H.; Schmuts, W.; Smith, L.J.; Willis, A.J.; 1982, Astron. Astrophys. Suppl. 47, 257.
Oke, J.B.; Bertola, F.; Capaccioli, M.; 1981, Astrophys. J. 243, 453.
Paczynski, B.; 1970, Acta Astronomica 20, 47.
Pence, W.; 1976, Astrophys. J. 203, 39.

Reimers, D.; 1975, Proceed. of 19 Colloque International
    d'Astrophysique, Liege, 369.
Renzini. A.; Buzzoni, A.; 1983, Mem. S.A.It. 54, 739.
Rood, R.T.; 1973, Astrophys. J. 184, 815.
Searle, L.; Wilkinson, A.; Bagnuolo, W.G.; 1973, Astrophys. J. 239, 803.
Seaton, M.S.; 1979, M.N.R.A.S. 187, 738.
Sil'chenko, O.K.; 1983, Soviet Astron. Lett. 9, 145.
Straizys, V.; Sviderskiene, Z.; 1972, Bull. Vilnius An. Obs. 35, 1.
Sweigart, A.V.; Gross, P.G.; 1976, Astrophys. J. Suppl. 32, 367.
Sweigart, A.V.; Gross, P.G.; 1978, Astrophys. J. Suppl. 36, 405.
Talbot, R.J.; Arnett, W.D.; 1971, Astrophys. J. 170, 409.
Thuan, T.X.; Gunn, J.E.; 1976, Pub. A.S.P. 88, 543.
Underhill, A.B.; 1980, Astrophys. J. 239, 220.
van der Hucht, K.A.; Cassinelli, J.P.; Wesselius, P.R.; Wu, C.C.; 1979,
    Astron. Astrophys. Suppl. 38, 279.
Whipple, F. L.; 1935, Harvard College Obs. Circ. 404, 1.
Wu, C.C.; Faber, S.M.; Gallagher, J.S.; Peck, M.; Tinsley, B.M.; 1980,
    Astrophys. J. 237, 290.

DISCUSSION

<u>Frogel</u> With regards to the clusters in the LMC: 1) Have you compared the numbers of stars in your models as a function of color and luminosity with the actual numbers that are observed? This is a very powerful test of the models. 2) The young LMC clusters have an E(B-V) of up to several tenths. Did you adjust your UV energy distributions for the large E(B-V) values?

<u>Barbaro</u> 1) No, we have not done this type of test. I agree with you that it can give sound constraints to the models. 2) While the cluster UV colours have been corrected for reddening, the spectra have been taken as they have been published (probably in my description I forgot mentioning this fact).

<u>Aaronson</u> I was wondering if you might comment on the rather remarkable claim we heard yesterday that the light of elliptical galaxies is dominated by a 5 - 10 billion year old population ?

<u>Barbaro</u> I can only say that our model of an elliptical galaxy ($t=1.6 \times 10^{10}$ yr, $\beta=10$) in the optical spectrum and UBV colours resembles the model of a generation $1 \times 10^{10}$ yr. old. This similarity does not mean that this generation is dominant, but is the consenquence of the superimposition of several generations. The most important of them, the generation $1.6 \times 10^{10}$ yr old, with the blue HB stars, produces more or less the same effect as a younger generation.

<u>Pickles</u> Do you think that models with maximum metallicity value Z=0.01 are metal rich enough to much observed galaxy feature strengths?
I find that the energy distribution of a solar metallicity (Z=0.017) 15 Gyr model is virtually identical to that of 10 Gyr model of Z=0.04, although the latter matches observed line strengths better. Would you agree with this?

<u>Barbaro</u> If one wants to reproduce the features of the spectrum, the maximum metal abundance of Z=0.01 is not suited, particularly for modelling elliptical galaxies. Probably this composition is sufficient

when the purpose is to model the general behaviour of the SED, although there can be the possibility, as you have mentioned, that an obseved SED can be fitted by models with different maximum Z.

Bruzual I think that your fits to the UV SED of globular clusters can be improved by a lot (for instance at 2420 A, 2640 break, 2800 Mg feature, etc.) if you use observed data (IUE) for late type stars. This stars do not emit "much" in the UV, but however they do emit some light, and the requirements of hot and intermediate temperature stars that you need in your models to reproduce the cluster and galaxy data may change when you include the correct late type UV SEDs.

Barbaro Although in my talk I have not explicitly mentioned, nevertheless for cool late stars (Te < 5500°K) we have used the observed spectra and for the UV region the IUE observations. Probably your suggestion extends also to cool stars with Te > 5500°K. I think that it would be interesting to apply this advice.

Mould I think you deserve our congratulations for completing this large task without deviating from theoretical precepts. It would be very useful to extend your models to higher spectral resolution and higher than solar metallicity to investigate the metallicity of Elliptical galaxies in the wavelength region 3000-4000 A, where the lack of Kurucz models cooler than 5000°K and UV excess are of lesser importance.

# PHOTOMETRIC EVOLUTION OF ELLIPTICAL GALAXIES IN THE COLOR-MAGNITUDE DIAGRAM

N. Arimoto
Observatoire de Paris-Meudon, 92190, Meudon, France
Y. Yoshii
Tokyo Astronomical Observatory, University of Tokyo, Mitaka,
Tokyo 181, Japan

ABSTRACT. Photometric evolutions of elliptical (E) galaxies are investigated by introducing galactic winds into the evolutionary population synthesis models. It is shown that E galaxies are one parameter family of their initial masses. E galaxies are composite stellar systems composed of stars with a wide variety of metallicity, and have the same age and the same initial mass function. The color-magnitude relation for E galaxies comes from their different average metallicities. Both color and luminosity evolutions of E galaxies depend strongly on their average metallicities and, hence, on their initial masses.

## 1. INTRODUCTION

From the point of view of chemical evolution, the initial activity of star formation and the subsequent multiple supernovae explosions are naturally expected in an E galaxy (Larson 1974, Ikeuchi 1977). Then, remaining gas will be ejected completely from a galaxy by induced galactic winds and the production of heavy elements will be ceased abruptly. Thereafter, photometric properties of a galaxy will evolve, without any additional star formations, in a way characterized by the evolution of stars with an average metallicity. Thus, the photometric evolution of an E galaxy will strongly depend on the star formation history in an early explosive era.

The observed color-magnitude (CM) relation of E galaxies shows that a blueness of the stellar content is tightly connected with an absolute magnitude (Visvanathan and Sandage 1977). Since the CM relation is at least partly due to the metallicity variations (Faber 1977, Visvanathan and Sandage 1977, Frogel et al. 1978), the history of star formation during initial active stage should be correlated with an initial mass of a galaxy. More precisely, star formation should last longer for a more massive galaxy to produce heavy elements much more effectively. Thus, E galaxies should have composite stellar populations whose metallicity distributions depend on the initial masses. Therefore, the single-burst population models with a constant metallicity (e.g., Tinsley 1978, Aaronson et al. 1978) are not justified for the photometric study of E galaxies.

To study why and how the average metallicity varies with a galactic mass and how the composite HR diagram of the stellar contents changes progressively along the CM relation, we introduce galactic winds into the composite population models constructed by Arimoto and Yoshii (1985; hereafter paper I).

This paper will briefly discuss work on the evolutionary population synthesis for E galaxies. A full discussion will appear elsewhere.

## 2. MODELS

Assuming that a galaxy is a homogeneous sphere and the gas ejected from stars is quickly mixed throughout a galaxy (one zone model), we have traced the chemical evolution during the initial activity of star formation. The computations are carried out with three fundamental parameters, i.e., the coefficient $\nu$ of star formation rate (SFR), the power index $\mu$ of initial stellar mass function (IMF), and the age $T_G$ of a galaxy. Full details of equations are given in paper I.

As for IMF, we use a power law form as $\phi(m) \propto m^{-\mu}$, with $m_\ell = 0.05 M_\odot$ and $m_u = 60 M_\odot$ for lower and upper stellar mass limits, respectively [Salpeter (1955) gave $\mu = 1.35$ for the local IMF].

We use a SFR proportional to the fractional gas mass $f_g$,

$$C(t) = \nu f_g(t), \text{ for } 0 \leqslant t \leqslant t_{GW}$$
$$= 0, \text{ for } t_{GW} < t$$

where $t_{GW}$ denotes the time when galactic winds are induced. Hereafter, $\nu$ is normalized by a value $\nu_0 = 6.08 \times 10^{-18}$ sec$^{-1}$, which corresponds to the SFR per unit mass in the solar neighbourhood provided that the present age of our galaxy is $15 \times 10^9$ yr.

As for a gas escape condition, we assume that all remaining gas in an E galaxy is ejected immediately when the thermal energy of supernovae remnants exceeds the binding energy of the gas (Larson 1974). Equations formulated by Saito (1979b) are used in computing the thermal evolution of the gas.

In order to include the effect that the metallicity of subsequently formed stars should change according to the chemical evolution into the photometric population synthesis, we assume that the stars born in the discrete interval $[(Z_{k-1} Z_k)^{1/2}, (Z_k Z_{k+1})^{1/2}]$ evolve along the track with $Z_k$. The subscript k runs from 1 to 5, provided that the interval centered on $Z_1$ and $Z_5$ are $[0, (Z_1 Z_2)^{1/2}]$ and $[(Z_4 Z_5)^{1/2}, \infty]$, respectively, where $Z_1 = 10^{-5}$, $Z_2 = 4 \times 10^{-4}$, $Z_3 = 10^{-3}$, $Z_4 = 10^{-2}$, and $Z_5 = 3 \times 10^{-2}$. The range of stellar mass is from $0.05 M_\odot$ to $60 M_\odot$. The sources of evolutionary tracks are given in paper I.

We also include the metallicity effect due to the line blanketing corrections. To convert the bolometric magnitude and effective temperature of a star into the absolute magnitude and various colors, we calibrate bolometric corrections and colors with effective temperature, metallicity (only for UBV), and luminosity class. The adopted calibrations are given in paper I.

## 3. STELLAR POPULATION IN A GIANT ELLIPTICAL GALAXY

In this section, to compare with the single-burst population models with a constant metallicity for giant elliptical (gE) galaxies (e.g., Tinsley 1978), we assume that $t_{GW}=10^9$ yr. We have computed the composite models with $\nu=50$, $T_G=15\times10^9$ yr, an initial galactic mass $M_G=10^{12}M_\odot$, and various $\mu$.

Figure 1 shows the Johnson's (1966) broad-band colors of the composite population models as a function of $\mu$. Contrary to the constant metallicity models, the composite models show a minimum for each integrated color as $\mu$ increases. This can be understood if we remember that each color is determined by competition between a giant-to-dwarf ratio and an average stellar metallicity. As $\mu$ increases, the giant-to-dwarf ratio decreases and, as well known, the integrated color shifts redward. While, an increase of $\mu$ also decreases an average metallicity of a galaxy, which shifts the color blueward.

Figure 1 shows that the composite model with $\mu\simeq1$ gives an excellent fit to the observed colors for an average gE galaxy. Apparent discrepancies on V-R and V-I colors seem not to be fatal, because these observed values would be underestimated (Tinsley 1978). It is unlikely that a shape of IMF depends significantly on $M_G$, therefore we assume hereafter that all E galaxies have the identical slope of IMF as $\mu=0.95$ regardless of $M_G$.

Figure 2 shows a distribution of stellar metallicity in the composite model with $\mu=0.95$. Most of the integrated light of a gE galaxy is dominated by super- metal-rich stars ([Fe/H]$\geqslant$0.3), but there is still non-negligible contributions from metal-poor stars ([Fe/H]<0). Tinsley (1978) and Aaronson et al. (1978) encountered the persistent discrepancy for UV light between their single-burst population models and observations. But the present result suggests that such discrepancy is natually removed if we include properly the metal-poor stars.

## 4. COLOR-MAGNITUDE RELATION

We have computed photometric properties of E galaxies for $M_G=10^9$, $4\times10^9$, $10^{10}$, $10^{11}$, $10^{12}$, and $2\times10^{12}M_\odot$, by considering a dependency of $\nu$ on $M_G$. We suppose that the initial activity of star formation had been induced by successive collisions of interstellar clouds in virial equilibrium. Then, assuming that a time scale of star formation is equal to a collisional time scale of clouds, we obtain a relation among $\nu$, $M_G$, and an initial radius $R_G$ of a galaxy as $\nu \propto M_G/R_G^2$. If proto-elliptical galaxies had obeyed the same mass-radius relationship as presently observed in galactic globular clusters and E galaxies; i.e., $R_G \propto M_G^{0.55}$ (Saito 1979a), above relation would reduce to $\nu \propto M_G^{-0.1}$. Hereafter, we use this relation provided that $\nu=60$ for $M_G=10^{12}M_\odot$. Throughout this section, the slope of the IMF and the age of a galaxy are assumed to be $\mu=0.95$ and $T_G=15\times10^9$ yr.

The resulting chemical and photometric properties of the composite models are given in Tables I and II, respectively. In Table I, $M_G$ and $T_G$ are given in units of $10^9 M_\odot$ and $10^9$ yr, respectively. Columns (5) to (8) give a time ($t_{GW}$) when galactic winds are induced in unit of $10^9$ yr, a metal abundance ($Z_g$) and a helium abundance ($Y_g$) of the gas at $t=t_{GW}$, and a ratio of an amount of mass loss to initial mass of a galaxy ($\delta M/M_G$). In table II, columns (2) to (7) give an absolute magnitude ($M_V$), broad-band colors U-B, B-V, V-R, V-I, and V-K, respectively. Columns (8) to (10) give an average stellar metallicity (<[Fe/H]>), a visible stellar mass-to-luminosity ratio ($M_*/L_B$), and a ratio

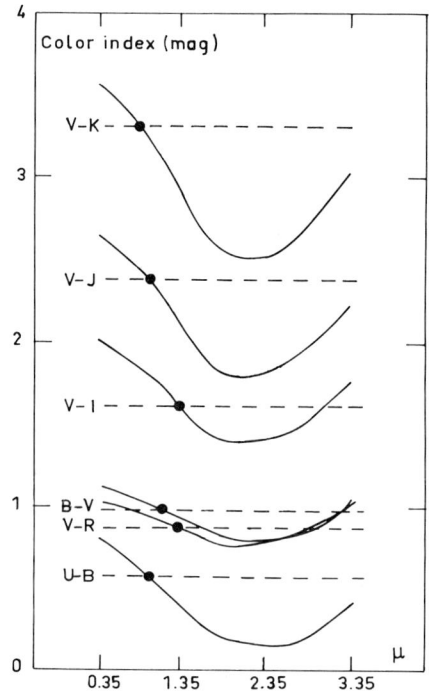

Fig.1. The integrated colors U-B, B-V, V-R, V-I, V-J, and V-K of the composite models at $15 \times 10^9$ yr. The lines represent loci of models with varying $\mu$. The dashed-lines represent observed colors for an average gE galaxy given by Tinsley (1978). Fitting points between the models and the observations are indicated by filled circles.

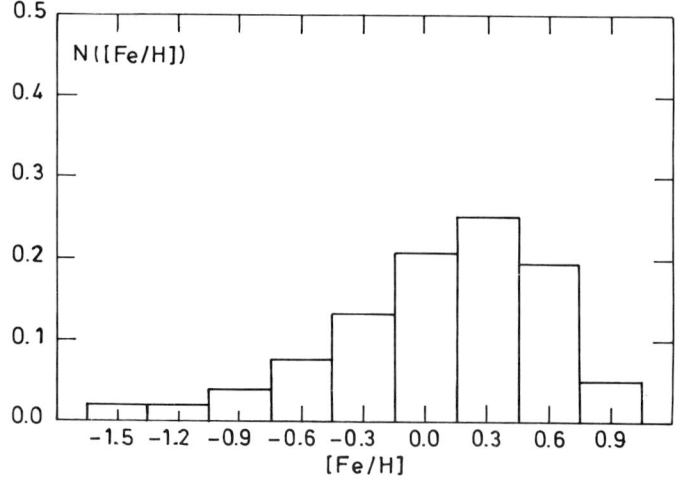

Fig.2. Distribution of stellar metallicity in the composite model with $\mu=0.95$. Relative number of stars is given as a function of stellar metallicity [Fe/H].

of mass of invisible remnants to present mass of a galaxy ($M_{rem}/M$).

Table I. Chemical properties of the models.

| $M_G$ | $\mu$ | $\nu$ | $T_G$ | $t_{GW}$ | $\log Z_g/Z_\odot$ | $Y_g$ | $\delta M/M_G$ |
|---|---|---|---|---|---|---|---|
| 1 | 0.95 | 120 | 15 | 0.020 | −0.20 | 0.23 | 0.71 |
| 4 | 0.95 | 104 | 15 | 0.050 | 0.27 | 0.23 | 0.51 |
| 10 | 0.95 | 100 | 15 | 0.085 | 0.52 | 0.24 | 0.36 |
| 100 | 0.95 | 80 | 15 | 0.355 | 0.91 | 0.31 | 0.08 |
| 1000 | 0.95 | 60 | 15 | 1.260 | 0.92 | 0.31 | 0.02 |
| 2000 | 0.95 | 56 | 15 | 1.770 | 0.91 | 0.30 | 0.01 |

Table II. Photometric properties of the models.

| $M_G$ | $M_V$ | U−B | B−V | V−R | V−I | V−K | <[Fe/H]> | $M_*/L_B$ | $M_{rem}/M$ |
|---|---|---|---|---|---|---|---|---|---|
| 1 | −14.1 | 0.17 | 0.78 | 0.73 | 1.30 | 2.43 | −0.74 | 6.7 | 0.62 |
| 4 | −16.0 | 0.32 | 0.88 | 0.82 | 1.50 | 2.78 | −0.42 | 9.0 | 0.58 |
| 10 | −17.3 | 0.41 | 0.92 | 0.85 | 1.58 | 2.92 | −0.20 | 10.7 | 0.56 |
| 100 | −20.2 | 0.52 | 0.98 | 0.89 | 1.67 | 3.06 | 0.15 | 12.3 | 0.49 |
| 1000 | −22.9 | 0.62 | 1.03 | 0.94 | 1.78 | 3.23 | 0.35 | 12.8 | 0.42 |
| 2000 | −23.7 | 0.63 | 1.03 | 0.95 | 1.82 | 3.29 | 0.38 | 12.9 | 0.40 |

The composite models show that the larger $M_G$ is, the larger is the binding energy of the remaining gas, and the later are induced galactic winds. Therefore, more massive E galaxies become more metal-rich stellar systems, because longer activity of star formation makes the chemical evolution proceed more effectively.

The integrated colors become redder for an E galaxy with the larger initial mass. The mass dependency is particularly evident in U−B and V−K colors, thus agrees with the observed wavelength dependency of the CM relation (Visvanathan and Sandage 1977, Frogel et al. 1978).

Figure 3 indicates the CM relation predicted by the composite models. The composite models reproduce the observed CM relation of E galaxies (Frogel et al. 1978) excellently. In Fig.3, average metallicity, defined by a luminosity mean, of each model is also indicated. The average metallicity changes from <[Fe/H]>=−0.74 to 0.38 for $M_V$=−14 to −24 mag. Thus, a difference in <[Fe/H]> is concluded as the reason why the blueness of stellar contents is connected with $M_V$.

We assume that materials in a galaxy are initially in virial equilibrium and that a galactic mass loss occurs instantaneously due to galactic winds. If the system recovers virial equilibrium after the mass loss, we obtain the relation between an increment of galactic radius $\delta R$ and an amount of mass loss $\delta M (<0)$:

$$\delta R/R_G = -\delta M/(M_G+2\delta M).$$

It must be noted that there is an upper limit for the amount of mass loss; thus $\delta M$ should be less than a half of $M_G$, otherwise the remaining stars would be no longer bound to a galaxy. Hence, strictly speaking, the composite models with $M_G \leq 4 \times 10^9 M_\odot$ cannot survive as E galaxies.

A total visible stellar mass-to-luminosity ratio ($M_*/L_B$) increases as a galaxy becomes luminous but is nearly constant for galaxies brighter than $M_B = -18$ mag. The composite models give systematically larger values of $M_*/L_B$ than the observed values of Michaud (1980). Such an apparent discrepancy, however, is easily removed without changing other predicted properties if we slightly increase the lower limit of stellar mass ($m_\ell$).

It must be noted that in E galaxies there is a significant amount of dark mass in a form of dead remnants. Therefore, if we include the invisible mass of remnants into the total mass of a galaxy, the total mass-to-luminosity ratio becomes nearly constant ($M/L_B \simeq 20$) regardless of $M_B$.

## 5. PHOTOMETRIC EVOLUTIONS OF ELLIPTICAL GALAXIES

We have traced photometric evolutions of E galaxies by computing the composite models for $T_G = 10^7$, $10^8$, $10^9$, $5 \times 10^9$, $10 \times 10^9$, and $15 \times 10^9$ yr. Other parameters are the same as those in Sect. 4. Generally speaking, elliptical galaxies evolve toward lower and right corner of the CM diagram (Fig. 4). But their photometric evolutions depend significantly on $M_G$. The luminosity of a massive galaxy becomes maximum when the luminosity decrements due to deaths of previously formed stars exceed the contributions of newly formed stars. Whereas, in a less massive galaxy, most of gas was ejected in the very early stage of evolution, therefore, the contribution of newly formed stars is very small and the luminosity decreases monotonously without attaining any maximum.

The evolution of integrated colors of an elliptical galaxy depends strongly on its average metallicity. The metal-rich galaxy evolves redward much more rapidly than ther metal-poor one. During early stages of evolution ($T_G \leq 10^9$ yr), however, massive galaxies were relatively bluer than less massive ones, because metal-rich and young massive stars contributed dominantly to the total lights (see paper I).

## 6. CONCLUSIONS

E galaxies are the one parameter family of their initial masses. E galaxies have the same age of $15 \times 10^9$ yr, have the same slope of IMF ($\mu \simeq 1$), but have the different average stellar metallicity.

The CM relation in E galaxies comes from their different average metallicities. If a galaxy had initially larger mass, the initial star formation lasted longer and the chemical evolution proceeded in much more extent. Thus, the gE galaxy has the larger average metallicity, the brighter magnitude, and the redder colors. The average metallicity changes from $<[Fe/H]> = -0.42$ to $0.38$ for $M_V = -16.0$ to $-24.0$ mag.

Both color and luminosity evolutions of E galaxies depend strongly on their average metallicities, and hence, on their initial masses. The CM relation changes significantly as galaxies evolve, which will have a great potential for studying distant E galaxies.

# ELLIPTICAL GALAXIES IN THE COLOR–MAGNITUDE DIAGRAM

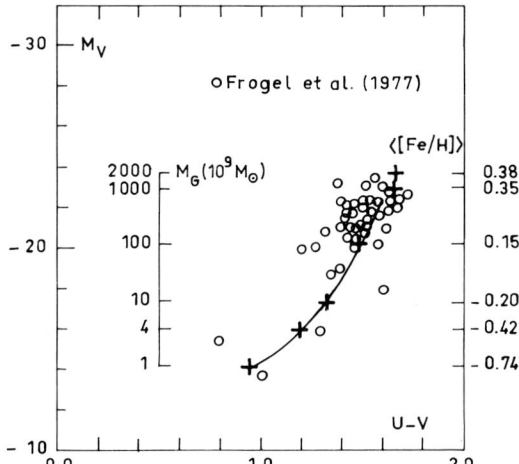

Fig.3. The CM relation in the ($M_V$, U–V) diagram predicted by the composite models at $15 \times 10^9$ yr. Crosses denote a series of models with varying $M_G$ and constant $\mu$ (=0.95). The value of $M_G$ is given in units of $10^9$ $M_\odot$. Open circles give the observational data by Frogel et al. (1978).

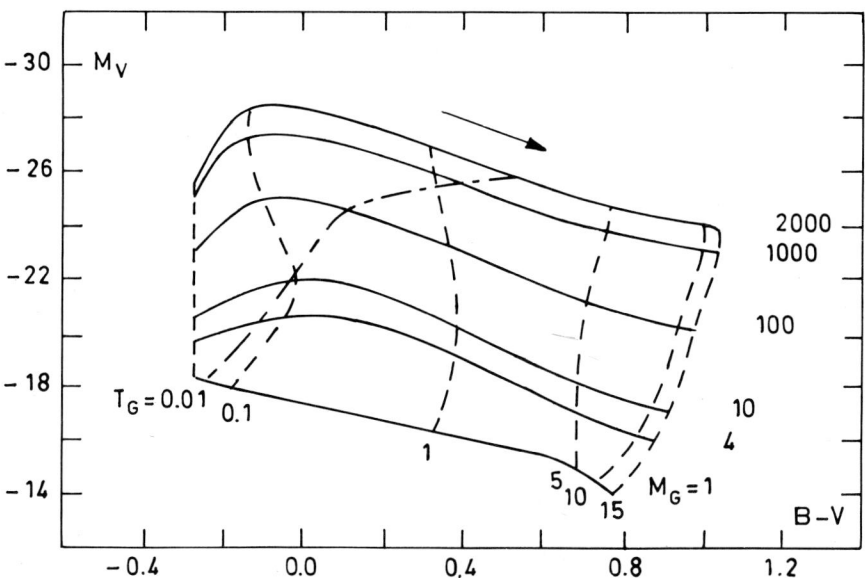

Fig.4. The evolutions of the composite models on the ($M_V$, B–V) diagram. Solid lines denote evolutionary tracks of the models with $M_G$. Dashed-lines denote isochrones of the models at age of $T_G$. $M_G$ and $T_G$ are given in units of $10^9$ $M_\odot$ and $10^9$ yr, respectively. A dot-dashed line denotes a phase when galactic winds are induced.

# REFERENCES

Aaronson,M.,Cohen,J.G., and Mould,J.: 1978, Astrophys.J.223,824.
Arimoto,N., and Yoshii,Y.: 1985, submitted to Astron.Astrophys.Suppl.
Faber,S.M.: 1977, In the Evolution of Galaxies and Stellar Populations, B.M.Tinsley and R.B.Larson (Eds), Yale University Observatory, New Haven, p.157.
Frogel,J.A., Persson,S.E., Aaronson,M., and Matthews,K.: 1978, Astrophys.J. 220,75.
Huchra,J.P.: 1977, Astrophys.J.Suppl.35,171.
Ikeuchi,S.: 1977, Prog.Theor.Phys.58,1742.
Johnson,H.L.: 1966, Ann.Rev.Astron.Astrophys.4,193.
Larson,R.B.: 1974, Monthly Notices Roy.Astron.Soc.169,229.
Michaud,R.: 1980, Astron.Astrophys.91,122.
Saito,M.: 1979a, Publ.Astron.Soc.Japan 31,181.
Saito,M.: 1979b, Publ.Astron.Soc.Japan 31,193.
Salpeter,E.E.: 1955, Astrophys.J.121,161.
Tinsley,B.M.: 1978, Astrophys.J.222,14.
Visvanathan,N., and Sandage,A.: 1977, Astrophys.J.216,214.

# DISCUSSION

Renzini: It would be extremely interesting to see how the integrated spectra of these model E galaxies look like, in particular because of their large spread in metallicity. Concerning the models themselves, did you consider also the nucleosynthesis of helium, and if so did you include this effect in the photometric modles?

Arimoto: Yes, we computed the nucleosynthesis of helium (see Table I). The models predict $\Delta Y_g/\Delta Z_g \simeq 0.5$. But, we did not include the effect of helium in the present photometric models. If we consider the nucleosynthesis of helium, it would have an effect opposite to that of metallicity.

Rocca-Volmerange: Did you take into account any variation in the photometric properties of stars with metallicity? More precisely how do you estimate colors and bolometric correction of metal-rich stars?

Arimoto: Yes, we did for metal-poor and normal metal stars ([Fe/H]≤0), but for metal-rich ones we assumed the calibrations for normal stars by Johnson (1966).

Koo: Your models are very impressive. Is there some practical way to explore spatial (radial) changes in metallicity of gE galaxies by adding your formulation to the work of dissapational galaxy formation models like those of Larson?

Arimoto: It will be possible to estimate the spatial change in metallicity required to reproduce the radial color gradients observed in some of gE galaxies by combining galaxy formation models with our photometric models.

Mould: The mass-metallicity relation appears to continue to lower masses than your disruption limit.

Aaronson: The dwarf spheroidals fit smoothly onto the mass-metallicity relation for larger ellipticals. Can you adjust your models to explain them?

De Jong: Is it possible that when galaxies are disrupted due to expulsion of gas in a galactic wind the dwarf spheroidals are created? If so your models would predict the metallicity of the stars in the fragments that are the progenitors of the dwarf spheroidals.

Arimoto: If kinetic energies of stars are dessipated effectively, the gas would be expelled <u>slowly</u> (not instantaneously). In this case, it would be possible that stellar velocities are always less than the escape velocity, because stellar velocities would decrease as a galaxy expand. Therefore, even if $\delta M/M_G \geqslant 0.5$, the remaining stars could be bound after a mass loss. However, to discuss the problem precisely, it is necessary to estimate how an ejection rate of the gas is.

Dressler: Have you included massive halos in your calculation of the mass loss due to winds? This would certainly have a large effect for low mass galaxies, for example, by preventing the galaxy from being unbound after a substantial mass loss. Presumably, such a galaxy would have a low central density and high M/L, as some observations indicate.

Arimoto: No, we haven't. But, I agree with you. Massive halos are one of the candidates to prevent less-massive galaxies from being disrupted.

Barbaro: Have you checked if your results depend on the assumption about the star formation rate?

Arimoto: No, we haven't. Colors of the model depend on the SFR. If we assume a lower value of $\nu$, the nucleosynthesis of metallicity becomes less effective and the initial activity of star formation lasts much longer. Thus, an average metallicity of the model decreases and younger stars becomes to contribute to the photometric properties, both of which shift the integrated colors blueward.

O'Connell: How do your predictions depend on parameters such as the lower mass limit of your IMF?

Arimoto: The model properties, except $M_*/L_B$ and $M_{rem}/M$, don't depend on the lower stellar mass limit ($m_\ell$). A larger value of $m_\ell$ decreases $M_*/L_B$ and increases $M_{rem}/M$.

O'Connell: Dressler asks me whether it is possible to reconcil your models with the synthesis results I discussed yesterday. I couldn't answer his question without understanding your models in great detail. So let me just ask you what adjustments you would have to make in your parameters as input assumptions in order to reproduce the observed colors and metallicities of gE galaxies at an age of 8 Gyr rather than 15 Gyr? (It appears that there is not much difference in your diagram between isochrones at 10 and 15 Gyr.)

Arimoto: The 8 Gyr model will predict bluer colors than the 15 Gyr one if other input parameters are the same. A decrease of $\mu$ shifts the colors redward, but it also increases largely the average stellar metallicity of the model. To keep the metallicity unchanged, a decrease of $\nu$ is also required, which shifts the colors blueward again. Therefore, it is rather difficult to reproduce the observed colors and metallicities of gE galaxies by the 8 Gyr model.

Aaronson: Could an 8 Gyr model with a red AGB reproduce the colors as well as your 15 Gyr model with no AGB?

Arimoto: The 8 Gyr model with a red AGB might reproduce the observed colors, because required adjustments for $\mu$ and $\nu$ would be smaller than the case without a red AGB.

Bruzual: From your $M_V$ vs U-V diagram it is not clear to me that 15 Gyrs is any better than 10 or 12 Gyrs. In your conclusions you stated that galaxies are 15 Gyrs old. How do you choose this age? Do you think that all normal ellipticals have the same age? Can you allow for a dispersion in the age of E galaxies? What are your best estimate of <age> and $\sigma_{age}$ of E galaxies?

Arimoto: The model properties become insensitive to $T_G$ if $T_G \geqslant 10$ Gyr. Thus, 10 or 12 Gyr may also be the age of E galaxies. But I beleive that E galaxies are as old as our own galaxy which is older than 15 Gyr according to the age estimations of galactic globular clusters. The CM relation of the coeval models agrees quite well with the observed slope of distributions of E galaxies on the CM diagram. The scatter in colors along the CM relation could be explained by an intrinsic dispersion in the efficiency of SFR ($\nu$). Therefore, our best estimations for normal E galaxies are <age>=15 Gyr and $\sigma_{age}=0$.

Mould: 1) Would you agree that it is inappropriate to determine the ages of E galaxies from the U-V colors of your models, because of the uncertainties in the U-V colors of super-metal-rich stars? 2) How are your models dependent on the assumption that star formation ceases suddenly at $t=t_{GW}$?

Arimoto: 1) Yes, SMR stars have large contributions to the integrated colors. If SMR stars are not properly taken into account, the resulting ages would be overestimated. But to calibrate the ages of E galaxies in terms of U-V, or other colors, effects of mixing length parameter on evolutionary tracks and contribution of horizontal branch stars should also be considered. At least both of these would partly cancel the effect of SMR stars. 2) We have assumed no star formation after $t=t_{GW}$, but presumably the gas shed from low mass stars would begin to accumulate and the star formation would again progress. Then, after a short period, the second burst of supernovae explosions would occur. Thus, if we include such intermittent star formations, the composite model would give an evolutionary track which makes loopes on the (U-B,B-V) diagram.

## IV.    EMPIRICAL POPULATION SYNTHESIS


R. W. O'CONNELL
Analysis of stellar populations at large lookbacks

A. PICKLES
Population synthesis and epochs of star formation in NGC 1316 (Fornax A)

B. ROCCA-VOLMERANGE, B. GUIDERDONI
Far-UV stellar populations of S0 galaxies

F. BERTOLA
The UV energy distribution of elliptical galaxies

R. NESCI
UV spectra of normal ellipticals

ANALYSIS OF STELLAR POPULATIONS AT LARGE LOOKBACKS


Robert W. O'Connell
Astronomy Department
University of Virginia
Charlottesville, VA 22903  USA


ABSTRACT. The capabilities of optimizing spectral synthesis for interpretation of stellar populations are reviewed with emphasis on the growing evidence that the light of nearby E/S0 galaxies and Sb bulges is dominated by a population significantly younger than expected from the study of globular cluster color-magnitude diagrams. The implications of these results for the study of high redshift galaxies are considered. The observational requirements and analytical tools necessary for proper interpretation of populations in galaxies viewed at large lookbacks are discussed in the context of the capabilities of the Hubble Space Telescope and the 10-m telescope.


1. INTRODUCTION

My purpose in this paper is to propose an observational agenda for interpreting the evolutionary state of stellar populations viewed at large lookback times and to review the implications of recent studies of nearby galaxies for such systems. The discussion will center on the technique of optimized spectral synthesis, which has been applied to integrated light observations by a number of workers over the past 15 years.

I am motivated by the accumulating evidence that evolutionary effects are readily detectable in a significant fraction of galaxies in rich clusters at relatively modest redshifts $z \sim 0.7$ (Butcher and Oemler 1978 and 1984, Dressler and Gunn 1983, Schild 1985). Observations of individual, more distant galaxies, mostly radio-selected (Lilly and Longair 1984, Djorgovski and Spinrad 1985), yield trends of color and magnitude which are consistent with the $\mu = 0.5$ evolutionary models of Bruzual (1983). In these calculations the initial gas complement is consumed in an e-folding time of only 1.4 Gyr. However, the models also assume that all material shed by stars on the giant branch is recycled into new stellar populations, a process which dominates the restframe near-UV spectrum which is redshifted into the visible/near IR region observed. The color and magnitude trends observed therefore imply considerable star forming activity at intermediate epochs. Without high resolution imaging, one cannot decide whether or not this activity is confined to disk populations; evidence to be discussed in Sec. 3 suggests that at least some of it may occur in spheroidal populations.

Spectroscopic evidence of strong bursts of star formation (Dressler and Gunn 1983, Butcher and Oemler 1984) confirms one's expectation that these late stages of the assembly of galaxies may be dominated by stochastic events (collapses of subclusterings, galaxy collisions and mergers, infall of circumgalactic material, etc.) which are likely to be difficult to model from first principles. We will be dependent on detailed analysis of the observations for a sound understanding of these processes. With the advent of unprecedentedly powerful instruments, such as the Hubble Space Telescope (HST) and the new generation of large ground-based telescopes, this is a good time to consider the analytical tools and observational requirements appropriate to this task.

## 2. SPECTRAL SYNTHESIS TECHNIQUES

There is by now a large literature of spectral synthesis studies with which most attendees at this conference will be reasonably familiar. I will therefore touch only on those aspects of technique which are of most interest for the problem of galaxies at high redshifts. Further details are available in another review (O'Connell 1985).

The basic approaches to current synthesis technique were described by Tinsley (1968), Spinrad and Taylor (1971), and Faber (1972), although there is still no universally-adopted standard in the field. The features distinguishing the various methods in current use are the photometric system employed, the component "library", the stellar evolutionary constraints assumed, and, most importantly, the synthesis algorithm. The salient characteristics of the photometric system are the wavelength range; the spectral resolution, $R = \lambda/\Delta\lambda$ ; N, the number of data points; and P, the photometric precision achieved. Precision is not ordinarily considered a defining characteristic of a photometric system, yet it is often a determining factor in the outcome of a synthesis problem. The component library is a file of spectral energy distributions (SED's), obtained on this photometric system, of stars or other sources which will be combined to produce the synthetic SED. The library's usefulness is governed mainly by the volume of stellar parameter space (mass, age, chemical content) which it covers thoroughly. The most widely used set of stellar evolutionary constraints is derived from the Yale isochrones (Ciardullo and Demarque 1977), with adjustments for mixing lengths and with other important phases of evolution, such as the asymptotic giant branch (AGB) and horizontal branch (HB), added using a mixture of theoretical and empirical evidence (e.g. Aaronson et al. 1978, Rabin 1980).

There are two distinct approaches to the incorporation of stellar evolutionary constraints. In "evolutionary" synthesis, described by

Bruzual and Arimoto elsewhere in this volume, a particular evolutionary scenario and a set of stellar evolutionary constraints are adopted and used to populate the model color-magnitude diagram (CMD). The corresponding composite SED is then calculated. It is governed by a small number of parameters, most commonly age, metallicity (Z), the slope of the initial mass function, and the e-folding time for star formation. Internal evolutionary constraints are not varied, and there is usually no attempt to adjust the population parameters to obtain optimal fits to individual observed SED's.

"Optimizing" synthesis may be less familiar. Here, the internal evolutionary constraints and population parameters are permitted to vary within preset limits in order to achieve best agreement with observed composite SED's. This might be regarded the inverse of evolutionary synthesis: it derives the distribution of stars in the CMD, subject to the condition of astrophysical plausibility, which best fits an observed SED. Goodness of fit and the uncertainties in derived parameters may be explicitly evaluated.

Optimizing studies usually employ linear or quadratic programming algorithms, which are widely available owing to their commerical applications. (Karmakar's algorithm, a recent improvement in linear programming, has the potential to treat problems of up to one million variables at speeds up to 50 times faster than current techniques.) These permit the inclusion of inequality constraints, which are a natural and powerful means of expressing limits on our knowledge of the evolution of single stars and of stellar systems.

The advantages of optimizing techniques for spectral synthesis include the following: (i) allowance for the considerable uncertainties, both theoretical and empirical, in stellar evolution, which, as discussed by Renzini and Chiosi elsewhere in this volume, are typically 30% in a given parameter; (ii) automatic inclusion of known stellar types (e.g. blue stragglers, AGB stars, post-AGB stars, etc.) not accurately predictable by current evolutionary scenarios; (iii) inclusion of arbitrary star formation histories (e.g. unconstrained mixtures of single-generation clusters representing random bursts of star formation) or initial mass functions; (iv) first order compensation for important stellar types missing from the library; and (v) improved sensitivity to the remarkably subtle distinctions between composite SED's which may imply significant differences in populations, as vividly illustrated recently by Burstein et al. (1984) and Rose (1985). These advantages generally permit optimizing algorithms to fit data to observational precision (typically 1-3%).

In the next two sections I wish to review the capabilities of current synthesis procedures with particular reference to an astrophysical problem of direct relevance to galaxies viewed at large lookbacks.

## 3. INTERMEDIATE AGE POPULATIONS IN LOCAL E/S0 GALAXIES

There is now good evidence that the nuclei of E/S0 galaxies contain a significant "warm" stellar component, by which I mean one which is bluer than expected for an old (13-16 Gyr), metal rich (Z $\sim$ 2-3 $Z_\odot$) population. In this section I wish to argue that the most plausible interpretation of the warm component is as the main sequence turnoff of an intermediate age (5-10 Gyr) population.

### 3.1 The Relationship of Globular Clusters to E Galaxies

From the time of Baade's (1944) original definition of population types, it has generally been assumed that both globular clusters and elliptical galaxies are nearly coeval, ancient stellar systems. SED distinctions which emerged later (Morgan and Mayall 1957, Baum 1959) were thought to reflect a difference in mean chemical content but not age; globulars and E galaxies appeared to exhibit a smooth continuum of photometric properties with the galaxies being the high metallicity extension of the globular sequence (Faber 1973). Recent high precision studies of line strengths and infrared colors in Galactic and M31 globulars, however, have demonstrated that this continuity is illusory (Frogel et al. 1980, Rabin 1980, O'Connell 1983a, Burstein et al. 1984, Burstein 1985, Rose 1985). Clusters and galaxies differ significantly, for instance, in their Mg I/Hβ/CN relation (Burstein et al. 1984) and in their Sr II/Fe I/CN/H I relation (Rose 1985). An age difference is a leading candidate for the origin of this behavior. Regardless of the ultimate interpretation, however, the conventional view that globular and E galaxies represent a continuum of single generation systems which differ only in their mean metal abundance must be abandoned. It is important to appreciate that it was predominantly this apparent continuity and the large ages obtained from cluster CMD's, rather than any direct evidence on ellipticals, which led to the interpretation of elliptical populations as 15 Gyr old.

### 3.2 Evolutionary Predictions of Broad Band Colors

Over the past decade, increasingly sophisticated evolutionary synthesis models have been developed to predict the broad band colors of E/S0 galaxies (Tinsley and Gunn 1976, Whitford 1977, Tinsley 1978, Aaronson et al. 1978, Frogel et al. 1980, Rabin 1980, Bruzual 1983). These studies are in general agreement that the UBV colors of luminous E/S0 galaxies (U-B $\sim$ 0.55, B-V $\sim$ 0.95) are consistent with ages of 10-13 Gyr for a population with Z = $Z_\odot$. However, it is well established that luminous E/S0 galaxies have nuclear Z $\sim$ 2-3 $Z_\odot$ (Faber 1977) and exhibit only small color gradients (Sandage and Visvanathan 1978, Strom and Strom 1978, Wirth and Shaw 1983, Davis et al. 1985), which indicate that metallicities remain higher than solar to large distances from the nucleus. Blanketing corrections for metal rich stars remain uncertain, especially in the U band, but for a 15 Gyr-old population with $Z = 2\ Z_\odot$ Rabin's (1980) models indicate that the integrated colors will be (U-B) $\sim$ 0.95 and (B-V) $\sim$ 1.10. These are significantly redder than observed in normal E/S0 systems, and it is evident that some component much warmer than the turnoff stars of an old, metal rich population is a major contributor to their light. Tinsley (1978), for example, suggested blue stragglers or a minority metal-poor population as likely candidates.

## 3.3 Optimising Synthesis for M32

Optimising studies of high resolution data have independently confirmed the presence of warm components but provide better information on their nature. The best case is that of M32, for which all optimizing syntheses based on photoelectric data agree that stars as blue as (B-V) = 0.50 (F8 V equivalent spectral type) are major contributors (Spinrad and Taylor 1971, Faber 1972, Pritchet 1977, O'Connell 1980, Keel 1983). O'Connell (1980) pointed out that M32's SED is entirely consistent with a $Z = Z_\odot$, age = 5 Gyr population, which would have a main sequence turnoff at F8 V.

Strong, independent confirmatory evidence that M32 experienced vigorous star formation only $\sim$5 Gyr ago has recently become available. Burstein et al. (1984) have demonstrated that M32's 2500 Å/5500 Å color and H$\beta$ index are consistent with an F8 main sequence turnoff. Rose's (1985) 2 Å-resolution spectral classification system confirms that a sharp truncation of the main sequence of M32 occurs around F8 and that residual star formation, blue stragglers, or hot HB stars cannot account for the warm component. Further, his luminosity-sensitive Sr II/Fe I index demonstrates that M32 is much more strongly dwarf-dominated than the metal rich globular clusters (a 5$\sigma$ effect). This not only rules out a major contribution from MRGC-type populations in M32 (including the possibility that a red HB supplies the F8 starlight) but indicates a turnoff age significantly less than globular clusters.

The young component in M32 could be regarded as anomalous, perhaps resulting from its tidal interaction with M31, except that M32's SED is entirely normal for E/S0 galaxies of its metallicity (Sandage and Visvanathan 1978), a fact most recently confirmed in a principal component analysis of 20 Å resolution data by Gregg (1985).

## 3.4 Optimizing Synthesis for Luminous E/S0 Galaxies

The nuclei of luminous E/S0 galaxies and Sb spirals like M31 and M81 are more difficult to deal with than M32 because available libraries do not match their higher Z's as well. A 15 Gyr population with $Z = 2\ Z_\odot$ should have a main sequence turnoff at (B-V) $\sim$ 0.85, according to the corrected Yale isochrones. However, the turnoffs derived in numerous synthesis studies based on photoelectric data for gE's, M31 and M81 (Spinrad and Taylor 1971, Faber 1972, O'Connell 1976, Pritchet 1977, Taylor and Kellman 1978, Wu et al. 1980, Peck 1980) are significantly bluer, in no case being redder than (B-V) $\sim$ 0.73 and averaging (B-V) $\sim$ 0.63. These models contain very little light at V from stars hotter than the turnoff. Perhaps the most thorough recent study is by Pickles (1985), who performed a differential analysis of 17 E/S0 galaxies in the Fornax cluster. His models explicitly allowed an arbitrary mixture of solar, metal-rich, and metal-poor populations, and he was able to re-derive the expected metalliity-luminosity relation. His turnoffs for gE nuclei average (B-V) $\sim$ 0.70, implying that the last major epoch of star formation was about 8 Gyr ago. Likewise, Rose's (1985) luminosity-sensitive index again suggests dwarf dominance, and therefore relatively warm turnoffs, in gE's.

By contrast, Gunn et al. (1981, hereafter GST) in a study of 7 gE's and M31 advocate "combination" models, in which a component hotter than F0 is added to an old population with turnoff at (B-V) $\sim$ 0.80, corresponding to an age of $\sim$12 Gyr. The nature of the hot component is not entirely clear, and systematic residuals in the near UV persist in even their best models, but GST argue that the component is most plausibly interpreted as upper main sequence stars associated with residual continuing star formation.

The origin of the discrepancy between GST and other workers appears to be that the GST models involve major contributions from a giant branch (derived from solar neighborhood field stars) which is bluer by $\sim$0.3 mag in (B-V) than would be normal for the age and metallicity assigned. In order to fit the infrared continuum, a cool turnoff must then be selected. But because the turnoff dominates at short wavelengths, this leaves an ultraviolet deficiency which must be made up by adding the hot component. It is possible to design strong observational tests for this component. Recent limits on hot starlight in gE's near 4000 A (Rose 1985) and the fact that far UV data suggests little continuing star formation (Faber 1983, Bohlin et al. 1985) appear to rule out the combination models favored by GST.

Thus, though the evidence is not as stong as for M32, nearby E/S0 nuclei and Sb bulges also appear to be dominated by intermediate age populations. Gregg (1985) has recently developed a good case for 1-5 Gyr populations in many S0 <u>disks</u> as well, based on departures from the mean spectral sequence for E/S0 nuclei.

## 3.5 Summary and Implications for High Redshift Galaxies

The existence of warm components in E/S0 galaxies and the distinctions between globular clusters and galaxies are well established by the integrated light studies discussed above independent of their interpretation. Whether or not the warm components represent intermediate age populations, they must significantly affect the evolution of galaxy colors at moderate lookbacks. However, the preponderance of the evidence does indicate that the light of nearby E/S0 galaxies and Sb bulges is dominated by a population 5-10 Gyr old, significantly younger than the metal poor globular clusters in our Galaxy. This interpretation is very sensitive to the metallicity assigned to the galaxies; for a given turnoff near (B-V) = 0.60, a factor of two increase in Z implies a 50% decrease in age. The weakest links in the chain of argument therefore appear to be the Z determinations for the galaxies and the calibration of the turnoff/Z/age relation by theoretical isochrones.

I think it is significant that the CMD studies reviewed by Mould at this conference independently show that many ostensibly old systems, including nearby dwarf spheroidals and the bar and halo of the LMC, contain intermediate age populations. Indeed, the combined evidence suggests that the Local Group experienced widespread episodes of active star formation 3-8 Gyr ago.

One therefore expects that many spheroidal ancestor systems will appear as active star formers at $z \sim 1$. It should be emphasized that the synthesis results do not imply that galaxies formed only 5-10 Gyr ago or that "primeval" systems should be observed at $z \sim 1$. Integrated light observations are sensitive primarily to the last major epoch of star formation, and it is difficult to place limits on earlier evolution. The oldest stars in all the nearby E/S0 systems dominated by intermediate age populations could in fact be 15 Gyr old. Viewed at a lookback of 5 Gyr, a spheroidal system which had formed stars at a constant rate starting 15 Gyr ago would have a (B-V) $\sim$ 0.50 in the restframe (Larson and Tinsley 1978); photometrically, it would resemble a current-epoch spiral and would not have a conspicuously peculiar SED.

However, these last events in the assembly of galaxies are likely to vary greatly in character from galaxy to galaxy or cluster to cluster, depending chiefly on the amount of gas remaining to fuel star formation and local dynamics. Some efficient systems, perhaps in the densest environments, will have completed star formation quickly and will form an upper envelope of red, "quiescent" galaxies at $z \sim 1$ (Sandage 1973, Oke 1983, Hamilton 1985). The color distribution of spheroidal ancestors below this envelope should change significantly over a lookback of 8 Gyr, however. Morphologies determined by the Hubble Space Telescope (HST) for distant blue galaxies will also shed light on the problem. The ancestors of local E/S0's with intermediate age populations will, of course, not necessarily resemble current-epoch spheroidals; one expects many disturbed morphologies, perhaps similar to accretors like N1275 or mergers like N7252 (Schweizer 1983).

## 4. CAPABILITIES OF OPTIMIZING SYNTHESIS

Considering the emphasis on "nonuniqueness" in synthesis analyses which one can find in the literature, the degree of unanimity in the studies cited above may seem surprising. In fact, I think too much has been made of the nonuniqueness problem by focussing on the fine details of synthesis solutions rather than their global properties. Fundamentally, nonuniqueness is a product of the fact that stellar SED's are not sufficiently independent of one another in a mathematical sense. It affects all methods of studying integrated spectra, evolutionary and optimizing synthesis alike. Its best known symptoms are the gaps or negative components appearing in the CMD of optimizing synthesis solutions obtained without astrophysical constraints (e.g. Williams 1976), which occur when adjacent components in the CMD are traded off by the algorithm to achieve slightly improved fits to the data.

However, it has been demonstrated by a number of workers (Faber 1972, Williams 1976, Turnrose 1976, Pritchet 1977, Pickles 1985) that the imposition of relatively mild smoothing constraints in the CMD renders optimizing solutions reasonably stable against observational errors, deficiencies in the libraries, or poor initial guesses at the solution. External tests of optimizing synthesis, such as prediction of IR or UV spectra from optical data (Aaronson, Frogel and Persson 1978; Ciani et al. 1984) or consistency between synthesis solutions and known CMD's (Christensen 1972), are also very encouraging. The largest uncertainties in the method appear to lie in our understanding of the details of stellar evolution (e.g. Renzini and Buzzoni, this volume), in the translation of these into numerical constraints on solutions, and in the difficulty of generating either empirically or theoretically a complete component library--all problems extrinsic to the synthesis technique itself.

It must be emphasized, however, that success with the technique cannot be trivially achieved. Considerable care in establishing a photometric system and library, in developing appropriate astrophysical constraints, in formulating answerable questions, and in estimating errors in solutions is necessary. The situation treated in Sec. 3 is one in which the composite SED is strongly dominated by populations with a relatively small range in age and Z. In this circumstance, careful application of optimizing synthesis to high quality data appears able to yield accuracies of about $\pm 0.15$ in log t or log Z. Because of the extrinsic uncertainties mentioned above and the strong interaction between population components in determining the composite SED, it is not possible to determine these two quantities from only a few measured parameters. The more comprehensive studies in Sec. 3 were characterized by: coverage from 3400 A or below to $2.2\mu$; $R \sim 200$ at 4000 A; $N \gtrsim 50$; a library containing $\sim 50$ component types; and $P \sim 1-5\%$ in relative fluxes throughout the spectrum. These are formidable requirements, but it is not obvious that a substantially smaller investment of observing time

# ANALYSIS OF STELLAR POPULATIONS AT LARGE LOOKBACKS

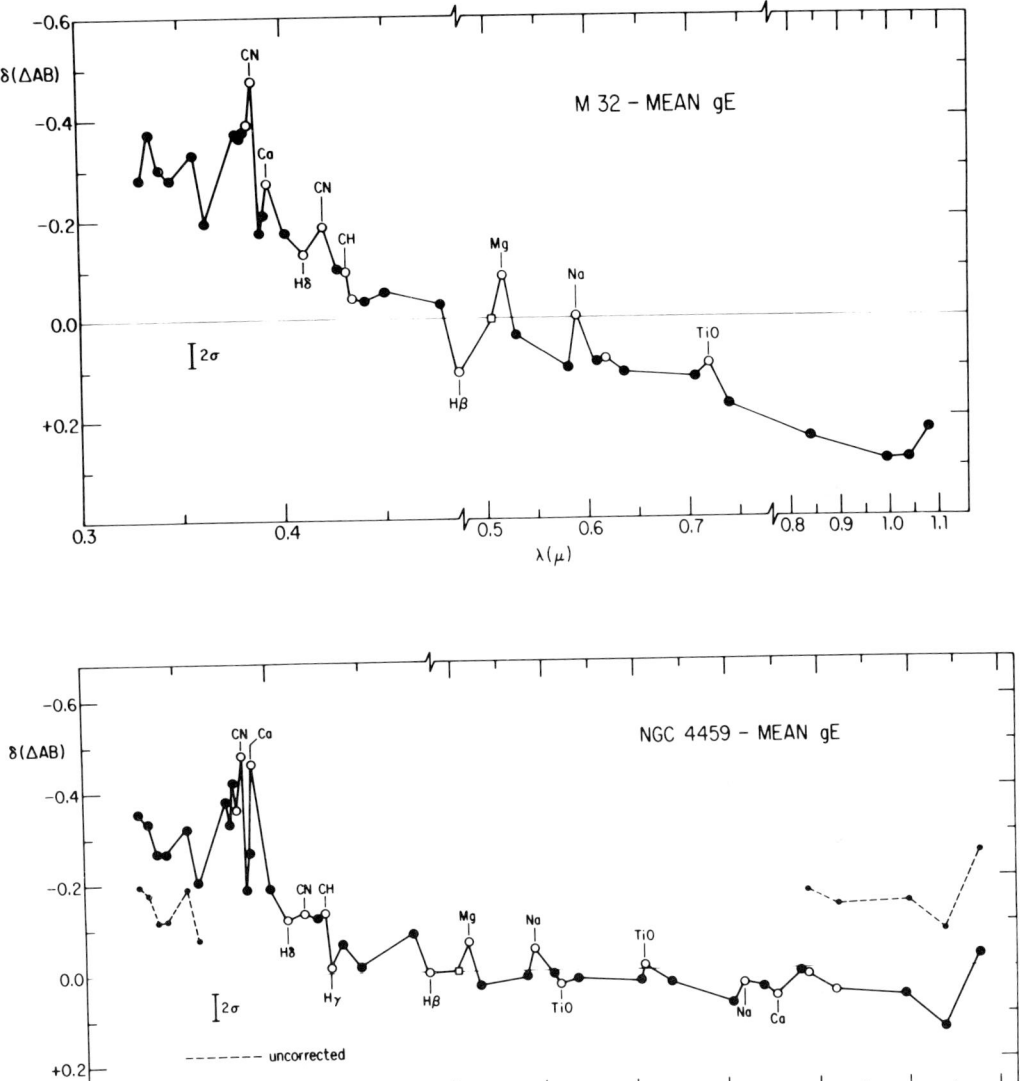

FIGURE 1: Differential energy distribution diagrams, expressed in magnitudes and normalized at 5050 Å (open square), for M32 (Fig. 1a) and N4459 (Fig. 1b) compared to a mean gE SED. Foreground and internal reddening have been estimated and removed as in O'Connell (1980); the dashed lines in Fig. 1b indicate the uncorrected values for N4459. Open circles indicate positions of strong absorption features. Negative values indicate that the object plotted is brighter than the mean gE relative to 5050 Å; spectral features "in emission" are stronger in the gE.

will permit a satisfactory outcome.

One encounters a more complex situation where the spectrum contains comparable contributions from a wide range of populations. This case applies, for example, to the "dilute elliptical" spectra observed in some clusters at $z \sim 0.5$ (Dressler and Gunn 1983, Butcher and Oemler 1984), which seem to represent old galaxies with a recent ($\lesssim$ 1 Gyr old) burst of star formation. In a situation where there are M important stellar populations each characterized by Q parameters (e.g. age, Z, IMF slope, etc.), one requires a minimum of $N = (Q+1)M - 1$ independent observational parameters to specify a complete solution. Unfortunately, because of the lack of independence of stellar SED's and their slowly varying changes with wavelength, increasing the number of data points in a given region of the spectrum does not necessarily provide additional information. A longer wavelength baseline is more useful than a higher density of data points. Thus, a formal requirement for, say, 9 independent data points may actually demand 30-40 points spread across the entire accessible spectrum. Even then, uncertainty in the final solution will exceed that in the single population case.

The most succesful optimizing approach to multi-population spectra is to create a library containing a wide range of strongly-constrained subpopulations (e.g. $Z = Z_\odot$ star clusters with ages from 0.001 to 10 Gyr) and seek solutions consisting of an arbitrary mixture of these (e.g. O'Connell 1980, GST, Pickles 1985). As an example, consider the case of N4459, an S0 galaxy in the Virgo cluster which is one of a class of blue-nucleated E/S0's first pointed out by Tifft (1969). Perhaps 15% of the cluster E/S0 objects studied by Sandage and Visvanathan (1978) fall in this class. N4459 can be regarded as a mild, local example of the "dilute E" phenomenon. In Fig. 1 I compare the SED's of M32 and N4459 (10" aperture) to a "mean gE" SED (O'Connell 1976). Both objects exhibit a near UV excess of about 0.4 mag and general line weakening. For M32, the blueness extends to the near IR; what we see here is predominantly a metallicity-driven shift in the position of the giant branch and turnoff, reflecting the factor of 2-3 difference in Z between M32 and gE's. N4459's near IR continuum resembles a gE's, indicating a similar giant branch temperature; the UV excess then presumably originates in a separate, younger component. The figure indicates that the first signs of young components in an old object will be a strengthening of the UV continuum and weakening of the strong metallic lines, particularly below 4300 A, rather than strengthening of the Balmer series, as is commonly supposed.

Optimizing synthesis demonstrates quantitatively that a single generation population is not consistent with the N4459 spectrum. Successful models include an old (8 Gyr) background population with an 11% contribution at V from generations $\lesssim 1$ Gyr old with normal IMF's. The best models fit to observational precision (0.017 mag). The implied star formation history is illustrated in Fig. 2, where it is

ANALYSIS OF STELLAR POPULATIONS AT LARGE LOOKBACKS 331

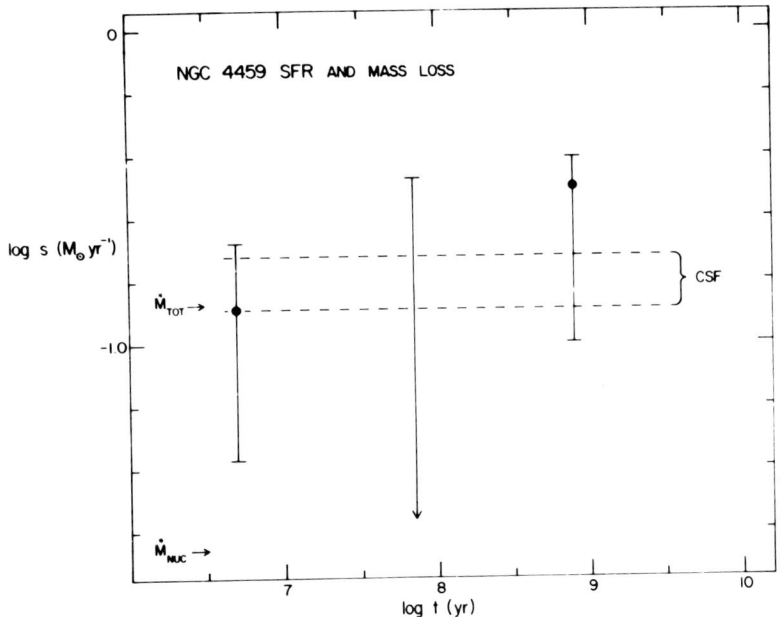

FIGURE 2: The history of recent star formation in N4459 from optimizing synthesis models. Dots indicate nominal rates for young generations selected from a possible 6 generations younger than 3 Gyr, normalized to the 10" region observed. Error bars are $3\sigma$ limits. Dashed lines indicate $3\sigma$ limits for a model assuming continuous star formation during the past 3 Gyr; this model fits less well than the multiple burst case. At the left are indicated giant branch mass loss rates for the entire galaxy and for the nucleus only. All quantities are calculated as in O'Connell (1983b).

compared to the rate of return of material to the ISM by mass loss from the giant branch (also calculated from the synthesis models). We see that whereas nuclear mass loss is insufficient to fuel the star formation, the rate of loss from the entire galaxy agrees well with the observed SFR. N4459 possesses an X-ray corona (Forman et al. 1985), and the results are in good agreement with the picture in which the heated material shed by stars throughout the galaxy ultimately cools and flows into the central regions, forming stars (Nulsen et al. 1984). Since only a small fraction of E/S0's have N4459-like color anomalies, this process is evidently not universal, at least not with a normal IMF.

It should be emphasized that the important photometric distinctions between N4459 and M32, or between the blue nucleus of M33 and metal poor globular clusters (O'Connell 1983b) for example, would not be clearly detectable with observational precisions worse than about 5-10%. The synthesis studies of M32 and the recognition of distinctions between globular clusters and galaxies disussed in Sec. 3 required precisions of 1-2%. High precision in the data base <u>and</u> in the fits emerging from the synthesis procedure are key elements in the

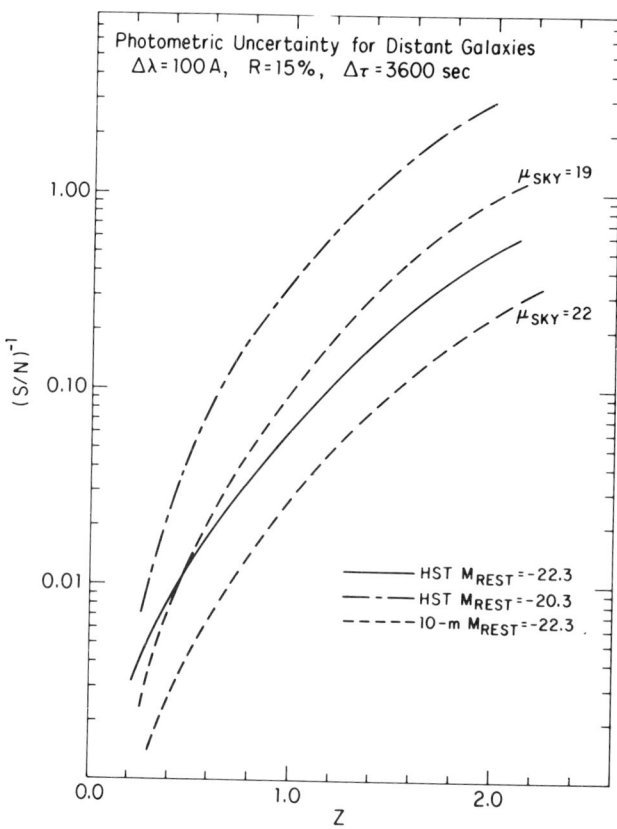

FIGURE 3: Photometric precision achievable on distant galaxies as a function of redshift. Assumptions include: $H_o$ = 75 km/sec/Mpc, $q_o$ = 0.25 (implying a lookback time at z = 1 of 6 Gyr, or 60% of the age of the universe); a spectrometer yielding 100 Å spectral resolution with a net throughput of 15%; a one hour integration time. The only noise source considered is photon statistics. For the Hubble Space Telescope (HST) I assume a sky background of 23 mag/square arcsec, applying for 5000-11000 Å, and a 2" diameter entrance aperture, corresponding to the 1/2 light diameter of a high redshift gE (Sandage 1972). For a ground-based 10-m, I assume a 5" entrance aperture to avoid differential refraction; the sky background surface brightnesses of 22 and 19 mag/square arcsec correspond to 5500 Å and 8500 Å, respectively. Calculations are done for 5500 Å, but the results will apply to other wavelengths for the given monochromatic sky brightness and absolute magnitude (per unit wavelength) at those wavelengths if the N/S ratio is scaled by $(5500 \text{ Å}/\lambda)^{0.5}$. M = -22.3 corresponds to the absolute V magnitude of the brightest gE galaxies; M = -20.3 corresponds to objects near M*.

interpretation of stellar populations using integrated spectra.

## 5. APPLICATION TO GALAXIES AT LARGE LOOKBACKS

It is evident from the foregoing that our understanding of stellar populations in high redshift systems will be most strongly limited by the observational precision achievable. At $z = 1$ the most luminous gE galaxies will have integrated apparent magnitudes of $\sim 22$ in the redshifted V band, but more typical E galaxies (near $M^*$, Bahcall 1977) will be $\sim$24th magnitude.

Figure 3 illustrates the observational precisions likely to be achievable in the relatively near future on high redshift systems with state-of-the-art instrumentation in space or on the ground. For objects with $z \gtrsim 0.5$, photometry will be strongly sky-limited at most wavelengths for either space or ground-based telescopes. The very strong effects of sky background level and of source redshift are evident in the figure. It indicates that the 5-10% precision necessary for population studies should be achievable on the brightest gE's to redshifts near $z = 1$ with the HST or a ground-based 10-m but that more typical objects near $z = 1$ would require excessive integration times. However, adequate precision would be possible on galaxies with more normal luminosities to $z \sim 0.6$. Since the sky background drops quickly toward shorter wavelengths, better results can be obtained below 5000 A for objects which are bluer than $(B-V) \sim 0.5$.

One concludes that the sheer photon-collecting power necessary for detailed population studies at high redshifts will be available in the relatively near future. An observational program and an analysis method must be developed together, however, with the opportunity for iterations of both. Experience suggests the following guidelines for this process:

(i) One must formulate questions about populations which are <u>answerable</u> by the synthesis technique. This is more difficult than it appears, and the only way to judge whether a question is answerable is by numerical simulations including artificial data with characteristics expected from the real data.

(ii) An automatic, optimizing synthesis technique is preferred which permits a large range of population constraints to be explored. Given the generally lower precision of the data expected, it is probably best to construct a library of single-generation clusters characterized by age, metallicity, and (perhaps) IMF. For generations younger than 1 Gyr, large intervals of $\sim 0.5$ in log t and log Z are probably acceptable. Finer intervals are more appropriate for older generations. Development of reliable library components (probably theoretically) for $Z > Z_\odot$ is essential, as is a complete error

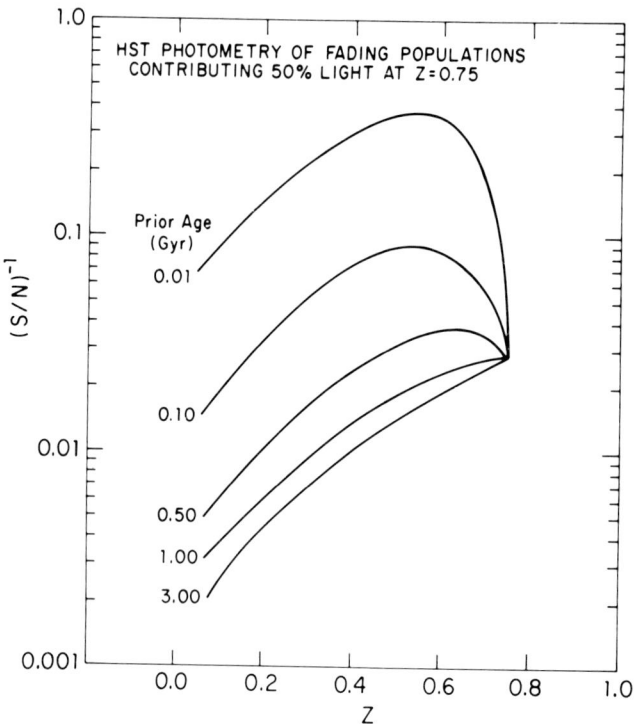

FIGURE 4: Photometric precisions as a function of redshift, calculated as in Fig. 3, for the descendents of bursts of star formation which occurred prior to the lookback time corresponding to z = 0.75 (5 Gyr). Luminosity decay of the burst populations in the V band is taken from the models of Larson and Tinsley (1978). The burst is assumed to produce 50% of the integrated light at z = 0.75. The background population is assumed to be unevolving.

analysis mechanism for the resulting solutions.

(iii) The actual photometric system employed will depend on iteration of (i) and (ii), but desirable goals appear to include: wavelength coverage (not necessarily continuous) of 1200-22000 A in the restframe; $P \sim 5-10\%$ in relative fluxes; and $R \gtrsim 100$, at least in spectral regions containing the most information (e.g. the 3500-4300 A region sensitive to the population differences in Fig. 1). This clearly implies long integrations and multiple-object spectrometers, regardless of the telescope employed.

(iv) Use of differential synthesis (simultaneous analysis of many galaxies in one cluster or many clusters) is desirable wherever feasible.

(v) As Figure 4 suggests, some problems concerning the populations of very distant systems may be solvable by observation at more readily accessible intermediate redshifts because of the very strong dependence of photometric precision on redshift and the relatively slow luminosity fading of bursts of star formation. For the particular example shown, observations of a descendent galaxy at $z < 0.2$ will yield better S/N on populations in a $z = 0.75$ ancestor which were at least 0.1 Gyr old at $z = 0.75$ (in the absence of subsequent star formation in the descendent) than would direct observation of the ancestor.

Overall, there are grounds for optimism that a relatively detailed understanding of stellar populations at large lookbacks can be achieved with foreseeable instrumentation, providing that workers (and time allocation committees) are willing to make the considerable investments in observing and development of analysis programs necessary for success.


REFERENCES

Aaronson, M., Cohen, J. G., Mould, J., and Malkan, M. 1978, Ap. J., 223, 824.
Aaronson, M., Frogel, J. A., and Persson, S. E. 1978, Ap. J., 220, 442.
Baade, W. 1944, Ap. J., 100, 137.
Bahcall, N. 1977, Ann. Rev. Astr. Astrophys., 15, 505.
Baum, W. A. 1959, Publ. Astron. Soc. Pacific, 71, 106.
Bohlin, R. C., Cornett, R. H., Hill, J. K., Hill, R. S., O'Connell, R. W., and Stecher, T. P., 1985, submitted to Ap. J.
Bruzual, A. G. 1983, Ap. J., 273, 105.
Burstein, D. 1985, Publ. Astron. Soc. Pacific, 97, 89.
Burstein, D., Faber, S. M., Gaskell, C. M., and Krumm, N. 1984, Ap. J., 287, 586.
Butcher, H. R., and Oemler, A. 1978, Ap. J., 219, 18.
Butcher, H. R., and Oemler, A. 1984, Nature, 310, 31.
Christensen, C. G. 1972, Ph.D. Thesis, California Inst. of Technology.
Ciani, A., D'Odorico, S., and Benvenuti, P. 1984, Astron. Astrophys., 137, 223.
Ciardullo, R. B., and Demarque, P. 1977, Trans. Astr. Obs. Yale Univ., 35.
Davis, L. E., Cawson, M., Davies, R. L., and Illingworth, G. 1985, A. J., 90, 169.
Djorgovski, S. and Spinrad, H. 1985, preprint
Dressler, A., and Gunn, J. E. 1983, Ap. J., 270, 7.
Faber, S. M. 1972, Astron. Astrophys., 28, 109.
Faber, S. M. 1973, Ap. J., 179, 731.
Faber, S. M. 1977, In Evolution of Galaxies and Stellar Population, eds. B. M. Tinsley and R. B. Larson (New Haven: Yale University

Observatory), p. 157.
Faber, S. M. 1983, Highlights of Astron., ed. R. M. West, 6, 165.
Forman, W., Jones, C., and Tucker, W. 1985, Ap. J., 293, 102.
Frogel, J. A., Persson, S. E., and Cohen, J. G. 1980, Ap. J., 240, 785.
Gregg, M. 1985, Ph.D. Thesis, Yale University.
Gunn, J. E., Stryker, L. L., and Tinsley, B. M. 1981, Ap. J., 249, 48.
Hamilton, D. 1985, Ap. J., in press.
Keel, W. C. 1983, Ap. J., 269, 466.
Larson, R. B., and Tinsley, B. M. 1978, Ap. J., 219, 46.
Lilly, S. J., and Longair, M. S. 1984, M.N.R.A.S., 211, 833.
Morgan, W. W. and Mayall, N. U. 1957, Publ. Astron. Soc. Pacific, 69, 291.
Nulsen, P., Stewart, G. and Fabian, A. 1984, M.N.R.A.S., 208, 185.
O'Connell, R. W. 1976, Ap. J., 206, 370.
O'Connell, R. W. 1980, Ap. J., 236, 430.
O'Connell, R. W. 1983a, Highlights of Astron., ed. R. M. West, 6, 147.
O'Connell, R. W. 1983b, Ap. J., 267, 80.
O'Connell, R. W. 1985, in prep. for Publ. Astron. Soc. Pacific.
Oke, J. B. 1983, in Clusters and Groups of Galaxies, eds. F. Mardirossian, G. Giuricin, M. Mezetti (Dordrecht: Reidel), p. 99.
Peck, M. 1980, Ap. J., 238, 79.
Pickles, A. J. 1985, Ap. J., in press.
Pritchet, C. 1977, Ap. J. Suppl., 35, 397.
Rabin, D. M. 1980, Ph.D. Thesis, California Inst. of Technology.
Rose, J. 1985, A. J., in press
Sandage, A. R. 1972, Ap. J., 173, 485.
Sandage, A. R. 1973, Ap. J., 183, 711.
Sandage, A. R. and Visvanathan, N. 1978, Ap J., 223, 707.
Schild, R. 1984, Ap. J., 286, 450.
Spinrad, H. and Taylor, B. J. 1971, Ap. J. Suppl., 22, 445.
Strom, K. M. and Strom, S. E. 1978, A. J., 83, 73.
Taylor, B. J. and Kellman, S. A. 1978, Ap. J. Suppl., 37, 101.
Tifft, W. G. 1969, A. J., 74, 354.
Tinsley, B. M. 1968, Ap. J., 151, 547.
Tinsley, B. M. and Gunn, J. E. 1976, Ap. J., 203, 52.
Tinsley, B. M. 1978, Ap. J., 222, 14.
Turnrose, B. E. 1976, Ap. J., 210, 33.
Whitford, A. E. 1977, Ap. J., 211, 527.
Williams, T. B. 1976, Ap. J., 209, 716.
Wirth, A. and Shaw, R. 1983, A. J., 88, 171.
Wu, C. C., Faber, S. M., Gallagher, J. S., Peck, M., and Tinsley, B. M. 1980, Ap. J., 237, 290.

# DISCUSSION

LEQUEUX: Can you assume, e.g. from DDO photometry, a metallicity independent of other factors or do you work from a guess and loop the system to get a final metallicity?

O'CONNELL: Metallicity and age must be simultaneously determined. You can use a mixture of library populations of different ages and metallicities and allow the algorithm to optimize the mixture, as Pickles has done. You can also estimate metallicity by examining residuals in spectral features from the best-fitting models using a library of solar metallicity. Estimates obtained in various ways by various authors tend to agree that the nuclei of gE galaxies have $Z \sim 2-3\ Z_\odot$.

RENZINI: As a general rule in an old population, colors, temperatures, and then integrated spectra, are more sensitive to metallicity than they are to age. This applies for instance to the turnoff color, for which a difference of, say 0.3 dex in [Fe/H] is equivalent to a difference of 3-4 Gyr in age. With this in mind, I wonder whether single metallicity model elliptical galaxies are adequate. Could the younger component you find actually be a metal-poor component? Indeed, I cannot see how a chemically homogeneous elliptical could be formed.

O'CONNELL: You are quite right that the age deduced from the main sequence turnoff in a population is a very strong function of the assumed metallicity. I think it was not adequately appreciated until the last 5 years how much the high metallicities assigned to giant ellipticals would reduce their derived turnoff ages. This emphasizes the importance of accurate determination of galaxy metallicities.

With regard to the age results I discussed, I doubt that they are strongly affected by a possible _mixture_ of metallicities (though they are certainly sensitive to the _mean value_ adopted). Pickles explicitly included a mixture of three metallicities in his models. In M32, a metal poor component of sufficient size to produce the observed F8 turnoff light would also dominate the giant light, leading to much weaker line strengths than observed.

While I certainly expect that E galaxies contain a finite range of metallicities, I don't believe this effect is responsible for the warm turnoffs now being detected.

ROCCA-VOLMERANGE: We confirm your analysis about the galaxy M32 from a far-UV spectrum with the IUE satellite (Johnson, 1979). As you can see in our paper of this section, the long wavelength (LWR Camera) spectrum corresponding to 2000-3000A agrees perfectly well with the spectrum of an F8 V star, obtained from the IUE spectral Atlas (NASA Newsletter, 1983)

O'CONNELL: That is a nice result.

AARONSON: Faber, Burstein and collaborators have a recent paper where they show the M31 globulars do not form a continuum with the galactic globulars. Do you have an explanation for this effect, and do you think the same sort of thing may be responsible for the differences between the globulars and the ellipticals?

O'CONNELL: I don't have an explanation. Burstein et al. slightly favor a younger age for the M31 clusters, but there are other possibilities as well. In his thesis, Christensen found it impossible to match the spectra of metal rich M31 clusters well with Galactic metal rich or poor stars. On the other hand, M32 can be well fitted with such stars. I think at the present that the M31 clusters are harder to understand than the E galaxies!

AARONSON: How well does a cluster like M67 integrated up fit M32?

O'CONNELL: Not very well. Rose has looked at this problem and sees evidence of blue straggler light at 4000A in M67 which is not present in M32.

WHITFORD: Can you comment on the nearly equal contributions of M dwarfs and M giants in the near infrared to the synthesis of the high resolution spectrum you showed? Isn't this sensitive to the IMF, and does the IMF lead to any difficulty with the mass-luminosity ratio?

O'CONNELL: The model illustrated was by Andrew Pickles and I would ask him to comment on it.

PICKLES: The constraints used on the main sequence are the simplest possible--that the numbers of stars per mass interval be non-decreasing ($x > 0$ in Tinsley's notation). This constraint does permit jumps in the M dwarf luminosity function but has no effect on the determination of the main sequence turnoff group. M/L ratios derived in this way are overestimated however. More detailed fits to the near infra-red region (700-1000nm) indicate that the total M dwarf contribution of 800nm is 10% or less. Recent observations of the Wing-Ford band at 991nm show that the contribution at 1000nm from latest M dwarfs (Wolf 359 type) is 3% or less. This implies a power law IMF slope of Salpeter value ($x=1.35$) or less.

EISENHARDT: Was the number of free parameters the same in fitting M33 and globular clusters with a metal rich population? Could the globulars be fitted with a metal rich population and more free parameters?

O'CONNELL: Details of these models were published in the Astrophysical Journal in 1983. The ability of the method to distinguish between old metal poor and young metal rich poplations does not appear to depend on the specific assumptions adopted for the models. The question addressed here was whether it was possible to fit metal poor globular cluster SED's by a combination of metal rich clusters of various ages, and it is clear that the answer was "no". Christensen studied the general synthesis of globular clusters in greater detail. I believe his results agreed with mine, namely that one can't fit high precision spectra of metal poor globulars with metal rich stars. High observation precision is, however, crucial for this problem.

NEWBERRY: (1) Can you comment on the effects of both Galactic and internal extinction on the population synthesis solution?

(2) In the solution space, how independent of metallicity and age is the extinction solution obtained from line strengths?

O'CONNELL: In most of the ellipticals studied so far extinction is not much of a problem. It can be estimated as part of the optimization procedure and agrees well with independent estimates of foreground extinction where available. The method is based on the fact that extinction affects continuum colors but leaves line strengths unchanged. In Sc galaxies extinction estimated from synthesis is consistent with Balmer emission line decrements (Turnrose). Most of these estimates use unconstrained models, and my impression is the results will not depend heavily on the makeup of the library, as long as it covers a wide temperature range.

AARONSON: Do you think there is any conflict between the deep color-magnitude diagrams of the Baade's window fields we saw a few days ago and the idea that ellipticals are predominantly intermediate age?

O'CONNELL: There are two areas where we need to clarify the situation in the window CMD results: (1) a very metal rich, old giant branch ought to fall well to the red of the 47 Tuc isochrone in the CMD rather than superpose on it as it does in the figure shown by Whitford; (2) one should use metal rich isochrones to compare to the turnoff seen in the deep CMD. These will be significantly fainter than $Z_\odot$ isochrones; the bulge turnoff could be consistent with a 5 Gyr metal rich population. (Whitford agreed with these remarks.)

MOULD: (1) Can you avoid a blue turnoff by adding a low metallicity tail to the abundance distribution in the form of globular cluster light in the model?

(2) The outer parts of the dE companions of M31 have less than 10% of this 5 Gyr population, or we would see their AGB stars. What color gradient is required to make this result consistent with yours?

O'CONNELL: (1) No. Very metal poor light with a strong horizontal branch will lead to an increasingly blue energy distribution as one moves to shorter wavelengths. You can't add enough MP light to avoid the intermediate age turnoff and yet be consistent with the SED below 4500A or with line strengths. In the case of M32 this is demonstrated by the close similarity of the IUE spectrum to a late F star, as commented on by Rocca-Volmerange.

(2) We haven't studied the two systems you worked on (N147 and 205). N205 shows evidence of a very recent burst of star formation in its center. It seems that the dwarf systems of the local group exhibit a wide range of SF histories, and one wouldn't necessarily expect that the nuclei of all local dwarfs will exhibit 5 Gyr turnoffs. It is important, however, to perform synthesis analysis on the outer parts of E galaxies or on their integrated light (though this is technically difficult). I might remark that Seitzer and I have recent CCD imagery on M32 which resolves its outer parts, so a direct comparison to your observations will be possible soon.

DE YONG: If the required precision that you quote for the photometry of $\sim 0.03$ mag. is differential, i.e. relative between points at two different wavelengths in the spectrum, it implies that your population synthesis fits are extremely sensitive to galactic and internal reddening.

O'CONNELL: The figure refers to flux ratios normalized, say, at 5500A. Yes, these ratios are sensitive to reddening, which must therefore be determined. This can be done by the synthesis technique itself or by external means. Most of the galaxy results I discussed are unlikely to be affected by mis-estimated reddening.

DE YONG: You express some confidence in the possibility to also model spirals in this way, but I think that must be very difficult in view of the fact that populations of different age, and therefore different colors, have different scale heights and will therefore be differently affected by the dust in the galaxy. Also, the youngest stars may still be completely obscured by dust in the molecular clouds from which they form.

O'CONNELL: There is no question that, especially, active star forming galaxies are much more complicated than any astronomer would prefer. Younger stars would be preferentially obscured, and if there were sound theoretical estimates of differential obscuration, this could be taken into account in the modeling procedure. It appears, however, that regions such as the nuclei of most Sc galaxies are not subject to such effects (e.g. Turnrose's analysis) insofar as the components which dominate the optical light are concerned. The only case that comes to mind where effects such as you describe are well established from optical data is that of M82. In the brightest star clusters there dust seems well mixed with the stars, and the observations sample a volume corresponding to about one optical depth at any wavelength.

DE YONG: At the time that ellipticals are still experiencing major star formation you have to include the effect of the absorption and reddening by dust in the models to correctly predict and/or interpret the observed spectrum/colors. I would think that this must strongly affect your conclusion that ellipticals show major star formation 5 Gyrs ago.

O'CONNELL: The observations of current epoch, nearby ellipticals are not strongly affected by reddening. But you are certainly right that galaxies viewed at lookbacks corresponding to the last major epoch of star formation could suffer significant effects of extinction, especially in the UV. This would not only redden their spectra but would produce selection effects which might lead one to underestimate the frequency of active star formers at a given epoch.

PICKLES: A bright elliptical spectrum can be fitted with a 'single burst' population of turnoff age about 8 Gyr and metallicity roughly 2-3 times solar; I find it difficult to interpret this as a 'formation age' however. If, as seems likely, there is a large range of metallicities in ellipticals, with metal-weak stars and stars as metal-rich as 5 or more times solar possibly, then could these facts reasonably imply extended periods or recurrent bursts of star formation over several billion years?

O'CONNELL: Yes, I certainly agree that the synthesis analyses are sensitive mainly to the last significant epoch of star formation and do not preclude an extended period of star formation preceding that. It is hard to place limits on earlier star formation in systems dominated by these intermediate age populations.

DRESSLER: If the spectral energy distributions (SEDs) of the nuclear regions of nearly ellipticals are typical of the outer regions as well, then your model predicts that star formation 5-8 Gyr ago dominates the present day integrated SED. Observations of rich clusters at Z=0.4-0.6 (a look back time of $\sim$6 Gyr, H =50) show that the majority of galaxies have the colors and spectra of nearby E or S0 galaxies, that is, they show little or no sign of star formation within 1 Gyr of the epoch of observation. Could you comment on this apparent contradiction with your interpretation of the SEDs of nearby ellipticals?

O'CONNELL: I expect we will want to address this problem in greater detail over the next couple days, but let me make several remarks now:

(1) The results on intermediate age turnoffs do not depend on the details of the synthesis method employed (e.g. the optimizing algorithm) and have been reached by several different groups using different data sets and approaches to the problem. They are consistent astrophysically with the CMD results reviewed by Mould concerning an epoch of active star formation $\sim$5 Gyr ago in the Local Group.

(2) I believe the weakest links in the argument are the mean metallicity assigned to the galaxies and subsequent age-dating derived from the theoretical isochrones. If the results are conclusively proven wrong by direct observations at high redshift, I would look first to these areas to explain the discrepancy.

(3) The results strictly refer only to the nuclei of nearby objects. However, in view of the generally small color gradients found in E's, I doubt that results for the integrated light would be much different, although this problem should definitely be examined soon.

(4) The results do not exclude an extended period of star formation prior to the turnoff epoch. This would inevitably be a stochastic process, especially if strong interaction with the environment were involved. It seems evident from the literature that some component of the galaxy population is exhibiting evolution at redshifts $\sim 1$, and it is not obvious to me that the ancestors of the kind of ellipticals observed in the local neighborhood (out to Virgo) can be excluded from that component.

BRUZUAL: Have you tried to predict the observed colors at $z \sim 0.5-1$ of all E-galaxies in which the turnoff is at $\sim 5$ Gyrs? I am afraid that the predicted colors may be too blue. If the turnoff is that young it makes no sense to speak about non-evolving S.E.D.'s from $z=0$, to $z=0.7$. It would be interesting to test the predicted colors against the observations as a function of $z$.

O'CONNELL: I have not made detailed models for the color evolution of objects like M32, since these would depend on the history of star formation prior to the turnoff age, on which we have little direct information. The most conservative assumption would be an extended period of relatively constant star formation prior to the turnoff age. This would imply that M32 would have a color at a lookback $\sim 5$ Gyr similar to a current-epoch spiral galaxy. Considering the range of plausible evolutionary scenarios, and the possibly important effects of dust in the rest wavelength UV (emphasized by de Yong), it is my impression that the color evolution claimed by Spinrad and Djorgovski, Butcher and Oemler, or Lilly and Longair is not inconsistent with this turnoff age for M32. The sampling effects I mentioned earlier could also importantly influence the colors observed at high redshifts. Of course, if the Space Telescope is able to demonstrate that no elliptical galaxies at lookbacks of 5-10 Gyr have spectra similar to current epoch spirals, then I would agree there is a serious problem with the interpretation of M32.

FROGEL: For an age of 5-8 Gyr the IR light will be dominated by luminous AGB stars, probably M's. This implies rapid luminosity evolution in the IR.

O'CONNELL: I believe you are right, and there seem to be several claims in the literature that such luminosity evolution is observed at redshifts of ~1. I believe Lilly and Longair have examined the influence of the AGB in detail.

NEWBERRY: I've made a number of evolutionary population models using single burst, constant, and exponentially decreasing star formation rates. Using a common, e.g., Salpeter, IMF it turns out that most of the <u>main sequence</u> light in the visual comes from the main sequence turnoff corresponding to that for a single age burst model. So it would seem difficult to determine much about the history of the star formation rate in a population synthesis approach.

O'CONNELL: I'm not sure I quite understand the implications of your models. If the last episode of star formation contributes a large fraction of the light, then I certainly agree it is difficult to obtain good information on earlier star formation. However, subsequent, smaller bursts of star formation are straightforward to detect, if they are above a certain threshold, especially against the cool background light of a population older than a few Gyr.

POPULATION SYNTHESIS AND EPOCHS OF STAR FORMATION IN NGC 1316 (Fornax A)


Andrew Pickles
Kapteyn Laboratorium
Postbus 800
9700 AV Groningen
Netherlands


Population synthesis models are presented for the galaxy NGC 1316 (Fornax A), with the main emphasis being placed on deriving epochs of star formation in this disturbed and possibly merged system. IRAS and radio results indicate that the nucleus of this dusty galaxy is optically obscured, and that the observed optical light is emitted along a line of sight which becomes opaque about 10 arcsec in front of the nucleus. The star formation history qualitatively indicates extended periods or recurrent bursts of star formation, with strong current star formation as well. The detailed synthesis results depend on the internal reddening, and results are tabulated for a range of reddenings encompassing the best estimate of $E(B-V) = 0.13$ mag.

1. INTRODUCTION

The double-lobe radio source Fornax A has been mapped at 1415 MHz by Ekers et. al. (1983) who found a bridge of radio emission connecting the lobes but offset from the central weak radio source. The optical galaxy NGC 1316, which is coincident with the central radio source, has been the subject of an intensive morphological and spectroscopic study by Schweizer (1980, 1981), who described it as a small, bright core within a large system of shells and ripples. Both the radio and optical structure has been ascribed by these authors to a past history of merger activity.
    Optical structure is visible on deep photographs out to radii of about 10 arcmin, and NGC 1316 was an early and well studied example of the class of galaxies with shells (Malin and Carter, 1983). Shells are visible on contrast enhanced photographs as sharp, low surface brightness edges in the galactic light distribution, and are believed to be stellar in origin (Malin and Carter 1980). The shell phenomenon has been interpreted as being due to the merger of a disc system with a more massive sheroid (Quinn, 1984), and this interpretation for NGC 1316 is consistent with the detection of a radial velocity step at a location coincident with its innermost shell (Bosma, Smith and Wellington, 1985).
    The technique of population synthesis is used here in an attempt to determine the history of star formation in this complicated system, where multiple epochs of star formation are likely.

## 2. INTERNAL REDDENING

The synthesised spectrum must fit the galactic optical energy distribution as well as the spectral lines, and the process is therefore complicated by the presence of large quantities of dust in NGC 1316. Dust is visible out to radii of about 1 arcmin on short exposure photographs published by Schweizer (1980, 1981) and Bosma et. al. (1985), and it is not surprising that NGC 1316 appears as a relatively strong source in the IRAS point source catalogue (1985). The IRAS point source flux strengths of NGC 1316 are 0.3, 0.3, 3.1 and 7.4 Jy at 12, 25, 60 and 100 microns respectively.

### 2.1 IRAS observations

The good correlations between infrared emission due to dust and neutral hydrogen distribution (Boulanger, Baud and van Albada, 1985), and between optical absorption and HI density (Bohlin, Savage and Drake, 1978), indicate that it is possible to estimate the optical absorption from the measured infrared flux strengths, provided that the spatial distribution of the dust is known. Boulanger et. al. (1985) derive the optical absorption as:

$$A_V = 0.45 \ I_{100} \quad \text{mag.} \qquad (1)$$

where $I_{100}$ is the 100 micron flux strength in Jansky per square arcmin.

This result is strongly dependant on temperature however, and extinction values which are roughly three times higher than this have been obtained from comparisons of optical star counts and interstellar infrared cirrus maps (de Vries and Le Poole, 1985). The difference appears to be entirely due to temperature, with the dust grains being up to 10 K hotter in the region studied by Boulanger et. al. (1985).

The grain temperature is in principle derivable from the ratio of the 60 and 100 micron flux strengths, assuming that these are generated in the same region. This calculation assumes thermal equilibrium between grains of roughly equal size, but it now seems clear that there is a broad spectrum of grain sizes, with large grains dominating the 100 micron emission, and smaller grains contributing predominantly to the 60 micron emission. The observed flux ratios do not therefore establish an unambiguous temperature, but may be used with caution as a comapritive indicator. The observed 60 to 100 micron flux ratio closely resembles that of the Boulanger et. al. (1985) sample, and we therefore choose to adopt their extinction estimate here. Adoption of a lower temperature for the dust grains would increase the emount of extinction calculated from the IRAS fluxes.

### 2.2 Spatial Extent of the Dust and Infrared Emission

Infrared images of NGC 1316 were obtained with the Chopped Photometric Channel (CPC) instrument on board IRAS, and these have been examined to determine the spatial extent of the infrared emission. The CPC images are approximately circular, with full width half maxima diameters of 1.6 and 1.4 arcmin at 100 and 50 microns respectively. These diameters are those expected for a point source however, due to the measured IRAS-CPC beam characteristics (Wesselius et. al. 1985), and indicate that the source is unresolved by CPC (and that the point source fluxes may be reliably used). This sets an upper limit of 50-80 arcsec on the diameter of the source at both 50 and 100 microns, although a different distribution at these wavelengths cannot be excluded. Dust is certainly present on larger scales, but the CPC results indicate that the 100 micron flux, and hence the bulk of the dust distribution, is concentrated at the nucleus.

VLA observations of NGC 1316 at 4.9 and 1.4 GHz show the central radio component to have a spatial extent of about 20 arcsec, with dual opposing jets in the centre (Geldzahler and Fomalont 1978, Fomalont and Geldzahler 1984). We assume here that this dimension represents the typical scale-length of the dust distribution, or at least (for our purposes) a lower limit. The optical extinction, calculated from equation 1 above, is listed in Table 1 for dust distribution diameters in the range 20 - 80 arcsec (2 - 8 kpc for an assumed distance of 20 Mpc).

Table 1    NGC 1316:   Dust Scale-length and Internal Reddening

| Dust Diameter (arcsec) | Filling factor | $I_{100}$ (Jy.arcmin$^{-2}$) | $A_V$ (mag.) | $E(B-V)$ (mag.) |
|---|---|---|---|---|
| 20 | 1.0 | 84.8 | 38.2 | 11.6 |
| 50 | 0.06 | 13.6 | 6.1 | 1.9 |
| 80 | 0.02 | 5.3 | 2.4 | 0.8 |

If most of the dust is concentrated within 20 arcsec ( 2 kpc for an assumed distance of 20 Mpc), then the optical extinction there is nearly 40 magnitudes. Lower, but still large, extinctions are indicated if the dust distribution is somewhat more extended. The CPC results do not exclude the possibility that the 50 micron flux is more concentrated than that at 100 microns, and hence that the dust temperature is higher and the extinction lower in the nucleus. This would have the effect of making the extinction flatter as a function of radius however, and we therefore conclude that the centre of NGC 1316 is optically obscured for all reasonable dust spatial distribution scalelengths. The dust is probably concentrated within 20 arcsec, and obscuration most probably extends out to radii of about 10 arcsec (about 1 kpc).

This conclusion provides a simple explanation for the peculiar dynamics observed in the nucleus of NGC 1316. Bosma et. al. (1985) found a low value of 230 km.s$^{-1}$ for the central velocity dispersion in NGC 1316, confirming unpublished results by Visvanathan. Thsi value is 30 km.s$^{-1}$ lower than the central velocity dispersions measured in the (brightest)

Fornax ellipticals NGC 1399 and NGC 1404 (Pickles and Visvanathan 1985) for example, although NGC 1316 is more than one magnitude brighter than these galaxies. The anomalously low velocity dispersion measured in NGC 1316 is naturally explained if it refers mainly to stars more than 1 kpc distant from the nucleus.

This conclusion also implies that the blue luminosity of NGC 1316 is underestimated (by about 10%, cf. Young 1976), and that corrections to the measured colours (Griersmith 1982) are large, although these corrections can only be determined from accurate surface photometry.

## 2.3 Synthesis estimate of Internal Reddening.

Dust is visible on photographs of NGC 1316 out to radii of 1 arcmin (6 kpc) and, although the infrared and radio observations show it to be mainly concentrated towards the centre, it is clear that there is also extended dust with a patchy distribution. We make the simplifying assumption that most of the dust giving rise to the infrared flux is so concentrated in the centre, that this region is entirely optically obscured. The observed spectrum then consists of starlight integrated along a tube of patch extinction, which increases very sharply near the centre, and is reddened by only a small fraction of the total dust.

Extinction affects the energy distribution, but not the spectral lines, and synthesis results should therefore be optimised when the correct amount of dereddening is applied. The results show that the fit is optimzed for values of assumed reddening in the range $0.10 < E(B-V) < 0.15$ mag, and this is adopted as the best estimate that can be derived from the data. A standard $1/\lambda^2$ extinction law has been assumed, but the uncertainties are larger for this determination of internal reddening than would be the case for determining Galactic foreground reddening for a dust-free system (cf. O'Connell 1980). Synthesis results are therefore also presented for reddening values outside of this range.

## 3. OPTICAL SPECTRUM

Spectra at a resolution of 1.5 nm in the wavelength range 360 - 800 nm were obtained in 1982 with an intensified reticon photon counting instrument at the cassegrain focus of the Mount Stromlo Observatory 1.9m telescope. Additional spectra at 1nm resolution in the wavelength range 800 - 1000 nm were obtained with the RGO CCD camera at the f/8 cassegrain focus of the 3.9m Anglo Australian Telescope. The spectra were flux calibrated by reference to flux standard stars observed concurrently, and combined to form one spectrum appropriate to the central 6 arcsec of NGC 1316. Comparison with large aperture photometry is difficult for this galaxy with strong colour gradients, but the spectra should be photometrically accurate to the same level of accuracy (4%) as was achieved for normal ellipticals observed concurrently (Pickles and Visvanathan 1985).

The combined, flux calibrated spectrum was dereddened for five representative values of $E(B-V)$: 0.05, 0.10, 0.13, 0.15, 0.20 mag, and synthesised for each of these separately.

## 4. SYNTHESIS LIBRARY AND TECHNIQUES

The synthesis library used here consists of standard stellar spectra, and fixed combinations of these constructed to represent the light due to single epoch (isochrone) stellar populations of specified age, metallicity and mass function (assumed power law) slope. The standard stellar spectra were observed in the same way as the galaxy spectra and are described in Pickles (1985a). The fractional contributions of each of these to the 'isochrone' spectra were calculated from the revised Yale isochrone tabulations (Green, Demarque and Ciardullo 1985). For this work the Yale tabulations were interpolated logarithmically to a helium abundance of $Y = 0.23$, and adjusted to a ratio of mixing length to pressure scale height of $\alpha = 1.6$ via the algorithm given in Twarog (1980). The isochrone spectra were constructed for two values of the mass function slope ($s = 1$, 2.35 for $dN/dm \, \alpha \, m^{-s}$), for each of three ages (5, 10 and 15 Gyr) and each of three metallicities ($Z = 0.004, 0.017, 0.04$), giving 18 isochrone spectra in total.

Figure 1 shows the adjusted Yale isochrones in the theoretical HR diagram, where isochrones of ages 5, 10 and 15 Gyr are plotted for each selected metallicity. The location of several stellar standard groups whose spectra were used to construct the isochrone spectra are also shown in this diagram. Complete details of this procedure are given in Pickles (1983).

The isochrone spectra do not include contributions from horizontal branch giants, and the late red-giant contributions are underestimated due to adoption of a slightly incorrect colour-temperature relation for these cool groups. These giant groups and groups representing newly formed upper main sequence stars have therefore also been included in the synthesis library as free parameters. This gives a library of 32 groups in which lower main sequence, subgiant and most giant branch constriants are completely specified according to the Yale isochrone tabulations, but which is uncostrained towards either ongoing star formation or horizontal branch contribution, and which can adjust the late red giant branch as necessary to best fit the observations.

The synthesis technique optimises the difference between the observed spectrum and the synthesised spectrum, where the latter consists of the summed, unconstrained contributions from isochrone and stellar standard spectra in the synthesis library. The program synthesises line and continuum information concurrently, and a weighting function is included to emphasise features of particular importance (eg. Mgb, NaD and H$\beta$), and to exclude regions of low photometric accuracy (atmospheric absorption bands). The programming technique uses a non-linear, constrained optimising program XRQP from the Numerical Optimisation Centre (Bartholomew Biggs 1979), and is fully described in Pickles (1985b).

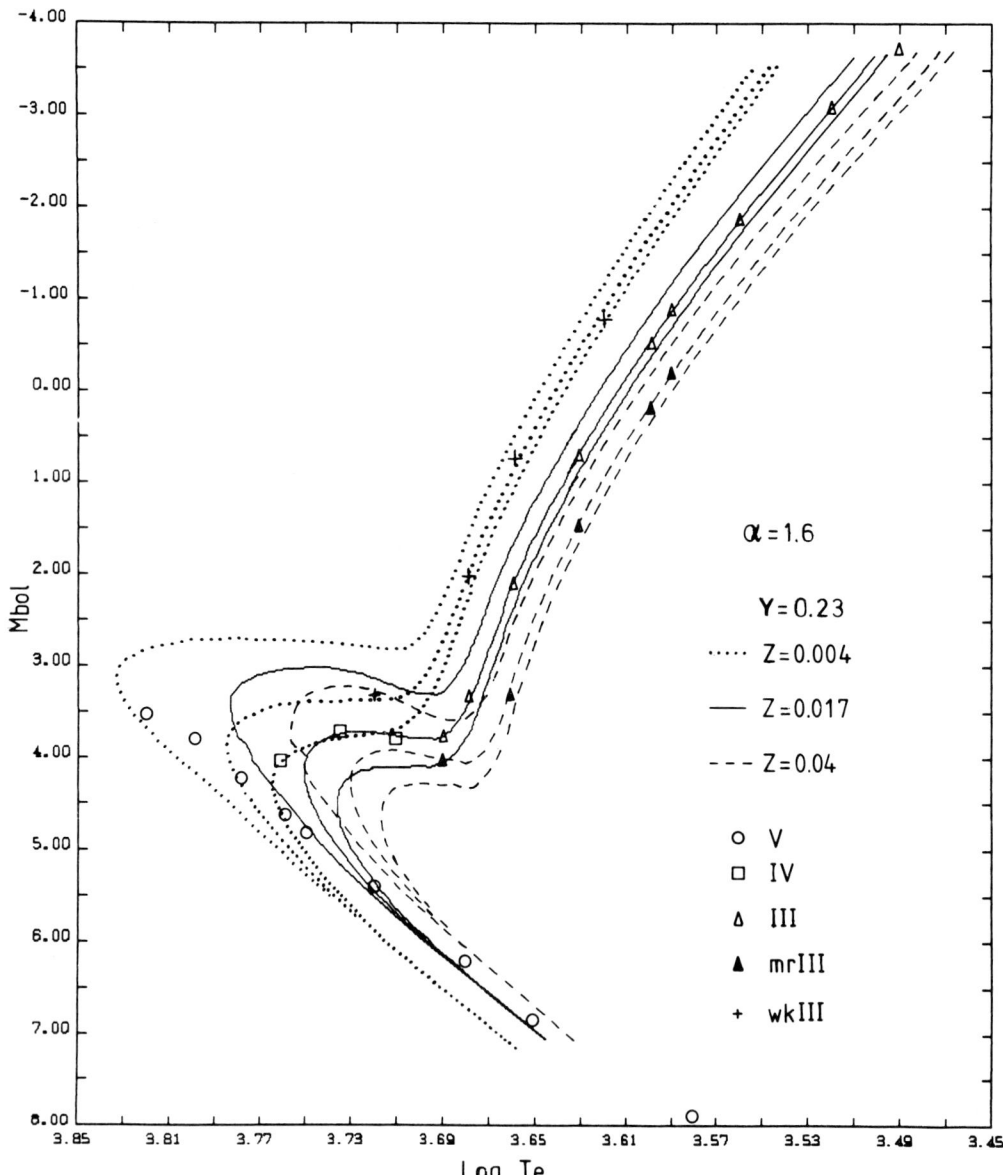

Figure 1. Interpolated Yale Isochrones are shown for three ages of 5, 10 and 15 Gyr at each of three metallicities. The figure also shows the locations of several standard stellar groups whose spectra are used to form the isochrone spectra used in the syntheses.

## 5. SYNTHESIS RESULTS AND DISCUSSION

The results are displayed graphically in Figure 2 for the value of assumed internal reddening which optimises the fit achieved. Figure 2 shows the observed, flux calibrated spectrum of NGC 1316 dereddened by $E(B-V) = 0.13$ mag. as the thin line normalised to $F(\lambda) = 100$ at 545 nm. The thick overlay represents the synthesised spectrum of NGC 1316, and is the sum of the labelled population components shown as spectral traces in the lower part of the figure. Each of the component spectra shows the relative light contribution of that component as a function of wavelength. The components R05, R10 and R15 refer to metal-rich ($Z = 0.04$) populations of age 5, 10 and 15 Gyr respectively. The mass fraction contributed by each component is also indicated in the figure.

Table 2 lists, for selected values of the assumed internal reddening in column 1, the Cousins BVI colours of the dreddened spectra (central 6 arcsec) in columns 2 and 3, the weighted mean photometric error of the fit in magnitudes (col. 4), the derived mass to visual light ratio in solar units (col. 5), and the relative V light and mass contributions of each significant component in columns 6 to 10.

Table 2  NGC 1316 Composite Isochrone Synthesis Results as function of $E(B-V)$.

| $E(B-V)$ (mag.) | Colours (B-V) | $(V-I)_c$ | Merit (mag.) | $M/L_V$ | % V Light and Mass Contributions | | | | | |
|---|---|---|---|---|---|---|---|---|---|---|
| | | | | | R05 | R10 | R15 | B | M III | |
| 0.00 | 0.91 | 1.58 | 0.0467 | 6.9 | . | . | 90.1 | 7.3 | 2.7 | V light |
| | | | | | . | . | 100.0 | 2.7E-2 | 1.5E-3 | Mass |
| 0.05 | 0.86 | 1.51 | 0.0429 | 6.8 | . | . | 89.6 | 9.3 | 1.2 | V light |
| | | | | | . | . | 100.0 | 3.6E-2 | 7.0E-4 | Mass |
| 0.10 | 0.81 | 1.45 | 0.0417 | 6.2 | 6.3 | 12.7 | 69.5 | 10.9 | 0.6 | V light |
| | | | | | 4.5 | 10.1 | 85.3 | 4.8E-2 | 3.9E-4 | Mass |
| 0.13 | 0.78 | 1.42 | 0.0416 | 5.7 | 26.8 | 4.3 | 56.7 | 11.6 | 0.7 | V light |
| | | | | | 20.8 | 3.7 | 75.5 | 5.5E-2 | 4.6E-4 | Mass |
| 0.15 | 0.76 | 1.39 | 0.0417 | 5.3 | 39.7 | . | 47.6 | 12.0 | 0.7 | V light |
| | | | | | 32.7 | . | 67.3 | 6.0E-2 | 5.4E-4 | Mass |
| 0.20 | 0.71 | 1.32 | 0.0425 | 4.4 | 65.5 | . | 20.1 | 13.4 | 1.0 | V light |
| | | | | | 65.3 | . | 34.6 | 8.1E-2 | 9.0E-4 | Mass |

The results indicate that the solution is optimised for values of internal reddening in the range $0.10 < E(B-V) < 0.15$ mag, where the fit achieved is comparable to the expected photometric errors of about 4%. The extra contribution from late M giants is small in all fits, and

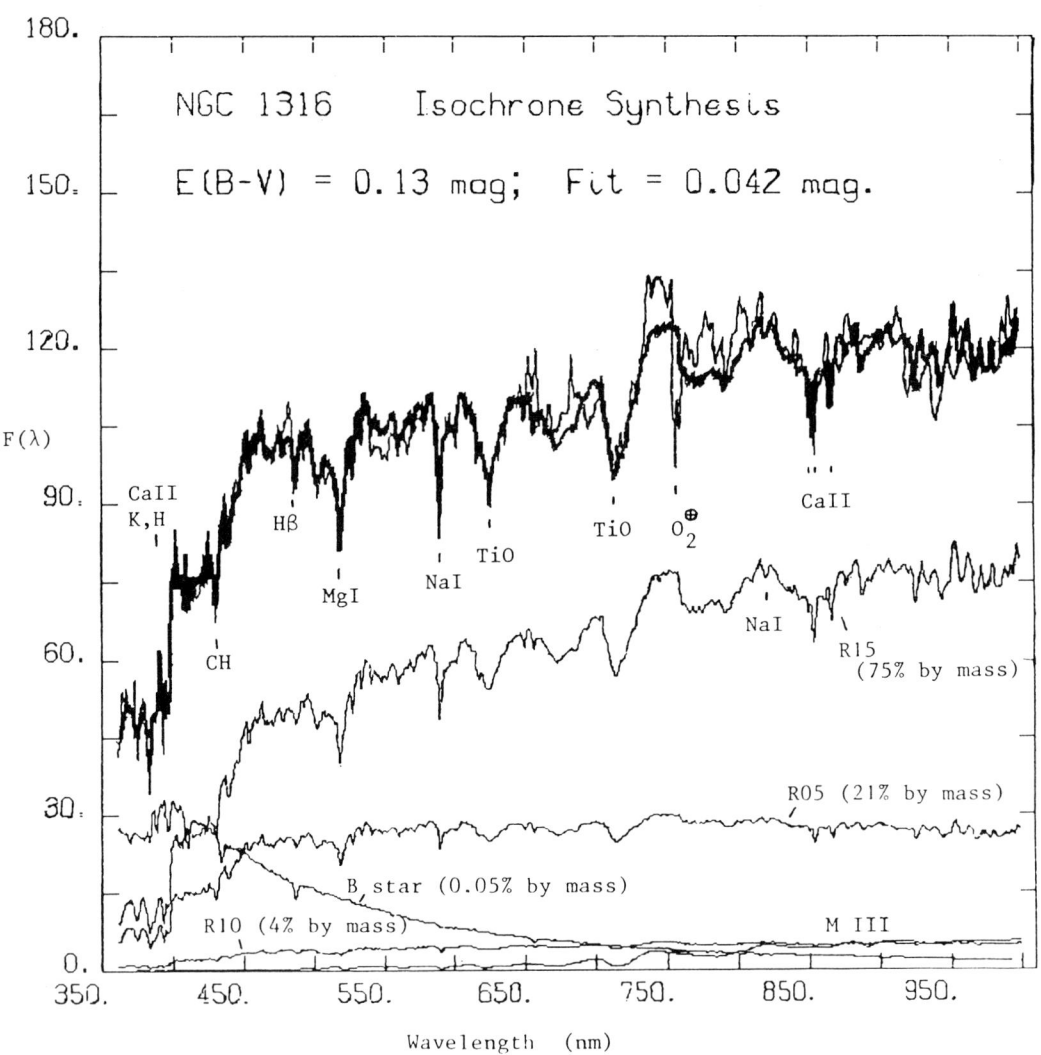

Figure 2. The observed spectrum of NGC 1316 dereddened by E(B-V)=0.13 mag. is shown as the thin trace normalised to 100 at F($\lambda$)=545 nm. The thicker overlay is the synthesised spectrum, and is the sum of the component spectra included in the lower part of the figure. Each component is labelled with its mass contribution, as discussed in the text.

represents an acceptable minor perturbation on the rigid constraints used in forming the isochrone spectra. This small extra contribution is necessary to properly match the observed TiO band strengths.

Photometry by Griersmith (1982) shows NGC 1316 to be significantly bluer in (U-V) than normal ellipticals, even without corrections for internal reddening, and the large contribution from B stellar group spectra is not surprising. The synthesis library contains horizontal branch giant spectra appropriate to stars cooler than 10,000 K, but only the hotter main sequence B stellar spectra are utilised. The blue flux could be interpreted as being due to very blue (10,000 K and hotter) horizontal branch giants, but this interpretation leads to astrophysical problems for such a strong-lined and metal-rich galaxy. The simplest interpretation is that this component is due to upper main sequence stars, and therefore indicates present day star formation.

The B star component provides about 10% of the V light, and about 0.05% by mass in the syntheses (assuming they are main sequence stars). These values can be compared with those of about 20% of the V light and about 0.2% by mass due to main sequence B stars in the present day mass function appropriate to the solar neighborhood (Miller and Scalo 1979). This comparison indicates that ongoing star formation may account for a large fraction of the galaxy mass in the nucleus (which is of order $10^{10}$ M$_\odot$).

For low values of the internal reddening (E(B-V)$\leq$ 0.05) only the oldest metal-rich isochrone model is used (because it is the reddest), together with a contribution from B stars. These fits are worse than expected due to photometric errors alone, and show that a purely old population with current star formation is not a good fit to the data.

The fits achieved for high values of assumed internal reddening (E(B-V)$>$ 0.15) are dominated by young (5 Gyr) metal-rich models, with contributions due to current star formation as well. These fits are also worse than expected, and indicate that the observed spectrum has been overcorrected for reddening.

The best fits (0.10$<$E(B-V)$<$ 0.15) indicate a predominantly old (15 Gyr) metal-rich population, corresponding to the epoch of globular cluster formation (VandenBerg 1983), but with large mass and light contributions from 10 and 5 Gyr metal-rich populations also. The significant differences between these solutions illustrates the strong dependence of synthesis solutions on galactic energy distributions, which depends here on the assumed internal reddening. The synthesis library should be extended to include isochrone spectra at younger and more closely spaced age intervals, but the best fits clearly show the existence of star formation over extended periods of time, either continuously or in bursts.

The synthesis results are commensurate with the proposition that NGC 1316 is a giant metal-rich elliptical which has undergone a recent merger with a system containing gas and dust, leading to a period of strong central star formation which is still continuing. IUE spectra of NGC 1316 do NOT show a turnup in the ultra-violet flux below 180 nm (as seen in some ellipticals), but this is not surprising considering the strong concentration of dust in NGC 1316. The synthesis results are quantitatively dependent on the applied correction for internal reddening and limited by the age resolution of the present synthesis library, but qualitatively show that significant populations of intermediate age (5 to 10 Gyr) are present in addition to a basic old population of globular cluster age.

I thank P. Wesselius and T. de Jong for assistance in interpreting the IRAS/CPC data, and the organisers at Erice for arranging such a pleasant and interesting conference.

REFERENCES

Bartholomew-Biggs, M.C., 1979, Technical Report No. 105, Numerical Optimisation Centre, Hatfield Polytechnic.
Bohlin, R.C., Savage, B.D. and Drake, J.F., 1978, Ap.J., 224, 132
Bosma, A., Smith, R.M. and Wellington, K.J., 1985, MNRAS, 212, 301
Boulanger, F., Baud, B. and van Albada, G.D., 1985, Astr. Ap. 144, L9
Ekers, R.D., Goss, W.M., Wellington, K.J., Bosma, A., Smith, R.M. and Schweizer, F., 1983, Astr. Ap. 127, 361
Fomalont, E.B. and Geldzahler B.J., 1984, BAAS, 16, 411
Geldzahler, B.J. and Fomalont, E.B., 1978, Astron. J., 83, 1047
Green, E., Demarque, P. and Ciardullo, R., 1985, Yale Trans. (in press).
Griersmith, D., 1982, Astron. J., 87, 462
The IRAS Point Source Catalogue, 1985, prepared by Joint Science Working Group (Washington D.C., U.S. Govt. Printing Office)
Malin, D.F. and Carter, D., 1980, Nature, 285, 643
_____, 1983, Ap.J., 274, 534
Miller, G.E. and Scalo, J.M., 1979, Ap.J. (Suppl.), 41, 513
O'Connell, R.W., 1980, Ap.J., 236, 430
Pickles, A.J., 1983, Thesis, Australian National University
_____, 1985a, Ap.J. (Suppl), (in press)
_____, 1985b, Ap,J. (in press)
Pickles, A.J. and Visvanathan, N., 1985, Ap.J. (in press)
Quinn, P.J., 1984, Ap.J., 279, 596
Schweizer, F., 1980, Ap.J., 237, 303
_____, 1981, Ap.J., 246, 722
Twarog, B.A., 1980, Ap.J., 242, 242
VandenBerg, D.A., 1983, Ap.J. (Suppl) 51, 29
de Vries, C.P. and Le Poole, R.S., 1985, Astr. Ap. (in press)
Wesselius, P.R., Beintema, D.A., de Jonge, A.R.W., Jurriens, T.A., Kester, D.J.M., van Weerden, J.E., de Vries, J. and Perrault, M., 1985, 'The IRAS-DAX Chopped Photometric Channel Explanatory Suppl.'
Young, P.J., 1976, Astron. J., 81, 807

## QUESTIONS

RENZINI: You have shown a case in which this galaxy is reconstructed in terms of one unique metallicity and three age groups. What happens if you keep only one age (say, 15 Gyr) and several metallicities (from metal poor to super metal rich)? How much worse is the fit in this case?

PICKLES: The mean metallicity of NGC 1316 is at least 2-3 times solar, which corresponds to the most metal-rich spectra and models that I have. To fit a range of metallicities with this mean would require models of 5-10 times solar, which I eagerly await. The best fit with existing metallicities and a single old (15 Gyr) age to a spectrum dereddened by $E(B-V) = 0.10$ mag. is poor, with a merit function value of 0.051 mag.

MADORE: You have shown solutions for a broad range of reddenings which are input into your models, and yet O'Connell has indicated that reddening is output as a fairly well determined parameter. Could you comment on the difference between these two apparently contradictory views of population synthesis?

PICKLES: There is none. The reddening is determined by O'Connell's method. The synthesis fits are optimised for reddening values over a range of about 0.05 mag. in $E(B-V)$, and the lack of a SHARPER minimum probably indicates that the standard extinction law assumed here is not correct for internal rather than foreground reddening. Most of the uncertainty is due to the fact that the reddening here is internal.

BRUZUAL: What happens if you do the fit to the spectrum of NGC 1316 in the way that R. O'Connell fitted the spectrum of M32? Since this is a dust rich galaxy with some evidence of star formation, one might expect to find a younger turnoff than he did for M32. Could you comment on this?

PICKLES: The main sequence turnoff group, derived in syntheses with a purely stellar synthesis library, is a function of assumed internal reddening, and goes from a late G dwarf turnoff for $E(B-V) = 0.05$ to an early G dwarf turnoff for $E(B-V) = 0.15$ mag. The extra contribution due to upper main sequence stars is similar to the solutions discussed here. My only comment is that I don't believe that the stars in galaxies are all coeval.

KOO: Among giant ellipticals for which you have used the isochrone fitting technique, what are the OLDEST ages or components that you have found? And what percentage of gE light comes from the oldest components? Comment - There is a rather disturbing inconsistency between the claims for a major portion ($\sim$50%) of gE mass being formed at 5-8 Gyr, and the observations of large CaII breaks (K and H lines - which are independent of reddening) in high redshift galaxies, which are comparable to those found in gE today.

PICKLES: The relative age contributions to bright Fornax ellipticals for which I have performed Isochrone syntheses are:

    NGC 1404      82% 5 Gyr,      18 % 15 Gyr,
    NGC 1399     100% 5 Gyr,
    NGC 1374      73% 5 Gyr,      27 % 10 Gyr,

where these contributions are almost entirely metal-rich. I would like to point out that the colours of a 5 Gyr metal-rich population are very similar to those of a 10 Gyr solar abundance population. The H and K break at 400 nm is strong in a 5 Gyr isochrone model, which has a roughly solar mass main-sequence turnoff. The H and K break would still be visible in a 2 Gyr old population, with a mid-F dwarf main sequence turnoff.


AARONSON: I'm a bit surprised that NGC 1316 comes out dominated by a 15 Gyr population, given how messed up this galaxy is, and the fact that normal ellipticals were argued as being dominated by an intermediate age population. But a larger reddening would increase the contribution of younger stars, yes?

PICKLES: The short answer is yes, but I don't think the reddening can be much higher than $E(B-V) = 0.15$ mag. If mergers are an important feature in the evolution of ellipticals, then these galaxies will reflect their environment. Perhaps mergers happen quicker, and ellipticals emerge quicker, in dense clusters such as Coma or some of the high redshift clusters under study now. The process may take longer in less concentrated clusters like Fornax, and even longer in galaxies like NGC 1316 which lies on the periphery of the Fornax cluster.

# FAR-UV STELLAR POPULATIONS OF S0 GALAXIES*

B. Rocca-Volmerange[1,2] and B. Guiderdoni[1]

[1] Institut d'Astrophysique, 98 bis bld Arago, F-75014
   PARIS, FRANCE
[2] Laboratoire René Bernas, Bat. 108, Université Paris XI,
   BP 1, F-91306 ORSAY, FRANCE

ABSTRACT   IUE spectra of S0 galaxies are analyzed in correlation with their gas content. For two respectively gas-rich (NGC 5102) and gas-poor (NGC 3115) S0 galaxies, a preliminary stellar population synthesis with IUE Atlas shows that the old population (A or G spectral type) is surprisingly uniform and of one single type. It contributes mostly to the 2000Å–3000Å flux and also partly to the 1500Å–2000Å range. For the two galaxies, far-UV excesses are low and strongly increasing with shorter wavelengths.

## 1. INTRODUCTION

Interest in observing S0 galaxies in view to analyze present or past star formation is obviously high for several reasons. From their disk component as well as their bulge or nucleus, such galaxies can give clues to their origin and so the galaxy formation and possibly to the origin of UV-excess, detected in early-type galaxies, owing to their possible intermediate morphological type between spiral and elliptical galaxies and to their large HI gas content range (Van Woerden et al., 1982). If S0 galaxies in clusters are actually favorite objects to study processes which could affect galactic disks, such as a stripping process in the past as we present in this conference (Guiderdoni and Rocca-Volmerange and references therein), we here only intend to compare from a preliminary far UV spectral synthesis some estimates of UV excesses in nuclei of two gas-deficient and gas-rich S0 galaxies. According to several authors (Deharveng et al., 1982, Bertola et al., 1982, Nesci and Perola, 1985), UV excesses of early-type galaxies are characterized by an important variation from one galaxy to another one. We suggest that most part of this variation may be attributed to known causes : different contributions of the old stellar population in nuclei, present star formation in the disk or shells in correlation with their gas content, etc ..., and for a comparison of far-UV

---

\* Based on observations and archives of the International Ultraviolet Explorer.

excesses, we have to estimate and substract this contribution to reach an "actual" UV excess in nuclei.

## 2 "OLD" STELLAR POPULATION IN THE FAR UV LIGHT

By old stellar population, we mean the evolved stars (including A type) mainly contributing to the visible light and which does not by definition create far-UV excess. By using an IUE library, we want to estimate the actual contribution of such an old population to UV light.
Visible spectra or integrated light of inner part for giant early-type galaxies have been found quite similar by several authors (Morgan, 1958, Oke and Sandage, 1968, Whitford, 1978). A stellar synthesis population model (O'Connell, 1976) associated to Iben's (1967) models of internal structure give for such nuclei a bulk of stars formed 8 to 11 Gyrs ago. Best model (type C) does not include normal K giants (different from Gunn et al., 1981) and their turn-off group is G0-5V with an age of $9 \; 10^9$ yrs. Contribution of K3-4V group is also important. If galaxies are smaller or having suffered gas exchange with environment inducing bursts of star formation, the turn-off group may appear younger until A spectral type (5 Gyrs old).

From the IUE stellar atlas (NASA Newsletter, n°22), a stellar population synthesis would be likely fruitful since stellar spectra are obtained with identical apertures and calibrations as the galactic ones. In a preliminary phase, we assume that the stellar population of nuclei is uniform and roughly having a spectral type defined from their optical colors U-B and B-V or visible spectra. In a previous paper (Rocca-Volmerange and Balkowski, 1984), we already gave such an average spectral type of the two different S0 galaxies the optical data of which fit respectively a type A3 for the gas-rich one NGC 5102 and G2 for the gas-poor one NGC 3115.

Figures 1 and 2 present a comparison of the dereddened ($E_{B-V}=0.07$ for NGC 5102) galactic IUE spectrum of each galaxy and the stellar IUE spectrum of spectral type deduced from visible data.

Our main conclusions are :
. The stellar population emitting in the 2000Å-3000Å wavelength range is quite uniform and essentially corresponds to one only spectral type (A2 for NGC 5102 and G2 for NGC 3115) about similar to visible.
. In the two galaxies, this "old" population of A or G type is relatively young : this means an age of 5 to 10 Gyrs.
. NGC 3115 presents a net depression around 2000 Å which is not an extinction law feature due to the lack of dust but attributed to a decreasing emissivity of the stellar population.
. The contribution of the A type population is noticeable in the 1500Å-2000Å wavelength range.

# FAR-UV STELLAR POPULATIONS OF SO GALAXIES

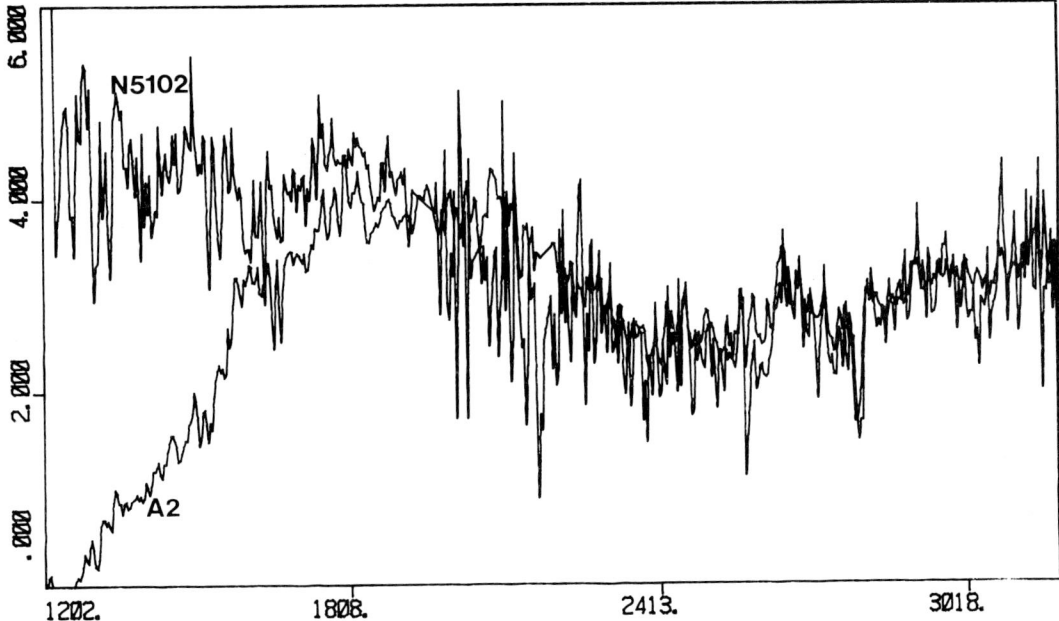

Fig. 1 A comparison of the dereddened IUE spectrum of the SO galaxy NGC5102 (Images n° 12239, 16417, 15886) with an A2V star from the IUE NASA Atlas (images n° 11235, 9855). Galactic flux is $10^{-14}$ erg $(s. \, \text{Å}. \, cm^2)^{-1}$

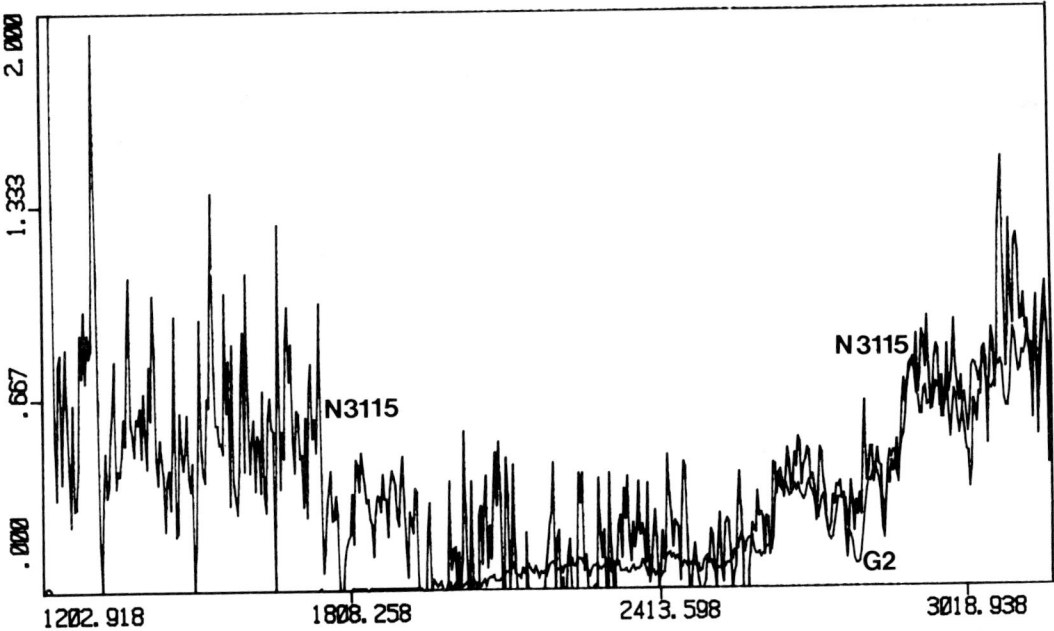

Fig. 2 A comparison of the IUE spectrum of the SO galaxy NGC3115 (images n° 9383, 7488, 10673) with a G2IV star from the IUE NASA Atlas (images n° 4760, 9864). Galacti flux is $10^{-14}$ erg $(s. \, \text{Å}. \, cm^2)^{-1}$

## 3 THE ACTUAL FAR UV EXCESS

From such previous spectral fits, the UV excess becomes the difference between the dereddened galactic spectrum and the old stellar population composite spectrum (or the average stellar spectrum). Then :
. UV excesses are very faint for the two galaxies. They are strongly increasing from 2000 Å to 1200 Å corresponding to very hot stellar temperatures.
. Some OB star lines are evidenced in the gas-rich galaxy NGC 5102 and so UV excess must be estimated from a more accurate population synthesis.
. Energy fraction of UV excess in NGC 3115 does not exceed a few percent of the total energy emitted by the galaxy.

## 4 DISCUSSION

From our selected S0 galaxy sample of IUE observations and archives, the most interesting ones are the two different galaxies NGC 5102 and NGC 3115, respectively gas-rich and gas-poor. Both of them have a noticeable S/N ratio while the others have either only signal in the long wavelength range (NGC 3998, NGC 4382, NGC 2784 and IIZW 67) or no signal in the two wavelength ranges (NGC 1316, NGC 1553).

As a first result, the most important stellar population which is emitting in the 3000Å-1500Å light is evolved, surprisingly uniform and of one spectral type, with a low dispersion. Such a spectral type (A or G according to the galaxy) corresponds with the turn-off group's as defined from a stellar population synthesis in visible (O'Connell, 1976) and confirmed by spectra and colors of these galaxies in UBV bands. Then at lower wavelengths, the youngest A or G populations dominate while colder groups such as K-dwarfs are not detected. This can be interpretated in terms of star formation history that we are presently doing (Rocca-Volmerange and Guiderdoni, in preparation).

Secondly, we may note the depression around 2000Å observed in the gas-poor and consequently dust-poor galaxy NGC 3115, essentially due to the lack of emissivity of the stellar old population and to absence of UV-excess at this wavelength. We have to notify that using such a feature to deredden any UV flux would be incoherent because presence of dust generally goes with presence of gas which, at its turn, produces young stars strongly emitting at 2000Å, as we can see on Fig.1.

Thirdly while NGC 5102 is a very atypical galaxy, due to its gas and bright nucleus (Pritchett, 1979) and is in the same group than the very peculiar galaxy Centaurus A (it could be an "E+A" starburst galaxy (Dressler and Gunn, 1983), easily explained by the presence of gas), NGC 3115 is more interesting with regard to the UV excess. Even when the stellar G spectrum is substracted from the galactic spectrum, an actual UV excess is observed which can only be attributed to very hot stars or nebular emission rapidly decreasing at 2000 Å.

No comment about the origin of the UV excess can presently be done but the extreme far-UV excesses, the faintness of which prevents to prelude massive star formation from residual gas, will be studied in a similar way for S0 and elliptical galaxies with various gas contents.

REFERENCES

Bertola, G., Capaccioli, M., Oke, J.B., 1982, Ap.J., **254**, 494
Deharveng, J.M., Joubert, M., Monnet, G., Donas, J., 1982, A.A., **106**, 16.
Dressler, A., Gunn, J.E., 1983, Ap.J., **270**, 7.
Gunn, J.E., Stryker, L.L., Tinsley, B.M., 1981, Ap.J., **249**, 48.
Iben, I., 1967, Ap.J., **147**, 624.
Morgan, W.W., 1958, Science, **128**, 1147.
NASA Newsletter n°22, ISSN 0738-2677
Nesci, R., Perola, G.C., 1985, A.A., **145**, 296.
Pritchett, C., 1979, Ap.J., **231**, 354
O'Connell, R.W., 1976, Ap.J., **206**, 370.
Oke, J.B., Sandage, A., 1968, Ap.J., **154**, 21.
Rocca-Volmerange, B., Balkowski, C., 1984, Proceedings of the Fourth European IUE Conference, ESA SP-218, p.69
Van Woerden, H., Van Driel, W., Schwartz, U.J., 1982, IAU Symp. n°100, 99.
Whitford, A.E., 1978, Ap.J., **226**, 777.

## G. BRUZUAL

In your spectrum of NGC 3115, there is almost no flux around 2200 Å. There are no stars which emit high at this wavelength and because of this reason this type of feature has been attributed in the past to absorption by dust (e.g. M81, Bruzual, Peimbert, Torres-Peimbert, Ap.J., 1982). Could you comment on this ?

## B. ROCCA-VOLMERANGE

As I said in my contribution, NGC 3115 has no gas and so dust appears to be little probable. The depression at 2000 Å is quite normal from Fig. 2 as a decreasing effect of the stellar population : the old one strongly emitting at 3000 Å does no more emit at 2000 Å and the UV-excess strong at 1200 Å (Rocca-Volmerange and Balkowski, 1984) strongly decreases at 2000 Å.

## C. BALKOWSKI

I would like to point out that NGC 5102 is a very peculiar galaxy with a relatively large HI content and a type A optical spectrum. In a spectroscopic survey of 56 S0 galaxies (Balkowski, Alloin, Ledenmat, 1985) there are only 2 other S0 galaxies with a type A spectrum.

# THE UV ENERGY DISTRIBUTION OF ELLIPTICAL GALAXIES

Francesco Bertola
Istituto di Astronomia, Università di Padova (Italy)

The IUE satellite, which is now completing its eighth year of successful operation, has brought us interesting and unexpected information on the UV energy distribution of early-type galaxies. Almost fourty ellipticals or bulge dominated disk galaxies have been so far studied, in most cases both in the short (1150 A - 2000 A) and in the long (1800 A - 3200 A) wavelength range. The observed objects are listed in Table I.
The following observational results have been established (Bertola et. al. 1980; Oke et al. 1981; Bertola et al. 1982):
i) Shortward of about 2000 A the energy distribution $F_\lambda$ exhibits a rising branch with a gradient similar to that of a black body at about 30000 K. This fact indicates the presence of a hot component in addition the stellar cold component dominating the visual spectrum.
ii) The hot component is of stellar origin since the UV luminosity profile perpendicular to the dispersion closely matches the visual profile.
iii) The level of the rising branch, when the spectrum is normalized in the visual region, is not constant, but varies by more than a factor 10 from NGC 221, where the rising branch is barely present, up to NGC 6166, the galaxy with brightest UV emission (Fig. 1).
Several attempts have been made to match the rising branch using different types of stellar components. Nesci and Perola (1985) were able to reproduce the observed energy distribution by using either young main sequence stars or horizontal branch stars (with bimodal temperature distribution) or accreting white dwarfs. Therefore they were unable to discriminate among the different components.
Using the approach of differential spectral synthesis, a good fit of the UV rising branch in elliptical galaxies with high UV flux has been obtained by adding the contribution of post asymptotic giant branch stars in the phase of planetary nebula nucleus to the energy distribution of low UV flux galaxies (Fig. 2). However almost the same result can be obtained adding a component characteristic of ongoing star formation (Bertola et al. 1985).
At a first glance the fluctuations in the UV level of elliptical galaxies seemed to be random, thus favoring the idea that young stars were responsible for the rising branch. However a relationship later found by Faber (1983) between the UV level and the magnesium line-strength index

TABLE I

Early-type galaxies observed with IUE

| Object | Morph. Type | Range |
|---|---|---|
| NGC 205 | E5 | Short/Long |
| NGC 221 | E2 | Short/Long |
| NGC 584 | E4 | Short |
| NGC 936 | S0 | Long |
| NGC 1023 | S0 | Long |
| NGC 1052 | E4 | Short/Long |
| NGC 1316 | S0 | Short/Long |
| NGC 1399 | E1 | Short |
| NGC 1404 | E1 | Short/Long |
| NGC 1407 | E0 | Short |
| NGC 1553 | S0 | Short/Long |
| NGC 2768 | E6 | Short |
| NGC 2784 | S0 | Short/Long |
| NGC 3115 | S0 | Short/Long |
| NGC 3379 | E1 | Short/Long |
| NGC 3998 | S0 | Short/Long |
| NGC 4111 | S0 | Short |
| NGC 4125 | E6P | Short/Long |
| NGC 4278 | E1 | Short/Long |
| NGC 4350 | S0 | Short/Long |
| NGC 4374 | E1 | Short/Long |
| NGC 4382 | S0 | Short/Long |
| NGC 4406 | E3 | Short/Long |
| NGC 4472 | E2 | Short/Long |
| NGC 4486 | E0 | Short/Long |
| NGC 4494 | E1 | Short |
| NGC 4552 | E0 | Short/Long |
| NGC 4621 | E5 | Short |
| NGC 4649 | E2 | Short/Long |
| NGC 4697 | E6 | Short/Long |
| NGC 4742 | E4 | Short |
| NGC 4762 | S0 | Short/Long |
| NGC 4853 | S0 | Short/Long |
| NGC 4889 | E4 | Short/Long |
| NGC 5102 | S0 | Short/Long |
| NGC 5128 | S0 | Long |
| A 1795 | E | Short |
| PKS 1404-26 | S0 | Short |
| NGC 5846 | E0 | Short |
| NGC 6166 | E2 | Short/Long |

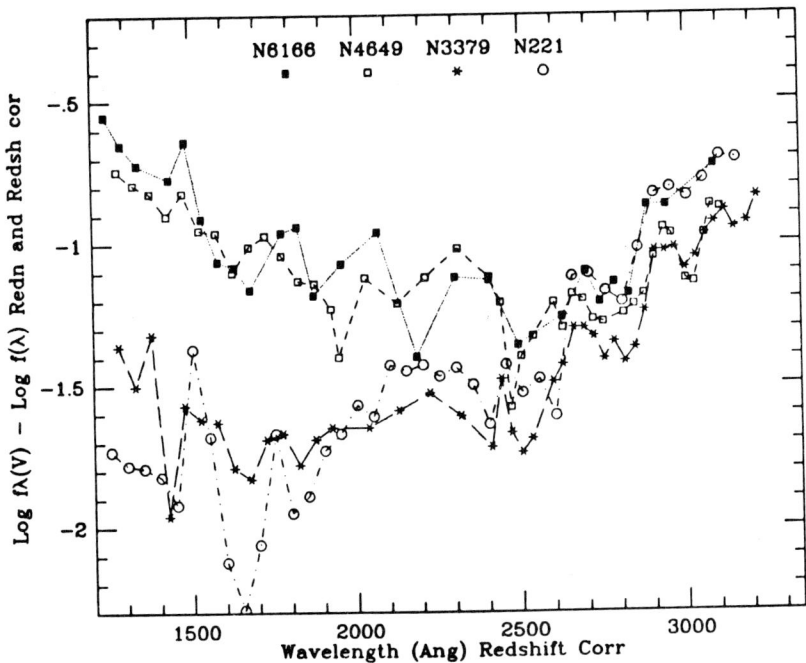

Fig. 1: The UV energy distribution for a representative sample of elliptical galaxies.

Mg2, weakened the above possibility. An extensive study of the early-type galaxies observed with IUE (Bertola et al. 1985) confirmed this relationship (Fig. 3), which requires careful analysis. The majority of representative points indicates a trend in the sense that more metal-rich galaxies emit more in the ultraviolet. Four points corresponding to objects with clear sign of ongoing star formation fall outside it. The scatter in the relationship seems to be real. Since the most deviating points toward blue correspond to galaxies with emission lines, a way to interpret the scatter is that ongoing star formation tends to displace representative points from the upper envelope of the relationship. Still the meaning of the variation of Mg2 with UV color remains unexplained.

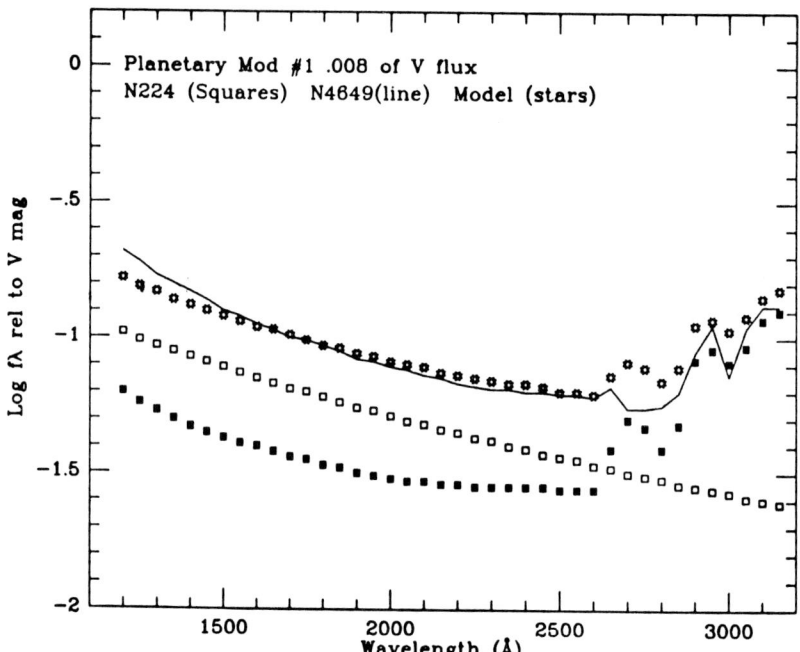

Fig. 2: A fit to the spectrum of NGC 4649 by adding a planetary nebula nucleus component to the energy distribution of the bulge of M 31. The model uses black-body spectra at appropriate temperatures, luminosity weighted by the lifetimes of the star at those temperatures.

An interesting case among the ellipticals studied with IUE is that of the cD galaxy NGC 6166 which has the highest level of the rising branch. For this galaxy Gregg (1985) has found that the best fit of the optical spectrum can be obtained by adding 4% of the light (at 5550 A) produced by a young population to the spectrum of the mean population of a normal elliptical of the same metallicity. This addition to the UV spectrum of a normal elliptical gives a rather satisfactory fit of the UV spectrum of NGC 6166. However in order to optimize the fit an enhancement of the UV ligth by adding a component of 0.075% O6 stars, whose presence is unexpected and requires independent confirmation, is required.

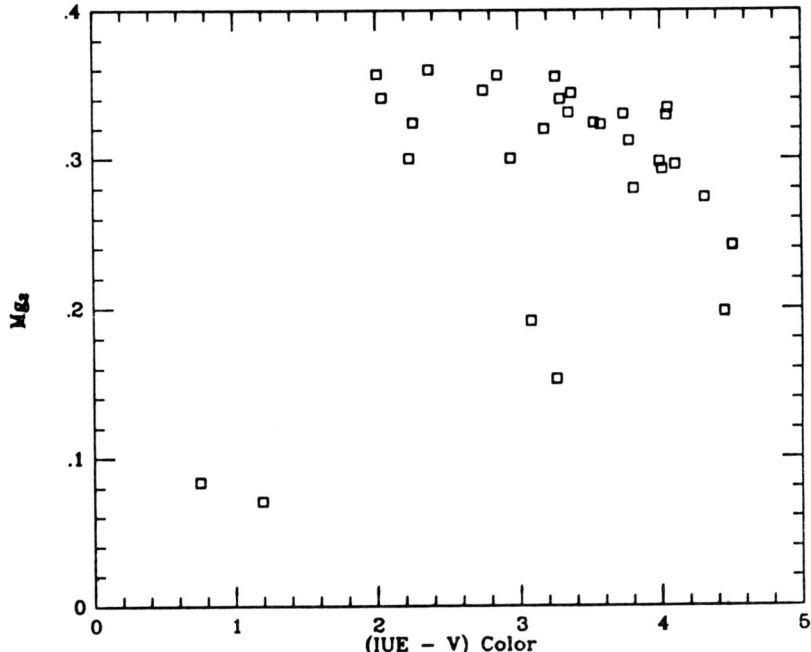

*Fig. 3*: The relationship between the magnesium line-strength index and the (IUE-V) color index. The IUE magnitude is derived from the mean flux in the 1150 A - 2000 A region.

REFERENCES

Bertola, F., Capaccioli, M., Holm, A.V. and Oke, J.B. 1980, Ap. J. (Letters) 237, L65

Bertola, F., Capaccioli, M. and Oke, J.B. 1982, Ap.J. 254, 494

Bertola, F., Burstein, D., Buson, L.M., Faber, S.M. and Lauer, T.R. 1985, in preparation

Faber, S.M. 1983, Highlights of Astronomy, 6, 165, Ed. R.M. West, Reidel Publ. Comp., Dordrecht

Gregg, M.D. 1985, Ph.D. Thesis, Yale University

Nesci, R. and Perola, G.C. 1985, Astr. & Astrophys. 145, 296

Oke, J.B., Bertola, F. and Capaccioli, M. 1981, Ap.J. 243, 453

# UV SPECTRA OF NORMAL ELLIPTICALS

R. Nesci
Istituto Astronomico
via Lancisi 29
Roma, Italy

Since the results of this study are already amply described in a paper published elsewhere (Astron. Astrophys. 145, 296), we present here only the main aspects of it.

Using a population synthesis technique we have been able to reproduce fairly well the shapes of the UV spectra of some ellipticals using in turn 3 kinds of hot stellar populations: HB stars, accreting white dwarfs in binary systems and young stars.

The HB stars solution is attractive because such stars are expected to be present in an old stellar population like that of an elliptical. We found however that these stars should be distributed in the HR diagram in a quite peculiar way, grouped into two separated clumps around log Te 3.75 and 4.30. Stellar evolution theories up to now do not account for such a distribution and the bimodal HB morphologies shown by a few galactic globular clusters are different from this one. Furthermore, stocastic fluctuations in the number of HBstars (some millions) are not likely to give an acceptable explanation of the wide spread (0.6 dex) in the UV excess shown by ellipticals. Two stellar populations with quite different metallicities might well produce this HB morphology, but in this case strong UV galaxies should be rather metal poor, which is not observed. We conclude therefore that if HBstars are actually responsible for this excess, then an effective mechanism must exist which is capable of producing the HB morphology indicated by our models with a large spread in the number ratio of the stars in the two groups.

The active white dwarfs (AWD) solution looks attractive in so far as it predicts a rate of Type I Supernovae which is not in disagreement with the observed one (very poorly known anyway). However, as in the case of the HB stars, the number of AWD's in a galaxy should be too high (several thousands) for an explanation of the spread in the UV excess in terms of statistical fluctuations to be acceptable.

The young stars solution, if their formation is assumed to occur in bursts, seems to offer a rather natural explanation both of the intensity and of the variance of the UV excess. However it requires a very high efficiency of the star formation mechanism and a mass function much flatter than the one derived for our own galaxy.

In conclusion, we feel that the source of the short-UV fluxes of elliptical galaxies cannot be surely ascribed to any of the three populations considered, and that other stars like nuclei of planetary nebulae or hot subdwarfs are worth being explored. In our opinion however, it is evident from our simulations that a population synthesis approach dealing only with the general shape of the energy distribution can hardly give an unambiguous solution: the only reliable way to identify the nature of the hot stars in ellipticals would therefore be to measure the equivalent widths of suitable absorption lines, a task for the Space Telescope in the near future.


DISCUSSION

Aaronson: Is the UV spatially extended in all ellipticals, so you can rule out any non-thermal component?

Nesci: Yes, it is.

Bruzual: A comment. I think that in the region from 3000 to 4500 A the different populations that you are adding may change the shape of the spectrum. Since in these bright galaxies you have very good observed spectra, a synthesis from 1200 to 5000 A, say, would allow you to set more constraints on the possible evolutionary schemes. Clearly both the UV and the visual spectra should be taken through the same aperture.

Nesci: The contribution to the blue-visual should not be very large for any of the populations. However I agree with you that population synthesis should be performed on as large a spectral baseline as possible.

Comte: 1) It seems difficult, in a system where the available gas comes only from mass loss from red giants, and is probably of quite low density, to get such efficiencies in production of main sequence stars, and also so high (30 $M_o$) cutoff masses of the IMF!
2) The active white dwarfs stars are likely to be X-ray emitters in the final accretion stage and during the burst. It is possible from Iben's models to predict an X-ray flux and to compare this with the limits of diffuse X-ray emission set by observations (Einstein, Exosat)?

Nesci: I agree with you for the first point, but one of my purposes was just to see what efficiency was required (and what upper limit for the IMF) for the young stars formation in order to explain the UV excesses observed! For the second point, it is not possible simply from Iben's paper to derive the X-ray flux of the AWD's: surely it would be a very nice test for this stellar population, but I don't know if the theory of X-ray binaries is developed enough as to make such a prediction.

Eisenhardt: There seems to be a feature in many of the IUE spectra near 1750 A, which does not appear in any of your models: do you have any idea if this is a real feature, or what it might be caused by ?

Nesci: The spectra I showed were obtained by Bertola, Capaccioli and Oke, and are an average of IUE data over about 50 A, due to the very low signal level. I asked Bertola about this feature time ago, and his opinion is that it is just an artifact of the background subtraction. Generally speaking, I think that we are asking too much from IUE when we look for emission lines in normal ellipticals at the distance of the Virgo cluster.

Kunth: The number of young mai- sequence stars you derived from your models of NGC 4472 is very high. More than 1000 O-type stars are required in both your continuous and burst star formation models. Similar rates of star formation are found in blue compact galaxies and give rise to very strong emission lines (H alpha, forbidden lines ...) that should also be observed here. The fact that they are not observed would tend to rule out this hypothesis, unless you can argue that you can have massive star formation without surrounding neutral and ionized gas!

Nesci: The aim of my work was just to see, from the number of stars required, which kind of stars were more likely to produce the UV excess of ellipticals: you cannot do it with less than 1000 hot stars. I didn't compute the expected intensity of the emission lines of the related H II regions, but surely I should expect them to be present. Do not forget that, if overshooting is to be taken into account as Chiosi has shown, the number of stars required should drop by a factor 3 or 4. Anyway I am not favouring any of the possible solutions that I have explored: in my opinion more things must be learned about the later stages of stellar evolution in order to understand the UV flux of elliptical galaxies.

Renzini: If the UV light from ellipticals is interpreted as due only to young stars then a fit to observed UV fluxes gives the current rate of star formation (having adopted some IMF). Then the fact that all observed ellipticals emit UV photons would imply that all of them are currently forming stars, and the average of the derived star formation rates (SFR) should be adopted as the representative SFR in ellipticals over the past several billion years. If so, the optical part of the spectrum should be considerably affected, and I wonder whether the optical part of the spectrum can actually rule out the presence of stars younger than, say, several billion years, at the level which would be impied by this interpretation of the UV flux.

Nesci: I did not consider the optical spectrum in this work, but surely I will include it in a next population synthesis code. As far as I know there is evidence of the hot stellar component also in the optical.

## V. GALAXIES AT LARGE LOOKBACK TIMES

A. DRESSLER
Studies of cluster galaxies at large lookback times

D. HAMILTON
Observational tests for galaxy evolution

P. R. M. EISENHARDT
Colors of 3CR and first-ranked high redshift galaxies

B. GUIDERDONI, B. ROCCA-VOLMERANGE
Evolution of disk galaxies in high-redshift clusters

D. C. KOO
Quests for primeval galaxies: a review of optical surveys

P. R. F. STEVENSON, T. SHANKS, R. FONG
New observations of galaxy number counts

# STUDIES OF CLUSTER GALAXIES AT LARGE LOOKBACK TIMES

Alan Dressler
Mt. Wilson and Las Campanas Observatories of the
Carnegie Institution of Washington
813 Santa Barbara Street
Pasadena, CA 91101-1292

## 1. INTRODUCTION

In our attempts to understand the spectral evolution of galaxies, we are fortunate indeed to have the ability to look back in time and observe galaxies as they were billions of years ago. Perhaps in no other discipline is it possible to gain such a direct view to history. The galaxies we seek to study are remote, their light faint, and thus only recently has it become technically feasible to sample the spectra of normal luminosity galaxies at lookback times of five billion years or more.

What do we seek to learn from such studies? From the point of view of the galaxy researcher, we seek to find out how and when spheroids were formed as ellipticals and the centers of disk galaxies. Likewise, we would like to know the timescales of formation and growth of stellar disks, and details of the manufacture and maintenance of the gaseous disks that feed them. Pieced together, the answers to these general questions should allow us to understand the range in morphology and the relation of morphology to environment that is observed for present-epoch galaxies.

These questions can be recast from the macroscopic to the microscopic perspective of stellar evolution. Our aim, then, is to chart the histories of star formation in these galaxies, as well as the composition and variation with time and location of the interstellar medium. It might also be hoped that the development of active nuclei in galaxies could be tied in. All of these factors are undoubtedly linked to stellar and gas dynamics, which must also be understood if galaxy evolution is to be successfully modeled.

The dream of the ambitious researcher goes even beyond this, driven by a belief that understanding the evolution of galaxies will lead to a knowledge of the nature of the primeval universe that gave birth to the galaxies we see today.

## 2. METHODOLOGY

The two approaches to the study of galaxy evolution mirror the difference in outlook mentioned above. From the point of view of structure (and dynamics), observing the development of galaxy morphology is the most direct way to learn how present-day galaxies evolved from their ancestors. We are fortunate that the Hubble Space Telescope will soon give us a view to galaxy morphology at cosmologically interesting distances, a view hitherto obscured by the earth's atmosphere. Until that time, however, our knowledge of the structures of distant galaxies is limited to little more than an estimate of how concentrated or diffuse a galaxy is, and sometimes the detection of an unusually prominent disk.

Spectrophotometry of distant galaxies, on the other hand, has become well within the capabilities of large ground-based telescopes. These studies began many decades ago with the work of Sandage, Kristian, Westphal, Oke, Gunn, Spinrad, and others, in the crusade to find the value of $q_o$ by means of the Hubble diagram. As has been discussed elsewhere in these proceedings, the first-ranked cluster galaxies that formed the bases of these studies are probably unrepresentative of galaxies on the whole. Nevertheless, it is interesting to note that Oke and Wilkinson (1978), and Lilly and Longair (1984), among others, have demonstrated that there is little color or spectral evolution of these galaxies since a redshift of $z \sim 0.5$, and possibly back as far as $z \sim 0.8$.

Therefore, it came as somewhat of a surprise, I think, when Butcher and Oemler (1978) announced their findings that two distant clusters, Cl 0024+16 ($z = 0.39$) and the cluster containing 3C 295 ($z = 0.46$) had much larger fractional populations of blue galaxies, presumably star-forming galaxies, than similar clusters at low redshift. Clusters of galaxies provide a significant advantage over field galaxies for this type of study, because the uncertain interpretation that follows the detection of rare, abnormal galaxies is simplified when a representative sample of galaxies at a common distance have been observed.

Butcher and Oemler estimated that 40-50% of the galaxies in these two distant clusters had colors that were bluer than what is expected for E or S0 galaxies. Although their observations were not questioned, their interpretation, that these were likely to be normal spirals, and their conclusion that a large amount of evolution had been observed, were questioned in a literature debate that has since lost most of its import. Later work by Butcher and Oemler (1984), and Couch and Newell (1984a,b) has confirmed for a larger sample (including low redshift clusters) that the fraction of blue objects does rise from only about 5% for low-z concentrated clusters to 20% or more at $z \sim 0.5$. The scatter is enormous, but a clear trend seems to be present. These more modest fractions compared to the original work reflect the tightening of membership criteria by Butcher and Oemler, who now require a galaxy to be brighter than $M_V < -20$ and within the radius that contains 30% of the cluster galaxies. To be considered blue, a galaxy must also have a rest frame B-V at least 0.2 magnitudes bluer than the reddest galaxies in the cluster.

The case for spectral evolution based on such broadband colors looks promising; nevertheless, spectrophotometry has a crucial role to play in both verifying and understanding this trend. This is true for several reasons. First, contamination of the cluster populations by projected field galaxies (or superposition of two clusters) becomes more of a problem with increasing distance; therefore, redshifts are necessary to validate a large blue fraction of true cluster members. Second, broadband colors must be K-corrected to recover the distribution of flux in the rest frame, and these require assumptions about the flux distribution itself that are better not made. Ultimately, of course, spectroscopy with resolution of 20 Å or better provides the best data for analysis of the stellar populations that make up the galaxy.

## 3. A SUMMARY OF SPECTROPHOTOMETRY IN THREE DISTANT CLUSTERS

Jim Gunn and I have been following this spectroscopic approach with the PFUEI CCD camera/spectrograph at the prime focus of the 200" Hale telescope. We have obtained substantially complete data for three clusters, the two original ones studied by Butcher and Oemler, and the "red" cluster studied by Koo (1981), and are continuing on with several other clusters at redshifts $0.4 < z < 0.8$. Our program includes two-color broadband photometry for all galaxies in a 25 sq. arcminute region and low resolution ($\sim$15 Å) spectroscopy for some 20-40 objects. We also image each cluster with a 100-Å wide interference filter centered on [O II] $\lambda\lambda 3727$ in the rest frame of the cluster in order to select out unusually strong emission associated with AGNs and intense star formation.

The results for the first three clusters show a wide range in cluster populations that does not bolster one's confidence that a simple model will be found. The field of 3C 295 was found to be rather heavily contaminated by projected field galaxies (Dressler and Gunn 1983) that accounted for more than half of the blue galaxies apparently belonging to the cluster. Nevertheless, the 20% that remain is a significant increase over the $\lesssim$5% found by Butcher and Oemler for low-z concentrated clusters. These six blue cluster members do not have the spectral characteristics of normal spirals, however. Three of them have spectra typical of active galactic nuclei (AGNs) and three are what I will refer to as post-starburst galaxies (PSGs), i.e., galaxies that have had a large increase in the rate of star formation which has apparently subsided. These are discussed in more detail below.

Dressler, Gunn, and Schechter (1985) found that the Cl 0024+16 was less contaminated and thus an even higher fraction of $\sim$25% of blue galaxies is indicated. In contrast to the 3C 295 cluster, however, the spectra of these blue galaxies resembled those of low-z, star-forming spirals. Again, several AGNs were found, but in this cluster no PSGs have been turned up.

Recently, Gunn and I have obtained spectra (with redshifts) for 29 galaxies in Cl 0016+16 (z = 0.54), a cluster found by Kron and touted by Koo to have a very small blue population, despite its higher redshift. We confirm Koo's claim that among the brighter galaxies, at least, there are no very blue galaxies like those found in the two clusters discussed above. Among the few marginally blue objects, however, are two PSGs which suggests that the cluster has had some activity. Because of its greater redshift, we will have to work even harder to reach the faint limits to which we have surveyed Cl 0024+16 and the 3C 295 cluster. Thus, new data are needed to be completely certain that this cluster is as poor in blue galaxies as its low-redshift counterparts.

We conclude from the results for these first three clusters that there is evidence for spectral evolution to the extent that galaxies active in star formation, AGNs, and PSGs are found in greater frequency than in low-z concentrated clusters in which these types are rare. This tentative conclusion, the confirmation of which awaits better statistics, is discussed more fully in Dressler, Gunn, and Schneider (1985). For the purposes of this workshop, it is more relevant to turn aside and examine the spectra of these distant galaxies.

## 4. A CLOSER LOOK AT THE SPECTRA OF GALAXIES IN HIGH-Z CLUSTERS

For the purpose of discussing the spectral evolution of galaxies, it is convenient to separate the spectra into four general categories:
(1) elliptical-like (2) post-starburst (3) AGN and (4) spiral-like. All of these types are represented in today's clusters as well; so, the case for evolution rests primarily on the question, "How many?"

### 4.1 The Reddest Galaxies

It is important to remember that, even in Cl 0024+16, the best case for a cluster with a significantly evolved population, ∼75% of the members have colors that lie within $0^m.2$ of the reddest galaxies. In fact, E and S0 galaxies at the present epoch show a smaller scatter than this, roughly half, but the errors in photometry for the higher redshift objects make this too fine a point to pursue. This broad definition of red galaxies would include M31 and M81, two local spirals, so it is best to remember that this color definition is not one-to-one connected with morphological type. Perhaps the best that can be said is that present-day galaxies that are dormant or weak in star formation fall into this category and, therefore, galaxies in these distant clusters that have similarly red colors were also relatively dormant for at least several billion years before the epoch of observation.

Gunn and I have measured the strength of the 4000 Å break in many of the reddest cluster members, and conclude that up to $z \sim 0.5$ there is no significant difference in this feature compared to present-epoch ellipticals. This suggests that even passive evolution has not been observed in these galaxies, which is consistent with the expectation that the effect in this wavelength region will be small up to $z = 0.5$. It may be difficult to improve on the accuracy of such measurements since there appears to be a real scatter in the strength of the 4000 Å break in both low-z and high-z samples. This could be due to differences in metal abundance or residual star formation; regardless, it seems to be present in both.

The fact that such a large fraction of the galaxies in distant clusters have elliptical-like spectra is a challenge for models which invoke large amounts of star formation in most ellipticals within the last 5-10 x $10^9$ years. Unfortunately, no quantitative analysis has yet been made to check the consistency of this idea with the data in hand. At this point, we can only conclude that most of the galaxies in the distant clusters studied to-date have spectra and colors that are indistinguishable from those of elliptical, S0, and Sa galaxies in low-z clusters.

## 4.2 Post-Starburst Galaxies

The 3C 295 cluster was first to show prominent examples of galaxies that have nearly elliptical-like spectra (a fairly red continuum, clear 4000 Å break, and metal lines like H & K of Ca II and the G band), but also contains relatively strong Balmer lines. Two examples found in Cl 0016 +16 are shown in Figure 1 (see also, Figs. 5 and 6 of Dressler and Gunn 1983). These galaxies have rest frame B-V colors of 0.70-0.80, a color typical of a G7 dwarf, but the H$\gamma$ and H$\delta$ have equivalent widths of 7-8 Å rather than 2-3 Å found in these stars. Conversely, their strong Balmer lines are typical of an F5 dwarf, but the B-V color of such a star is a much bluer 0.45. Similarly, these spectra are notable in that they do not match those of low-z, star-forming spirals, which, at the appropriate color, have much weaker Balmer lines. Also clearly missing in the distant objects is the [O II] emission and higher UV continuum characteristic of ongoing star formation.

Gunn and I found that the spectra can be matched exactly (within the observational errors) by adding light from A1 and A5 dwarfs to the spectrum of a giant elliptical, in the ratio of equal contribution from A stars and elliptical at 4400 Å. Thus, we originally referred to these objects as "E + A" spectra.

In order to see if such a model is consistent with a coeval stellar population, we compared these E + A spectra to those of Magellanic Cloud clusters obtained by L. Searle and H. Smith. A fair agreement was found with "Type V" clusters which have Balmer lines of 7-8 Å and B-V colors of 0.6-0.7. These are noticeably too blue, but one cannot rule out the possibility that dust is important in the distant galaxies. Spectral synthesis models by A. Manduca reproduce the Searle and Smith Type V clusters with a single, coeval starburst 2-3 billion years old. If we allow two coeval populations, the perfect match described above can be reproduced, with the addition of a $\sim 1 \times 10^9$-year-old population, dominated by A stars, to a substantially older (>5 x $10^9$ year old) population whose light comes mainly from K giants. These Manduca models are certainly not ideal in that they do not include post-AGB stars or the very metal-rich stars that are likely to be found in the distant galaxies; nevertheless, they provide a rough sketch of the types and ages of the required stellar components.

As a further embellishment, we considered whether this two-population model is consistent with truncation of star formation in a spiral that had formed stars at a constant or exponentially declining rate for some 11 x $10^9$ years (assuming a cosmic age of 16 x $10^9$ years and a look-

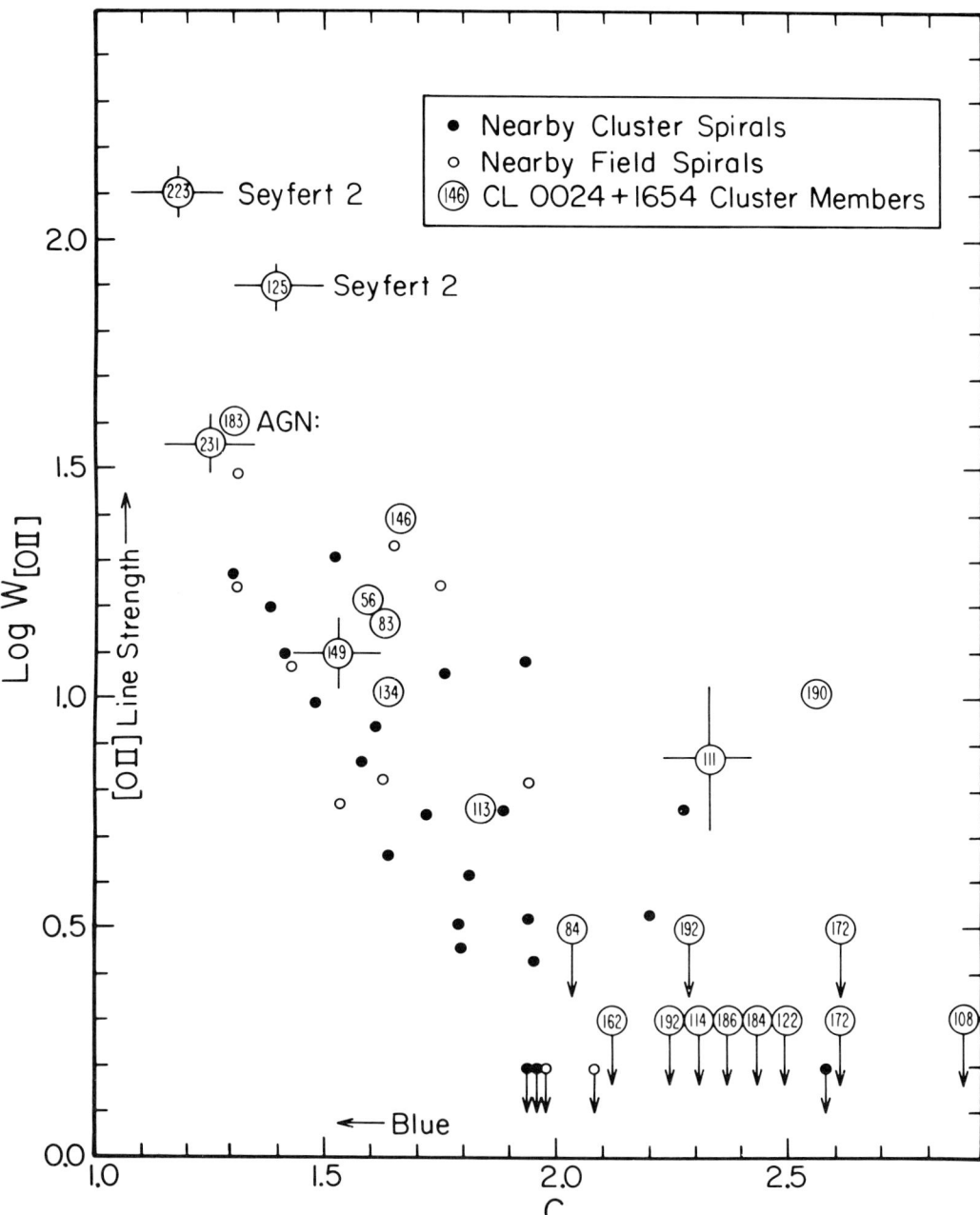

Fig. 2. The relation between galaxy color and the strength of [O II] $\lambda\lambda 3727$ emission. Plotted is C, the continuum slope determined from the spectra from 4100-5000 Å vs. the log of the equivalent width of [O II]. The similarity of the relationship for the nearby and distant samples suggests that the blue cluster members of Cl 0024+16 are also spirals which experienced extended epochs of star formation.

back time of about $5 \times 10^9$ years). We found that stopping star formation would, of course, account in the disappearance of the UV light and [O II] emission in such galaxies; however, even in the case of constant star formation that ends $1 \times 10^9$ years before the observation, the Balmer lines never reach equivalent widths greater than $\sim 5$ Å because they are diluted by the light from previous generations of older stars. This means that a true burst is required, in which the star formation rose above the past average. A reasonable model to produce equal contributions at 4400 Å of the old and young components is a burst that occurred roughly $1 \times 10^9$ years before the epoch of observation. The strength of the Balmer lines constrains this time scale within a factor of two either way. Before this time, the past average star formation is $\sim 10$ solar masses per year, but during a burst that lasts $\lesssim 5 \times 10^8$ years, about 10-20% of the galaxy's mass is converted to stars, at the rate of $\sim 40$ solar masses per year. (This estimate assumes a Salpeter luminosity function that stretches to very low-mass stars; if stars of a solar mass or more are the only ones produced in the burst, the fraction of mass converted and the rate are an order of magnitude less.) If this were the only such starburst that the galaxy underwent, there would be only subtle signs in the spectrum of the galaxy some $5 \times 10^9$ years later. Thus, in this picture the observed spectrum is the result of a burst of star formation when the star-formation rate rose at least a factor of two above the past average, but then dramatically subsided. I will refer to such a galaxy as a "post-starburst galaxy," or PSG.

These types of spectra are found at low-z also; in our 1983 paper, we give several examples. The difference is a question of frequency of occurrence. (Of course, there might also be a difference in the morphological types in which this activity is found, but we are unable to make such a comparison until the HST images become available. It is probably significant that many of the nearby examples are S0 galaxies.) We have now found about 10 PSGs among the $\sim 100$ good spectra we have accumulated during this project. I have recently made a cursory check of the 1095 spectra of galaxies in low-z clusters that Steve Shectman and I have collected, and I estimate that only a few percent at most have PSG spectra. Thus, the $\sim 10\%$ fraction seems to be a significant increase. I suspect that the true fraction in the distant clusters could be even higher, since these intermediate-color objects are probably underrepresented in our sample. R. Ellis and his collaborators have also found other examples of what appear to be post-starburst galaxies.

Perhaps the best analogs to these distant PSGs are 7 spiral and irregular galaxies in the Coma cluster that have blue colors and strong Balmer lines but very low H I contents (Bothun and Dressler 1985). They differ from the distant examples in that they are less luminous, on average, by a magnitude or two, and that they show signs of continuing star formation. Bothun and I estimate that the observed rate of star formation ($\sim 5$-15 solar masses per year) cannot be supported for more than $\sim 10^9$ years due to the low H I content, so that if and when star formation ceases, these galaxies too will exhibit the spectral characteristics of the PSGs. We speculate in this paper that ram pressure on the disk of a galaxy infalling into a cluster may actually stimulate star formation at the same time the lower density ISM is stripped away, thus producing a burst that will end abruptly. Perhaps this same fate awaited the more luminous spiral galaxies that fell into the hot intracluster gas of 3C 295.

### 4.3 Active Galactic Nuclei

Among the spectra that have emission lines, those due to active nuclei (AGNs) are the easiest to recognize. The emission lines are usually strong compared to the continuum, and have ratios of [O II], [O III], and H$\beta$ that are typical of gas at high excitation. Sometimes the Balmer lines are thousands of kilometers per second wide, a certain sign of the nuclear activity found in Seyfert 1 galaxies and quasars.

A classification scheme such as that of Baldwin et al. (1981) provides very good discrimination of AGNs from, for example, high excitation, nuclear H II regions. Unfortunately, some of the diagnostic emission lines are too weak to be measured in the relatively crude spectra of distant galaxies. Gunn and I therefore have adopted a scheme that discriminates solely by the ratios of [O II], [O III], and H$\beta$ that should be nearly as good as long as the metal abundance of the galaxy in question is greater than 1/3 solar. In any event, all AGNs qualify for the sample, but there may be some slight contamination by H II regions if the galaxies have metal-poor components. These same criteria have been applied by Dressler, Thompson, and Shectman (1985) to the low-z cluster sample mentioned above. The result is that only $\sim$1% of the low-z cluster sample are AGNs by these criteria. This $\sim$1% figure is to be directly compared to a fraction of 8/70 = 11% AGNs (selected by the same criteria) for four high-z clusters. If we can believe these small number statistics, we again see a dramatic increase in the fraction of these rare galaxies in high-z clusters compared to their low-z counterparts.

The fact that the low-z field contains a fraction of AGNs several times higher than the low-z clusters suggests that AGN activity and star formation may be linked, perhaps simply through galactic gas content. (Recall that the fraction of star-forming spirals is considerably higher in the field than in clusters.) I note in passing that our preliminary results suggest that this is true in the distant clusters as well. Those with a significant fraction of blue galaxies seem to have AGNs; the two clusters without AGNs are the reddest in our sample.

Because AGNs are rare objects, it is important to have an efficient search technique if representative samples are to be acquired. Toward this end, Gunn and I have incorporated into our program imaging through narrow-band filters centered on redshifted [O II]. This method is very effective and efficient since it covers an entire field at once. A similar program for low-z clusters would be very valuable.

### 4.4 Other Emission-Line Galaxies

In addition to the easily recognized AGN spectra described above, there are many low-excitation, emission-line spectra in our sample of distant cluster galaxies. These systems are characterized by a prominent [O II] emission and comparable or weaker H$\beta$, with even weaker [O III]. Such spectra are characteristic of H II regions with near-solar metal abundance, so they are typical of luminous late-type spirals. Since we have no morphological information yet for the distant galaxies, Gunn and I have used a spectrophotometric test to see if these distant objects are also likely to be normal spirals. The test rests on the rather good correlation between [O II] emission-line strength and integrated color, shown in Figure 2. The data shown were obtained for spirals in relatively nearby clusters where morphological types can be reliably assigned, and the spectra were taken though very large apertures (16", or roughly 10 kpc) to insure that the nearby galaxies were sampled in integrated properties as is necessarily the case for the distant objects. (Such spectral data for which large apertures have been used to study the integrated properties of present-epoch galaxies for comparison with distant samples is very sparse, and much needed.)

In addition to its value as a diagnostic for studying star-formation properties of distant galaxies, the relation shown in Figure 2 has important implications for the process of star formation in present-day spirals. This is because the diagram actually compares the "instantaneous" star-formation rate (averaged over $\lesssim 10^7$ years), as indicated by the [O II] flux from H II regions, with the long-term ($10^{8-9}$ years) average, which is measured by the continuum flux contributed by A and F stars. The fact that there is a relatively well-behaved relation between these two quantities suggests that normal spiral galaxies do not often pop on and off like firecrackers. If they underwent drastic changes in their rates of star formation on timescales of $\lesssim 10^{8-9}$ years, there should be a much larger scatter in the diagram than is observed. For example, the PSG galaxies discussed earlier fall significantly below the observed relation. As yet, there are not quantitative estimates as to what kind of limits can be placed on the oscillation of star-formation rates in such spirals, but this type of diagram will be quite helpful.

Cl 0024+16 is the only distant cluster we have studied thus far that has a large population of galaxies with low-excitation, emission-line spectra. The majority of its 25% blue fraction appears to be galaxies of this type. Examples of these and two AGN spectra in Cl 0024+16 can be found in Figure 3 of Dressler, Gunn, and Schechter (1985). The relation between color (determined by the continuum slope) and [O II] strength for these galaxies is compared to the low-z spiral sample in Figure 2. A general agreement with the trend for low-z spirals is seen in the sense that bluer galaxies have proportionally stronger [O II] emission. This implies that they, too, could be normal spiral galaxies, but I hasten to introduce a note of caution at this point. There seems to be a marginal trend that these distant objects are a factor of two more active in star formation than low-z spirals of similar color. It is difficult to be certain that this is not merely a calibration problem, but, if genuine, it means that the star-formation rates are quite large in most of the observed "spirals" of Cl 0024+16. As one goes to higher and higher rates of star formation, the relation in Figure 2 becomes more degenerate in the sense that the timescales along the two axes become more similar. That is, for a very active spiral, the color may be dominated by quite hot, young stars whose life-

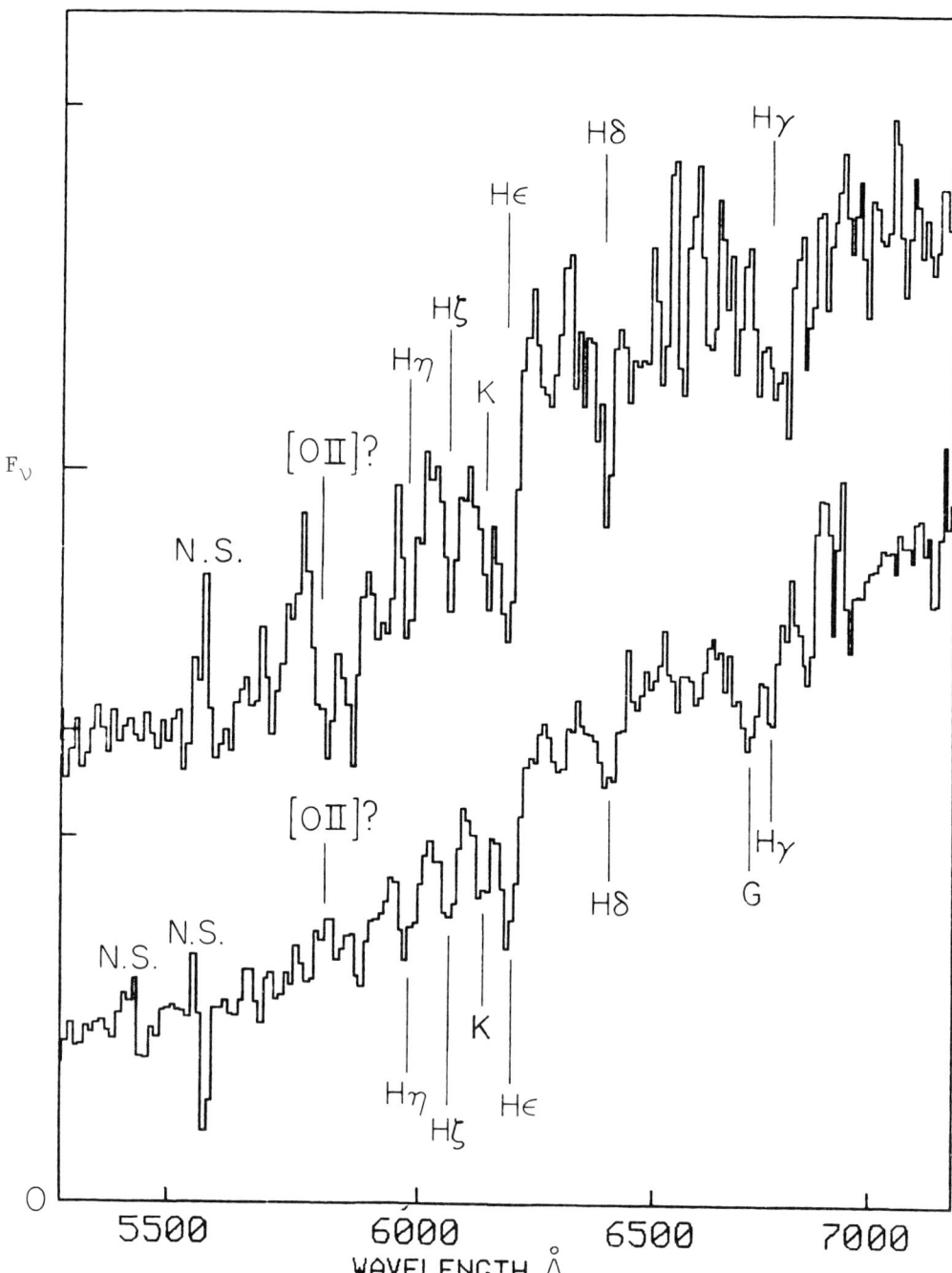

Fig. 1. Two additional examples of PSGs found in Cl 0016+16. The Balmer absorption lines are prominent, but metal lines are also seen. Note that no clear [O II] emission is seen in these examples, as is common in the class. (N.S. refers to poor subtraction of night-sky emission.)

times are not much longer than a burst of star formation in the system. Thus, fairly strong starburst activity may populate this diagram in a way that is roughly consistent with the relation for the bluest, most active, low-z spirals. This leads me to wonder if these distant galaxies may not be normal spirals after all, but instead are gas-rich galaxies falling into the ICM that have been stimulated into more active star formation in the manner discussed in Bothun and Dressler (1985). This is a crucial link because, if the latter interpretation is true, the PSGs may simply be later stages of these starburst galaxies. This in turn would suggest a fairly simple interpretation of all the results to date: (1) a subclump of gas-rich spirals fall into a distant cluster and are stimulated to one final burst of star formation (e.g., Cl 0024+16); (2) the gas is exhausted and the activity fades away leaving PSG galaxies (e.g., the 3C 295 cluster); and (3) the cluster slowly returns to a dormant (red) appearance, awaiting a further infall of gas-rich galaxies (e.g., Cl 0016+16).

As a postscript to this unbridled speculation, I note that this simple picture leaves at least two critical points unexplained. First, why does there appear to be an increased number of AGNs? Presumably, this has something to do with the ram pressure and high star-formation rates driven by it, which somehow manages to feed a central collapsed object. Second, why aren't there examples of such activity in present-epoch clusters? The first answer is that there may be, for example, DC 0248-53 (Dressler, Thompson, and Shectman 1985), and the low luminosity galaxies that appear to be falling into Coma. If this activity is more rare at the present epoch, or generally limited to lower luminosity galaxies, it could be because gas contents of spirals are significantly lower in today's spirals compared to their ancestors. This could significantly lessen the strength of the burst of star formation, and also make stripping of the infalling galaxies more effective.

## 5. SUMMARY

Up to redshifts $z \sim 0.5$, most of the galaxies in clusters are as red as would be expected for passively evolving systems that formed most of their stars early. On the other hand, there are a number of types of galaxies--PSGs, AGNs, and spirals with robust star formation--that are relatively rare in present-epoch, concentrated clusters, but appear at as much as an order-of-magnitude greater frequency in the distant clusters. This certainly suggests that some spectral evolution has been observed for a small but significant fraction of the cluster population, but to the question of what drives this evolution, the data provide no clear answer. If we are observing only the running of the cosmic clock, the supposition that galaxies were more active in the past and are "running down," we must wonder at why the populations in all three well-studied clusters are so different. This variety probably suggests some sort of environmental influence, or a clock that is synchronized to cluster evolution rather than cosmic time. It is in this context that observations of field galaxies at high redshift are so important. With their very different environments, they offer a way to disentangle the cosmological effects from the environmental ones. Studies of distant field galaxies are less straightforward because, as mentioned earlier, a range of distance and luminosity is sampled simultaneously, making the sample more subject to selection biases. Nevertheless, these types of programs, like the one being carried out by Kron, Koo, and Windhorst, are vital to the understanding of the evolution of cluster galaxies as well.

## 6. FUTURE WORK

It should go without saying, but never does, that samples of distant galaxies with good spectrophotometry are woefully small. This is presently one of the major limitations to understanding spectral evolution, at least from an observational point of view. From the presentations of O'Connell and Pickles at this workshop, it seems obvious that spectrophotometry of cluster galaxies at $z \sim 1$ will provide a definitive result concerning the importance of late star-formation in elliptical galaxies.

On the other hand, there are other important approaches for which data is at least as scarce. Most notable of these is the limited wavelength coverage of both distant galaxies and low-z galaxies to which model spectral energy distributions are to be compared. It is clear that the UV and IR are going to provide unique diagnostics of the history of star formation. Lilly and Gunn (1985), and Thuan (1985) have begun to investigate the near-IR $(1-2\mu)$ fluxes of many of the distant objects, and have found color anomalies relative to low-z cluster galaxies. These types of studies will be particularly valuable in assessing the contribution of intermediate-age populations, although we need both theoretical as well as observational work in this area. We have heard reports at this conference by Lonsdale and De Jong of results from the IRAS satellite which show the great potential for studying star formation in the far infrared. One hopes that SIRTF will have sufficient sensitivity to detect galaxies at cosmologically interesting differences that are undergoing modest, though not extraordinary bursts of star formation.

The reports by Bertola and Nesci on the IUE observations of elliptical galaxies show how varied and perplexing this new area of exploration is going to be. One can imagine that HST will, through its much gerater UV sensitivity, bring this to a mature field and provide a powerful tool for evaluating claims of "recent" star-formation in these seemingly dormant galaxies. A new photometric study by Ellis <u>et al.</u> (1985) suggests that the far-UV fluxes of galaxies in Cl 0016+16 are abnormally bright compared to expectations and some observations of low-z galaxies. When questions of aperture corrections and low-z templates can be thoroughly answered, as HST will do, this type of approach should provide another good way to examine star-formation histories.

By broadening our wavelength coverage, we will be able to better separate different epochs of star formation. When such observations are available for a wide range of lookback times, understanding the spectral evolution of galaxies should be within our grasp.

## REFERENCES

Baldwin, J.A., Phillips, M.M., Terlevich, R.: 1981, Publ. Astron. Soc. Pac., 93, 5.
Bothun, G.D., Dressler, A.: 1985, preprint, Astrophys. J., submitted.
Butcher, H., Oemler, A., Jr.: 1978, Astrophys. J., 219, 18.
Butcher, H., Oemler, A., Jr.: 1984, Astrophys. J., 285, 426.
Couch, W.J., Newell, E.B.: 1984a, Astrophys. J. Suppl., 56, 143.
Couch, W.J., Newell, E.B.: 1984b, in preparation.
Dressler, A., Gunn, J.E.: 1983, Astrophys. J., 270, 7
Dressler, A., Thompson, I.B., Shectman, S.A.: 1985, Astrophys. J., 288, 481.
Ellis, R.S., Couch, W.J., MacLaren, I, Koo, D.C.: 1985, preprint.
Koo, D.C.: 1981, Astrophys. J. (Letters), 280, L43.
Lilly, S.J., Gunn, J.E.: 1985, preprint.
Oke, J.B., Wilkinson, A.: 1978, Astrophys. J., 220, 376.
Thuan, T.X.: 1985, preprint.

## DISCUSSION

O'CONNELL : Is the definition of your PSG class based mainly on the strength of the Balmer absorption lines or on the absence of emission lines? I seem to recall a number of spectra in Humason, Mayall, and Sandage with strong Balmer absorption and not terribly strong emission, though these may have been mainly in the central areas of long-slit spectra.

DRESSLER : I consider the archetype PSG spectrum to be one with strong Balmer absorption and negligible emission, though in practice, we do not have sufficient spectral resolution to rule out [O II] emission at the level of several angstroms. This however, is far less than what would have occurred during the burst itself. The objects that Bothun and I are studying do have fairly strong emission and therefore are not what I would consider PSGs.

MADORE : You suggest that a single burst might consume 20% of the mass of the galaxy. Given a reasonable efficiency of star formation, does this not imply that there can only be one or two bursts in the lifetime of the galaxy, and that prior to the burst, the gas fraction must have been exceedingly high?

DRESSLER : Yes, unless only high-mass stars are formed in the burst, in which case, these constraints are greatly relaxed. If the 10-20% figure is correct, however, one must conclude that gas fractions in galaxies were somewhat higher in the past, since today only low-luminosity dwarf galaxies have gas fractions of 20% or more.

AARONSON : How strongly can you rule out continuous or exponentially decaying star-formation models for the "E + A" galaxies?

DRESSLER : If we form stars at a constant rate for $10^{10}$ years using the Manduca models, the Balmer lines reach a maximum of $\sim 5$ Å as the main sequence burns down to A stars, which is clearly weaker than what we find. Of course, exponentially declining rates produce even weaker Balmer absorption features. Independent of the models, I think you can believe this if you have looked at integrated spectra of typical spirals of the appropriate color. The Balmer lines are just not that prominent due to dilution from K and M giants in the old disk.

RENZINI : It would be interesting to see whether these post-starburst galaxies have any significant excess in the near IR, as one would expect for the AGB contribution in a $\sim 1$ Gyr population.

DRESSLER : I believe Thuan has been studying this issue.

O'CONNELL : First, just a comment on Thuan's results. Thuan finds strange JHK colors for galaxies in clusters at $z \sim 0.5$, which he does interpret as AGB effects. Second, are the clusters you observe very rich, dense clusters comparable to Coma? And, in such clusters at low redshifts, how much fainter are the brightest spirals than the brightest ellipticals? Blue objects seem to appear at relatively bright magnitudes in your color-magnitude diagrams for distant clusters.

DRESSLER : Yes, these clusters are comparable to Coma in density and richness. In such low-redshift, concentrated clusters, few spirals are within a magnitude of the brightest galaxies, and most are two magnitudes or more down. You are correct that in clusters where the blue excess is seen, there are many that are only about a magnitude fainter than the brightest members. It is important to remember, however, that this does not imply that these galaxies are as massive, since the implied M/L for a burst or an active spiral is much lower than for a normal elliptical. The color-magnitude diagrams for these distant clusters with significant blue populations are probably comparable to present-epoch, irregular clusters of galaxies which also contain spirals active in star formation.

COMTE : Concerning the low-z cluster sample of 1095 spectra, can you say something about any possible correlation between the fraction of emission-line objects and the cluster core density of galaxies, or any population/density related parameter you have?

DRESSLER : We have not looked at any correlation with core density. We did look for a correlation of emission-line frequency with local density and found no significant trend. Such trends would be hard to find since there are only 78 emission-line objects in the cluster sample.

KUNTH : What is the fraction of elliptical galaxies which is generally found with signs of [O II]$\lambda\lambda 3727$ emission?

DRESSLER :  For our sample it is only about 3%, but remember that we require a relatively strong emission of ∼4-5 Å.  I believe that much higher fractions, ∼20-30%, have been found by Nelson Caldwell when the threshhold has been dropped to ∼1 Å.

# OBSERVATIONAL TESTS FOR GALAXY EVOLUTION

Donald Hamilton
National Optical Astronomy Observatories
Cerro Tololo Inter-American Observatory
P. O. Box 26732
Tucson, Arizona, U.S.A. 85726-6732

**Abstract.** A discussion of recent optical observational work using the look-back time method is presented. This discussion includes brief outlines of the techniques of and problems associated with broad-band counts of galaxies, broad-band photometric comparisons, and spectrophotometric comparisons. The importance of the spectrophotometric comparisons will be emphasized as well as the most recent work in this field. The evidence for evolution has not been solid.

## 1. INTRODUCTION

The idea that the rest-frame radiometric properties of a presumed homogeneous sample of galaxies will change with time is not new (Hubble and Tolman 1935). These changes occur in both color and in absolute luminosity. Unfortunately, color or spectral evolution which could be measured directly, is observationally uncorrelated with luminosity evolution, at least in the optical region alone. Luminosity evolution is the critical unknown in the determination of the deceleration parameter.

The technique of looking for correlations of some signature of evolution with redshift (or look-back time), or equivalently, studying the star-formation history of galaxies, is usually referred to as the look-back time method. It can be divided roughly into three categories: broad-band counts of galaxies, broad-band photometric comparisons, and spectrophotometric comparisons. All of these techniques to some extent rely upon the assumption of the homogeneity of the sample. By homogeneity, it is meant that the galaxies which comprise a sample, if all born at the same point in space-time, would have identical properties, to within some small dispersion.

The changes of these properties with redshift (or equivalently time) are manifestations of the changes in the character of the typical star which comprises one of the test galaxies. These changes measure the star-formation history of galaxies.

This overview of spectral evolution is restricted to recent observational work involving the look-back time method as mentioned above. This discussion will not involve the Butcher-Oemler effect as it is thoroughly discussed by Dressler elsewhere in these proceedings.

## 2. BROAD-BAND GALAXY COUNT SURVEYS

### 2.1 The Use of Galaxy Counts as an Evolutionary Probe

It is expected that any significant change in the comoving luminosity density of galaxies will reveal itself as an alteration in the slope of the number of galaxies as a function of apparent magnitude. If galaxies were brighter in the past than they are now then more galaxies would be counted in a given magnitude interval and hence yield a steeper slope. Galaxy counts are dependent upon the luminosity function of each type of galaxy, the K-corrections for each type or class, and the redshift distributions of each class. Also, galaxies of different classes would have different rates of evolution (see, e.g., Bruzual 1981). Claims for evolution have to be placed in the context of the significance of the discrepancy relative to the derived no-evolution case.

The principal assumptions in galaxy count studies are 1) proper segregation of galaxies according to type and luminosity; 2) the extrapolation of nearby galaxy counts to faint limits is a valid approach and is without systematic error; and 3) there is no systematic problem associated with the observables.

Assumption 1) above, as normally conducted, is heavily model dependent. It is imperative to disentangle the effects of evolution intrinsic to the sources and evolution in the type of object sampled. This can only be accomplished by redshift measurements.

The problem with assumption 2) is that accurate local galaxy counts are difficult, since it entails surveying a large area and obtaining reliable total absolute magnitudes for at least a statistically meaningful subsample. The survey which has come the closest in fulfilling these requirements is that of Kirshner et al. (1978, 1979, 1983). This survey is usually used to normalize the faint galaxy-count data at the bright-end. A major problem in this normalization procedure, which can be easily rectified, is that the color transformation between the magnitude systems of the bright counts and that of the faint counts is poorly known.

Assumption 3) would appear as a systematic increase or decrease in the measured magnitude of a galaxy as a function of the sampling depth, due to the nature of the definition of a magnitude. Clearly, for proper interpretation, total magnitudes, which are independent of depth and redshift, need to be used. The more easily determined isophotal magnitudes are much more difficult to properly interpret in any conventional and straightforward manner. Also, it must be kept in mind that since number-magnitude surveys are ususally conducted in a fixed bandpass, those objects which have spectral energy distributions different than that of a typical object will be selectively removed or enhanced according to the K-corrections. Intrinsically faint blue galaxies have low K-corrections as compared with red, giant elliptical galaxies and therefore will dominate the counts for surveys conducted in bandpasses blueward of about 7500Å.

## 2.2 Some Recent Galaxy Count Surveys

Galaxy count surveys have been conducted by a number of different groups. Kron (1978) obtained counts in two colors (J and F) as a function of the apparent magnitude $0.5(J+F)$. This survey did reveal an anomalously high slope at the faint end: with an onset of about $0.5(J+F)$ = 21. Kron defined a large-aperture magnitude and it is basically a measurement of the total light. The Kron counts are entirely consistent with the no evolution case, given that the observed counts vary from field to field, and that there are errors in the no-evolution extrapolation. Kron has stressed this point on several occasions.

A survey which relied much upon Kron's technique and even his data base, was that of Koo (1981). Koo's major contribution was to add bandpasses in the ultraviolet and in the far-red. By using four-colors, it is possible to qualitatively estimate the type of galaxy and its redshift. Hamilton (1985a) used this technique to estimate the redshifts of galaxies for his spectrophotometric program. The interpretations of Koo (1981) are strongly model dependent and the effect of unproven or incorrect assumptions could be substantial. The bright-end normalization discussed above is poorly determined and the interpretations depend heavily on its accuracy.

Another survey of galaxy counts, that of Tyson and Jarvis (1979), and subsequent Bell Labs data, will not be considered here because these surveys were not properly calibrated. Also, the segregation between stars and galaxies even at bright magnitudes for the Bell Labs data is known to be poor (Koo and Kron 1982).

Shanks et al. (1984) have obtained galaxy counts using COSMOS, based on plate material obtained with both the UK Schmidt and the AAT. They claimed to have detected strong evolutionary effects in both the blue counts and in the color distributions. However, using the same reduction technique for red bandpass data, they concluded that the evolutionary effects are much less. Their counts in the blue and in the red are for the same fields, hence to a first approximation, they are the same galaxies. This discrepancy is an important one as it hints that something other than color or luminosity evolution is involved.

Hall and Mackay (1984) have conducted a deep CCD survey of several areas in both the blue and in the red. They used the drift scan technique for this survey and so the precision of flat-fielding is much more accurate than for surveys conducted in the normal guided exposure mode. They concluded that to a red magnitude of 25.0, the slope of the counts is 0.4 ± 0.1. This slope is the no-evolution value.

It can be concluded that counts of galaxies to faint limits are entirely consistent with the no-evolution case. The claims for evolution in the past have usually vanished when a proper analysis of both the technique and the interpretation have been conducted.

## 3.0 BROAD-BAND PHOTOMETRIC COMPARISONS

This technique consists of obtaining multicolor broad-band photometry of a presumed homogeneous sample of galaxies and with some assumption about or derivation of the K-corrections. A comparison is then made between the observed colors of the sample of galaxies at the high-redshift end and those which are predicted by the use of the K-corrections applied to the low-redshift sample. If there are any deviations, then one explanation might be that the discrepancy is due to something other than redshift only, e.g., color evolution. In other words, the anomalous K-corrections are redshift (or time) dependent. This technique has been used by at least two independent groups: Kristian, Sandage, and Westphal (1978), and Schneider et al. (1983). The primary purpose of these studies was the construction of the Hubble diagram and they were based on observations of first-ranked cluster galaxies. Since neither of these groups found any anomalous K-corrections, to their redshift limit of about $z \sim 0.3$, it can be safely concluded that the effect of color evolution is very small to this redshift. The early results of the (g-r) color-redshift relation from the Schneider et al. program were reported in Gunn (1982). In this review, the observed relation based on photometry of first-ranked cluster galaxies was shown to deviate significantly from the predicted relation thus implying color evolution. Apparently, only a zero-point error in (g-r) was involved which was corrected in the final presentation by Schneider et al. (1983).

Problems of broad-band comparisons, which could affect proper interpretation include poor photometric calibration, reddening, metallicity variations, and poorly determined K-corrections. More importantly, colors are not very good age discriminators for composite stellar systems.

This technique can still yield useful results if enough faint rich clusters can be located. With current photographic technology, clusters with redshifts between $z \sim 0.8$ and $z \sim 1.0$ can be straightforwardly discovered.

## 4. SPECTROPHOTOMETRIC COMPARISONS

A much more straightforward technique than those outlined above, is to obtain good signal-to-noise ratio spectrophotometry of a homogeneous sample of galaxies. Spectrophotometry has several advantages in that 1) a redshift is obtained, 2) monochromatic magnitudes can be defined and therefore the need for K-corrections can be avoided altogether, and 3) accurate indices and line-strengths can be derived. These quantities can be defined so that their correlations with extraneous and indeterminate parameters such as reddening and metallicity could be minimized or at least better understood.

The principal disadvantage of the technique is that only a limited number of objects can be observed. However, the information content in low signal-to-noise ratio spectrophotometry is still far greater than in high signal-to-noise ratio broad-band photometry. Also, assuming that the sample selection is well understood (which is rarely the case), the final interpretation is usually straightforward.

### 4.1 Signatures of Spectral Evolution

An observable needs to be defined which is sensitive to evolutionary effects. The primary evolutionary effect is that the character of the typical star will become earlier the larger the look-back time (bluer and brighter main-sequence turn-off). The effect of a bluer turn-off will quite naturally depend on how much light is produced by the main-sequence relative to that of the giant branch. The latter is well known to be basically insensitive to the mass of the star which feeds it. This bluing effect will be more dramatic in the ultraviolet. A sensitive observable is the amplitude of the 4000Å break (Spinrad 1980; Bruzual 1981; Hamilton 1985a). It is also a practical indicator because it is essentially monochromatic, hence the effects of reddening and improper calibrations are minimal. The 4000Å break is a high contrast feature and is much less sensitive to the signal-to-noise ratio of the data than other possible indicators such as the strength of the H$\beta$ absorption. The sensitivity of the 4000Å break to changes in the stellar population have been described both by Bruzual (1981) and by Hamilton (1985a). Also, Hamilton (1985a) has presented an analysis of the break based on spectrophotometry of stars and of local galaxies.

### 4.2 Sample Selection

Before any observational program is begun, it is imperative that the potential biases be evaluated; selection effects afflict nearly every observational program in cosmology. In the past, the traditional selection criteria have been rich-cluster morphology or the strength of the radio flux density. These two methods still form the basis for modern observational programs. A new dimension to the problem has recently been introduced: selection of galaxies according to their optical colors. The major observational programs which have been based on these criteria will be discussed below.

## 4.3 Spectrophotometric Surveys of First-Ranked Cluster Galaxies

Oke (1971) compared the spectral energy distributions of three galaxies with redshifts ranging from $z \sim 0.2$ to $z \sim 0.5$. He found no discernable difference between the spectral energy distributions of these galaxies.

The first major search for spectral evolution was by Wilkinson and Oke (1978) using the spectrophotometric data base of Gunn and Oke (1975). Unfortunately, as pointed out by Wilkinson and Oke, the dispersion amongst the spectral energy distributions of their program galaxies was too large to make any definitive conclusions about evolution possible. However, they did set an upper limit to the amount of evolution that could have taken place in the objects at high redshifts.

The latest in this style of work has been presented by Oke (1984). The selection criterion of Oke and collaborators was the morphological appearance of the cluster in one of four different types of surveys conducted. Oke's sample is based almost exclusively on first-ranked cluster galaxies. He obtained a measure of the 4000Å break and basically found no change to a redshift of about $z \sim 0.8$.

## 4.4 Spectrophotometric Studies of Radio Galaxies

In recent years observational studies of galaxies selected by their radio flux density such as the 3CR sample have been quite popular (Lilly and Longair 1982; Lilly, McClean, and Longair 1984; Djorgovski, Spinrad, and Marr 1985; Djorgovski and Spinrad 1985). The principal reason for selecting radio galaxies is that a distant sample of galaxies can be obtained. As the recent work of Djorgovski and Spinrad (1985) has shown, redshifts for 3CR galaxies up to $z \sim 1.8$ have been readily measured.

The problems associated with the use of radio galaxies in cosmological studies are formidable: potential correlations between the optical flux density with that of the radio flux density, non-thermal components diluting thermal light, correlations of the strength of the non-thermal contribution with the strength of emission lines (hence detectability), and changes in the type of galaxy as a function of apparent depth all help to preclude straightforward interpretation. It has been noted (van den Bergh 1975), that bursts of star formation and the existence of a radio source are correlated. Any amount of residual star formation will alter such observables as the 4000Å break (Bruzual 1981; Hamilton 1985a). It would be better to choose a sample of galaxies whose star-formation history was well described by current ideas and concentrate any observational effort on these objects. It is not possible to uniquely model the evolutionary behavior of galaxies with random or episodic star formation with the constraints of present-epoch observations.

From the studies of Djorgovski et al. (1985), and Djorgovski and
Spinrad (1985), it is clear that there is an abrupt change in the
morphology of the distant 3CR radio galaxies at about z ~ 1. At this
redshift, changes in the spectral properties of the sample are also
evident. Could it be that the 3CR double-lobed sample is really
inhomogeneous and that there are at least two distinct populations of
galaxies the 3CR survey selected?

Because of the bewildering number of nearly intractable problems
associated with the use of radio galaxies in cosmological studies, any
far-reaching conclusions based on such data should be treated with
circumspection. The problems mentioned above have been pointed out on
several occasions by Spinrad (1980), Smith (1977), and Djorgovski
and Spinrad (1985).

## 4.5 The Spectrophotometric Survey of Color-Selected Galaxies

In an attempt to control (or at least to understand) selection effects
more thoroughly, Hamilton (1985a), in a major new survey, selected
galaxies according to their redness relative to the typical galaxy
(most common) at the same brightness level. Much of the design of
Hamilton's observations was based on the previous work of Wilkinson and
Oke (1978) and of Spinrad (1980).

The typical galaxy at magnitudes fainter than about g ~ 21, is considered
to be blue. These galaxies (blue usually implies low luminosity, see,
e.g., Hamilton 1982) are easily selected in surveys conducted in the
blue because their K-corrections are significantly lower than those
of giant elliptical galaxies. By choosing red galaxies, Hamilton has
narrowed the selection of objects to those which could be described as
having formed from one massive burst of star-formation. If some
galaxies had additional star-formation subsequent to their initial
turn-on, then these would be bluer. Hamilton's assumption is that
there exists some galaxies that are describable as having formed from
only one episode of star-formation, and that these are ellipticals.

The definition of redness used by Hamilton depends upon the apparent
magnitude and was chosen so as to sample approximately the same part of
the spectral energy distribution of the galaxy, irrespective of redshift. The color-indices are formed from the **ugri** photometric survey
(Hamilton 1983, 1985a). For objects with a red magnitude between 19 and
21, the (g-r, r-i) diagram was used: objects with (g-r) > 1.5 and with
(r-i) > 1.7 were selected. For objects with red magnitudes greater than
21, only those objects with (r-i) > 2.0 were selected. These definitions
were chosen so that the selection color was redder than (u-g) which is
almost equivalent to the 4000Å break in information content. A possible
bias might occur, if the 4000Å break was strongly correlated with these
red-color indices. However, based on an analysis of nearby galaxies,
this does not seem to be the case (Hamilton 1985a).

It is believed that the red-color selection criterion yields elliptical-
type galaxies. If Hamilton's survey had included Virgo, then NGC 4472
would have been selected for the spectrophotometric program. Galaxies
redder than giant elliptical galaxies are quite rare. Even such
galaxies as M82 and MARK 231 are still bluer in optical broad-band
colors than the classical giant ellipticals such as NGC 4472 and NGC
4889. Also, the low-redshift sample of the spectrophotometric program
only contains those objects which would be described as elliptical
galaxies, based on their morphological appearance on the survey material.

Hamilton (1985a) obtained long-slit spectrophotometry about the 4000Å
break for approximately 25 galaxies with redshifts from z ~ 0. to z ~
0.8. All of the faint galaxies were observed with the Cryogenic Camera
on the Mayall telescope. The latest version of the evolution diagram is
presented in Figure 1. Also included are some data from Spinrad
(1984), which used the same instrument. The one galaxy that was
observed both by Hamilton (1985a) and by Spinrad (1984) had identical
measurements of the 4000Å break and so to a first approximation no
transformation was needed.

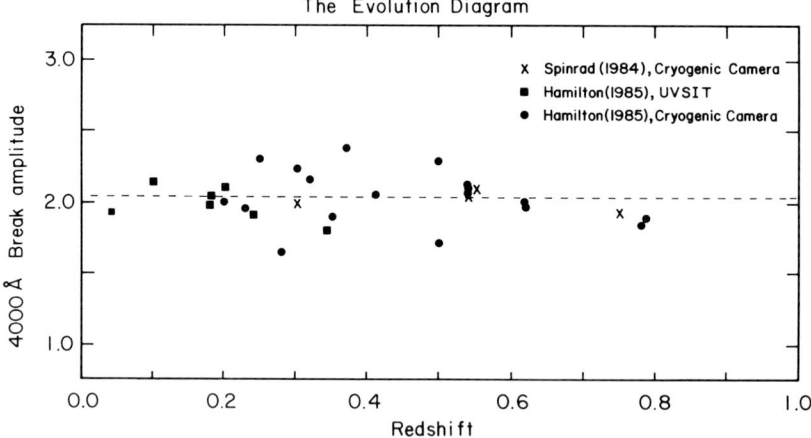

Figure 1: The 4000Å break amplitude versus redshift diagram
for data from the red-selected galaxy set of Hamilton (1985a)
and some additional data from Spinrad, which were chosen for
various reasons. The error in a break measurement is about 0.1.

The data of the evolution diagram imply that spectral evolution is less
than 7% at the very highest redshift galaxies surveyed (z ~ 0.8).
But it is clear that there seems to a definite trend towards bluer
break amplitudes at the high-z end. The best fitting model of Hamilton
(1985a) predicts that the break amplitude will have a value between 1.0
and 1.2 at z ~ 1. Observations are being planned to reach absorption-
line redshifts of z ~ 1.

Criticism about Hamilton's approach has usually been based on the mistaken belief that he has chosen the very reddest galaxies. This is most certainly not the case (Hamilton 1985a). The selection criterion is intended to select the characteristic elliptical galaxy from a large background of intrinsically faint blue galaxies. An effort was made to select galaxies so as to sample the elliptical distribution without bias. A more thorough analysis of the red galaxy sample will be presented in the next paper in the spectral evolution series (Hamilton 1985b).

The typical galaxy in a rich cluster of galaxies, whether it be near (e.g. Coma), or distant (e.g., Cl 0016+16, $z = 0.54$) is red relative to the field galaxy distribution. The red-selected galaxies of Hamilton are in general no redder that those found in these rich galaxy clusters. However, a full and proper analysis of the selection criterion of the red-selected set cannot come about until the photographic survey is completely calibrated.

## 5. CONCLUSIONS

Three major observational techniques for the study of the spectral evolution or star-formation history of galaxies have been reviewed. The problems associated with each technique have been briefly outlined. Spectrophotometry of individual sources is the most straightforward and powerful technique.

Although it appears that color or spectral evolution is negligible, this does not rule out strong luminosity evolution. It has been the hope that the measurement of spectral evolution would allow the determination of the amount of luminosity evolution, however, the correlation is at best a second-order one. The spectrophotometric surveys have been primarily of late-type galaxies (Morgan sense) and galaxy counts have been based on field galaxies which are spirals for the most part. The evolutionary behavior of elliptical galaxies (those formed in a single, short-burst of star-formation) will probably be more noticeable with look-back time than that of spiral galaxies. It is not possible to directly measure luminosity evolution since cosmology is also involved.

The evidence for spectral evolution in recent years has not been solid. Many of the claims for detection have vanished when a proper analysis of both the interpretations and of the techniques involved have been conducted. Also, there has been a large body of evidence for little or no evolution which has not been taken seriously. It is of interest to note that the results of Oke (1984) and those of Hamilton (1985a) agree very well, yet they have distinctly different selection criteria. Finally, it can always be assumed that any observed effect in observational cosmology has several tractable explanations, the least likely one in the end is usually the most obvious one in the beginning.

# REFERENCES

Bruzual A., G. 1981, Ph.D. thesis, University of California, Berkeley.
Djorgovski, S., and Spinrad, H. 1985, preprint, University of California.
Djorgovski, S., Spinrad, H., and Marr, J. 1985, proceedings of the Special IAU Colloquium New Aspects of Galaxy Photometry, ed. J.-L Nieto, Lectures in Physics Series, Springer Verlag, in press.
Gunn, J. E. 1982, in Astrophysical Cosmology, Proceedings of the Vatican Study Week on Cosmology and Fundamental Physics, eds. H. A. Bruck, G. V. Coyne, and M. S. Longair (Specola Vaticana), 233.
Gunn, J. E., and Oke, J. B. 1975, Astrophys. J., **195**, 255.
Hall, P., and Mackay, C. 1985, Mon. Not. R. astr. Soc., **210**, 979.
Hamilton, D. 1982, Pub. Astr. Soc. Pacific, **94**, 754.
Hamilton, D. 1983, in IAU Colloquium 78, Astronomy with Schmidt-type Telescopes, ed. M. Capaccioli, 461.
Hamilton, D. 1985a, Astrophys. J., **297**, 000 (Paper I).
Hamilton, D. 1985b, Paper II, in preparation.
Hubble, E. P., and Tolman, R. C. 1935, Astrophys. J., **82**, 302.
Kirshner, R., Oemler, A., and Schecter, P. 1978, Astron. J., **83**, 1549.
Kirshner, R., Oemler, A., and Schecter, P. 1979, Astron. J., **84**, 951.
Kirshner, R., Oemler, A., and Schecter, P., and Shectman, S. 1983, Astron. J., **88**, 1285.
Koo, D. 1981, Ph.D. thesis, University of California, Berkeley.
Koo, D., and Kron, G., Astron. and Astrophys., **105**, 107.
Kron, G. 1978, Ph.D. thesis, University of California, Berkeley.
Kristian, J., Sandage, A., and Westphal, J. 1978, Astrophys. J., **221**, 383.
Lilly, S., and Longair, M. S. 1982, Mon. Not. R. astr. Soc, **199**, 1053.
Lilly, S., McClean, I., and Longair, M. S. 1984, Mon. Not. astr. Soc., **209**, 401.
Oke, J. B. 1971, Astrophys. J., **170**, 193.
Oke, J. B. 1984, in Clusters and Groups of Galaxies, eds. F. Mardirossian, G. Giuricin, and M. Mezetti (Dordrecht: Reidel), 99.
Schneider, D., Gunn, J., and Hoessel, J. 1983, Astrophys. J., **264**, 337.
Shanks, T., Stevenson, P., and Fong, R. 1983, Mon. Not. R. astr. Soc, **206**, 767.
Smith, H. 1977, in IAU Symposium 74, Radio Astronomy and Cosmology, ed. D. L. Jauncey (Dordrecht: Reidel), 279.
Spinrad, H. 1980, in IAU Symposium 104, Objects of High Redshift, eds. G. O. Abell and P. J. E. Peebles (Dordrecht: Reidel), 39.
Spinrad, H. 1984, private communication.
Tyson, A., and Jarvis, J. 1979, Astrophys. J. (Lett), **230**, L153.
van den Bergh, S. 1975, Ann. Rev. Astron. Astrophys., **13**, 217.
Wilkinson, A., and Oke, J. B. 1978, Astrophys. J., **220**, 376.

# DISCUSSION

**R. O'Connell:** I am somewhat surprized by the scatter of the 4000Å break you show for local galaxies (see Fig. 10 of Hamilton 1985a). This seems to amount to a range of over 0.5 mag. My impression was that local E galaxies brighter than $M_V=-20$ had more homogeneous spectral properties. An excellent determination of the 4000Å break scatter at the current epoch is important in judging possible evolutionary effects (to larger or smaller 4000Å break values) at high redshifts.

**D. Hamilton:** I was also surprized to see such a large scatter. Our impressions may be due to tighter correlations derived from analyses of galaxies in rich clusters of galaxies. Some of the scatter for local galaxy data may be due to local supercluster effects and to aperture effects. Also, one should note that the first-ranked galaxy data of Oke also show a lot of scatter.

**A. Dressler:** Gunn and I also see a spread in the break amplitude for bright ellipticals in a given cluster. I would have thought that this was due to metallicity differences, which according to $Mg_2$ measurements by Terlevich et al. (1981) and my own study of Coma and Virgo ellipticals, are quite large at a given absolute magnitude.

**D. Koo:** 1) To what redshifts can you distinguish S0 from gE? 2) I have heard that the discs of S0's are redder than the colors of average gE. Will this affect you results? 3) If the 4000Å break distribution is not extremely narrow, an effect equivalent to the Scott effect might result in your finding a null result when in fact the median, mode, or mean of the 4000Å break distribution among field gE has changed by more than the 7% change you see at $z \sim 0.8$.

**D. Hamilton:** 1) The segregation between E's and S0's will clearly depend upon inclination angles/intrinsic ellipticities, but given ideal conditions, i.e., edge on S0 versus E0, then $z \sim 0.4$, otherwise $z \sim 0.2$ is a safe lower limit. 2) It will only affect my results if there is some redshift dependent selection, which is possible at the lower redshifts, but unlikely beyond $z = 0.3$. Also, I have obtained 4000Å break amplitude measurements (based on IDS spectrophotometry) of several points along a radius in the galaxy NGC 6703, which is claimed by some to be a face-on S0. Based on these points, there is a very dramatic bluing trend away the nucleus. 3) In response to your comment, first of all there are few distributions in cosmology which are extremely narrow; secondly, if your point is that there might exist a correlation between the selection criterion and the 4000Å break amplitude, I have demonstrated that it does not exist for at least local galaxies. Finally, an effort was made to sample uniformly the perceived elliptical restframe color distribution.

**G. Bruzual:** Do you give any significance to the behavior of the 4000Å break in the serendipitous sources that you showed in the lower panel of your figure (see Fig. 13 of Hamilton 1985a)? Why do they behave differently in your diagram?

**D. Hamilton:** The behavior of the 4000Å break amplitude for the serendipitous sources can be explained entirely by the color-luminosity effect; or these are lower luminosity sources which were close to the line of sight of the primary program objects. They behave quite differently because they were selected in effect by apparent magnitude, and the probability of finding a galaxy increases with depth.

COLORS OF 3CR AND FIRST-RANKED HIGH REDSHIFT GALAXIES

Peter R. M. Eisenhardt
National Optical Astronomy Observatories
950 N. Cherry Ave.
Tucson, Arizona 85726

ABSTRACT. Forty radio selected and thirty nine optically selected giant elliptical galaxies of known redshifts ranging from 0.019 to 1.2 have been observed simultaneously in four colors from 0.4 to 2.3 microns. No evidence for differences between the colors of radio and optically selected galaxies is found. The infrared H-K colors are best fit by a passively evolving model with little residual star formation. The optical-IR colors are contained by a red envelope well fit by a passively evolving model, but some galaxies show strong blue deviations for $z > 0.4$. This behavior is most easily explained by episodes of star formation involving small fractions of the total number of stars.

1. INTRODUCTION

The most direct way to study spectral evolution of galaxies is to observe them at high redshift. Given a sample which extends to large lookback times, we ideally want accurate UV to IR spectrophotometry as a function of time. However, high z galaxies are faint objects, and time is a function of $H_o$, $q_o$ and z, so we settle for color as a function of redshift. We then address questions such as: How much stellar evolution is needed to reproduce these colors? When did galaxies form? Is galaxy formation and evolution dependent on local conditions? And of course, can we learn anything about cosmology (i.e. $H_o$ and $q_o$)?
   Previous studies have naturally focused on the brightest normal galaxies known, first ranked cluster galaxies, which are invariably ellipticals, and have remarkably constant luminosities (Sandage 1972b). While nearby giant ellipticals are fairly homogeneous, with no evidence for evolution from BVR photometry for redshifts < 0.4 (Kristian, Sandage, and Westphal 1978), there is considerable controversy about evolution in more distant objects. At these redshifts, morphological classification is impossible, but by analogy with nearby examples, the first ranked galaxies in clusters and 3C radio galaxies are assumed to be giant ellipticals, or at least their progenitors. At optical wavelengths, some authors have found evidence for evolution at redshifts as low as 0.4 (Butcher and Oemler 1978,1984; Lilly and Longair 1984; Djorgovski and Spinrad 1985; Eisenhardt 1984) while others see little or none up to z=0.9 (Oke 1983; Hamilton 1985; Koo 1985). In the infrared there is also controversy. Lebofsky (1981) and Lilly and Longair (1984) found no observable evolution in IR <u>colors</u> with

redshift (though they do detect luminosity evolution), while Puschell, Owen, and Laing (1982) measured anomalously blue IR colors for high redshift objects. Ellis and Allen (1983) found IR J-K colors for a sample of optically selected galaxies differ from those of radio galaxies, while Thuan et al. (1984) find these colors are independent of radio power.

IR photometry of high redshift galaxies is essential to monitor the normal stellar population (Lebofsky 1981), but the effects of evolving stellar populations are greatest in the UV (Bruzual 1981), because the main contribution here is from quickly evolving hot stars. Since UV light emitted by high redshift galaxies is observed at optical wavelengths, both optical and IR observations of high redshift galaxies are needed. Such observations, when made at all, have usually been done on different telescopes, on different instruments, through different apertures (Grasdalen 1980; Lilly and Longair 1984; Thuan et al. 1984).

With such considerations in mind, a study of the optical and IR colors of radio and optically selected giant ellipticals was begun, using the Simultaneous Photometer for Infrared and Visible light (hereafter SPIV). SPIV is naturally suited to accurate color measurements, because it measures its four bands through the same aperture at the same time. The optical/ IR colors in particular have not been previously measured with the same aperture. The other major goal addressed here is to <u>directly</u> compare radio and optically selected galaxies, for this has only been done in the optical for $z < 0.4$ (Sandage 1972c; Kristian, Sandage, and Westphal 1978). Discussion of luminosity evolution and $q_o$ is deferred to that by Lebofsky and Eisenhardt (1985).

## 2. OBSERVATIONS

The design of SPIV was first presented in Eisenhardt (1982). SPIV uses three dichroic filters to divide light from a common aperture in the telescope focal plane into four colors with bandpasses (in microns) of: 0.42 to 0.7 ($V_B$); 0.7 to 0.95 ($I_B$); 1.45 to 1.8 (H); 1.97 to 2.27 (K). The H and K bands are detected by liquid helium cooled InSb diodes. $I_B$ is detected by a helium cooled Si diode, and $V_B$ by an uncooled EMI 9658R photomultiplier tube.

Observations were made on 11 nights at the MMT on Mt. Hopkins with a 7 arcsec aperture, and 3 nights each at the UAO 2.3m on Kitt Peak and 1.54m on Mt. Bigelow with a 20 arcsec aperture, from June 1983 to March 1984. Typically five standards and ten to fifteen galaxies were observed per night. The observing procedure is fully described in Eisenhardt (1984).

Radio galaxies were taken from the catalog of Burbidge and Crowne (1979) and various studies by Spinrad and collaborators. The optically selected galaxies are first ranked cluster members. Those with $z > 0.5$ are from a survey for faint clusters of galaxies conducted over the past decade by Gunn, Hoessel and Oke (Oke 1983). Information on these was kindly presented in advance of publication. In all, data from 79 galaxies is presented, 40 of them radio galaxies. The main selection criterion was that redshift be known.

Two important possible selection effects here are: 1) A bias for bluer (in the rest frame), richer clusters as redshift increases; 2) Favoring emission line galaxies. To forestall these problems, Gunn, Hoessel and Oke used red sensitive plates, and obtained redshifts from fitting to standard galaxy spectra. However, such procedures may discriminate against the unexpected. The 3C sample has nearly complete identification and redshift measurements, so the selection effects for it are quite different. Comparing radio and non-radio galaxies is therefore very useful.

Because calibration to a standard magnitude system is complicated, particularly for the very broad $V_B$ and $I_B$ bands, the results here are in the SPIV instrumental system. Reduction and calibration are thoroughly discussed in Eisenhardt (1984). For reasons discussed there, all colors are referred to the H magnitude. The reduced observations on the SPIV instrumental system are shown in Figures 1 and 2, with open circles representing optically selected galaxies, and filled circles radio galaxies. The colors have been corrected for atmospheric and galactic extinction. In a few cases where an object was not detected in one or more bands, two sigma upper or lower limits are shown.

## 3. ANALYSIS

The similarity of the colors of radio and non-radio galaxies in Figures 1 and 2 is striking. The distributions appear the same at all redshifts in all colors, and statistical tests confirm this (Eisenhardt 1984), as was found in Lebofsky and Eisenhardt (1985). This finding has important consequences regarding the optical/IR properties of radio galaxies, and implies selection effects are not responsible for the deviations from the no evolution predictions.

### 3.1 Predicted Colors.

Because the K correction depends entirely on the choice of a standard elliptical galaxy spectrum (I use that of Lebofsky and Rieke 1985), I treat it as a "zero order" (no evolution - NE) theoretical prediction for comparison with the observed colors. This NE prediction is shown as the upper solid line labelled "a" in Figures 1-4. Note this NE model contains no free parameters: it was calculated without reference to the observations.

Evolutionary models were taken from Bruzual (1981) because they pay particular attention to the UV, where evolutionary changes are greatest, and cover a wide range of SFR. Because only three spectral points beyond one micron are tabulated in Bruzual's spectra, and these points maintain the same relative fluxes independent of model and time (except for a brief period of $10^8$ years following the epoch of galaxy formation), I decided to append the NE spectrum to his model spectra, normalizing at one micron. This is reasonable since the epoch of galaxy formation is almost certainly at redshifts well beyond those observed here. The assignment of redshift to time requires the choice of the formation redshift $z_f$; and of $H_o$ and $q_o$. I chose $q_o=0.5$ because of the inflationary universe models, and $H_o=80$, with $z_f$ a variable.

The predictions for several representative models are plotted in Figures 3 and 4. Mu models have an exponentially decaying SFR. Mu represents the fraction of the total mass that has formed stars after one Gyr. In c models the SFR is constant for the first Gyr, and zero thereafter. All models assume the Salpeter IMF. Bruzual (1981) found that mu=0.6 to 0.7 models best reproduced the observed properties of ellipticals. The $z_f=2$ model is included in an attempt to reproduce the blue $V_B$-H colors, not because it it is thought to be a realistic description of the global SFR in ellipticals.

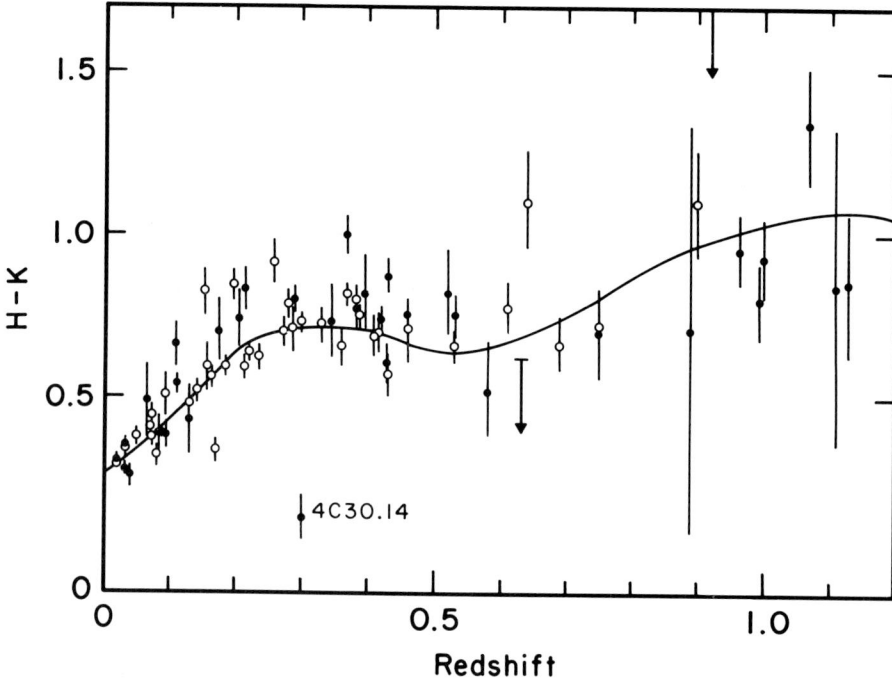

Figure 1. Observed H-K Color vs. Redshift. Open circles are optically selected galaxies, filled circles radio selected. Arrows mean only upper or lower limits were measured. The solid line is the K correction, or no evolution prediction.

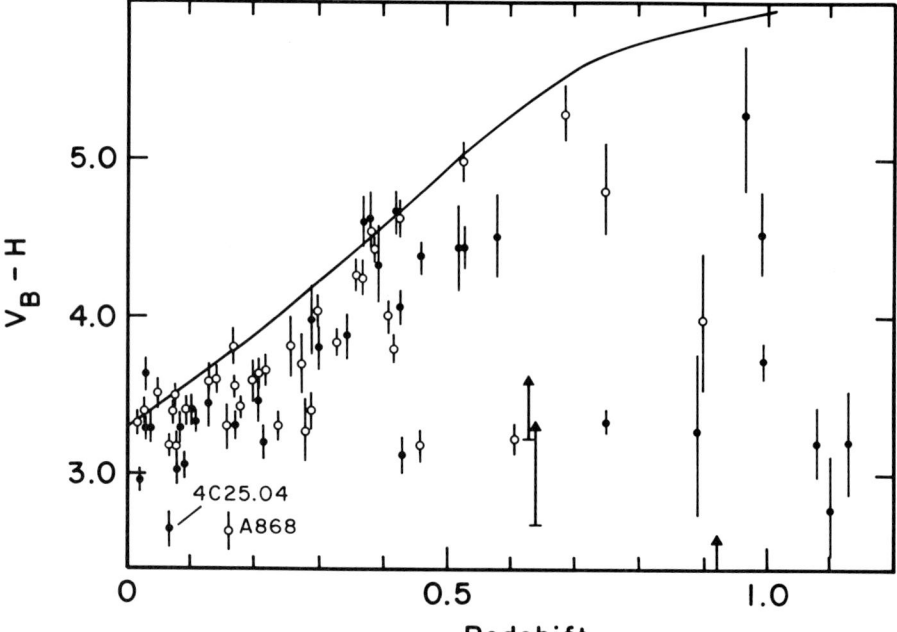

Figure 2. Observed $V_B$-H Color vs. Redshift. Symbols same as in Figure 1.

# COLORS OF 3CR AND FIRST-RANKED HIGH REDSHIFT GALAXIES

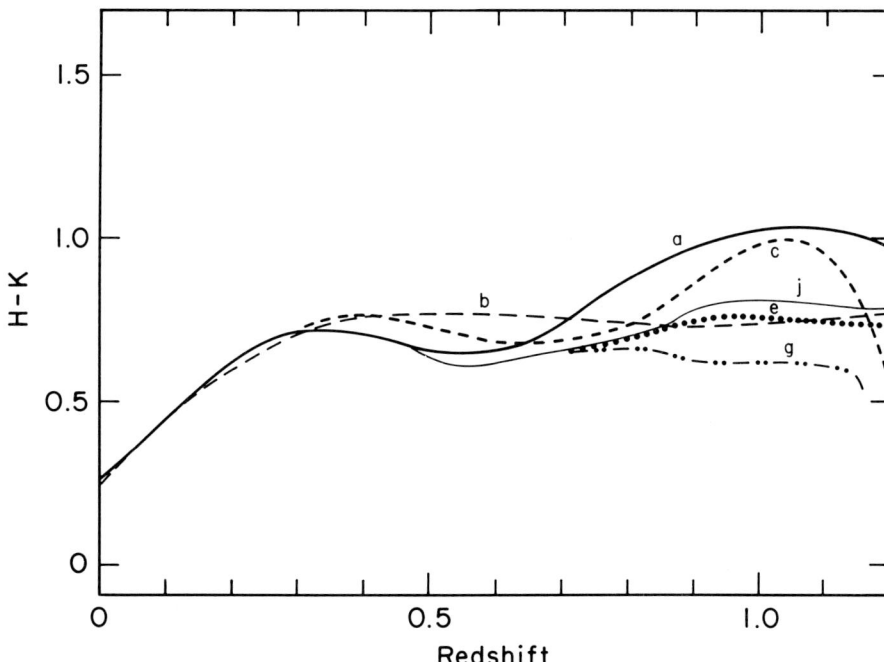

Figure 3. Predicted HK vs. Redshift. (a) NE model; (b) 3rd order fit to data; (c) 5th order fit to data; (e) $z_f=20$, mu=0.7 model; (g) $z_f=2$ c model; (j) $z_f=5$, mu=0.5 model. For e and g $H_o=80$ and $q_o=0.5$, for j they are 50 and 0.

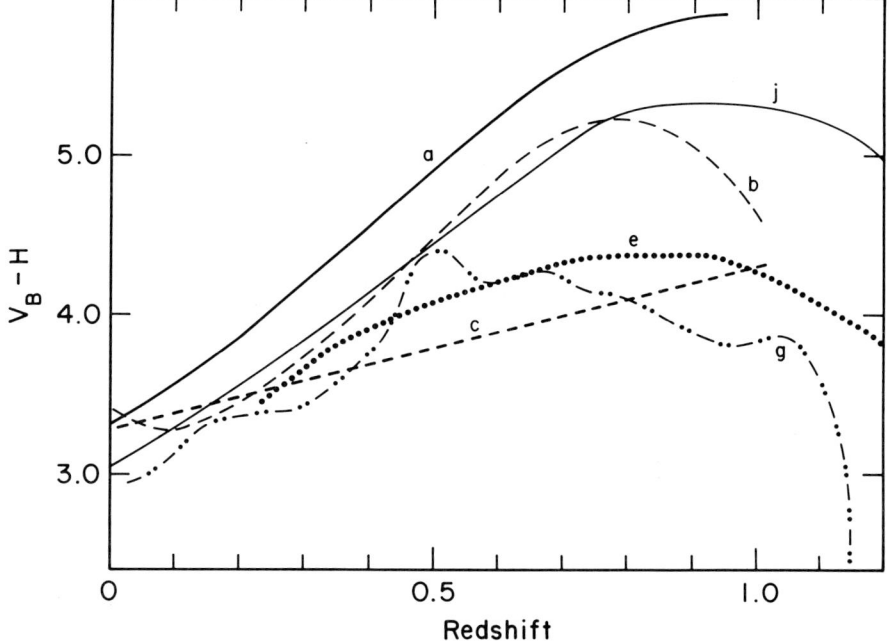

Figure 4. Predicted $V_B H$ vs. Redshift. As for Fig. 3, except (c) is 1st order.

## 3.2. Comparison with Observed Colors

### 3.2.1. H-K.
The agreement in H-K between observation and prediction is generally excellent. 4C30.14 is much bluer than other objects. Observational error is unlikely as it was observed twice. Because 4C30.14 is a high frequency radio source, and its image is not obviously extended, it may be an N galaxy.

Third and fifth order polynomial fits to the data are plotted in Figure 3. Several red points were omitted in calculating these polynomials. The excess redness in these objects is most likely due to a nonthermal source, unless it is caused by reference beam problems. The IR colors depend only slightly on evolution. All models are somewhat bluer at z=1 than the NE curve, in agreement with the observations. Very recent formation epochs ($z_f < 3$) appear to be ruled out, and larger values of mu fit the observations better than the low mu models, though this difference is slight. These results are in accord with the conventional picture of elliptical galaxy formation, with most stars formed over a short period at an early epoch. Evolution in this scenario is passive (c models) with little or no active star formation at the redshifts observed here.

### 3.2.2. $I_B$-H.
The $I_B$-H color is troublesome. The scatter is large, and agreement with predictions is poor even at low z. As I have been unable to rule out instrumental error or to confirm these phenomena with independent data, these colors are not presented here. A full discussion is given in Eisenhardt (1984).

### 3.2.3. $V_B$-H.
Two phenomena are evident in Figure 2: 1) The NE curve represents an upper (red) envelope to the data; 2) At z=0.4 the data, which had been tracking the NE curve with moderate scatter, show a sudden large deviation towards the blue. Rather than follow a particular path in the color-redshift plane, however, the data beyond z=0.4 are scattered over a two magnitude range, bounded on the red by the NE curve. Two blue points with z < 0.4 deserve special mention: 4C25.04 showed erratic behavior at $V_B$ when it was observed; and A868 has another galaxy 12 arcseconds to the south, making a reference beam problem virtually certain.

It is clear that no single model can fit the data beyond z=0.4. To determine how well the predictions match the low redshift points and the upper red envelope at high redshift, a third order polynomial, shown as the dashed line b in Figure 4, was fit to the data, excluding points which were substantially bluer than this envelope. The form of this fit is similar to the NE prediction, but consistently bluer. The c model is actually a better fit to the upper red envelope of the $V_B$-H data than the NE model, implying that the V flux has been underestimated relative to that at H. The mu models cannot produce the red colors of the high redshift galaxies for $H_o$=80 and $q_o$=0.5. This demonstrates the powerful effect of even a small rate (5% for mu=0.7, $z_f$=20, at z = 1) of residual star formation on $V_B$. More time is needed than the cosmological parameters chosen allow for a mu=0.7 model to be sufficiently red at z= 1. This is confirmed by curve j in Figure 4, which used $H_o$=50, $q_o$=0.

Either low values of mu or very recent formation redshifts are needed to produce the blue envelope of the $V_B$-H observations. Hot blue stars are required, and these are only available in the models if there has been recent star formation. The H-K data, however, suggest this has not happened on a global scale, and the $V_B$-H data are not so blue as to require that the majority of stars formed recently. This is best illustrated by the c model with $z_f$=2. In this model star formation continues at a constant rate up to z=1.16, and then terminates. The dramatic effect of the onset of star formation is shown by the precipitous blueward plunge of this model for z > 1.1, in all colors.

A possible explanation for the $V_B$-H observations, then, is that some elliptical galaxies had bursts of star formation at or just prior to the redshifts observed, but the majority of stars formed long ago. A starburst forming roughly $10^7$ stars in total would suffice, which would not significantly affect the overall galaxy properties. The conditions leading to such a starburst could well vary from one galaxy to the next, giving rise to the observed scatter. It is hard to see why this process should begin at a redshift of 0.4, however.

## 4. COMPARISON WITH OTHER OBSERVERS

Others have found excess blue light beginning at redshifts near 0.4 (Lilly and Longair 1984, Butcher and Oemler 1978). DeGioia-Eastwood and Grasdalen (1980) suggested this is simply due to the variation observed in the UV spectra of nearby ellipticals, as these wavelengths are redshifted into the V band. However, Bertola, Capaccioli and Oke (1982) show that this variation only occurs below 2500 Å. Because $V_B$ is not sensitive below 4200 Å, these variations should not affect the observations for redshifts < 0.7. On the other hand, Gunn, Stryker and Tinsley (1981) suggest these local UV variations are indeed caused by a young star population, just as proposed here. Other explanations include hot stars which are not young, such as blue stragglers or horizontal branch stars. If these are part of the normal evolutionary sequence, however, it is difficult to explain the wide range of $V_B$-H observed.

Lilly and Longair (1984) find essentially the same results as are presented here. This implies the radio galaxies observed here are not systematically biased by unknown selection effects. The lack of difference between radio and non-radio galaxies strongly suggests the latter are also representative. It is therefore curious that Oke (1983) states that there is no evidence for changes of colors with redshift for their optically selected sample, since the non-radio galaxies observed here are drawn from this same sample. In fact the statement is at odds with the data presented in Oke's Figures 4 and 5, and with an earlier statement that there is a hot UV source which varies from galaxy to galaxy. It is true that the scatter in color means there is no well-defined correlation of color with redshift, and when interpreted in this way there is no contradiction between Oke's statement and the findings presented here.

Hamilton (1985) has studied a sample of relatively isolated galaxies which show no evidence for evolution to z=0.8. This result is not in disagreement with those found here, since galaxies with essentially unevolved colors have been observed here, in both the optically and radio selected samples. Note also that Lilly and Longair (1984) find a positive correlation of [OII] 3727 emission with blue color, and that Hamilton detects no emission in any of his galaxies.

There can be little doubt that some galaxies at redshifts near one have negligible populations of young stars, while others show blue colors which are clearly extended, as shown by CCD images (Lilly and Longair 1984; Djorgovski and Spinrad 1985). The difference is not attributable to radio emission. Hamilton's results suggest cluster richness may be a factor, but his red selection criteria are likely to have eliminated blue field galaxies.

DISCUSSION:

Whitford: For $z > 1$ are not the rest frame "visual" magnitudes in the satellite UV? What is the source for the observed magnitudes?

Eisenhardt: The predictions for observed V magnitude at high redshift are based on IUE observations of nearby ellipticals.

Hamilton: You claim that in your $V_B$-H diagram the nonthermal part contributes little; could you remind me please of the basis of this claim and at what level are you sensitive.

Eisenhardt: There are two arguments against a significant nonthermal contribution in the 3CR galaxies: 1) CCD images obtained by Lilly and Longair and Djorgovski and Spinrad are extended in blue light, indicating the UV excess is not due to a point source. Djorgovski and Spinrad estimate a limit of < 10% nonthermal contribution. 2) My measurements of optically selected galaxies show no difference in any color at any redshift from the 3CR galaxies, implying that any non-thermal source contribution is unrelated to radio power.

Acknowledgements: Marcia Lebofsky played an essential role throughout this investigation, as did George Rieke. I thank Gustavo Bruzual for allowing me to use unpublished results of his models, and George Djorgovski and Don Hamilton for assistance in using them. The support of the NSF is acknowledged.

REFERENCES

Bertola, Capaccioli, and Oke, J.B. 1982, Ap.J. **254**, 494.
Bruzual, G. 1981, Ph.D. Thesis, Univ. of California at Berkeley.
Burbidge, G. and Crowne, A.H. 1979, Ap.J.Supp. **40**, 583.
Butcher, H. and Oemler, A. 1978, Ap.J. **219**, 18.
——————————. 1984, Ap.J. **285**, 426.
DeGioia-Eastwood, K. and Grasdalen, G.L. 1980, Ap.J. **239**, L1.
Djorgovski, S.G. and Spirad, H. 1985, submitted to Ap.J.
Eisenhardt, P.R.M. 1982, Proc. S.P.I.E., **331**, 434.
——————————. 1984, Ph.D. Thesis, Univ. of Arizona.
Ellis, R.S., and Allen, D.A. 1983, M.N.R.A.S. **203**, 685.
Grasdalen, G.L. 1980, in IAU Symposium No. 92, "Objects of High Redshift," eds. G. Abell and P.J.E. Peebles (Reidel, Dordrecht) p. 269.
Gunn, J.E., Stryker, L., and Tinsley, B. 1981, Ap.J. **249**, 48.
Hamilton, D. 1985, Ap.J. in press.
Koo, D. 1985, private communication.
Kristian, J., Sandage, A., and Westphal, J.A. 1978, Ap.J. **221**, 383.
Lebofsky, M.J. 1981, Ap.J. **245**, L59.
Lebofsky, M.J., and Eisenhardt, P.R.M. 1985, submitted to Ap.J.
Lebofsky, M.J., and Rieke, G.H. 1985, in preparation.
Lilly, S.J., and Longair, M.S. 1984, M.N.R.A.S. **211**, 833.
Oke, J.B. 1983, in "Clusters and Groups of Galaxies," eds. F. Mardirossian, G. Giuricin, and M. Mezetti (Dordrecht:Reidel) p. 99.
Puschell, J.J., Owen, F.N., and Laing, R.A. 1982, Ap.J. **257**, L57.
Sandage, A. 1972b, Ap.J. **178**, 1.
——————————. 1972c, Ap.J. **178**, 25.
Thuan, T.X., Windhorst, R.G., Puschell, J.J., Isaacman, R.B., and Owen, F.N. 1984, Ap.J. **285**, 515.

EVOLUTION OF DISK GALAXIES IN HIGH-REDSHIFT CLUSTERS

B. Guiderdoni[1] and B. Rocca-Volmerange[1,2]

[1] Institut d'Astrophysique, 98 bis bld Arago, F-75014 PARIS, FRANCE
[2] Laboratoire René Bernas, Bat. 108, Université Paris XI, BP 1, F-91306 ORSAY, FRANCE

ABSTRACT  We propose an evolutionary scenario for spiral galaxies in rich, compact clusters. Interactions with the environment, probably dominated by the processes in the hot gaseous intergalactic medium, remove the gas from the disks of blue objects infalling into cluster cores from outer layers. Predictions are made for UBV colors of cluster spirals versus the redshift z, accounting for cosmological and evolutionary effects, by means of a model of spectrophotometric evolution. We show that spirals stripped at ages corresponding to $z \simeq 0.5$ very rapidly acquire redder colors and an earlier appearance. The fraction of blue objects in compact and open clusters, observed by Butcher and Oemler, may possibly be interpretated.

## 1. EVOLUTION OF SPIRAL GALAXIES IN A CLUSTER ENVIRONMENT

There is now strong evidence for environmental influences on the dynamical and photometric evolution of galaxies (see the recent review by Dressler, 1984). The disk galaxies are particularly sensitive to various processes of gas removal, of which the most effective ones appear to be the evaporation and ram-pressure stripping in the hot intergalactic medium (IGM) in rich, compact clusters (Gunn and Gott, 1972, Cowie and Songaila, 1977, Giovanelli and Haynes, 1984, Haynes et al., 1984) and the cumulative effect of tidal interactions during two-bodies encounters in groups and loose clusters (Spitzer, 1958, Toomre and Toomre, 1972, Farouki and Shapiro, 1981) Moreover, an aggressive environment rapidly disrupts the large HI disks detected in some galaxies (Bosma, 1981, Sancisi, 1983), as well as any hypothetical gaseous halo that might be the reservoir for star formation in disks (Larson et al., 1980).The earliest spirals are more sensitive to these environmental influences (Guiderdoni and Rocca-Volmerange,1984 and 1985a).

From a statistical study of the HI content and far-UV to visible colors of Virgo Cluster spirals, and an analysis of the observational data by means of a model of photometric evolution (Guiderdoni and Rocca-Volmerange, 1985a,b), we showed that the redder colors and the deficient gas content of such spirals, as well as the large scale properties of the cluster, are consistent with one or two complete, recent gas removals, occurring at age $t > 1/2\ t_g$, where $t_g$ is the present age of the disks, and associated with the crossing of the IGM in the inner degrees of the core with a time scale $\leq 1$ Gyr. Most of these deficient spirals are evolving towards the anemic class of Van den Bergh, 1976. A slight drift of the morphological types in the spiral sequence is predicted because of the increase of the bulge/disk ratio due to the fading of the disk. Objects with colors, average surface brightness and bulge/disk ratio of S0s can form from effective, permanent gas removal if it begins not later than $t \simeq 1/2\ t_g$. In fact, cluster S0s have smaller bulge/disk ratios and fainter bulge magnitudes than "field" S0s (de Souza et al., 1984). On the other hand, infall of late-type galaxies from outer layers provides the core with new, gas-rich, blue objects (Tully and Shaya, 1984).

The properties of a number of nearby clusters show the same general trend (Giovanelli and Haynes, 1984). Nevertheless, this scenario of rapid, effective interaction is still controverted (Bothun et al., 1984, Kennicutt, 1983a, Kennicutt et al., 1984). In any way, the unavoidable fate of spiral galaxies in cluster cores seems to be a drastic depletion of their gas content and a decrease of the star formation rate which results in redder colors and an earlier appearance. In an aggressive environment, the cessation of star formation is predicted and observed to occur earlier than in the "field" where the gas-consumption time scales deduced from the observed present SFRs and total gas contents are from, say, $\simeq 3$ Gyr to $\simeq 10$ Gyr (Larson et al., 1980, Kennicutt, 1983b). Field spirals might be refueled by some infall of inter/protogalactic matter, which is inhibited in a dense environment.

## 2. A MODEL OF SPECTROPHOTOMETRIC EVOLUTION FOR HIGH-REDSHIFT GALAXIES

This evolutionary scenario for cluster spirals predicts that high-redshift clusters, corresponding to a significant look-back time with respect to the age of the disks $t_g$, are likely to harbour a population of objects bluer than those in nearby clusters, and perhaps to show a higher spiral/S0 fraction if some of the S0s originate in strongly stripped spirals. Butcher and Oemler, 1984, used photometry of 33 clusters with $0.003 \leq z \leq 0.54$ to probe the evolution up to a look-back time of, say, $\simeq 6$ Gyr with $H_0 = 50$ km s$^{-1}$ Mpc$^{-1}$ and $q_0 = 1/2$. They found that the fraction $f_B$ of blue galaxies (defined as $<(B-V)>(E,S0)-(B-V) \geq 0.20$ in the rest frame of the galaxy) in the inner core strongly increases with the redshift. Additionally, nearby rich clusters harbour much less blue objects than the "field" and their spirals are redder. These have also earlier types, according to Gisler, 1980, and most of them have a low surface brightness and an anemic appearance (Wirth and Gallagher, 1980).

We here intend to use a model of spectrophotometric evolution of galaxies to predict the colors of high-redshift spirals, either with a "closed-box" evolution or with strong interactions resulting in a complete, permanent gas removal. The age of the Universe is hereafter $t_0$ = 13 Gyr. The disk galaxies are assumed to form at redshift $z \simeq 4$ and we have presently $t_g$ = 12 Gyr. According to our previous models of evolution, the stars form from the gas and evolve in the HR diagram along theoretical stellar tracks corresponding to a solar metallicity. The synthetic spectrum of the galaxy is a linear combination of 28 stellar spectra (1200 Å<$\lambda$<10680 Å) from the IUE Ultraviolet Spectral Atlas, 1983, and Gunn and Stryker, 1983, accounting for the various spectral types and luminosity classes. In our model, the colors are calculated in the UBV system from this redshifted spectrum, and account both for evolution and cosmological effects in standard Friedmann models. The adopted stellar initial mass function (IMF) gives a good fit of the Solar Neighborhood. The star formation rate (SFR) varies as $\tau_*(t)$ = a exp -at for early-type galaxies (a=1 for a typical S0) or $\tau_*(t) = \nu\ M_{gas}(t)$ for spiral galaxies ($\nu$=0.2 for a typical Sb).

## 3. RESULTS AND DISCUSSION

Fig. 1 and 2 respectively show the predicted apparent colors U-B and B-V versus the redshift z for the typical S0 and Sb galaxies. Additional metallicity effects can explain the reddest and bluest objects of the whole morphological sequence. Internal extinction corrections and the light from ionized regions can also increase the color range. The evolution of such a spiral galaxy after complete, permanent stripping from age $t_s$ = 6 or 8 Gyr is represented by dotted lines on fig. 1, 2 and must be understood as the limit case of several short strippings occuring with the short period defined by the small crossing time of the rich, compact clusters observed by Butcher and Oemler. It is easy to see that these recently-stripped spirals rapidly evolve towards redder colors, comparable to those of a Sa or a S0 (see also Guiderdoni and Rocca-Volmerange, 1985b). It is worthwhile to notice that the predicted whole color range continuously increases from z=0. to z$\simeq$0.5, and is rapidly crossed by the evolving, stripped galaxies. So the population of clusters with z$\simeq$0.5 is expected to have a very wide color distribution.

Fig. 3 shows the predicted apparent magnitude $m_V$ versus the redshift z. The typical S0 and spiral galaxies are scaled to respective apparent magnitudes 9.85 and 10.85 at the distance of the Virgo Cluster ($d_{VC}$ = 20 Mpc). The permanent stripping results in only a moderate fading of $\Delta m_V \leqslant 1$ mag.

Thus a drastic interaction with the environment, probably the hot IGM, from redshift $z \simeq 0.5$ to the present, makes a blue object rapidly acquire red colors and even the anemic or S0 appearance. In conclusion, the fraction of blue objects in compact and open clusters is the product on the one hand of the various interaction processes,

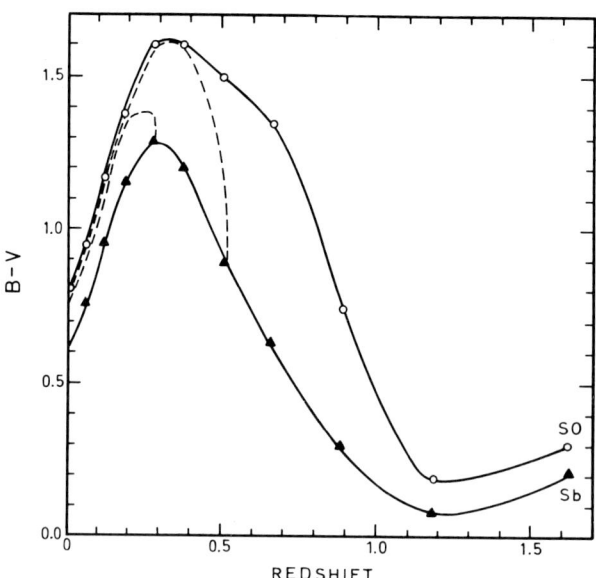

Fig. 1 : Predicted apparent U-B versus redshift z for typical S0 ($\tau_*(t) = 1.0 \exp -1.0\, t$, full line with open circles) and Sb ($\tau_*(t) = 0.2\, M_{gas}(t)$, full line with filled triangles) The evolution of a spiral after complete, permanent stripping occuring at redshift $z_S = 0.51$ or $z_S = 0.28$ is denoted the two dotted lines.

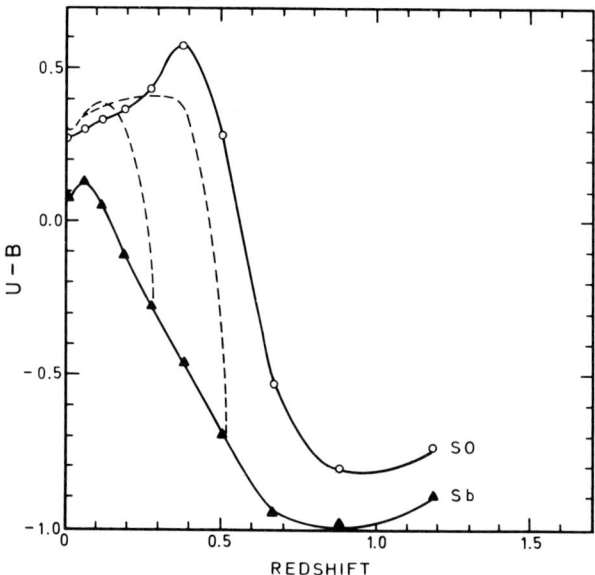

Fig. 2 : Predicted apparent B-V versus redshift z. Symbols as in fig. 1.

the effectiveness of which depends on the kind of environment and can lead to an acceleration of the evolution towards the cessation of star formation, and on the other hand of the probably clumpy arrival of blue, healthy objects from the collapse of the outer layers (Rivolo and Yahil, 1983, Tully and Shaya, 1984).

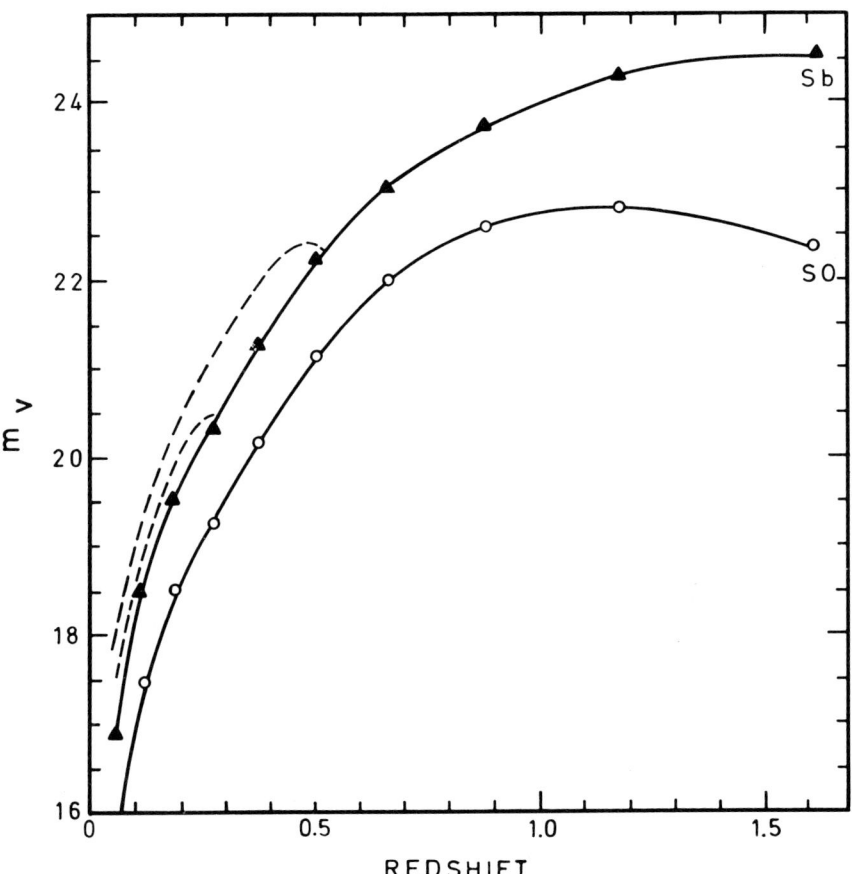

Fig. 3 : Predicted apparent magnitude $m_V$ versus redshift z. Symbols as in fig. 1. The typical S0 and Sb are respectively scaled to apparent magnitudes 9.85 and 10.85 at the distance of the Virgo Cluster (m-M=31.5).

REFERENCES

Bosma, A., 1981, A. J., **86**, 1825
Bothun, G.D., Schommer, R.A., Sullivan, W.T., 1984, A.J., **89**, 466
Butcher, H., Oemler, A. Jr, 1984, Ap.J., **285**, 426
Cowie, L.L., Songaila, A., 1977, Nature, **266**, 501
De Souza, R.E., Vettolani, G., Chincarini, G., 1984, preprint
Dressler, A., 1984, Ann. Rev. Astron. Astrophys., **22**, 185
Farouki, R., Shapiro, S.L., 1981, Ap.J., **243**, 32
Giovanelli, R., Haynes, M.P., 1984, preprint
Gisler, G.R., 1980, A.J., **85**, 623
Guiderdoni, B., Rocca-Volmerange, B., 1984, ESO Workshop on "The Virgo Cluster of Galaxies", Garching
Guiderdoni, B., Rocca-Volmerange, B., 1985a, Astron. Astrophys., in press.
Guiderdoni, B., Rocca-Volmerange, B., 1985b, submitted
Gunn, J., Gott, J., 1972, Ap.J., **176**, 1
Gunn, J.E., Stryker, L.L., 1983, Ap.J. Supp. Series, **52**, 121
Haynes, M.P., Giovanelli, R., Chincarini, G.L., 1984, Ann. Rev. Astron. Astrophys., **22**, 445
IUE Ultraviolet Spectral Atlas, NASA Newsletter n°22, 1983
Kennicutt, R.C., 1983a, A.J., **88**, 483
Kennicutt, R.C., 1983b, Ap.J., **272**, 54
Kennicutt, R.C., Bothun, G.D., Schommer, R.A., 1984, A.J., **89**, 1279
Larson, R.B., Tinsley, B.M., Caldwell, C.N., 1980, Ap.J., **237**, 692
Rivolo, A.R., Yahil, A., 1983, Ap.J., **274**, 474
Sancisi, R., 1983, in Internal Kinematics and Dynamics of Galaxies, IAU, p. 55.
Spitzer, L., 1958, Ap.J., **127**, 17
Toomre, A., Toomre, J., 1972, Ap.J., **178**, 623
Tully, R.B., Shaya, E.J., 1984, Ap.J., **281**, 31
Van den Bergh, S., 1976, Ap.J., **206**, 883
Wirth, A., Gallagher, J.S., 1980, Ap.J., **242**, 469.

## T. DE JONG

Based on the same observational material, you seem to arrive at opposite conclusions from Kennicutt (1983) who claims that the observed colors, Hα fluxes and hydrogen masses are incompatible with recent ram-pressure stripping of the gas from spirals in Virgo. Could you explain to me how this is possible and tell me who of you I should believe ?

## B. ROCCA-VOLMERANGE

Our studies are not based on the same observational materials. Kennicutt's sample is strongly biased towards late-type spirals ($4 \leq T \leq 10$) which are faintly deficient and have redder colors more or less attuned to their mild deficiency. On the contrary, early-type spirals ($0 \leq T \leq 3$) which have a strong weight in our conclusions have a very low gas content for their type and colors and are strongly distinguishable from "field" early-type spirals.

## C. BALKOWSKI

I would like to make some remarks concerning the HI deficiency of the Virgo Cluster to complete what you said :
1) The non-deficiency of Virgo X spirals has already been pointed out by Chamaraux et al. (1980), Virgo X is low density subgroup with more late type spirals than in the Virgo Cluster center.
2) I would like also to remind that we found a correlation between the HI deficiency and the distance to the cluster center (Chamaraux et al., 1980 ; Van Gorkom et al., 1984).
3) There is no bright Sc galaxy in a $3°$ radius circle centered on the cluster center so it is normal that you find that the Sc are not HI deficient.
4) A recent analysis of the HI content of the S0 galaxies inside and outside the Virgo Cluster show that the S0 are much more HI deficient than the early or late spirals (Balkowski et al., 1984) showing that the stripping is more efficient in early-type systems.

Concerning the HI distribution of the S0 galaxies, Van Woerden et al. (1983) have shown that they tend to present HI rings at the border of the optical discs, this is particularly the case for NGC 4262 which is one of the few S0 detected in the Virgo Cluster. As the stripping is more efficient in the outer part of the disc, this could explain the large deficiency of the S0 galaxies in the Virgo Cluster (Balkowski et al., 1984).

## B. ROCCA-VOLMERANGE

We have argued that the flatness of the radial gradient of deficiency for $R > 6°$ and its steepness in the core, together with the existence of spirals with small HI disks in the inner degrees are evidence for rapid ram-pressure stripping occurring only in the center of the Virgo Cluster.

We have found a number of deficient late-type spirals in Virgo I. So the type effect is not only a spatial effect. There might be also the alternative possibility of a rapid type drift after a complete stripping. We have predicted $\Delta T \approx -2$ from the evolution of the bulge/disk ratio.

## J. FROGEL

Since color changes rapidly but not $m_V$ during stripping, the luminosity change at a $\lambda$ other than V should be rapid. So if you want to see luminosity evolution, what you have said implies that V is not the wavelength to look at.

## A. DRESSLER

Stripping spirals to make S0s is a much more attractive explanation than before if, only early type spirals are affected, as you have found in Virgo and I have found for the Giovanelli-Haynes sample. This explanation does little, however, to explain the existence of most S0 galaxies, which are in lower density environments where ram-presure sweeping seems unlikely

Secondly, according to models by Gunn and myself (Ap.J., 270,7) simply stripping old spirals, even active Sc's, does not result in a strong (W~8 Å) Balmer line phase. According to our calculations, this requires a burst of star formation to overcome the accumulated light of older stars, although stripping spirals that are much younger than $10 \times 10^7$ years in age might also work.

## B. ROCCA-VOLMERANGE

According to your second comment, it appears that the conclusions are strongly affected by the assumed ages of the disks. We are presently addressing the question of the Balmer lines by means of our model of spectrophotometric evolution.

QUESTS FOR PRIMEVAL GALAXIES: A REVIEW OF OPTICAL SURVEYS


David C. Koo
Space Telescope Science Institute
3700 San Martin Drive
Baltimore, Maryland 21218
U.S.A.


ABSTRACT. Optical surveys explicitly designed to detect continuum or line emission from individual primeval galaxies are reviewed. In addition to such searches, a variety of recent redshift surveys of very faint quasars, radio sources, and galaxies are also useful in constraining the epoch, number, and brightness of galaxies during their birth. The serendipity discoveries of a few high-redshift objects with narrow emission lines are tantalizing. Presently the strongest constraints come from long-slit spectroscopy of blank sky to search for Lyman α emission. The lack of positive detections suggests that galaxy formation occurred at redshifts $z > 5$, a result consistent with the lack of primeval galaxies among samples of quasars fainter than $B = 20$ and with the mild evolution predicted by models that fit faint galaxy counts and colors.


1. INTRODUCTION

The last review of primeval galaxies (PG) was by Davis (1980) and it emphasized galaxy-formation models rather than observations. In contrast, the aim of this review will be to summarize the current status of numerous and diverse surveys which constrain the nature of PG. We will limit the scope to optical measurements designed to detect positive flux from individual objects. Despite their importance, excluded are studies of intervening matter seen as absorption lines in background quasars (see review by Boksenberg and Sargent 1983) as well as measurements of extragalactic background emission in X-rays (Setti and Woltjer 1982), ultraviolet (Paresce and Jakobsen 1980), optical (Toller 1983), and infrared (Boughn, Saulson, and Uson 1985 and McDowell 1985).
  For this review, PG are defined as <u>high-redshift</u> galaxies undergoing their <u>initial stage of star formation.</u> PG are presumably the youthful counterparts to galaxies like our Milky Way or giant ellipticals, which are believed to be about 10 or more billion years old. Galaxies such as I Zw 18, though possibly "young" if they are being observed during their initial burst of star formation (Searle and

Sargent 1972), are not of high redshift and thus not a PG by our definition. By our definition, PG must exist, though they may be extremely faint or even invisible in the optical.

PG searches are worthy of intensive efforts for many reasons. When PG are eventually discovered, their redshifts will directly establish the epoch of galaxy formation, which alone has profound implications for theoretical models of galaxy formation (Binney 1977). The range of redshifts, on the other hand, will reflect the extent to which the birth of galaxies has been coeval and may give clues to the relative amplitudes on different mass scales of the initial fluctuation spectrum, especially if cluster formation epochs are also known. The spatial distribution of PG are useful measures of clustering at high redshifts; if the explosive scenarios for galaxy formation of Ostriker and Cowie (1981) or Ikeuchi (1981) are correct, we might even find arcs or chains of galaxies undergoing birth. Physical parameters of the formation process, including the mass and luminosity function of PG; the temperature, density, dust content, and metallicity of gas; initial mass function and star formation rate; etc., are all in principle derivable from the observed brightness and spectral features. Of course, information from other wavelengths as well as from the morphology of PG will add vital clues to our knowledge of star, galaxy, and active-galaxy formation.

Granting that searches for PG deserve attention, where can we expect to find them? The following Table I shows the range of values predicted by different models of PG. If you are not the gambling-type of observer, the implied uncertainty would be a nightmare; if you are, a dream. Historically, the theoretical foundations were far less chaotic.

TABLE I. Range of Predicted Properties from Published Models of Primeval Galaxies

| | | |
|---|---|---|
| Size: | <1 arc sec | >30 arc sec |
| Brightness: | 18 mag | >30 mag |
| Num. Density: | 1 deg$^{-2}$ | >100 arcmin$^{-2}$ |
| Redshift z: | <2 | >10 |
| Wavelength: | X-ray | Radio |

The "Golden Age" of PG research began with the seminal paper by Partridge and Peebles (1967). They predicted that PG would be observable as a high surface density of very-red images of large size (5 to 30 arcsec) and uniform surface brightness. The extended size was a result of their assumption that the major burst of star formation occurred when protogalactic gas clouds stopped expanding with the universe at redshifts typically between 10 and 30 with sizes 15 kpc. The redness was a result of redshifting the rest-frame ultraviolet light emitted by massive, hot stars to the optical. These theoretical models were the prime motivations for the PG surveys made by Partridge

(1974) and Davis and Wilkinson (1974), which were explicitly designed to look for <u>large, red</u> PG.

Within less than a decade, and already hinted by an unpublished paper by Weymann in 1966, theoretical predictions took a 180 degree turn by suggesting that PG would instead appear <u>small and blue</u>. This "Renaissance" of PG research was largely based upon the work of Larson (1974), who emphasized the importance of dissipation in the galaxy formation process. Dissipation results in more compact objects and longer, later formation times which favors lower redshifts and thus bluer colors than non-dissipational models. Combined with the newly popular view of a more open universe (Gott et al. 1974), these models predicted smaller and fainter PG consistent with the null results of surveys by Davis, Partridge, and Wilkinson. PG models by Meier (1976), Kaufman (1975, 1976), Sunyaev, et al. (1978), and Tinsley (1977 to 1980) made specific suggestions for finding PG: 1) PG may be masquerading as quasars 2) UBV two-color plots can separate PG from other types of faint objects 3) surveys should be made for objects with strong Lyman $\alpha$ emission lines or Lyman continuum breaks and 4) faint blue galaxies are PG. All these suggestions were taken seriously and provided the basis for several explicit searches for PG by Koo and Kron (1980, 1982), Koo (1981), Mackay (1985), and Turner, Gunn, and Sargent (1981).

Today a new era of PG research has begun. On the theoretical side, galaxy formation models may need to be more complicated and to include the effects of mergers and interactions of galaxies; the presence of dark matter (baryonic or not) and Lyman $\alpha$ clouds; the consequences of large-scale (adiabatic) fluctuations and biassed galaxy formation; the disturbances to intergalactic or intragalactic gas caused by energy output by active nuclei or by the heating and metal production of very massive-stars; etc. On the observational side, Charge Coupled Devices (CCD) have extended the frontier in the optical; infrared imaging arrays are becoming a reality; VLA and Westerbork routinely reach sub-mJy flux levels; gravitational lenses, as "telescopes" of Nature, further magnify the distant cosmos; the Hubble Space Telescope will soon be launched; space telescopes sensitive to the X-ray, ultraviolet, and far-infrared add new dimensions to deep surveys; etc. All these developments are relevant to PG models and observations. In the following sections, however, we will concentrate on optical surveys completed or already underway.

## 2. OPTICAL SURVEYS DESIGNED TO DETECT PG

### 2.1 Techniques

Almost all techniques available to the optical observer can be exploited in searches for PG. We begin by discussing the method containing the least spectral information, measurements of the total optical flux from an unresolved background. Taking advantage of a "dark" cloud to provide a foreground reference, Matilla (1976) detected a background of $12 \pm 4$ $S_{10}$ units, where an $S_{10}$ unit is a

TABLE II. Explicit Searches for Primeval Galaxies

| Type | Tele. Size (m) | Method/Data | Area (arcmin$^2$) | Flux Limit | Result | Reference |
|---|---|---|---|---|---|---|
| **Large Red PG** | | | | | | |
| " | 3 | Direct Imaging B, V, R, plates | 360 | R = 21.7 | 7 Cand. | Partridge (1974) |
| " | 2 | Aperture Phot. Red Fluctuations | 6 | R = 21-24 | Null | Davis and Wilkinson (1974) |
| " | 4 | Direct Imaging R, I CCD | 18 | R = 24 | 1 Cand. | Loh and Wilkinson (1979) |
| **Compact Blue PG** | | | | | | |
| Quasars | 4 | Multicolor Phot. Astrometry, U, B, V, I plates | 2000 | B = 23 | ≤200 deg$^{-2}$ | Koo and Kron (1982) Koo, Kron, and Cudworth (1985) |
| Faint Galaxies | 4 | Multicolor Phot. U, B, V, I plates | 1500 | B = 23.5 | ≤1000 deg$^{-2}$ | Koo (1981) |
| **Redshifts of PG Candidates** | | | | | | |
| Peculiar Galaxies | 5 | Spectroscopy SIT | 60 gal | B = 21 | Null | Turner, Gunn, and Sargent (1981) |
| Faint Quasars | 4 | Spectroscopy Cryocam | 50 obj | B = 22.5 | Null | Koo and Kron (1985) |

TABLE II. Explicit Searches for Primeval Galaxies (Continued)

| Type | Tele. Size (m) | Method/Data | Area (arcmin$^2$) | Flux Limit | Result | Reference |
|---|---|---|---|---|---|---|
| Blue Radio Galaxies | 4 | Spectroscopy Cryocam | 10 obj. | R = 21 | Null | Kron, Koo, and Windhorst (1985) |
| **Lyman α Surveys** | | | | | | |
| Lyman α | 4 | Narrow-band imaging Broadband imaging Image tube plates | 113 4 qso | V = 23.5 | Null Null | Mackay (1985) " |
| Lyman α Lyman cont. | 4 | Slitless Spec. I plates | 4000 | R = 20.6 | Null | Koo and Kron (1980) |
| " | 4 | Slitless Spec. R-I CCD | 200 | R = 22.5 | Null | " |
| Lyman α | 4 | Multiple Narrow-Band Imaging + Cryocam spect. | 6 | R = 26.2 | 10 Cand. | Hartwick and Pritchet (1985) |
| Lyman α Lyman cont. | 4 | Long-slit spect. of blank sky Cryocam spect. | 1 | R = 26.5 | Null | Cowie (1985) Koo and Kron (1985) |

measure of surface brightness equivalent to distributing the flux of a single A0 V star of 10th mag over a square degree (about $1.4 \cdot 10^{-9}$ ergs sec$^{-1}$ cm$^{-2}$ Å$^{-1}$ st$^{-1}$ at 5100 Å). This claim for a bright background generated tremendous excitement since the predicted values were closer to 1 $S_{10}$ (Peebles 1971); Tinsley (1977), for example, suggested that PG might be found at bright magnitudes B ≥ 20. The excitement was rapidly dampened with the report of a much smaller background, $1.0 \pm 1.2$ $S_{10}$, by Dube, Wickes, and Wilkinson (1977, 1979), whose experiment took great care in accounting for foreground emission. These low values were confirmed by Spinrad and Stone (1978), who also used a dark cloud as a reference in their spectroscopic measurements, and by Toller (1983) who obtained measurements with Pioneer 10. Although background measurements are in principle a good diagnostic of emission from PG, in practice such experiments do not provide strong constraints, since the foreground emission from the Earth's atmosphere, solar system, and Milky Way are more than 100 times brighter and the technique of merely counting faint galaxies already constrain their emission to several tenths of an $S_{10}$ unit to B = 26 (Toller 1983, Seitzer and Tyson 1985).

Beyond counting objects as a function of apparent brightness, we can add considerable spectral information and discrimination against low-redshift galaxies and stars by securing colors from broadband (~1000 Å FWHM) imaging. If Lyman α or other strong emission lines are to be detected, medium-band imaging, narrow-band imaging, and slitless spectroscopy all provide good depth with excellent areal coverage and moderate spectral resolution (50 Å to 200 Å FWHM). For final confirmation that specific candidates are indeed PG rather than quasars or other objects, slit spectroscopy with better resolutions (~15 Å) will probably be necessary. A major discriminant will be the presence in PG of narrow emission lines with velocity dispersions more typical of galaxies, i.e. ≤500 km sec$^{-1}$, rather than much broader lines seen in quasars. Since PG may exist in vast numbers (see Table I), one can in principle obtain long-slit spectra of only tiny regions (few square arcmin) of "blank" sky and hope to detect PG; even if individual PG overlap spatially, emission lines from each PG may be separated in wavelength from those of the others.

2.2 Surveys for Large Red PG

Table II summarizes published surveys and those still underway that are explicitly designed to find PG. We will later discuss surveys which might find PG by serendipity. Table II divides these surveys into four broad categories: 1) large red PG, 2) compact blue PG, 3) deep redshift surveys of PG candidates, and 4) PG with strong Lyman α or other emission lines.

The two pioneering PG surveys were those of Partridge (1974) and Davis and Wilkinson (1974). Partridge searched for extended red objects on photographic plates covering 0.1 deg taken in three broad bands, BVR, with the Lick 3-m telescope. To a limit of R = 21.7, only seven candidates passed the criteria of being large uniform-surface-brightness objects. R. Kron and the author have measured a spectroscopic redshift of z = 0.12 for the brightest candidate; the remaining six appear to be late-type (Sbc-Im) galaxies with redshifts z less than 0.3 on the basis of deeper 4-m multicolor photometry (Koo 1985a). Rather than attempting Partridge's method of searching for individual PG candidates, Davis and Wilkinson constructed a special large-aperture (1000 square arc sec) photometer to measure small (1%) fluctuations of the night sky due to the presence of a high surface density of extended red PG. Although their areal coverage was only 6 square arc min, the flux limit for individual objects could in principle be fainter than that of Partridge's experiment, if PG were sufficiently numerous; no positive fluctuations were detected.

Although technically not in the optical, and thus outside the scope of this review, the very recent 2 micron photometric survey for PG by Boughn, Saulson, and Uson (1985) is noteworthy as an experiment which extends the sensitivity of fluctuation measurements to the infrared and hence higher galaxy formation redshifts. PG were not found.

Since Partridge's survey, Loh and Wilkinson (1979) have continued the quest for red extended PG by using CCD detectors for direct imaging. One candidate was found several years ago but has yet to be confirmed spectroscopically.

## 2.3 Surveys for Compact Blue PG

Returning now to Table II, we note that several surveys have been designed to detect PG which are small and blue rather than large and red. Unless moderate-resolution spectra are obtained to secure identifications (see following section 2.4), these searches are useful mainly by providing solid upper-limits to the number-density and flux of possible PG. Multicolor photometry is used to isolate PG candidates from faint blue stars if quasar-like PG are being sought, or from the vast numbers of faint blue galaxies if slightly-fuzzy PG are the targets.

In the case that PG are masquerading as quasars (Meier 1976, Sunyaev et al. 1978, Bookbinder et al. 1980, Wyse 1985), the surveys of faint quasars by Koo and Kron (1982), Koo, Kron, and Cudworth (1985), and Koo and Kron (1985) place useful upper-limits to the number of such PG to B $\leq$ 23 (preliminary results have also been reported by Koo 1983a, 1983b, and 1986 ). Deep UBVI plates were used to detect all stellar-like objects which did not lie within the multicolor loci expected for Galactic stars. Such objects, limited to $\leq$300 deg$^{-2}$ to B = 23, were presumed to be good candidates for faint quasars and PG. After making spectroscopic, astrometric, and variability studies of a subsample of these quasar and PG candidates, we found that few are likely to be PG: after we exclude about one-third of the candidates because they are unresolved nearby (z $\leq$ 0.7) galaxies or Galactic stars, most of the remaining sample show variability or very broad emission lines more characteristic of genuine quasars than PG (Meier 1976, Sunyaev et al. 1978). Our best estimate of the quasar density is 150 deg$^{-2}$ to B = 22.5; if 20% of these are PG masquerading as quasars, we obtain an upper-limit of 30 deg$^{-2}$ for PG with stellar sizes. This 20% represents a conservative limit based upon our not finding any convincing case for narrow emission lines seen at $z \geqslant 1$ among three dozen spectra of quasar candidates.

Could the quasars themselves be a manifestation of PG? Until our faint quasar survey, the extrapolation of the very steep luminosity functions found for very-luminous quasars implied the possibility that all galaxies were undergoing a quasar event at high redshifts (Braccesi et al. 1980). The flattening of the apparent number counts of faint quasars and the turnover in the luminosity functions deduced from the redshifts of faint quasars indicate that this possibility is unlikely to be true. In fact, our observations are most consistent with the view that quasars (Seyfert 1 are included in this class) were on average much more luminous in the past but that the ratio of the number of quasars to that of galaxies today has been small ($\leqslant 1\%$) to redshifts of 2 to 3; beyond this, the number of quasars appear to drop rapidly.

Of course, PG could be compact and blue but not quite stellar in size as demanded by the faint quasar surveys. For example, massive galactic halos that undergo rapid star formation would appear extended (Kaufman and Thuan 1977). In this case, contamination by faint blue galaxies is severe and our limits are far less stringent. As before, we exploit the power of multicolor photometry, this time to segregate PG from nearby galaxies (Koo 1985 ). Since colors of galaxies at redshifts $z \geqslant 1$ have yet to be measured, we have relied heavily upon the models of the spectral evolution of galaxies, like those of Tinsley (1980) or Bruzual (1981), to estimate the colors of PG. If the model colors are accurate, only a small percent of faint galaxies lie in the predicted PG positions. Typical observed percentages range from 2% at B = 22.5 to 8% at B = 23.5; this translates to a total of about 1000 PG deg$^{-2}$ to B = 23.5. In contrast, galaxy count models that include evolution and use $\Omega = 1$ (e.g., Model C of Bruzual and Kron 1980 or Koo 1981) predict 20% at B = 22.5.

## 2.4 Redshift Surveys of PG Candidates

Another approach to finding PG has been to select candidates which can then be studied spectroscopically. We have already discussed one such survey in the previous section, in which the candidates were faint quasars. Another spectroscopic survey was that inspired through a suggestion by B. Tinsley at the 1977 Yale Conference on evolution of stellar populations that PG could be found even as bright as B = 20 if the redshift of formation was low ($z \leqslant 2$) and PG were very bright. The resulting redshift survey by Turner, Gunn, and Sargent (1981) is not well known but has been mentioned by Turner (1980) and Gunn (1982). Although the survey did not yield any PG, it was the first very-deep survey of faint non-radio galaxies. Redshifts from a SIT spectrograph on the 5-m were acquired for almost 60 faint galaxies, an impressive number even by today's standards with CCD's as detectors. R. Kron isolated the original total sample of 72 faint galaxies (B = 19 to 21) as PG candidates from an examination of three 4-m plates

taken with a blue filter. Since the image structures of PG are not
known, he decided to select a diverse range of morphological properties; the final sample includes very compact but slightly fuzzy
objects, large uniform surface brightness galaxies, apparently
disturbed systems with wisps and tails, and other objects with
peculiar shapes. Although a few redshifts were found to be in error
(see discussion of this sample by Koo 1985 ), their basic finding of
no PG has been confirmed by more recent deep redshift surveys (see
Section 3.3) and is consistent with other optical evidence which
suggests that evolution of field galaxies has been mild at best (see
reviews by Ellis 1982 and Kron 1982).

Another redshift survey of PG candidates by Kron, Koo, and Windhorst (1985) has been motivated by the proposal of Katgert, de Ruiter,
and van der Laan (1979) and van der Laan and Windhorst (1982) that the
faint (B ≳ 20) blue galaxies identified with low-flux radio sources
are the high-redshift counterparts to nearby radio sources seen as
luminous giant ellipticals. These counterparts are presumed to be
blue because of extensive star formation, which in turn would make
them bright enough to be visible at high redshifts; in other words,
these blue galaxies are PG. Multicolor photometry and spectroscopy of
several such blue candidates suggest that they have modest redshifts
generally less than $z' = 0.5$; the brighter blue galaxies almost always
exhibit peculiar morphology hinting of tidal interactions or even
mergers. Extrapolating to the blue radio galaxies too faint for spectroscopic redshifts and morphological classification, we suspect that
these are more likely to be the same population seen at higher redshifts rather than to be PG. This conclusion is also supported by the
correlation found between the compactness of the radio sources and
their color. Redshifts of fainter blue galaxies will settle the
question.

2.5 Surveys for PG using Lyman α Emission Lines

Since the problem of recognizing a PG is generally more severe than
that of detection, surveys which exploit the expected presence of
strong Lyman α emission lines (or other emission features or continuum
discontinuities, like the Lyman break at 912Å) are particularly powerful. Meier and Terlevich (1981) and Hartmann, Huchra, and Geller
(1984) have reported small Lyman α emission line fluxes in nearby
galaxies with extensive star formation, a finding consistent with the
presence of even a small amount of dust. Both works suggest that PG
may thus be difficult to detect in the Lyman α line; PG may, however,
not suffer such absorption during the earliest stages of star formation when sufficient metals and dust have yet to be produced. Also
encouraging has been the finding by Spinrad (private communication)
that the Lyman α emission lines found in the most distant radio
galaxies have equivalent widths up to 1000 Å! As already pointed out
by Partridge and Peebles (1967), 6% to 7% of the total luminosity of a
PG should be emitted in the Lyman α line alone. This translates to an
equivalent width of 70 Å in the rest frame or nearly 350 Å in the

observed frame at z = 4. Since the width of the line is expected to be narrow, say ≤600 km sec$^{-1}$ or 12 Å at z = 4, the line is almost 30 times stronger than the continuum! As previously mentioned, slitless spectroscopy, narrow-band imaging, or slit spectroscopy of specific objects or "blank" sky are all excellent techniques for both detection and recognition of this feature in PG.

The first two of such surveys were undertaken by Koo and Kron (1980). Both were mainly designed to detect Lyman α emission lines of PG (or quasars) in the near infrared (6000 Å to 9000 Å), corresponding to redshifts between 4 and 6.5, by using slitless spectroscopy with spectral resolutions of 100 Å. The first combined a "grism" (transmission grating attached to a prism) with hypersensitized infrared photographic plates taken at the 4-m prime focus. This setup reached a flux limit of $4 \cdot 10^{-15}$ ergs sec$^{-1}$ cm$^{-2}$ for emission lines from compact objects. The second survey replaced the plates with a CCD which reached a limit 6 times fainter but covered an area 1/20th as large (200 arcmin$^2$). Although several candidates were found, none proved to be PG when higher resolution spectra were acquired by H. Spinrad. At least for compact objects, these two surveys, when compared to those of Partridge (1974) or Davis and Wilkinson (1974), reached considerably fainter flux limits, covered more area, and possessed better recognition of PG.

Mackay (1985) has searched for PG by using image-tube plates with UBVRI broadband filters and five narrow band filters (100 Å each), the central one set to detect Lyman α of each of four quasars (0938+119, 1049-09, 1233-34, and OQ172). Each quasar was examined under good seeing conditions in hopes of detecting a Lyman α fuzz and a 3 arcmin diameter area surrounding each quasar was also searched to V = 23.5 for associated PG detected by their unusual colors. Both experiments gave null results but the idea is a good one that deserves further attention with the use of CCD's and higher redshift quasars.

If the redshift is not known, such narrow-band surveys require a large investment of exposures taken with many filters to obtain good spectral coverage, a technique suggested by Meier (1976) and Hogan and Rees (1979). Indeed Hartwick and Pritchet (1985) have already completed such a survey at CFHT using 100 Å bands covering 7300 Å to 9000 Å for a 6 arcmin$^2$ field. With CCD exposures of almost two hours per filter, their investment of over 35 hours makes this one of the deepest new PG surveys to date; their flux limit for detecting Lyman α is equivalent to a broadband continuum limit of R = 26. Most exciting of all is that they actually found 10 candidates! Within the last month (Feb. 1985), they have also used two nights of 4-m time at Kitt Peak National Observatory to check the nature of their candidates with the Cryocam system. We eagerly await their results.

A survey that in principle can achieve even fainter limits with a compromise in area is a long-slit survey of arbitrary patches of sky (Spinrad 1977, Hogan and Rees 1979, Koo and Kron 1980). The gain is made by using higher spectral resolution, say 12 Å, rather than the 100 Å resolutions of narrow-band filters. This gain is reduced at certain wavelengths, however, when one accounts for the difficulty of subtracting the noise of the strong emission lines of the night sky, especially in the near infrared. In the optimal case of working in smoother parts of the night-sky spectrum, one can reach a flux limit of $\sim 2 \times 10^{-17}$ ergs sec$^{-1}$ cm$^{-2}$ for detection of a single unresolved line in an hour exposure with a CCD spectrograph system like the Cryocam at Kitt Peak National Observatory. This system has a useable spectral range of 2500 Å, a resolution between 7 and 15 Å FWHM, and accomodates slits typically 2.5" x 300", i.e., 0.2 arcmin$^2$ of sky. This would translate to a broadband R ~ 26.5 for the object itself if indeed Lyman α contains 7% of the total flux.

Since 1981, R. Kron and the author have been using this setup to search for PG. To account for cosmic rays and other detector problems, at least two exposures are made, one slightly offset (10" to 30") from the other in position along the slit; in practice, a star visible on the TV acquisition systems is used as a reference. Areal coverage is extended by adding such pairs of exposures with slits oriented at new position angles but still centered on the reference star. The final coverage of the sky resembles the spokes of a wheel. A special "filter" is then applied to each pair of exposures (typically 2000 sec); these filters are designed to locate positive signals of approximately equal amplitude and having the offset spatial separation. Although analysis is still underway, a preliminary assessment of the data shows no evidence of line emission from numerous patches of sky.

Another long-slit approach has been taken by Cowie (1985), who is using auto-correlation techniques to extract PG signals from Cryocam exposures of an hour. This in principle is a powerful technique to extract overlapping spectra which all have a well-known and characteristic spectral distribution, namely strong Lyman α emission and a large Lyman continuum break. The data are still under analysis, but he estimates the limits to be $2 \cdot 10^{-17}$ erg sec$^{-1}$ cm$^{-2}$ for a total field size of one arcmin$^2$ (another 10 arcmin$^2$ is available).

2.6 Constraints on Models of Compact PG

What constraints do these surveys place on PG models? Fig. 1 of Davis (1980) is a useful graph that shows how the null results of various surveys limit the number-density of PG versus the flux from each object, which when multiplied together gives an average surface brightness. Our Fig. 1 is essentially an updated and rotated version of Davis' figure. Our R band refers to one centered on 7000 Å with equal sensitivity from 6000 Å to 8000 Å, which is a compromise between the actual bands adopted by different groups. To place the emission line surveys on the same scale, we have assumed that Lyman α is $(1 + z)70$ Å so that a line-flux limit at z = 5, i.e. around 7300 Å, is about 20% of the continuum R band limit that is actually plotted. Blue band limits for galaxies and quasars are assumed to be 1.5 mag and 0.5 mag, respectively, fainter than the equivalent R.

In addition to a variety of observations, Fig. 1 includes for illustration the range of values predicted by the dissipation models of Meier (1976) and revisions of the metal-production arguments given by Davis (1980). Without these revisions (see more detailed discussion by Cowie 1985), Davis would estimate, for formation redshifts $z \sim 5$, typical number densities, N, of 14,000 PG per $deg^2$, each with $R \sim 18$! This far exceeds the total number of observed galaxies. From the luminosity function suggested by Felten (1985), we adopt a fractional closure density 30 times smaller, $\Omega_L = 0.003$, by assuming an M/L of 3 and a luminosity density of $0.6 \times 10^8$ $L_\odot$ $Mpc^{-3}$ between $M_B = -20.5$ and $-22.5$ ($H_o = 50$ km $sec^{-1}$). Lowering the number density of galaxies by 2 to $0.001$ $Mpc^{-3}$ is compensated by our adoption of a low $q_o \sim 0.05$. Another major change is to presume that the ultraviolet light of PG would be absorbed by dust and reradiated to the infrared when the fraction of metals in the star-forming gas exceeds 1/100th solar. This value was estimated from finding an average optical depth $\sim 1$ when $2 \cdot 10^7$ $(R/10$ $kpc)^2$ $M_\odot$ of dust (with a mass absorption coefficient $\kappa = 10^5$ $cm^2$ $gm^{-1}$ in the UV) uniformly fills a sphere of radius R. The overall effect is to make PG appear fainter by a factor of 3000; our new value is marked as M5 in Figure 1, assuming the visible formation timescale is still close to $10^8$ yr. Yet the long-slit experiment still excludes this conservative estimate as well as much of the parameter space predicted by the models of Meier, but not by much. A more realistic timescale for

TABLE III. Symbols for Figure 1.

Redshift (z), Mass Density ($\Omega = 2$ $q_o$), Formation Timescale ($\tau$)

| | | |
|---|---|---|
| C1 | Galaxy Counts: | CCD (Seitzer and Tyson 1985) |
| C2 | Galaxy Counts: | CCD (Hall and Mackay 1984) |
| E | Extragalactic Background Light (Dube, Wickes, and Wilkinson 1977) equal to $\sim 1$ $S_{10}$ unit. | |
| G | PG Galaxy Candidates: UBVI Plates (Koo 1981) | |
| L | Long Slit: CCD (Cowie and Hu 1985, Koo and Kron 1985) | |
| M1 | $z = 5$ $\Omega = 2$ $\tau = 15 \cdot 10^8$ yr model (Meier 1976) | |
| M2 | $z = 2$ $\Omega = 2$ $\tau = 2 \cdot 10^9$ yr model (Meier 1976) | |
| M3 | $z = 7$ $\Omega = 0.06$ $\tau = 2 \cdot 10^9$ yr model (Meier 1976) | |
| M4 | $z = 16$ $\Omega = 0.06$ $\tau = 5 \cdot 10^8$ yr model (Meier 1976) | |
| M5 | $z = 5$ metals = 0.01 solar $\tau = 10^8$ yr $\Omega = 0.1$ $\tau = 3 \cdot 10^7$ yr (M5') $\tau = 3 \cdot 10^8$ yr (M5") $\tau = 3 \cdot 10^6$ yr (M5''') | |
| N1 | Narrow Band: CCD (Cowie 1985) | |
| N2 | Narrow Band: CCD (Hartwick and Pritchet 1985)* | |
| N3 | Narrow Band: Image-tube (Mackay 1985) | |
| Q | Quasar-like PG: CCD (Koo, Kron, and Cudworth 1985) | |
| R1 | Red PG: Photometry (Davis and Wilkinson 1974) | |
| R2 | Red PG: BVR Plates (Partridge 1974) | |
| R3 | Red PG: CCD Imaging (Loh and Wilkinson 1979) | |
| S1 | Slitless: I plates (Koo and Kron 1980) | |
| S2 | Slitless: CCD (Koo and Kron 1980) | |

*Limits assume 10 candidates are <u>not</u> PG.

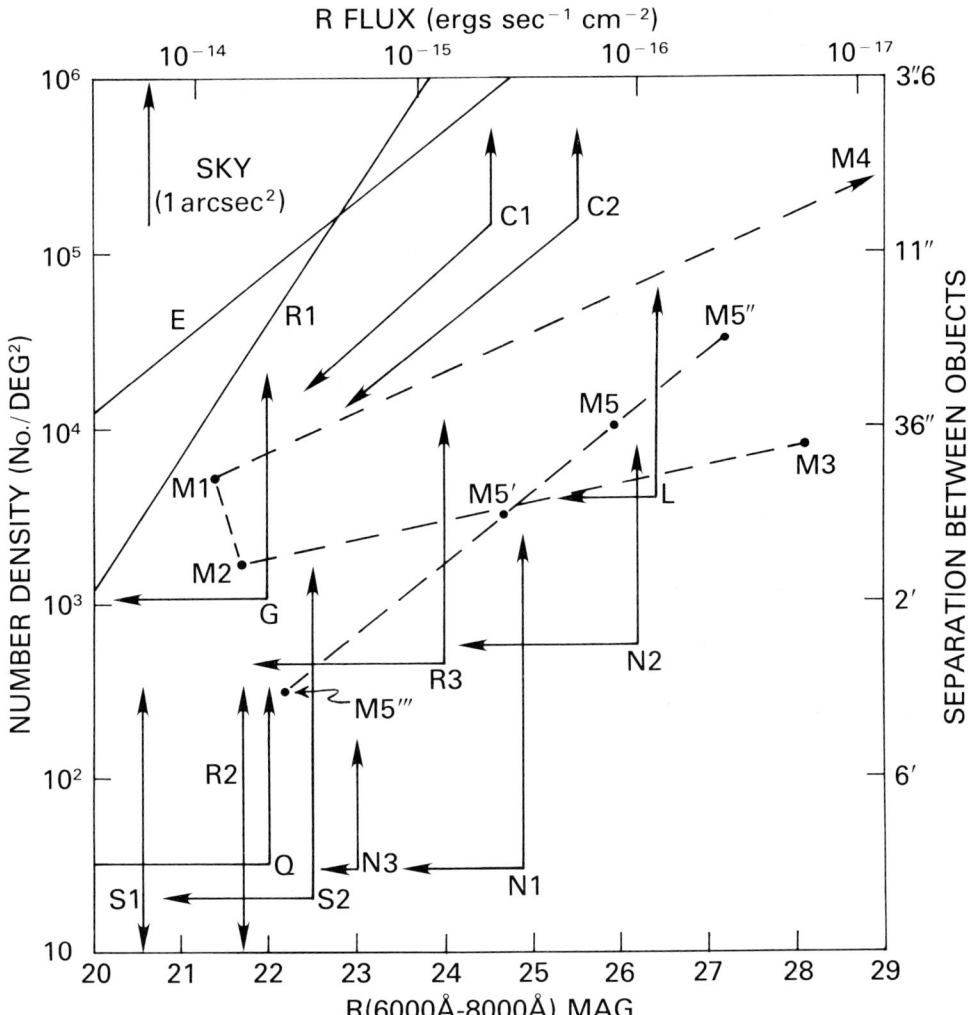

**Figure 1:** Surface density versus flux in 2000 Å wide band centered on 7000 Å. Observational limits (solid lines) exclude the parameter space to the upper-left. Models are marked with dashed lines. Table III identifies the labels. Surveys for Lyman α lines had R limits set to 5 times the line-flux limit.

forming 1/100th solar worth of metals is a few times $10^6$ yr (see M5'''), in which case brighter surveys provide stronger constraints. On the other hand, changing the timescale of formation to several times $10^8$ years (see point M5'') is enough to hide such PG. Adding intervening dust (Ostriker and Heisler 1984), increasing the sizes of PG, or enhancing the obscuration by internal dust by adopting small units or cells of star formation (since for a given amount of metals, the effective optical depth increases for smaller volumes) are other escapes from the conclusion that PG formation must have occurred at redshifts beyond 5.

## 3. OPTICAL SURVEYS USEABLE FOR PG SEARCHES

Many deep optical surveys that have been designed to explore problems unrelated to the search for PG can be very powerful and yet inexpensive methods to limit the properties of PG. Since the explicit searches already indicate that PG are at best faint and/or rare, PG, if found in the optical at all, may be the result of a serendipitous discovery. In general, any survey that limits the number and brightness of high redshift objects or the redshifts of faint objects can be useful for supplying data on PG. With large telescope time being so scarce, deep surveys designed only to look for PG with little guarantee of success would unlikely be allocated observing time.

### 3.1 Photometric Surveys

Deep photometric surveys of faint galaxies in the form of number counts are actually indirect probes of PG to the extent that models needed to interpret such data include high redshift galaxies undergoing extensive star formation. Unfortunately, unless additional information, such as colors or spectroscopic redshifts, are available, the contribution from PG to the counts are difficult to disentangle from lower redshift galaxies undergoing luminosity or density evolution. The total counts certainly provide solid upper limits to the surface density of PG to the depth of the survey.

The deepest surveys of galaxies are the broadband CCD images analyzed by Seitzer and Tyson (1985) and Hall and Mackay (1984). The latter survey reached a red limit of $R = 25.5$ and covered an area of 73 arcmin$^2$. The former survey reaches $B = 26$ for several fields totaling 130 arcmin$^2$ and includes good color information which show fainter galaxies to be bluer, a conclusion already known from brighter photographic surveys which use 4-m plates (Kron 1980, Koo 1981, Shanks et al. 1984). The total counts provide some but not stronger constraints on PG models (see Fig. 1), but when more color or hopefully redshift information becomes available for these very faint galaxies, the constraints on PG might reach interesting limits. Brighter galaxy counts ($B \leq 24$) and colors already suggest formation redshifts $z > 5$ in a low density universe (Bruzual and Kron 1980, Koo 1981).

Deep medium-band surveys, like those of Spillar and Loh (1984) of field galaxies or those of the Durham group of cluster galaxies (Couch et al. 1983), should provide better redshift discrimination than broadband surveys and should thus give more realistic upper limits to the number of PG. One danger, of course, is that PG may still be masquerading with colors similar to those of low redshift galaxies. In this case, narrow-band surveys, like that of faint field galaxies reported by Sharein at this conference, may identify PG by their strong emission lines. Similarly, Cowie (1985) has been searching for PG with deep images taken through a set of narrow band filters originally designed to examine the cooling flows in clusters of galaxies. As seen by the limits (N1) drawn in Fig. 1, this data complement well the long-slit surveys (L).

## 3.2 Slitless Surveys

Just as the slitless surveys of PG by Koo and Kron (1980) are useful in constraining the number of high-redshift quasars, the slitless surveys of quasars reported by Schneider, Schmidt, and Gunn (1983, 1984) are useful for finding PG. Their experiment uses a "grism"-CCD combination to search large areas of sky (3200 $arcmin^2$) with the Palomar 5-m telescope to a limit of R = 21.5. This is an improvement of a factor of 16 in area and with similar sensitivity to the CCD survey by Koo and Kron. So far no faint quasars with high redshift z > 2.7 have been found; presumably no PG have been discovered, either.

## 3.3 Redshift Surveys

The ideal experiments from which a serendipitous discovery of PG might be made are deep redshift surveys of faint objects. Especially promising would be quasar candidates which yield narrow emission lines or other indications of thermal emission when examined spectroscopically. So far, no PG have been found among the faint quasar surveys by Koo and Kron (1985), Marano et al. (1986) and Boyle et al. (1986). The latter survey is particularly impressive for it includes many hundreds of faint quasar candidates.

Redshift surveys of relatively bright field galaxies, like that of Turner et al. (1981) mentioned above (Sec 2.4), are unlikely to find many PG, but if very bright PG exist at all, even if very rare, they may be found in such surveys. Two such surveys are that of the Durham group as reported by Ellis (1983) and that by Koo and Kron as reported by Koo (1983a). Among a total sample of over two hundred random faint (B = 20 to 22) galaxies observed to date, none have yielded an unusually high redshift, say greater than z = 1.

In a similar manner, the deep redshift surveys of faint distant clusters (see review by Dressler at this conference) by Dressler and Gunn (1982, 1983), Dressler, Gunn, and Schneider (1985), Butcher and Oemler (1984), and Sharples et al. (1985) are all useful, since PG may exist among the non-cluster galaxies observed spectroscopically. Among a total of about 50 non-cluster members, none appear to be a PG.

Redshift surveys of faint optical identifications of radio sources provide another rich source of PG search material. Although the faint blue radio galaxies mentioned in Sec. 2.4 do not seem to be PG as suggested by Katgert et al. (1979), we cannot exclude the possibility that some of the very faintest blue galaxies are PG until more redshifts are measured. The redshifts $z > 1$ among the 3C radio galaxies acquired by Spinrad and co-workers over the last few years (Spinrad and Djorgovski 1984) give hope that some radio sources are pointers to unusual galaxies at high redshift. Of special interest would be galaxies which are extremely faint or even unidentified. As suggested by Sunyaev et al. (1978), PG may emit radio emission as well as optical; if all PG were also associated with an active nucleus, ultra-deep radio surveys reaching sub-mJy levels like those of Condon and Mitchell (1982), Windhorst et al. (1985), or Fomalont et al. (1984) may contain PG candidates.

3.4 PG Candidates ?

Prime suspects for PG are high-redshift objects possessing only narrow emission lines. A few such objects have already been discovered, all by accident as part of non-PG surveys. H1340 No. 10 with a redshift $z = 2.47$ and $V = 19.5$ was reported by Foltz et al. (1983) to have narrow Lyman $\alpha$ and CIV emission lines with FWHM of 1500 km sec$^{-1}$. A bona-fide PG is expected to have widths less than half of this, so this object may actually be more like a Seyfert galaxy.

Another very interesting candidate might be the gravitational lens system MG 2016 + 112 with $z = 3.27$ and $m = 22.5$ reported by Lawrence et al. (1984) and Schneider et al. (1985) to have five emission lines with widths that are unresolved (<1000 km sec$^{-1}$). This object was found as part of a radio survey for gravitational lens candidates and possesses a radio power 1000 times stronger than that of an average Seyfert. One of the optical components has been found to dim by 0.3 m over a 9.5 month interval. Both of these properties would argue against this object being an ordinary PG, but it can also be argued that this is not an ordinary quasar either, since the emission lines are very narrow.

Hazard (1985) has reported finding an object with $z > 2$ which has very narrow emission lines with widths less than 600 km sec$^{-1}$. He called it a high luminosity H-II region or young galaxy, i.e. a primeval galaxy. Its redshift is probably too low to represent the formation epoch of all galaxies, but this object may well be one of the few prototypes of PG available for detailed study.

## 4. CONCLUSION

This review has attempted to provide the reader with a flavor of the numerous optical approaches used in searching for PG. The diversity of techniques, which range from counting faint galaxies and quasars to the powerful spectroscopic surveys of blank sky, reflects our ignorance of the properties of such objects. Although many surveys have failed to find PG, the null results are still important in providing clues to where PG may be hiding and in constraining existing models of PG. Current limits are reaching interesting levels: robust estimates based upon the amount of observed metals in galaxies suggest that galaxy formation occurred at redshifts beyond ~5. The more detailed models of Meier are also close to being excluded for reasonable choices of the time-scale of initial star formation and deceleration parameter, $q_o$, for formation redshifts less than 5. Moreover, observations show a lack of PG among samples of faint (B > 20) quasars, whose number counts are already becoming more shallow with magnitude; thus large numbers of PG are not masquerading in size and color as quasars, a finding that also suggests galaxy formation occurred at redshifts higher than ~4. These and other models, however, are nearly 10 years old and do not reflect many recent developments in particle physics, for example, or in new scenarios for galaxy formation, interaction, and evolution. Even more tantalizing are hints that objects with properties close to those expected from PG, namely narrow emission lines, are being found at redshifts beyond 2. Several surveys, most notably that of Hartwick and Pritchet, may already have detected genuine PG but slit spectroscopy is needed for confirmation. What are the faintest or empty-field radio sources? What are the redshifts of the galaxies found in deep CCD images? Always lurking behind any extensive effort to find PG is the possibility that PG are invisible in the optical and that their discovery must await future deep infrared or X-ray surveys. The search, though becoming more demanding as brighter surveys fail, continues, often as part of other surveys that push the frontiers, for if PG are found, their detection promises to be one of the most exciting and important discoveries in observational cosmology.

## 5. ACKNOWLEDGEMENTS

Many people have generously provided information about their surveys before completion, including E. Turner, D. Hartwick, C. Pritchet, C. Mackay, H. Spinrad, and L. Cowie. Others have contributed to this review through their suggestions, discussions, and encouragement, including R. Kron, M. Fall, B. Partridge, and especially L. Cowie. To all I give my thanks and to the seekers of PG, best of luck.

## 6. REFERENCES

Binney, J. 1977, Ap. J., **215**, 483.
Boksenberg, A. and Sargent, W. L. W. 1983 in Proceedings of 24th Liege Astrophysical Coloquium, 'Quasars and Gravitational Lenses' p. 500.
Bookbinder, J., Cowie, L. L., Krolik, J. H., Ostriker, J. P., and Rees, M. 1980, Ap. J., **237**, 647.
Boughn, S. P., Saulson, P. R., and Uson, J. M. 1985 preprint.
Boyle, B. J., Fong, R., Shanks, T., and Peterson, B. A. 1986 Proceedings of Trieste Symposium, "The Structure and Evolution of Active Galactic Nuclei" ed. G. Giuricin, F. Mardirossian, M. Mezzetti, and M. Ramella (Reidel, Dordrecht).
Braccesi, A., Zitelli, V., Bonoli, F., and Formiggini, L. 1980 A. A., **85**, 80.
Bruzual, G. A. 1981 Ph.D. Dissertation, Univ. of Calif., Berkeley.
Bruzual, G. A. and Kron, R. G. 1980 Ap. J., **241**, 25.
Butcher, H. and Oemler, G. 1984 Nature, **310**, 31.
Condon, J. J. and Mitchell, K. J. 1982 A. J., **87**, 1429.
Couch, W. J., Ellis, R. S., Godwin, J., and Carter, D. 1983 M.N.R.A.S., **205**, 1287.
Cowie, L. L. 1985 private communication.
Davis, M. 1980 in IAU Symp. No. 92 on 'Objects of High Redshift', eds. G. O. Abell and P. J. E. Peebles (Reidel, Dordrecht) p. 57.
Davis, M. and Wilkinson, D. T. 1974 Ap. J., **192**, 251.
Dressler, A. and Gunn, J. E. 1982 Ap. J., **263**, 533.
    1983 Ap. J., **270**, 7.
Dressler, A., Gunn, J. E., and Schneider, D. P. 1985 Ap. J., in press.
Dube, R. R., Wickes, W. C., and Wilkinson, D. T. 1977 Ap. J. (Letters), **215**, L51.
    1979 Ap. J., **232**, 333.
Ellis, R. G. 1982 in Proceedings of Erice School on The Origin and Evolution of Galaxies' ed. B. J. T. Jones and J. E. Jones (Reidel, Dordrecht).
    1983 in IAU Symp. No. 104 on 'Early Evolution of the Universe and Its Present Structure' ed. G. O. Abell and G. Chincarini (Reidel, Dordrecht) p. 87.
Felten, J. E. 1985, preprint.
Foltz, C., Weymann, R., Hazard, C., and Turnshek, D. 1983 P.A.S.P., **95**, 117.
Fomalont, E. B., Kellermann, K. I., Wall, J. V., and Weistrop, D. 1984 Science, **225**, 23.
Gott, J. R., Gunn, J. E., Schramm, D. N., and Tinsley, B. M. 1974 Ap. J., **194**, 543.
Gunn, J. E. 1982 in 'Astrophysical Cosmology', Proceedings of the Vatican Study Week on Cosmology and Fundamental Physics, ed. H. A. Brück, G. V. Coyne, and M. S. Longair (Specola Vaticana, Vatican) p. 233.
Hall, P. and Mackay, C. D. 1984 M.N.R.A.S., **210**, 979.
Hartmann, L. W., Huchra, J. P., and Geller, M. J. 1984 Ap. J., **287**, 487.

Hartwick, D. and Pritchet, C. 1985 private communication.
Hazard, C. 1985 Proceedings of 1984 Manchester Conference on 'Active Galactic Nuclei' ed. J. E. Dyson.
Hogan, C. J. and Rees, M. J. 1979 M.N.R.A.S., **188**, 791.
Ikeuchi, S. 1981 P.A.S.J., **33**, 211.
Katgert, P., de Ruiter, H. R., van der Laan, H. 1979 Nature, **280**, 20.
Kaufman, M. 1975 Ap. and Space Sci., **33**, 265.
   1976 Ap. and Space Sci., **40**, 369.
Kaufman, M. and Thuan, T. X. 1977 Ap. J., **215**, 11.
Koo, D. C. 1981 Ph.D. Dissertation, Univ. of Calif., Berkeley.
   1983a in IAU Symp. No. 104 on 'Early Evolution of the Universe and Its Present Structure' ed. G. O. Abell and G. Chincarini (Reidel, Dordrecht) p. 105.
   1983b in Proceedings of 24th Liege Astrophysical Coloquium, 'Quasars and Gravitational Lenses' p. 240.
   1985 A. J., **90**, 418.
   1986 in Proceedings of Trieste Symp. on 'The Structure and Evolution of Active Galactic Nuclei' ed. G. Giuricin, F. Mardirossian, M. Mezzetti, and M. Ramella (Reidel, Dordrecht).
Koo, D. C. and Kron, R. G. 1980 P.A.S.P., **92**, 537.
   1982 A. A., **105**, 107.
   1985 in preparation.
Koo, D. C., Kron, R. G., and Cudworth, K. M. 1985 P.A.S.P., submitted.
Kron, R. G. 1980 Ap. J. Suppl., **43**, 305.
   1982 Vistas in Astronomy, **26**, 37.
Kron, R. G., Koo, D. C., and Windhorst, R. A. 1985 A. A., **146**, 38.
Larson, R. B. 1974 M.N.R.A.S., **166**, 585.
Lawrence, C. R., Schneider, D. P., Schmidt, M., Bennett, C. L., Hewitt, J. N., Burke, B. F., Turner, E. L., and Gunn, J. E. 1984 Science, **223**, 46.
Loh, E. D. and Wilkinson, D. T. 1979 (see private comm. in Davis 1980).
Mackay, C. D. 1985 private communication.
Marano, B., Zamorani, G., and Zitelli, V. 1986 in Proceedings of Trieste Symp. on 'The Structure and Evolution of Active Galactic Nuclei' ed. G. Giuricin, F. Mardirossian, M. Mezzetti, and M. Ramella (Reidel, Dordrecht).
Matilla, K. 1976 A. A., **46**, 77.
McDowell, J. 1985 in preparation.
Meier, D. L. 1976 Ap. J., **207**, 343.
Meier, D. L. and Terlevich, R. 1981 Ap. J. (Letters), **246**, L109.
Ostriker, J. P. and Cowie, L. L. 1981 Ap. J. (Letters), **243**, L127.
Ostriker, J. P. and Heisler, J. 1984 Ap. J., **278**, 1.
Paresce, F. and Jakobsen, P. 1980 Nature, **288**, 119.
Partridge, R. B. 1974 Ap. J., **192**, 241.
Partridge, R. B. and Peebles, P. J. E. 1967 Ap. J., **147**, 868.
Peebles, P. J. E. 1971 Physical Cosmology (Princeton University Press, Princeton) p. 63.

Schneider, D. P., Lawrence, C. R., Schmidt, M., Gunn, J. E., Turner, E. L., Burke, B. B., and Dhawan, V. 1985 preprint.
Schneider, D. P., Schmidt, M., and Gunn, J. E. 1983 B.A.A.S., **15**, 957. 1984 B.A.A.S., **16**, 488.
Searle, L. and Sargent, W. L. W. 1972 Ap. J., **173**, 25.
Seitzer, P. and Tyson, J. A. 1985 B.A.A.S., **17**, 580.
Setti, G. and Woltjer, L. 1982 in 'Astrophysical Cosmology' Proceedings of the Vatican Study Week on Cosmology and Fundamental Physics, ed. H. A. Brück, G. V. Coyne, M. S. Longair (Pontificiae Academiae Scientiarum, Vaticano) p. 315.
Shanks, T., Stevenson, P. R. F., Fong, R., MacGillivray, H. T. 1984 M.N.R.A.S., **206**, 767.
Sharples, R. M., Ellis, R. S., and Couch, W. J., and Gray, P. M. 1985 M.N.R.A.S., **212**, 687.
Spillar, E. and Loh, E. D. 1984 B.A.A.S., **16**, 488.
Spinrad, H. 1977 in 'Evolution of Galaxies and Stellar Populations' ed. B. M. Tinsley and R. B. Larson (New Haven, Yale) p. 301.
Spinrad, H. and Djorgovski, S. 1984 Ap. J. (Letters), **285**, L49.
Spinrad, H. and Stone, R. P. S. 1978 Ap. J., **226**, 609.
Sunyaev, R. A., Tinsley, B. M., and Meier, D. L. 1978 Comments Astrophys., **7**, 183.
Tinsley, B. M. 1977 Ap. J., **211**, 621.
1978 Ap. J., **220**, 816.
1980a Fund. of Cosmic Physics, **5**, 287.
1980b Phil. Trans. R. Soc. London, **A296**, 303.
Toller, G. N. 1983 Ap. J. (Letters), **266**, L79.
Turner, E. 1980 in 'Objects of High Redshift', IAU Symp. No. 92, eds. G. O. Abell and P. J. E. Peebles (Reidel, Dordrecht) p. 71.
Turner, E., Gunn, J. E., and Sargent, W. L. W. 1981 preprint.
van der Laan, H. and Windhorst, R. A. 1982 in 'Astrophysical Cosmology' Proceedings of the Vatican Study Week on Cosmology and Fundamental Physics, ed. H. A. Brück, G. V. Coyne, M. S. Longair (Pontificiae Academiae Scientiarum, Vaticano) p. 349.
Windhorst, R. A., Miley, G. K., Owen, F. N., Kron, R. G., and Koo, D. C. 1985 Ap. J., **289**, 494.
Wyse, R. F. G. 1985 preprint.

NEW OBSERVATIONS OF GALAXY NUMBER COUNTS

P.R.F. Stevenson, T. Shanks and R. Fong
Physics Department
University of Durham
England

1. INTRODUCTION

A well determined galaxy number-magnitude, n(m), relation contains important information on both galaxy luminosity evolution and cosmological world models and there have therefore been many recent attempts to observationally determine the form of n(m) to faint limits, in both blue and red passbands (Kron 1978, Tyson and Jarvis 1979, Peterson et al 1979, Koo 1981, Shanks et al 1984a; hereafter SSFM). However, the observational results from different authors in different fields, particularly in the blue passband, showed a spread in the observed n(m) relation at faint blue magnitudes ($B \sim 23^m$) of approximately one magnitude (see SSFM Fig. 11). More surprisingly SSFM also found that the form of n(m) seems to be not much better determined at bright ($B \sim 17^m$) magnitudes. This creates a problem for the faint count interpretation since the normalisation of the faint count models then become less certain.

SSFM suggested several reasons for these differences between different authors' counts. They could either be artificial, caused by zeropoint errors in the magnitude scales or by patchy Galactic obscuration; or they could be real, representing actual inhomogeneities in the galaxy distribution caused by large scale clustering. This latter possibility can be addressed by extending the area of sky on which the counts are based to determine the best overall average n(m) relation. In this paper we significantly extend the original SSFM survey area at bright limits using COSMOS (Stobie et al 1979) machine measurements of UKST plates and we discuss a more modest extension at faint limits using similar measurements of AAT plates.

2. DATA

2.1. UKST Data

The UKST surveys have now been extended over seventeen fields in the $b_J$ passband and cover an area of sky of $\sim 355$ square degrees, some thirty times larger than the original SGP field of SSFM. This data defines the n(m) relation between $16 \leqslant b_J \leqslant 20.5$ mag. Of these seventeen fields five lie adjacent to the original SSFM field at the SGP (J3721) and all six plates cover a total area of $\sim 110$ square degrees. The five new fields were calibrated using J3721 as a master catalogue (see SSFM for a detailed discussion of the calibration of this plate) and then comparing galaxy photometry on the small overlap areas. The counts for these fields have been presented by Shanks et al (1984 b; hereafter SSFM2) and show excellent agreement among themselves, suggesting that the original twelve square degree area of SSFM was representative in its number-count characteristics.

In order to strengthen this conclusion 11 more UKST $b_J$ fields have been studied which cover a further 245 square degrees of sky. Most (10) of these fields were chosen for the Durham galaxy redshift surveys (DRSS), 2 fields from the original Durham/AAT (DARS) Survey, and 8 new fields for the DARS extension. These fields are scattered across both the southern and northern galactic caps and should hence offer an excellent opportunity to check the representativeness of the SGP counts and the general isotropy of the galaxy distribution to relatively faint ($b_J \leqslant 21$ mag) limits.

All of the new UKST fields have been independently calibrated using CCD observations of galaxies. These observations not only allow the zero-point to be calculated to ± 0.1 mag accuracy, but also allow a test of the relative magnitude scale of the COSMOS machine to be made at bright magnitudes. It is important that no scale errors are present in the galaxy photometry because this would introduce an apparent change in the slope of the n(m) relation and hence affect the interpretation of the results (see Section 3). CCD magnitudes are plotted against COSMOS machine magnitudes for some bright galaxies on plates J3192 and J3390 in Figure 1. This comparison shows no evidence of scale errors between magnitudes of $16 < b_J < 18.5$ mag. Further work by Metcalfe (private communication) has shown that there are no scale errors to $b_J = 20$ mag on any of the eleven new fields studied here. The linearity of the magnitude scales on the SGP fields to these limits have been discussed in detail by SSFM and Stevenson (1985).

2.2. AAT Data

The AAT data of SSFM has now been extended to cover another two fields; the Pavo field (comprising plates J1634 and R1635) and a field centred on the distant cluster 0024 + 1654 (comprising plates J1747 and R1748). These plates enable the n(m) relation to be observed in the range $20.5 < b_J < 24$ mag, and $18.5 < r_F < 22$ mag. The total area of sky now surveyed to these depths is $\sim 1.6$ square degrees, a factor of four greater than the original SSFM area.

The calibration of the SSFM AAT field at the SGP (comprising plates J1888 and R1996) has been discussed in detail by SSFM and SSFM2. It was demonstrated there that the COSMOS magnitudes are linear and that they can be compared to pseudo-total magnitudes of the type used by Kron (1978). These results held in both the $b_J$ and $r_F$ passbands, in the magnitude ranges quoted above.

The Pavo field plate measurements and their CCD calibration have been described in detail by Stevenson et al (1985). It was shown there that the Pavo field magnitudes should also be close ($\leqslant 0.1$ mag) to total. It is very important that our magnitudes are close to total since the counts will be later compared to computer models based on total magnitudes.

The second newly studied AAT field has been calibrated using the 0024+1654 galaxy cluster photometry of Couch (1981). This photometry has recently been checked by CCD observations (Metcalfe; private communication) and a good agreement both in zero-point and relative magnitude scale was obtained. The magnitudes measured on this field can therefore also be compared to total magnitudes since Couch also measured Kron-type total magnitudes, in both $b_J$ and $r_F$ passbands.

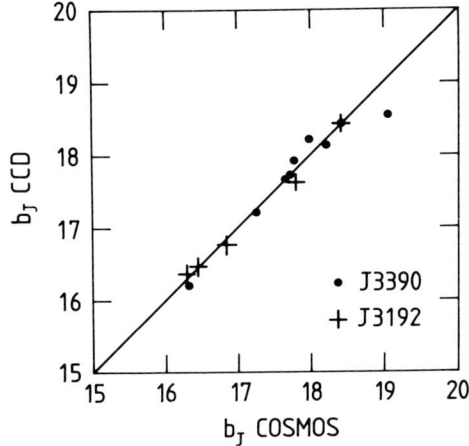

Figure 1. Comparison between COSMOS photographic galaxy photometry and CCD photometry for a sample of UKST galaxies.

## 3. OBSERVED COUNTS

### 3.1 UKST Counts

The $b_J$ galaxy counts in each of the eleven newly studied fields are shown in Fig. 2. It can be seen that in general they show very similar n(m) relations to the SGP counts of SSFM (dashed line in Fig. 2) with only three exceptions. Of these exceptions two fields (J3192, J3390) show an excess of galaxies at intermediate ($b_J \sim 18$ mag) magnitudes. On J3192 this excess is known to be at least partly caused by the Serpens-Virgo galaxy cloud, however on J3390 no large scale structure of this type has been reported. The excess is too large to be caused by uncertainties in the magnitude zero-point (see above and Fig. 1) and indeed if the counts were shifted faintwards in order to remove the observed excess they would be discrepant at faint magnitudes. It may therefore be possible that a super-cluster is also present in the J3390 area and this possibility is discussed in detail by Stevenson (1985). The other exception is that of plate J651 where the counts are seen to be consistently low at all magnitudes. This effect could most naturally be accounted for by absorption by dust in this field, assuming again that the zero-point is accurately determined. However, since we have no measure of whether there is absorption present in this field, we shall leave the counts as they stand.

The fact that only three out of 17 fields show significantly different n(m) relations to those at the SGP is excellent evidence that the original area of SSFM was representative in its galaxy counts. This is shown quantitatively in Fig. 3a where all UKST $b_J$ counts have been ensembled and plotted together with the original counts of SSFM. This result implies that the original SSFM UKST counts in the $r_F$ passband should also be representative of the true $r_F$ n(m) relation.

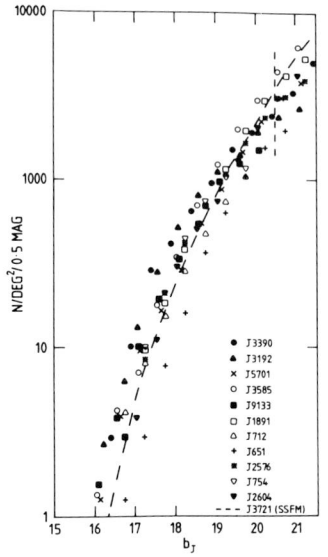

Figure 2:

Galaxy counts in 11 separate UKST fields. Also shown, represented by a dashed line, are the original SGP counts of SSFM. The vertical dashed line indicates the $b_J$ = 20ᵐ5 limit of reliability for the UKST counts.

## 3.2 AAT Counts

The AAT counts obtained from the two newly studied fields together with the counts of SSFM are shown in Fugure 3a and 3b for the $b_J$ and $r_F$ passbands respectively. Unlike the UKST counts there are still large discrepancies between the counts obtained in each field. However, this result was partly to be expected since at least one field (0024) was chosen for study because a large amount of absorption had been measured in this area by Couch (1981). These counts could therefore be used to confirm this measurement by seeing how low they are relative to the SGP counts (which come from an area of very little absorption, see SSFM).

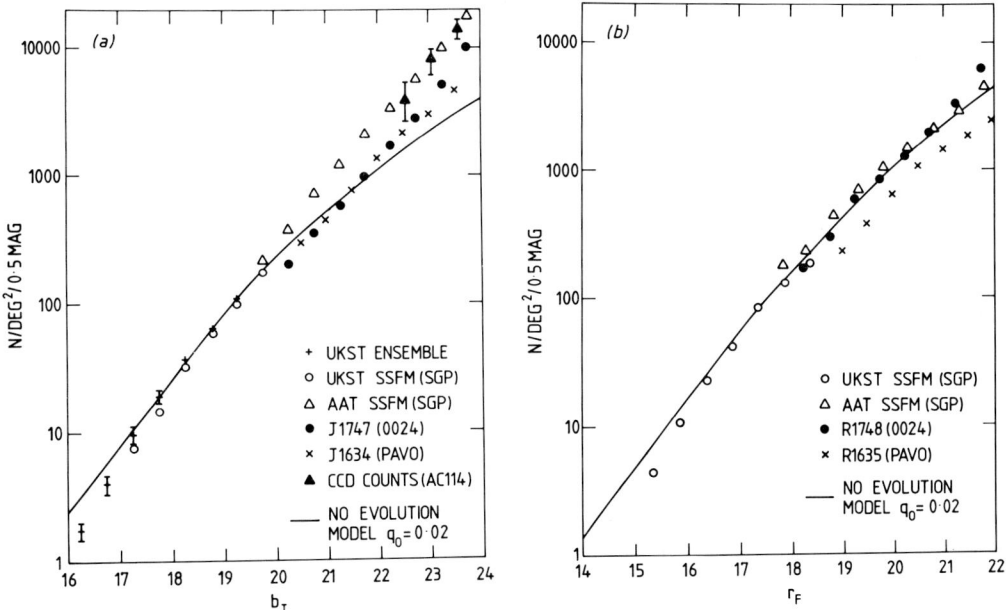

Figure 3: a) UKST and AAT differential galaxy number counts in the $b_J$ passband. Also shown for comparison is the no-evolution standard model of SSFM.

b) As for Figure 3a but in the $r_F$ passband.

The observed counts on the Pavo field can also be seen to be lower than those of the SGP by $\sim 0.7$ mag in the $b_J$ passband and $\sim 0.4$ mag in the $r_F$ passband. These discrepancies are too large to be caused by zero-point errors ($\pm 0.1$ mag) or isophotal effects and it has been argued by Stevenson et al (1985) that the effects of absorption was again the most likely cause. The fact that only half of the implied absorption is required in the $r_F$ passband is consistent with this idea. This amount of absorption is however larger than that expected from the usual galactic extinction law, but it is interesting to note that Couch (1981) has measured a significant amount of absorption ($\sim 0.5$ mag in $b_J$; estimated by comparing the field galaxy colour distributions to those near the poles) in a field only 4 degrees away from the Pavo field. The assumption will therefore be that an absorption of this amount is not unreasonable and for the purposes of interpretation the counts will be corrected accordingly.

As noted above, Couch (1981) has measured a significant absorption in the 0024 field ($\sim 0.8 \pm 0.3$ mag in $b_J$) and the counts in this field may be used to check this result. Indeed it is found that the observed n(m) relation in the $b_J$ passband in this field is lower than that of the

SGP by $\sim 0.6 \pm 0.1$ mag. In the $r_F$ passband the observed n(m) relation is only $\sim 0.1$ mag lower than at the SGP. However, to within the zero-point uncertainties ($\pm 0.1$ mag) and the uncertainty in the amount of absorption measured by Couch ($\pm 0.3$ mag), our results are consistent with Couch's. It can be seen from Figure 3a and b that if the 0024 counts were corrected for the absorption measured by Couch then a reasonable agreement with the SGP counts would be obtained.

As a final check on how absorption affects the form of the n(m) relation, counts have been obtained in another field (AC114) studied by Couch, where he has measured no absorption. The UKST counts from this field (plate J2604) are shown in Fig. 2 and agree with those of the SGP where very little absorption is present. Unfortunately no COSMOS measurements have yet been made of an AAT plate in this area. However, Couch and Pence (1985) have obtained counts to $\sim 24$ mag in $b_J$ using CCD observations. These counts are also plotted in Fig. 3a and are seen to give fairly good agreement at faint magnitudes with those of SSFM at the SGP.

Thus although at faint limits the counts are less well determined due to the effects of absorption in the two new fields studied here, if the counts are corrected by the amount of absorption measured independently by Couch (1981) then good agreement with the SGP counts is obtained. The counts of SSFM are therefore thought to be a reasonable representation of the true n(m) relation in the $b_J$ and $r_F$ passbands.

## 4. INTERPRETATION

The modelling of the n(m) relation has been discussed in detail by SSFM and SSFM2. It was shown that the number-count models depend chiefly on the K-corrections the assumed mix of galaxy types and the luminosity function. The parameters assumed here are the same as the standard model of SSFM.

It was also shown by SSFM that the normalization of the models at bright magnitudes ($b_J \sim 18$ mag) is very important, since this will determine how much evolution is required at faint magnitudes. The fact that the UKST ensemble agrees with the normalization of the SSFM no-evolution model prediction in Fig. 3a, in the range $17 < b_J < 19$ mag, is excellent evidence that the SSFM models were correctly normalized.

Since it was shown in the previous section that the observed n(m) relations of SSFM are a fair representation of the true n(m) relation at faint magnitudes the conclusions of SSFM still hold. These may be summarized as follows:

 i) The n(m) relation in the $b_J$ passband shows evidence for a large amount of luminosity evolution (see SSFM equation 15). Even assuming a fainter luminosity function for late type galaxies such as that used by Koo (1981), which decreases the average redshift of galaxies in the model and hence produces a steeper number-count slope, the model still only predicts 50 per cent of the observed galaxies at $b_J \sim 23$ mag. The difference between the data and the no-evolution model (see Fig. 3a) is too large to be explained by any non-evolving model.

 ii) In the $r_F$ passband the counts show evidence for smaller amounts of evolution. This result is expected since evolutionary models such as those of Bruzual (1981) predict that the luminosity of a galaxy should evolve faster at bluer wavelengths. In fact in the $r_F$ passband the effects of $q_o$ are quite significant and assuming that the red models are well determined (see SSFM), good constraints on possible combinations of evolution and $q_o$ can be obtained. Using single-burst evolutionary models of early-type galaxies (Tinsley 1978), which dominate the red counts, the n(m) models suggest that $q_o \lesssim 1$.

Acknowledgements

We thank the COSMOS group at the Royal Observatory, Edinburgh, especially Dr. H.T. MacGillivray for the assistance provided and the UK Schmidt and the Anglo-Australian telescope units for the use of photographic plates. We also wish to acknowledge Drs. W. Couch, W. Pence and N. Metcalfe for the use of their CCD data prior to publication.

References

Bruzual, A.G., 1981. Ph.D. thesis, University of California, Berkeley.
Couch, W.J., 1981. Ph.D. thesis, Australian National University.
Couch, W.J. and Pence, W., 1985. In preparation.
Koo, D.C., 1981. Ph.D. thesis, University of California, Berkeley.
Kron, R.G., 1978. Ph.D. thesis, University of California, Berkeley.
Peterson, B.A., Ellis, R.S., Kibblewhite, E.J., Bridgeland, M., Hooley, T., and Horne, D., 1979. Astrophys. J. Letters, 233, L109.
Shanks, T., Stevenson, P.R.F., Fong, R., and MacGillivray, H.T., 1984a. Mon.Not.R.astr.Soc., 206, 767.
Shanks, T., Stevenson, P.R.F., Fong, R., and MacGillivray, H.T., 1984b. Astronomy with Schmidt Type Telescopes, IAU Colloquium No. 78, ed. Cappaccioli, M., p499-505, Reidel, Dordrecht, Holland.
Stevenson, P.R.F., 1985. Ph.D. thesis, University of Durham.
Stevenson, P.R.F., Shanks, T., Fong, R., and MacGillivray, H.T., 1985. Mon.Not.R.astr.Soc., 213, 953.
Stobie, R.S., Smith, G.M., Lutz, R.K., and Martin, R., 1979. Image Processing in Astronomy, ed. Sedmak, G., Cappacioli, H., and Allen, R.J., Observatorie di Trieste, Italy, p48.
Tinsley, B.M., 1978. Astrophys. J., 222, 14.
Tyson, J.A., and Jarvis, J.F., 1979. Astrophys. J. Letters, 230, L153.

DISCUSSION

BRUZUAL: Could you remind us of how you take into account the luminosity evolution in your models? Is it the same for all galaxy types? How do you estimate the rate of luminosity evolution? For the high Z galaxies in your sample the colour evolution (neglected) can be important and, if included, may change the choice of cosmological parameters for your best fit.

STEVENSON: In the $b_J$ passband we fit simple empirical models for the change in galaxy brightness, $\Delta M$, as a function of redshift. This polynomial expression is then used as a modification to the assumed K-corrections. Due to uncertainties in the $b_J$ passband parameters (see SSFM) this empirical relation is not fitted to theoretical models. In the $r_F$ passband where the parameters are much better determined, the rate of luminosity evolution predicted by Tinsley's (1978) single-burst evolutionary models of early-type galaxies is assumed (which is assumed to be similar to first-order for E, So, Sab and Sbc type galaxies in $r_F$, with no-evolution assumed in later types). This theoretical relation can then be fitted, together with an appropriate value of $q_o$, to the empirical $\Delta M(Z)$ relation obtained from the observed $r_F$ counts.

KOO: In the last Erice Conference Ellis claimed that the normalization of Bruzual and Kron (1980) of 1.5 gal., between 14.5<J<15.5 was too high and should be lowered to 0.9 gal/mag. What normalization did you adopt?

STEVENSON: We also adopt the higher normalization of 1.5 gal. between $14.5 < b_J < 15.5$.

AARONSON: Since you are worried about reddening, have you tried to see if any of your fields lie near regions of infrared cirrus which IRAS found to be located at quite high galactic latitudes?

STEVENSON: We have looked at the IRAS data but have not finalized the results of this analysis at the present time.

## VI. CHEMICAL EVOLUTION OF GALAXIES AND MISCELLANEA

R. GÜSTEN
The chemical evolution of galaxies

M. TOSI, A. I. DIAZ
Nitrogen and oxygen evolution in nearby spiral galaxies

C. FORIERI
Effects of metal-dependent stellar models on the yield of nitrogen

M. A. SHAW, G. F. GILMORE
Surface brightness distributions in two edge-on spiral galaxies

THE CHEMICAL EVOLUTION OF GALAXIES

R. Güsten
Max-Planck-Institut für Radioastronomie
Auf dem Hügel 69
D-5300 Bonn 1
F.R.G.

SUMMARY. After a brief introduction into the evolution of spiral galaxies, the observational evidence for temporal and spatial abundance variations across the galactic disk is discussed. The theoretical framework of chemical evolution models is summarised, followed by a critical discussion of available constraints. Solutions applying to the special cases of the solar vicinity and the galactic abundance gradients are presented. We summarize evidence for a bimodal nature of star formation, and emphasize that if in regions and periods of high star formation activity the formation of low-mass stars is suppressed, a self-consistent model of star formation and chemical evolution of the Milky Way disk can be presented.

I. INTRODUCTION

The main purpose of this contribution is to review our concept of the chemical evolution of galaxies. There have been substantial improvements over the last few years of both the observational constraints and the theoretical framework of models describing the spatial and temporal gradients in the chemical composition of matter. Nevertheless the puzzle of chemical evolution is caused by numerous incompletely understood processes, with so many relevant input data not yet available (or uncertain) that we are far from possessing a generally accepted 'unique' solution. By necessity, this presentation is incomplete and, due to the huge amount of data available, is biased towards the characteristics of our own galaxy. For earlier reviews on the subject see Tinsley (1980) and Pagel and Edmunds (1981); those aspects of chemical enrichment that are related to radio astronomy (e.g. the isotope studies) have been summarised recently by Güsten and Mezger (1983). In Section II we present the general scenario of chemical evolution of spiral galaxies, and sum up the observational evidence for chemical enrichment of matter in Sect. III. The mathematical framework is introduced in Sect. IV, followed by a brief discussion of available constraints. In Sect. V we advocate a bimodal scenario of star formation, with high and low mass stars forming at different sites under different physical conditions. Model solutions for the evolution of the Milky Way disk are developed in Sect. VI, with special emphasis given to the question of what causes galactic abundance gradients across the disk.

## II. THE CONCEPT OF CHEMICAL EVOLUTION

Since the collapse of the protogalaxy $\sim$15 Gyr ago, stars have formed continuously out of the interstellar gas, forming first the spheroidal halo component (mass >10" $m_\odot$) and later the disk (6-7·10$^{10}$ $m_\odot$). Today, star formation is mainly confined to the narrow layer of interstellar gas (of 60-100 pc scale height) inside the solar circle ($R_{sc}$ = 10 kpc). Only a minor fraction (<10%) of the disk mass has been left in the interstellar gas phase, the bulk of the matter being locked up in long-lived lower mass stars and remnants. The release of nucleosynthesis products by shorter lived more massive stars (m >1 $m_\odot$) has changed the chemical composition of the gas from primordial values (Y(He) $\sim$0.22 by mass, 'metal'-free Z(A >4) $\sim$0) to Y $\sim$0.28 and Z $\sim$0.02 in the present-day local disk gas.

The network of processes that govern the chemical evolution of galaxies is summarized in Fig. 1. In the following, we focus mainly on the disk component, as our understanding of the evolution of the young Galaxy

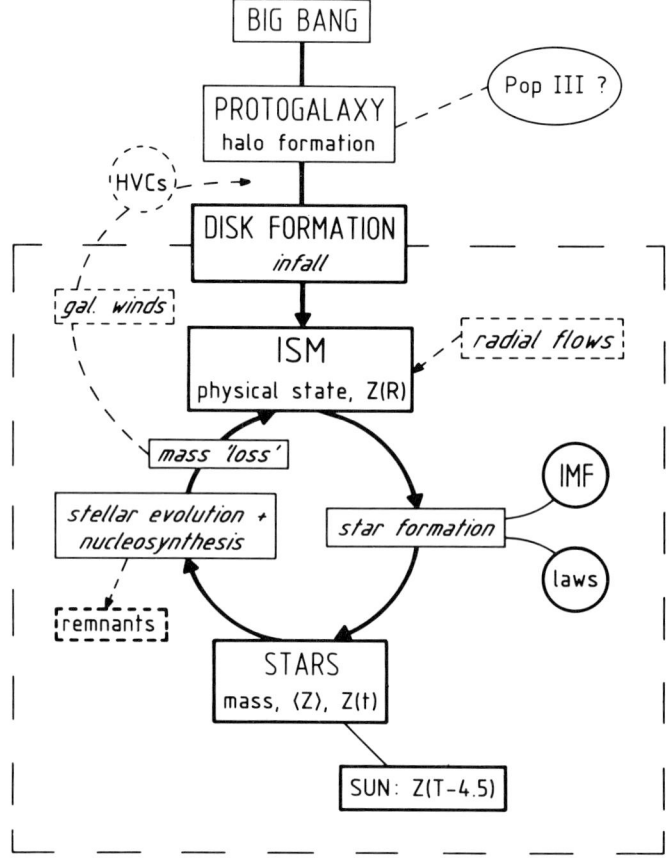

*Fig. 1. Network of processes governing the chemical evolution of galaxies.*

is only rudimentary. Obviously, star formation was very efficient during the collapse of the protogalaxy (lasting a few free fall times, $t_{ff} \lesssim 1$ Gyr), when much of the gas was transformed into 'halo' stars. Stellar nucleosynthesis processes during collapse led to rapid chemical enrichment of the gas co-moving with the stars ($Z \lesssim (0.1-0.2) \cdot Z_\odot$), as evidenced by the metallicity distribution of globular clusters and halo dwarfs (e.g. Searle and Zinn, 1978). Somewhat later, the disk was gradually built up from the settling (slightly enriched) halo gas. Note in this context, that there is little observational need for pregalactic enrichment from a hypothetical pop III generation of stars (Hartwick, 1983).

The cycling of matter between the stellar and interstellar phases is governed by the processes of star formation and evolution. Stars condense out of the interstellar medium (ISM) at a rate $\Psi(m_\odot \text{ yr}^{-1})$ with some initial mass spectrum $\phi(m)$, evolve during core hydrogen burning, then return their 'envelopes', partly enriched with nucleosynthetic products, to the ISM during the last few percent of their lifetime. Lifetimes range from a few $10^6$ yr for the most massive stars to >12-15 Gyr, the age of the disk, for masses $m \lesssim 0.9$ $m_\odot$. Low-mass remnants are left (white dwarfs with $<m> \sim 0.6$ $m_\odot$ for initial masses $m \lesssim 5-8$ $m_\odot$, and probably neutron stars $<m> \sim 1.4$ $m_\odot$ from more massive progenitors) that lock up some fraction of the disk matter permanently. Thus, since more and more gas is transferred into stars during galactic evolution, we expect the gas-to-total-mass ratio in the disk to decrease and the metallicity of the gas to increase steadily with time.

However, a real galaxy's life is more complicated. Important details of the star forming process are still unclear, and some crude parametrisation of its dependence on the physical environment (e.g. the average gas density) is unavoidable (Sect. V.2). Nucleosynthesis in particular of the rare nuclei occurs during late evolutionary phases (along the asymptotic giant branch) which are difficult to model. Hence their yields are generally uncertain (Sect. V.3) and sometimes not even available at all. Some 'observational' calibration in terms of a reliable evolution model is clearly needed (Sect. VI).

## III. ABUNDANCE GRADIENTS: EVIDENCE FOR EVOLUTION

Optical spectroscopy of emission nebulae has revealed in a number of nearby spiral galaxies abundance gradients similar to those across the Milky Way disk (Pagel and Edmunds, 1981), with lowest metallicity (O/H and N/H) in the outer disks, typical radial gradients $\sim 0.1$ dex kpc$^{-1}$ and highest abundances in their nuclei. In the Milky Way system, where optical studies are restricted to the solar vicinity due to interstellar extinction, radio techniques have turned into a powerful tool for exploring the global disk structure (isotope ratios). Complementary to such information on the present-day spatial abundance variations, the history of enrichment can be uniquely inferred from the metallicity distribution among local disk stars of different age.

## Stars in the Solar Vicinity: $Z(t, R_{\odot})$

(1) From color observations of long living G and M disk dwarfs with main sequence lifetimes exceeding the age of the disk, the *cumulative metallicity distribution* $S(Z)$ is derived, which gives the fraction of stars with metallicity greater than 'Z'. For more than two decades it has been known that this distribution is strikingly deficient in very metal poor stars (e.g. Pagel and Patchett, 1975; Mould, 1978): less than 10 percent of the stars have $Z < Z_{\odot}/3$. This paucity of metal-poor stars turned out to be a powerful constraint on the chemical evolution of the early disk, and as it is in conflict with predictions from simple evolution models (Sect. VI), has been denoted as the 'G-dwarf problem' in the literature (van den Bergh, 1962; Schmidt, 1963).

(2) Twarog (1980) fitted isochrones to Strömgren indices of more than 1000 F dwarfs in order to determine consistently their age and metallicity. The final *age-metallicity relation* (Fig. 2) rises with time approximately linearly at first but flattens during the last $\sim 5$ Gyr, the time since the solar system was formed. However, apparently 'calibration' is a great problem, and using a different set of isochrones, Carlberg et al. (1985) obtained a significantly flatter slope.

(3) The chemical composition of the early solar system is taken as a reference, assumed to be representative of the chemical composition of the ISM 4.6 Gyr ago.

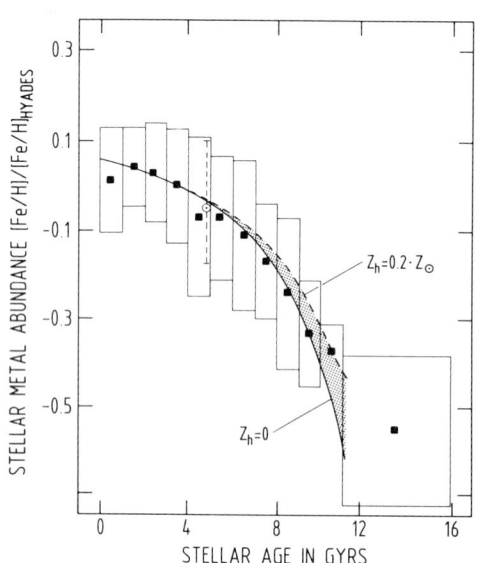

*Fig. 2. Age-metallicity relation as obtained by Twarog (1980) for local field F dwarfs (solar metallicity according to Pagel and Edmunds (1981)). Predicted AMRs are from models allowing for an exponentially decreasing infall rate of halo gas with metallicity $Z_h$ ($t_f \sim 5$ Gyr, $n=1$, $T=12$ Gyr).*

## The ISM across the Disk: $Z(T,R)$

Table I gives an updated compilation of the present-day abundances of the CNO nuclei and their isotopes across the Galaxy (see also Fig. 6). Abundances are given by relative number of atoms, disk gradients are defined by $\partial/\partial R\,[^{10}\log Z(R)]$ in [dex kpc$^{-1}$]. The order of presentation (solar system, solar vicinity, inner disk, and galactic center) is considered as an evolutionary sequence, with the most advanced population being that of the galactic nucleus. In column (6) the main tracer species (lines) are listed; for details of the observations, see the references (Güsten and Mezger, 1983).

**TABLE I**   COMPILATION OF SOLAR SYSTEM AND INTERSTELLAR ABUNDANCES

| Isotopic Species | Solar System | | Local ISM | Disk-Gradient [dex kpc$^{-1}$] | Galactic Center | | Observed Lines |
|---|---|---|---|---|---|---|---|
| $^2$H/H | $3.6(^{+1.0}_{-1.4}) \cdot 10^{-5}$ | (5) | ∿2 10$^{-5}$<br>5 10$^{-6}$ | positive | ? | (6,15)<br>(17) | molecular (HCN,DCN)<br>atomic (H,D) |
| $^4$He/H | 0.05-0.10 | (4) | ∿0.10 | -0.015<br>-0.020 | ? | (7)<br>(8) | radio recomb. (HeII)<br>optical (HeII) |
| $^{12}$C/H | $4.7(\pm 1.1) \cdot 10^{-4}$<br>$4.2(^{+1.3}_{-1.0}) \cdot 10^{-4}$ | (3)<br>(2) | $(3 \cdot 10^{-4})$ | <0 | ? | (8) | optical (CII) |
| $^{12}$C/$^{13}$C | 89±2 | (1) | 75(±5) | ∿0.05 | 20±5 | (9,18,19) | molecular (CO,H$_2$CO) |
| $^{16}$O/H | $8.3(\pm 2) \cdot 10^{-4}$<br>$6.9(\pm 1.3) \cdot 10^{-4}$ | (3)<br>(2) | $5(\pm 1) \cdot 10^{-4}$ | $-0.08 \atop -0.07 (\pm 0.015)$ | ? | (10)<br>(11) | radio recomb. (HII)<br>optical ([OII],[OIII]) |
| $^{16}$O/$^{18}$O | 490±25 | (1) | ∿400 | positive | ∿200 | (12,19,20) | molecular (H$_2$CO,CO) |
| $^{18}$O/$^{17}$O | 5.4 | (1) | 3.7 | ≤0.01 | 3.5(±0.2) | (13) | molecular (CO) |
| $^{14}$N/H | $1.0(\pm .2) \cdot 10^{-4}$<br>$.9(\pm .3) \cdot 10^{-4}$ | (3)<br>(2) | $4 \cdot 10^{-5}$ | -0.09(±0.015)<br>-0.095(±0.03) | ? | (11)<br>(16) | optical ([NII]) |
| $^{14}$N/$^{15}$N | 272 | (1) | 400 | (negative) | ∿1000 | (14,21) | molecular (HCN,NH$_3$) |

(1) Cameron (1980)
(2) Ross + Aller (1976)
(3) Lambert (1978)
(4) Hirayama (1971), Heasley + Milkey (1978)
(5) Kunde et al. (1982)
(6) Bruston et al. (1981)
(7) Thum et al. (1980)
(8) Peimbert (1978)
(9) Henkel et al. (1982)
(10) Mezger et al. (1979)
(11) Shaver et al. (1983)
(12) Henkel et al. (1979)
(13) Penzias (1981)
(14) Wannier et al. (1981)
(15) Penzias (1980)
(16) Binette et al. (1982)
(17) Vidal-Madjar et al. (1983)
(18) Henkel et al. (1985)
(19) Güsten et al. (1985)
(20) Schüller et al. (1985)
(21) Güsten + Ungerechts (1985)

As expected, the metallicity of the gas ($^4$He,C,N,O) increases with progressive 'astration'. This is most obvious for the galactic center data. The composition of the local disk gas, however, has changed only moderately since the time the solar system decoupled 4.6 Gyr ago. While the rare nuclei of C and O are progressively enriched relative to their main isotope with time (see $^{12}$C/$^{13}$C, and $^{16}$O/$^{18}$O), this is different for the nitrogen isotopes, whose ratio *increases* towards the galactic center. A quantitative analysis of these rare isotope data in terms of a reliable evolution model will provide important constraints on the 'cooking sites' of these nuclei (e.g. Güsten and Ungerechts, 1985). It is still controversial whether the dispersion in the abundances of stars, HII regions and molecular clouds (e.g. Fig. 6) are signposts of real inhomogeneities in the ISM due to incomplete mixing, or whether it simply reflects differences in e.g. the line formation process. Due to injection of processed stellar material and/or dilution by infalling halo gas, the ISM is locally inhomogeneous, but the timescales of the mixing processes are uncertain. If linked to the lifetime of giant molecular clouds, which is short (<10$^8$ yrs) compared with other galactic evolution timescales, azimuthal mixing seems a reasonable approach. However, in case that there were real strong abundance fluctuations White and Audouze (1983) have shown that the chemical enrichment of secondary nuclei could be severely affected.

## IV. SET OF EQUATIONS GOVERNING CHEMICAL EVOLUTION

The basic equations describing the chemical enrichment of the galactic disk, segmented into annuli $(R,R+dR)$, are

*(1) the net star formation rate*

$$\frac{\partial}{\partial t} M_* = \Psi(t) - \int_{m(t=\tau)}^{m_{up}} \Psi[t-\tau(m)]\phi(m)r(m)dm, \qquad (IV.1)$$

with $r(m)$ the mass of stellar material that is re-ejected into the ISM after the star's lifetime $\tau(m)$. $M_g(R,t)$ and $M_*(R,t)$ are the masses of the disk annulus $(R,R+dR)$ in form of gas and of stars, respectively. $\phi(m)$ is the initial stellar mass spectrum, and $\Psi(R,t)$ is the actual star formation rate which is assumed to go with the average total gas density (Schmidt, 1959) as

$$\Psi(R,t) = \nu \cdot \rho_g^n(R,t) \qquad (IV.2)$$

with $n \sim 1-2$. This is simply a convenient parametrisation without physical relevance, since stars are well known to condense out of the *molecular* gas phase. However, the processes that determine the molecular gas fraction are not yet understood, and obviously their efficiency varies with galactic radius (Sect. V.1).

*(2) the net gas consumption rate*

$$\frac{\partial}{\partial t} M_g = - \frac{\partial}{\partial t} M_* + A(R,t) - \frac{1}{R}\frac{\partial}{\partial R}(R \cdot V_{rad} \cdot M_g), \qquad (IV.3)$$

with $A(R,t)$ the gas infall rate which accounts for the gradual build-up of the disk by the settling halo gas. The last term accounts for possible radial flows in the plane of the disk.

*(3) the net enrichment rate of the gas*

$$\frac{\partial}{\partial t}(Z_i \cdot M_g) = -Z_i \cdot \Psi + \int_{m(t=\tau)}^{m_{up}} \Psi[t-\tau(m)]\phi(m) \cdot \left\{[r(m)-d(m)]Z_i[t-\tau(m)] + \Delta m(Z_i)\right\} dm$$

$$+ M_g \cdot V_{rad} \cdot \frac{\partial Z_i}{\partial R} + Z_h \cdot A(t,R). \qquad (VI.4)$$

$\Delta m(Z_i)$ is the newly synthesized mass of nuclear type 'i', returned to the ISM from evolved stars. For 'secondary' elements $\Delta m(Z_i) \propto Z_j[t-\tau(m)]$, with $Z_j$ the prestellar abundance, by mass, of the seed nucleus 'j' (Sect. V.3). $d(m)$ accounts for possible destruction of nucleus 'i' during stellar processing. A possible preenrichment of the infalling halo gas is included by $Z_h$. The yield $y_i$ of a stellar generation is defined as the mass of newly synthesized species '$Z_i$' per fraction of matter that is locked-up in long-living stars and remnants ($1-r \sim 0.6$)

$$y_i = \frac{\int \phi(m) \, \Delta m(Z_i) \, dm}{(1-r)} \tag{IV.5}$$

The integration of the above set of equations is mathematically rather simple, although fully time-dependent solutions can be obtained only numerically. In the instantaneous recycling approximation (IRA) where one assumes that the evolution of stars with lifetimes shorter than the age of the system is 'instantaneous' [$\Psi(t-\tau) \rightarrow \Psi(t)$, etc.], approximate analytical solutions are often possible. This IRA is a reasonable approach unless the system gets highly evolved (as the galactic nuclear bulge population, or elliptical galaxies) and as long as one does not want to study the enrichment with ejecta from long-lived stars.

The major problem arises from the often ill-determined initial conditions, such as the time-scale of disk formation, the laws of star formation etc. These uncertain parameters have to be, within 'reasonable' limits, adjusted to match the available observational constraints self-consistently.

## V. CONSTRAINTS ON THE CHEMICAL MODELS

### V.1. Characteristics of the Disk Component

Disk Formation

The gradual (and delayed) build-up of the disk by settling halo gas is a natural consequence of spiral galaxy formation (see e.g. Tinsley and Larson, 1978). Long-lasting infall was first invoked to explain the metallicity distribution among local disk dwarfs (see Lynden-Bell (1975) on the 'G dwarf problem', Twarog (1980) on the age-metallicity relation). To investigate whether infall is still a process chemically active today (by dilution), we compare the present-day rate $A(T)$ with the time-averaged rate of accretion $<A> \sim \sigma_D/T \sim 7$ $m_\odot$ $pc^{-2}$ $Gyr^{-1}$ ($\sigma_D$ the local surface density of matter, $\sim 75$ $m_\odot$ $pc^{-2}$, $T \sim 12$ Gyr). *Limits* on the actual rate, $A(R_\odot,T) \leq 1-2$ $m_\odot$ $pc^{-2}$ $Gyr^{-1}$ from X-ray observations (Cox and Smith, 1974) and from high-velocity cloud statistics (Oort, 1970), respectively, are small compared with $<A>$, suggesting that infall has in fact faded away. This is strengthened when one considers the uncertain nature of the HV-clouds. From their chemical composition, which within the uncertainty is consistent with local disk abundances, and their velocity field, HV clouds appear likely not to relate to early halo gas at all, but rather to represent the cooled and condensed debries of hot coronal material that has been ejected from the disk into the halo in accord with the galactic fountain model (van Woerden et al. (1985), Kaelble et al. (1985)). The accretion *flow pattern* is observationally poorly constrained. An exponentially decaying infall rate has been shown to match the chemical and kinematical properties of the local disk, with timescale $t_f \sim 5$ Gyrs for n=1 (Vader and de Jong, 1981; Lacay and Fall, 1983; see also Fig. 2). Infall versus galactocentric radius is often scaled to match the exponential disk profile ($R_D \sim 3.5$ kpc scalelength).

Disk Kinematics (radial flows)

The major effect of inward-directed flows is to enhance metallicity gradients across the disk. Physical processes which may drive radial flows are (1) accretion of low-angular-momentum gas, (2) viscosity in the ISM (from cloud-cloud collisions), and (3) dissipation of energy in response to the density wave. Order of magnitude estimates yield a local radial flow component $V_{rad} \sim (0.1-1.0)$ km s$^{-1}$. Observational limits are poor ($V_{rad}$ less than a few km s$^{-1}$), in part due to uncertainties in the LSR (see Lacey and Fall (1985) for references).

Distribution of Disk Matter

As in other late-type spirals, the bulk of the Milky Way matter is stellar, with less than 10 percent in the form of gas. The total surface density of the disk follows an exponential profile with scalelength $\sim 3.5$ kpc outside of R $\sim 5$ kpc, attains a maximum around 2-3 kpc and falls off toward the galactic center where the old bulge component dominates (Caldwell and Ostriker, 1981; Rohlfs and Kreitschmann, 1981). The local surface density is $\sigma_D \sim 70-80$ m$_\odot$ pc$^{-2}$, roughly half in the form of 'dark' (unidentified) matter (Bahcall, 1984). Most of the ISM is contained in cold clouds, which range from medium-dense low-mass objects with hydrogen mainly atomic to 'giant' molecular clouds with a significant fraction of their mass ($\sim 10^5$ m$_\odot$) found in high-density clumps ($\sim 10^{4-5}$ cm$^{-3}$). The distribution of HI surface density, as inferred from the 21 cm line, is roughly constant with galactic radius, $\sigma_{HI} \sim 2-3$ m$_\odot$ pc$^{-2}$ (Burton and Gordon, 1978), until it decreases exponentially at R $\geq 15$ kpc. The molecular gas is confined to a thin layer of dense clouds in a ring of emission between 4 and 8 kpc galactic radius. Since there is no molecular hydrogen transition observable in the radio range, estimates of the large-scale H$_2$ distribution have to be based on observations of the widespread CO molecule (for details and problems with the CO-H$_2$ conversion see e.g. Lequeux (1981), Güsten and Mezger (1983)). In Fig. 3 gas surface densities are given for molecular abundances [H$_2$]/[$^{13}$CO] $\sim 10^6$ (e.g. Solomon et al. 1979) and $4 \cdot 10^5$ [Dickmann (1978) from local dark clouds, and, independently, from the galactic $\gamma$-ray distribution by Bath et al. (1985)]. The local molecular gas surface density is 1-2 m$_\odot$ pc$^{-2}$. At the peak of the molecular ring, at R $\sim 4-6$ kpc, the mass of molecular gas exceeds the HI mass by a factor 2-4. Inside of R $\sim 4$ kpc both phases are deficient.

Important for constraining the degrees of astration is the fraction of matter left in the gaseous phase. Given the above numbers we compute a gas-to-total-mass ratio $\mu \sim 0.06-0.08$, roughly constant with galactic radius (Fig. 3).

# THE CHEMICAL EVOLUTION OF GALAXIES

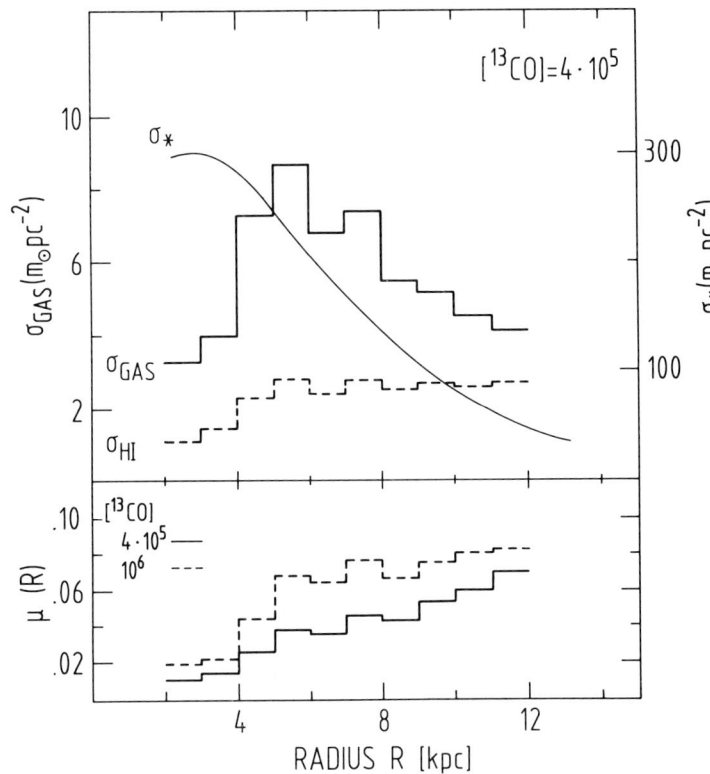

*Fig. 3. Surface density of matter and gas-to-total-mass ratio across the disk. The molecular gas content is derived for an $H_2/CO$ abundance $[^{13}CO] = 4 \cdot 10^5$ and $10^6$, respectively, with $[^{13}CO] \propto Z^{-1}$.*

## V.2 Star Formation in the Disk

### The Initial Mass Function (IMF)

From star counts in the solar vicinity, the present-day mass distribution $n(m)$ of main-sequence stars has been determined (e.g. Miller and Scalo, 1979; Lequeux, 1980). This is related to the initial mass function by

$$n(m) = \int \Psi(t) \phi(m) dt,$$

where $\Psi(t)$ is the star formation rate, and $\phi(m)$ describes the probability of forming a star of mass m in a newly born generation of stars, with

$$\int \phi(m) m \, dm = 1$$

For stars with ms lifetimes exceeding the age of the disk, $\phi(m) \propto n(m)$; for short-lived ($\tau_{ms} << T$) massive stars, $\Psi(T-\tau) \sim \Psi(T)$, thus $\phi(m) \propto n(m)/\tau(m)$. For the intermediate mass range, $1 \leq m \leq 2-3$ $m_\odot$, $\phi(m)$ has to be deconvolved from the history of star formation. IMFs commonly used in the literature are derived by (1) requiring a continuous transition between the upper and lower branches and (2) for a constant star formation rate.

Recent determinations are compiled in Fig. 4. Between 2 and 10 $m_\odot$, the IMF follows a power law $\phi(m) \propto m^{-\gamma}$ with $\gamma \sim 2.4$, close to the slope

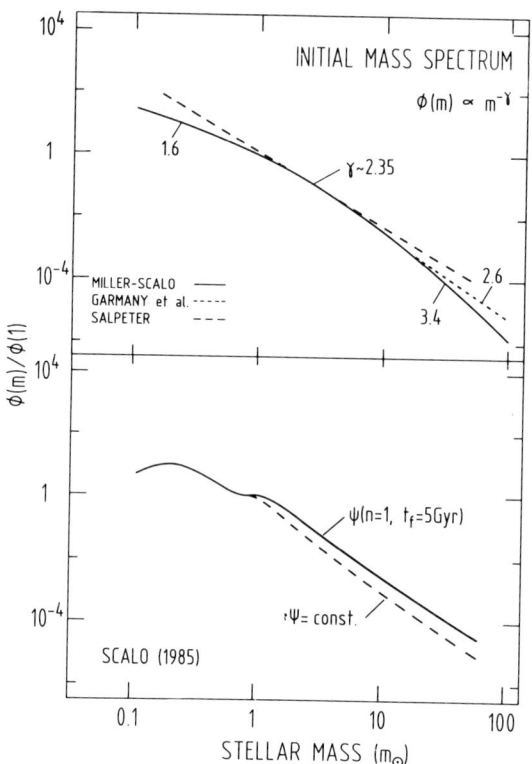

Fig. 4. 'Smoothed' initial mass functions for the local disk as derived for $\Psi(t) = const.$ according to Salpeter (1955), Miller and Scalo (1979), Garmany et al. (1982), and Scalo (1985). In (b) the effect of a decreasing SFR is demonstrated, causing a non-monotonic IMF with a prominent dip below $\sim 1$ $m_\odot$.

derived by Salpeter (1955) in his pioneering work. Below 1 $m_\odot$ it flattens ($\gamma \sim 1.6$) or even turns over, while above $\sim 20$ $m_\odot$ it may steepen considerably: $\gamma \sim 2.6$ (Garmany et al., 1982) to $\gamma \sim 3.4$ (Miller and Scalo (1979); Humphreys and Mc Elroy (1985)). IMFs given in Fig. 4 are smoothed, but there is increasing evidence for substantial fine-structure in the present-day mass function (Vereshchagin, 1982; Piskunov and Vereshchagin, 1984) with the most prominent gap around 1.5-2.0 $m_\odot$ (late A-stars). The mass range of the local IMF is from $\sim 0.1$ $m_\odot$ to $\sim 100$ $m_\odot$, with some evidence that $m_{up}$ decreases with decreasing galactic radius (Mezger, 1980; Panagia, 1980), likely due to the increased metallicity of the gas ($m_{up} \propto Z^{-0.5}$ (Kahn, 1974)). Data about the IMF outside the solar neighbourhood are rare. Only recently, Humphreys and Mc Elroy (1985) derived $\gamma \sim 3.0$ ($m \geq 15$ $m_\odot$) in the LMC and SMC, similar to the local slope. The same result seems to hold for the M33 spiral galaxy (Berkhuijsen, 1982).

## Evidence for a non-monotonic IMF and a bimodal nature of star formation

If we allow for a decreasing SFR (and hence some dependence on the amount of available interstellar gas, Eq. IV.2) the IMF derived from the present-day luminosity function *is* non-monotonic around m $\sim 1$ $m_\odot$ (see Fig. 4b). Usually a double-peaked distribution is considered 'unlikely', and the restriction of a monotonically decreasing IMF is used to argue for a rather constant SFR. Taken seriously, however, a double peaked IMF may indicate that star formation processes are bimodal (e.g. Smith et al., 1978) with high and low-mass stars forming at different sites under different physical conditions (Turner, 1984): The parental cloud of low mass stars (T associations in the local Taurus molecular cloud) is cold (T $\gtrsim 10$ K), clumpy and dispersed; star formation is scattered in small groups and spread out in time. High-mass star formation, on the other hand, is strongly clustered towards the warm (T $\sim$30 K) centrally condensed cloud cores (Orion, $\rho$-Oph), and apparently requires some external triggering (by small-scale shocks from local events, or by the global density wave via an enhanced cloud-cloud agglomeration rate).

What suppresses the creation of low-mass stars in sites of massive star formation is likely the strong dependence of the minimum fragment mass on the cloud temperature: $M_{min} \propto T^k$, $k \geq 2$, slightly dependent on details of the fragmentation process (energy cascade vs. angular momentum limited turbulence, Turner (1984), Larson (1985)). Once an early generation of massive stars has formed, their intense radiation field rapidly raises the ambient cloud temperature, and thus $M_{min}$ by an order of magnitude. With fragmentation into low-mass clumps being suppressed, the formation of high mass objects is favored and likely continues until the parent cloud will be dispersed after some $10^7$ yrs due to the energy/momentum release from dying (massive) stars. *Direct observational support* for an enhanced low-mass cut-off of the IMF in regions of high star formation activity is found toward star burst nuclei like M82 (Rieke et al., 1980), Arp 220 and NGC 6240 (Rieke et al., 1985), and from the detailed UBV color studies by Jensen et al. (1981) on the M83 spiral arm population.

Star Formation Rates

All tracers of star formation outside the solar neighbourhood are based upon characteristics of the most massive stars only (m $\geq$10 $m_\odot$); hence estimated rates depend sensitively on the slope of the IMF, the applied stellar atmospheres and the evolution of these stars. From UBV color photometry and $H_2$ emission measurements, Kennicutt (1983) obtains gas consumption rates into massive stars $\Psi$(m $\geq$10 $m_\odot$) $\sim$1 $m_\odot$ yr$^{-1}$ for normal Sc-Sd disks. Their FIR luminosity as measured by IRAS (de Jong et al., 1984) is consistent with these estimates. For a Miller-Scalo IMF this corresponds to a total SFR $\Psi$(0.1 $\leq$m $\leq$100 $m_\odot$) $\leq$10 $m_\odot$ yr$^{-1}$. Similar results hold for the Milky Way. From the galactic thermal radio background, Güsten and Mezger (1983) derive $\Psi \sim$7.5 to 11 $m_\odot$ yr$^{-1}$ for IMFs according to Garmany et al. and Miller-Scalo, respectively. Surprisingly, these estimates are quite comparable to the average past rate as derived from their disk masses. For the Milky Way, we find $<\Psi>_T \sim M_D/T(1-r) \sim 9$ $m_\odot$ yr$^{-1}$ (see Sect. V.1). Taken at face value, these results have been used to argue for a rather constant star formation rate, which then is difficult to reconcile with $\Psi \propto \rho_g^n$, n >0 (Eq. IV.2). In terms of the evolution scenario sketched above with exponentially decreasing infall the expected present rate is $\Psi_{exp}(T) \sim(0.2-0.5)$ $<\Psi>$ for n=2 and n=1, respectively.

A problem arises from the radial distribution of star formation across the disk, which is inconsistent with the galactic mass distribution (Fig. 5). While the average rate $<\Psi>$ (per 1 kpc wide annuli) is rather flat as a function of radius, the observed present-day rate $\Psi(T,R)$ attains a prominent maximum between 4 and 6 kpc with sharp fall-off inside 3-4 kpc and beyond R $\sim$8 kpc. Very similar discrepancies have been noted by Talbot (1980) and Jensen et al. (1981) for the nearby spiral M83. One possible implication may simply be that, as the observed SFR is derived from OB stars, while $<\Psi>$ and $\Psi_{exp}(T)$ are from the low-mass branch

of the IMF there are different efficiencies of forming high and low-mass stars across the galactic disk, with the lower masses being more and more suppressed with decreasing galactocentric radius. A likely solution to these puzzles is the above outlined concept of bimodal star formation as advocated by Güsten and Mezger (1983). If
(1) the large-scale galactic star formation is parametrized into two modes, an interarm population (that is simply proportional to the amount of available gas) and a spiralarm population (governed by the flow of gas through the density wave, $\psi^{sa} \propto (\Omega_p \cdot \Omega_R) \cdot M_{gas}$, and accounting for the formation of the bulk (>2/3) of the OB stars), and if
(2) the latter as site of active massive star formation is deficient in low-mass stars, a streadily increasing fraction of massive stars per stellar generation toward the inner disk is predicted, consistent with Fig. 5. This is because in the rest frame of the density wave ($\Omega_p \propto$ const.(R) is the speed of the spiral pattern, the angular velocity of the crossing gas is $\Omega_R \propto R^{-1}$), the weight of the high-mass favoring spiralarm mode increasingly dominates star formation in the inner disk. The total present-day galactic SFR is reduced by a factor $\gtrsim 2$, dependent on the low-mass cut-off of the high-mass mode of the IMF. As pointed out by Larson (1985b), another direct consequence of bimodal IMFs is that the predicted number of stellar remnants is significantly larger than in models with standard (monotonic) IMFs, and may therefore account for the amount of dark matter in the solar vicinity (Sec. V.I).

V.3  Sites of Nucleosynthesis

The enrichment timescale of the ISM is basically controlled by (1) the mass and hence lifetime of the stellar cooking site, and (2) the order (primary vs. secondary) of the synthesis process. Primary nuclei are generated in one stellar generation from primordial elements (maybe in a network of processes). The synthesis of secondary products is based on the ashes of an earlier generation, and is therefore naturally time-delayed (Sect. VI.1). We cannot enter here into a detailed discussion of the relevant nucleosynthesis processes. Generally, there is quite good agreement about the nuclear processes that synthesise the most abundant elements and their isotopes, but their astrophysical 'cooking sites' remain controversial. Table 2 is an updated compilation from Güsten and Mezger (1983) to summarise the relevant ingredients for models of chemical evolution. For reference, net bulk yields (Eq. IV.5) according to the monotonic Miller-Scalo IMF are given.

IV.  MODELS OF CHEMICAL EVOLUTION

The Simple Closed Model, and the Need for Infall

A convenient starting point is the so-called 'simple closed model', in which with the instantaneous recycling approximation the evolution of an initially unenriched (primoridal composition) isolated (no infall, no radial flows) test volume is followed. The model thus represents the extreme case of fast disk formation. For a primary nucleus (like $^{12}C$,

TABLE II   YIELDS OF NUCLEOSYNTHESIS PROCESSES

(from Güsten and Mezger, 1983)

| Species | Nucleosynthesis Process Production | Nucleosynthesis Process Destruction | Seed-nuclei | Type | Stellar Mass Range | Net Bulk Yields | Notes | References |
|---|---|---|---|---|---|---|---|---|
| $^2H$ | - | $^2D(p,\gamma)^3He$ | - | - | all | -0.6 | completely destroyed in stars of all masses | |
| $^4He$ | pp, CNO | $\alpha$-capture | H | prim. | MS | 1.3(-2) | hydrogen-burning, 50% in stellar winds (WR-stars) | 1,8,10 |
| | | | | | IMS | $\delta=0$  1.8(-2) $\delta=1.5$  1.9(-2) | dredge-up in Red Giants $\delta$-mixing length ratio | 2 |
| $^{12}C$ | triple $\alpha$' | $\alpha$-capture CNO | $^4He$ | prim. | MS | 1.5(-3) | Helium-burning, partly WR-stars | 1,3,10 |
| | | | | | IMS | $\delta=0$  3.9(-3) $\delta=1.5$  2.3(-3) | dredge-up in Red Giants | 2 |
| $^{13}C$ | incomplete CNO | CNO | $^{12}C$ | sec. (prim.) | IMS | $\delta=0$  4.4(-5) $\delta=1.5$  1.6(-4) 0.9(-4)prim. | incomplete CNO-burning in Red Giant envelopes low-temperature 'hot-bottom' burning | 2,4 |
| $^{14}N$ | CNO | $\alpha$-capture $^{14}N(\alpha,\gamma)^{18}F$ | $^{12}C(^{16}O)$ | sec. (prim.) | IMS | $\delta=0$  7.9(-4) $\delta=1.5$  3.1(-3) 2.1(-3)prim. | major sink in CNO-cycle; several dredging up phases along AGB, hot-bottom burning in convective envelope | 2 |
| $^{15}N$ | hot CN | CNO | $^{12}C, ^{14}N$ | prim. | MS | 2.5(-6) | hot CN-cycle ($T>2\cdot 10^8$) in (novae) or supernovae | 7,9,11 |
| $^{16}O$ | $^{16}O(\alpha,\gamma)^{12}C$ | $\alpha$-capture CNO | $^{12}C$ | prim. | MS | 6.4-9.2(-3) | helium-burning in massive stars | 1,3,8 |
| $^{17}O$ | incomplete CNO | (CNO) | $^{16}O$ | sec. (prim.) | (MS) | 2(-6) | enriched winds from WN-stars | 9 |
| | | | | | IMS | ~2(-5) | incomplete CNO-burning with deep mixing along AGB, no hot-bottom burning included | 4 |
| | | | | | | ? | hot CNO-burning in novae/supernovae | 6,7 |
| $^{18}O$ | $^{14}N(\alpha,\gamma)^{18}F(\beta^+)^{18}O(\alpha,\gamma)^{22}Ne$ | $^{14}N$ | | sec. | MS | ? | explosive He-burning (SN?) | 6 |
| | | | | | IMS | -3(-7)? | during quiet core He-burning in $m<5m_\odot$ destroyed(?) | 4,5 |

References:
(1) Arnett (1978)
(2) Renzini + Voli (1981)
(3) Woosley + Weaver (1980)
(4) Dearborn et al. (1978)
(5) Thielemann + Arnold (1978)
(6) Howard et al. (1971)
(7) Audouze et al. (1978)
(8) Maeder (1981)
(9) Woosley + Weaver (1980)
(10) Maeder (1982), Abbott (1982)
(11) Güsten and Ungerechts (1985)

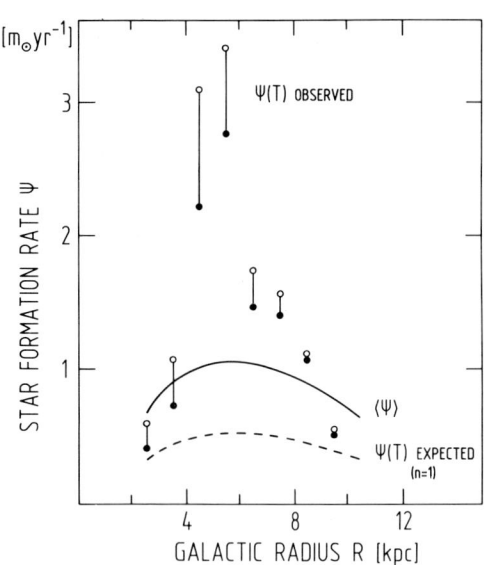

Fig. 5. Star formation rates (in 1 kpc wide annuli) across the disk. The observed rate is derived from the galactic thermal radio background (Güsten and Mezger, 1983). Estimates of the time-averaged rate <Ψ> are from the galactic disk mass distribution (Caldwell and Ostriker, 1981), from which the expected present-day rate is calculated in terms of the exponential infall model (Fig. 2).

16O) the set of equations VI.1-4 is then solved by

$$Z^P = y_i^P \cdot \ln \mu^{-1}, \quad (VI.1)$$

$\mu$ is the gas fraction ( (t=0)=1), $y_i$ the primary's yield according to Eq. IV.5. For secondary elements (such as $^{13}C$, $^{14}N$ etc.), whose synthesis requires the prestellar presence of (primary) 'seed' nuclei, the corresponding solution is

$$Z^S = \frac{1}{2} y_i^S y_i^P [\ln \mu^{-1}]^2 \quad (VI.2)$$

Hence, $Z^P/Z^S \propto Z^P$ is a direct measure of the gas metallicity. For a linear dependence of the SFR on density, $\mu(t)=\exp(-t/\tau)$, and primary nuclei (and $Z^P/Z^S$) enrich linearly with time.

This simple approach is known to fail in explaining important constraints. Regarding the metallicity of local field dwarfs, neither the paucity of metal poor stars is reproduced (unless we allow for pre-enrichment of the 'zero age' gas, cf. Tinsley, 1980), nor the flattening of the age-metallicity relation during the last ∼5 Gyrs (Fig. 2, Twarog, 1980). As $\mu(R)$ ∼const. (Fig. 3), no metallicity gradients across the disk are predicted. The strong peak of the present-day SFR in the inner disk is not explained. Finally, the detection of deuterium (which is easily destroyed during astration, Table 2) in the evolved galactic center gas argues against 'closed' processing (cf. Audouze, 1984).

The most natural modification (at least with respect to the galaxy formation process) is to allow for a gradual build-up of the disk by infalling halo gas (cf. Lynden-Bell, 1975). The effect of infall is that the metallicity gradients flatten due to dilution with unprocessed halo gas. This can be easily shown for the special case that infall constantly balances the net SFR (hence $M_{gas}$=const.) and Eqs. IV(1-4) yield

$$Z^P = y_i \left\{ 1 - e^{(1-\mu^{-1})} \right\} \quad (VI.3)$$

with $Z^P$ ∼$y_i$ for $\mu < 1$ (cf. Tinsley, 1980). During recent years several authors have demonstrated that reasonable (i.e. saturated) infall models nicely reproduce the chemical properties and kinematics of the local disk (Vader and de Jong (1981), Lacey and Fall (1983)). As an example, the numerical solution (without instant.recycling) for an accretion time scale $t_f \sim 5$ Gyrs (Eq. V.1) and n=1 (Eq. IV.2) is superimposed on the age-metallicity relation in Fig. 2.

The problem arises if we extend this local solution to the disk beyond the solar vicinity. If we assume that the infall pattern was similar across the disk ('$t_f$' is allowed to increase with radius), and simply scale the accretion rate acc. to Eq. (V.1) to match the disk mass distribution, no set of parameters can be found that self-consistently explains the strong gradients in $Z(R)$ and $\Psi(R)$, and the flatness of $\mu(R)$ (cf. Lacey and Fall, 1983). Remember that in radially closed systems (no radial flows), the latter is a direct measure of the degree of astration (evolution). Thus, some additional ingredient is needed to account for the observed disk gradients.

## What causes Abundance Gradients across the Disk?

There exist several suggestions for an additional free parameter: (1) the yield $y_i$ may increase toward the galactic center, (2) the effect of infall (dilution) changes systematically across the disk, and (3) inward directed gas flows in the plane of the disk may lead to enhanced astration in the inner Galaxy. I discuss these modifications briefly now.

'Pure' infall models ($V_{rad}=0$, const. yield) can predict acceptable gradients by specifying a radially decreasing ratio $k=(1-r)\Psi/A$, i.e. the effect of infall relative to the net star formation rate increases with R. For illustration, if we assume $k(t)$ and $A(t)$ constant, the IRA solution is

$$Z = y_i \cdot k \, [1-(\mu/\mu_o)^{k-1}] \qquad (VI.4)$$

which behaves asymptotically as $Z \sim y_i k$. $\mu_o$ is the initial gas fraction $\mu(t=0)$, at the time infall started. Thus adjusting k and $\mu_o$ one can obviously account for the observed gradients. Models of this class with high infall rates $A(R,T) \sim 5$ $m_\odot$ $pc^{-2}$ $Gyr^{-1}$ (constant with time and across the disk, see Sect. V.1), almost constant SFR $\Psi \propto \exp(-t/\tau)$, $\tau \sim 15$ Gyr (thus relaxing Eq. IV.2) and $\mu_o$ adjusted to fit the disk mass distribution are presented by Tosi (1982) and Diaz and Tosi (1984). However, in view of the observational constraints on the genesis of HV clouds (velocity fields, chemical composition - Sect. V.1) unsaturated infall models with high present-day accretion rates may be questioned.

Infall Models including Radial Gas Flows (Tinsley and Larson, 1980; Mayor and Vigroux, 1981; Lacey and Fall, 1985). Inward-directed gas flows of processed gas lead to enhanced astration of the ISM in the inner disk. For the special case $A=0$ (no infall) Eqs. (1-4) predict gradients (Tinsley, 1980)

$$\frac{\partial}{\partial r} (Z/y_i) \propto \frac{(1-r) \cdot \Psi}{M_g \cdot V_{rad}} \qquad (VI.5)$$

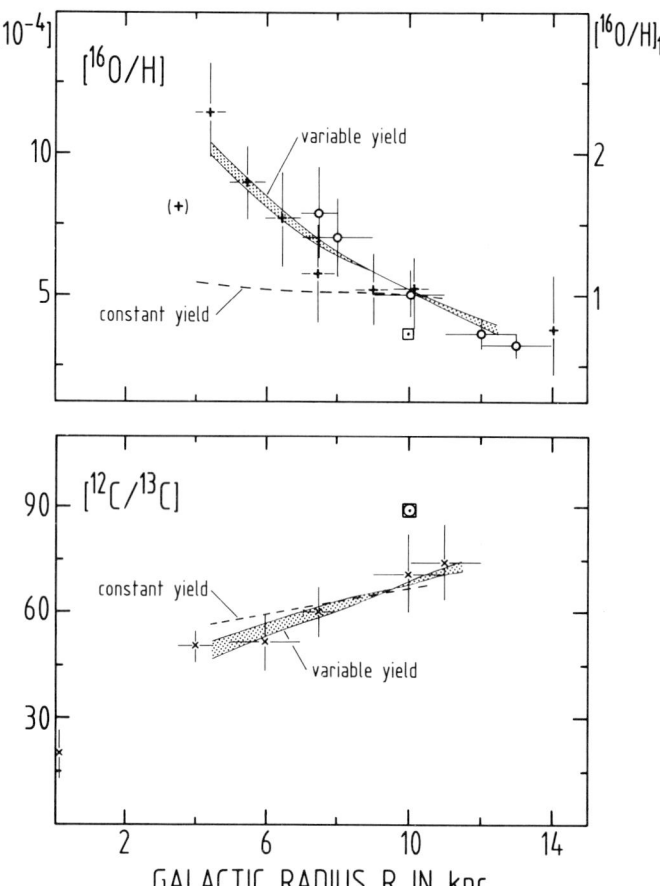

Fig. 6. Observed and predicted abundance gradients across the disk (updated from Güsten and Mezger (1983), for references see Table I).

From their recent models (incl. exponential infall), Lacey and Fall (1985) state that for moderate flow velocities ($V_{rad} \sim 1$ km s$^{-1}$, close to the observed limits (Sect. V.1)), acceptable abundance gradients can be achieved, with all the other above constraints roughly fulfilled.

Infall Models with Variable Yield. We advocated in Sect. V.2 that bimodal star formation is a natural consequence of double-peaked IMFs as derived from the present-day mass function for declining SFRs (Fig. 4). To resolve inconsistencies between the galactic star formation rates, we suggested that likely the IMF varies across the disk such that inside the the solar circle a steadily increasing fraction of massive stars per stellar generation is formed. Initially a variable IMF was introduced to explain the galactic abundance gradients. Assuming that the high-mass mode of star formation is governed by the flow of gas across the density wave [$\Psi \propto (\Omega_p - \Omega_R) \cdot m(H_2)$], Güsten and Mezger (1983) followed the chemical enrichment of the disk gas in Oxygen $^{16}$O and the carbon isotope ratios. Predictions of this model which otherwise is the saturated infall model described above (n=1, $t_f \sim 5$ Gyr, Sect. V.1), are shown to fit the observed constraints perfectly (Fig. 6). For this bimodal IMF, the net yield of a primary nucleus from massive progenitors varies approximately as

$$y_i(R) \propto (1 + \alpha \cdot \nu(R)), \qquad (VI.6)$$

with $\nu(R) = \Omega_R - \Omega_p / \Omega_\odot - \Omega_p$ ($\Omega_p \sim 13.5$ km s$^{-1}$ kpc$^{-1}$; $\Omega_R \propto R^{-1}$ roughly) and

$\alpha$ the ratio of massive star formation in the spiralarm to interarm population in the local bin ($\alpha \sim 1$; Güsten and Mezger, 1983).

Further advantages of a bimodal IMF have been presented recently by Larson (1985). If the hypothesis of a low mass star deficient star formation mode during periods of high star formation activity is generalized to earlier phases of galaxy evolution, the large number of stellar remnants (from massive progenitors) can account for the missing (invisible) mass in the local disk, and maybe in the galactic halo. Larson argues that problems with observed colors and $M/L_B$ ratios of spiral galaxies and also the increase of metallicity and $M/L_B$ with mass in giant ellipticals are explained in terms of bimodal star formation.

But clearly, more observational constraints (on spatial variations of the IMF, on IMF-sensitive abundance ratios across the disk, on limits on the radial flow velocity and on the accretion rate of halo gas) are required to discriminate between these competing models on the galactic evolution.

## REFERENCES

Abbott, D.C.: 1982, Astrophys. J. 263, 723
Arnett, W.D.: 1978, Astrophys. J. 219, 1008
Audouze, J.: 1984 in "Diffuse Matter in Space", eds. J. Audouze et al., Reidel
Audouze, J., Lequeux, J., Vigroux, L.: 1975, Astron. Astrophys. 43, 71
Bahcall, J.N.: 1984, Astrophys. J. 287, 926
Bath, C.L., Issa, M.R., Houston, B.P., Mayer, C.J., Wolfendale, A.W.: 1985, Nature 314, 511
Berkhuijsen, E.: 1982, Astron. Astrophys. 112, 369
Binette, L., Dopita, M.A., D'Odorico, S., Benvenuti, P.: 1982, Astron. Astrophys. 115, 315
Bruston, P., Audouze, J., Vidal-Madjar, A., Laurent, C.: 1981, Astrophys. J. 243, 161
Burton, W.B., Gordon, M.A.: 1978, Astron. Astrophys. 63, 7
Caldwell, J.A.R., Ostriker, J.P.: 1981, Astrophys. J. 251, 61
Cameron, A.G.W.: 1980 in "Nuclear Astrophysics", eds. C. Barnes, D. Clayton, D. Schramm, Cambridge
Carlberg, R.G., Dawson, P.C., Hsu, T., Vandenberg, D.A.: 1985, preprint
Cox, D.P., Smith, B.W.: 1976, Astrophys. J. 203, 361
Dearborn, D., Tinsley, B.M., Schramm, D.N.: 1978, Astrophys. J. 223, 557
de Jong, T., Clegg, P.E., Soifer, B.T., Rowan-Robinson, M., Habing, H.J., Houck, J.R., Aumann, H.H., Raimond, E.: 1984, Astrophys. J. 228, L67
Díaz, A.I., Tosi, M.: 1984, M.N.R.A.S. 208, 365
Dickman, R.L.: 1978, Astrophys. J. Suppl. 37, 407
Garmany, C.D., Olson, G.L., Conti, P.S., Van Steenberg, M.E.: 1981, Astrophys. J. 250, 660
Güsten, R., Henkel, C. Bartla, W.: 1985, Astron. Astrophys. (in press)
Güsten, R., Mezger, P.G.: 1983, Vistas in Astronomy 26, 159
Güsten, R., Ungerechts, H.: 1985, Astron. Astrophys. 145, 241
Hartwick, F.D.A.: 1983, Mem. Soc. A. It. 54, 51
Heasley, J.N., Milkey, R.W.: 1978, Astrophys. J. 221, 677
Henkel, C., Wilson, T., Downes, D.: 1979, Astron. Astrophys. 73, L13
Henkel, C., Wilson, T.L., Bieging, J.: 1982, Astron. Astrophys. 109, 344
Henkel, C., Güsten, R., Gardner, F.F.: 1985, Astron. Astrophys. 143, 148
Hirayama, T.: 1971, Solar Physics 19, 384
Howard, W.M., Arnett, W.D., Clayton, D.D.: 1971, Astrophys. J. 165, 495
Humphreys, R.M., Mc Elroy, D.B.: 1985, preprint
Jensen, E.B., Talbot, R.J., Dufour, R.J.: 1981, Astrophys. J. 243, 719
Kaelble, A., de Boer, K.S., Grewing, M.: 1985, Astron. Astrophys. 143, 408
Kahn, F.D.: 1974, Astron. Astrophys. 37, 149
Kennicutt, Jr. R.C.: 1983, Astrophys. J. 272, 54
Kunde, V., Hanel, R., Maguire, W., Gautier, D., Baluteau, J.P., Martin, A., Chedin, A., Husson, N., Scott, N.: 1982, Astrophys. J. 263, 443
Lacey, C.G., Fall, S.M.: 1983, M.N.R.A.S. 204, 791
Lacey, C.G., Fall, S.M.: 1985, preprint
Lambert, D.L.: 1978, M.N.R.A.S. 182, 249
Larson, R.B.: 1985a, M.N.R.A.S. 214, 379
Larson, R.B.: 1985b, M.N.R.A.S. in press
Lequeux, J.: 1980 in "Star Formation", 10th Advanced Course S.S.A.A., Saas-Fee
Lequeux, J.: 1981, Comments on Astrophysics 9, 117

Lequeux, J.: 1985 in 'Birth and Infancy of Stars', Les Houches, 1983
Lynden-Bell, D.: 1975, Vistas in Astronomy 19, 299
Maeder, A.: 1981, Astron. Astrophys. 102, 401
Maeder, A.: 1982, Astron. Astrophys. 120, 113
Mayor, M., Vigroux, L.: 1981, Astron. Astrophys. 98, 1
Mezger, P. G.: 1980 in "Radio Recombination Lines", ed. P. A. Shaver, Reidel
Mezger, P. G., Pankonin, V., Schmid-Burgk, J., Thum, C., Wink, J.: 1979, Astron.Astrophys. 80, L3
Miller, G.E., Scalo, J.M.: 1979, Astrophys. J. Suppl. 41, 513
Mould, J.R.: 1978, Astrophys. J. 220, 434
Oort, J.H.: 1970, Astron. Astrophys. 7, 381
Pagel, B.E.J., Edmunds, M.G.: 1981, Ann. Rev. Astron. Astrophys. 19, 77
Pagel, B.E.J., Patchett, B.E.: 1975, M.N.R.A.S. 172, 13
Panagia, N.: 1980 in "Radio Recombination Lines", ed. P. A. Shaver, Reidel
Peimbert, M.: 1978, IAU Symp. 84 "The Large Scale Characteristics of the Galaxy", ed. W.B. Burton, Reidel, p. 84
Penzias, A.A.: 1980, Science 208, 663
Penzias, A.A.: 1981, Astrophys. J. 249, 518
Piskunov, A.E., Vereshchagin, S.V.: 1984, Nauchn. Inf. Astron. Council 57
Renzini, A., Voli, M.: 1981, Astron. Astrophys. 94, 175
Rieke, G.H., Lebofsky, M.J., Thompson, R.J., Low, F.J., Tokunaga, A.T.: 1980, Astrophys. J. 238, 24
Rieke, G.H., Cutri, R.M., Black, J.H., Kailey, W.F., McAlary, C.W., Lebofsky, M.J., Elston, R.: 1985, Astrophys. J. 290, 116
Rohlfs, K., Kreitschmann, J.: 1981, Astrophys. Space Sci. 79, 289
Ross, J.E., Aller, L.H.: 1976, Science 191, 1223
Salpeter, E.E.: 1955, Astrophys. J. 121, 161
Scalo, J.M.: 1985, Fundamentals of Cosmic Physics, in press
Schmidt, M.: 1959, Astrophys. J. 129, 243
Schmidt, M.: 1963, Astrophys. J. 137, 758
Schüller, M., Henkel, C., Güsten, R.: 1985, in preparation
Searle, L., Zinn, R.: 1978, Astrophys. J. 225, 357
Shaver, P.A., McGee, R.X., Newton, L.M., Danks, A.C., Pottasch, S.R.: 1983, M.N.R.A.S. 204, 53
Smith, L.F., Biermann, P., Mezger, P.G.: 1978, Astron. Astrophys. 66, 65
Solomon, P.M., Scoville, N.Z., Sanders, D.B.: 1979, Astrophys. J. 232, L89
Talbot, R.J. Jr.: 1980, Astrophys. J. 235, 821
Thielemann, F.-K., Arnould, M.: 1979 in "Les Elements et leurs Isotopes dans l'Univers", XXII$^e$ Coll. Internat. d'Astrophys., Liège, p. 59
Thum, C., Mezger, P.G., Pankonin, V.: 1980, Astron. Astrophys. 87, 269
Tinsley, B.M.: 1980, Fundamentals of Cosmic Physics 5, 287
Tinsley, B.M., Larson, R.B.: 1978, Astrophys. J. 221, 554
Tosi, M.: 1982, Astrophys. J. 254, 699
Turner, B.E.: 1984, Vistas in Astronomy 27, 303
Twarog, B.A.: 1980, Astrophys. J. 242, 242
Vader, J.P., de Jong, T.: 1981, Astron. Astrophys. 100, 124
Van den Bergh, S.: 1962, Astron. J. 67, 486
van Woerden, H., Schwarz, U.J., Hulsbosch, A.N.M.: 1985 in "The Milky Way Galaxy", IAU Symp. 106, H. van Woerden et al. (eds.), Reidel, p. 387
Vereshchagin, S.V.: 1982, Sov. Astron. Lett. 8, 294
Vidal-Madjar, A., Laurent, C., Gry, C., Bruston, P., Ferlet, R., York, D.G.: 1983, Astron. Astrophys. 120, 58
Wannier, P.G.: 1980, Ann. Rev. Astron. Astrophys. 18, 399
Wannier, P.G., Linke, R.A., Penzias, A.A.: 1981, Astrophys. J. 247, 522
White, S.D.M., Audouze, J.: 1983, M.N.R.A.S. 203, 603
Woosley, S.E., Weaver, T.A.: 1980 in "Nuclear Astrophysics", eds.C. Barnes, D. Clayton, D. Schramm, Cambridge

DISCUSSION

*J. Lequeux:* Comment on [O/H] in Sun and local ISM. I think that it is not possible to alleviate the apparent discrepancy between the Solar and ISM $^{16}$O/H by saying that oxygen is hidden in grains. A substantial fraction of oxygen can only be hidden in ices around grains. Apparently these ices exist only deep inside molecular clouds and are not present in HII regions where the O/H is measured. However the O/H as derived from spectra of HII regions is model-dependent and systematic errors are possible.

*T. de Jong:* Since the evidence for infall of gas into the galactic plane other than due to the high-velocity clouds is basically lacking, I think that it would be a good idea to try to calculate models in which the expulsion of gas into the halo and the later cooling and fall-back into the galactic plane is included. This might lead to a slow build-up of the galactic disk which is indicated by much of the observational constraints.

*R. Güsten:* This depends critically on the cooling time of the coronal gas, which is difficult to estimate. For an injection of $\sim 3$ $m_\odot$ $yr^{-1}$ Salpeter (1985, Mitt. Astron. Ges. 63, 11) argues for a residence time of $\sim 3 \cdot 10^8$ yr to have $\sim 10^9$ $m_\odot$ warm HI in orbit above the disk.

*M. Tosi:* From J. Silk's talk I understand that where you form massive stars you immediately form also low and intermediate mass stars. Isn't this inconsistent with your bimodal star formation?

*R. Güsten:* I argued above that in sites of high (massive) star formation activity the strong radiation field rapidly increases the gas temperature and hence the minimum fragment mass in the parental cloud, thus suppressing the formation of low mass stars (see Sect. V.2, and Larson (1985) for details).

# NITROGEN AND OXYGEN EVOLUTION IN NEARBY SPIRAL GALAXIES

M.Tosi[1] and Angeles I.Diaz[2]

[1] Osservatorio Astronomico Universitario
C.P.596, I-40100 Bologna, Italy

[2] Royal Greenwich Observatory
Herstmonceux Castle, Hailsham, U.K.

## INTRODUCTION

Reliable data on nitrogen and oxygen abundances are available for many HII regions both in the Galaxy and in nearby spirals and a plot of the N/O ratio vs O/H should provide, in principle, some indication on the primary and/or secondary origin of nitrogen. There is a general trend of N/O increasing with O/H if all the data for individual HII regions are considered together. However, the N/O distribution is much flatter within a given galaxy (see e.g. Pagel and Edmunds 1981). This classical diagram has received different interpretations. Pagel and Edmunds (1981) suggest that varying amounts of primary nitrogen are required to reproduce the data if standard assumptions on stellar and galactic evolution are made. On the other hand, by releasing some of these assumptions Serrano and Peimbert (1983) are able to reproduce the general trend of the diagram with purely secondary nitrogen.

Although there have been claims that the origin of nitrogen is solely primary (Tomkin and Lambert 1984), the stellar evolution theory predicts that some secondary production of nitrogen must exist in all stars through the CNO cycle, while primary nitrogen can be produced only in stars experiencing the envelope burning phase during which dredged-up material is processed (Renzini and Voli 1981). In order to put some limits on the contribution of primary nitrogen, we have computed numerical models for the evolution of the spirals for which sufficiently reliable data exist (namely the Galaxy, M31, M33, M51, M83, M101, NGC2403, NGC6946, and IC342). To avoid ad hoc assumptions, we have adopted for our models standard nucleosynthetic predictions and kept the number of free parameters down to two: star formation and infall rates. The star formation is assumed to be exponentially decreasing with time (e-folding time $\tau$) and the infall F is taken to be constant and uniform (see Diaz and Tosi 1985 for details). Our models are computed without instantaneous recycling approximation, in order to take into account the actual lifetimes of stars. This is especially important for this study since nitrogen and oxygen are mainly produced by stars of different mass ranges.

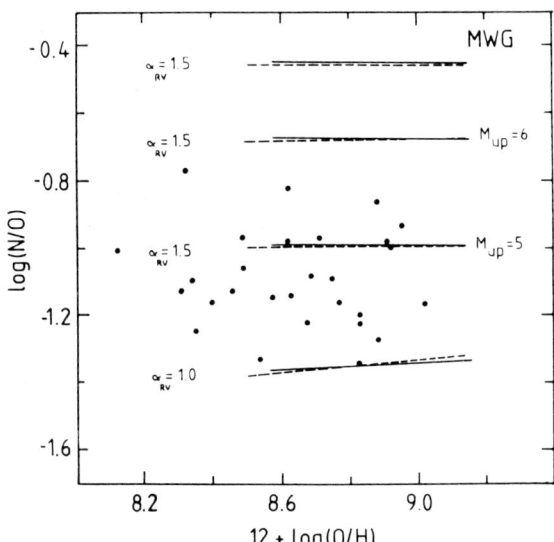

Fig.1. The Galaxy: Logarithmic N/O vs O/H distribution as derived from HII region observations (dots) and from theoretical models (thick and dashed lines). See text for details.

NUCLEOSYNTHESIS PRESCRIPTIONS

For massive stars ($M \geq 10\ M_{\odot}$) we have adopted Arnett's (1978) results on explosive nucleosynthesis but scaling his initial mass-helium core mass relation in order to take mass loss into account (Maeder 1981, 1983). The contribution to the interstellar medium enrichment by massive star winds has also been taken form Maeder (1981, 1983).

For low and intermediate mass stars we have adopted the yields derived by Renzini and Voli (1981) for a mass loss rate corresponding to $\eta = 1/3$ and different mixing lengths $\alpha_{RV}$. Different primary nitrogen contributions are predicted by Renzini and Voli depending on the adopted value of $\alpha_{RV}$: the higher this value, the larger the expected primary contribution, since the mixing length sets the lower mass limit for hot-bottom burning processes. The upper mass limit for these processes to take place is that of carbon ignition in degenerate cores, whose standard value is $M_{up} = 8\ M_{\odot}$ although lower values have recently been suggested (Bertelli et al. 1985, Castellani et al. 1985, Renzini et al. 1985).

RESULTS

Fig.1 shows the logarithmic N/O vs O/H distribution in the Galaxy as derived from Shaver et al. (1983) observations of HII regions and from the models of ours which best fit the observed oxygen distribution. The top and bottom sets of models have been computed with the standard

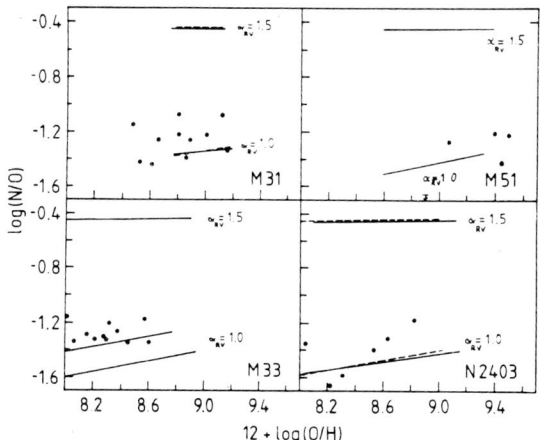

Fig.2. Logarithmic N/O vs O/H diagram for M31, M33, M51 and NGC2403. Symbols are as in Fig.1.

$M_{up}=8\,M_\odot$ and $\alpha_{RV} = 1.5$ and 1.0 respectively. The corresponding results for $\alpha_{RV} = 0.0$ and 2.0 are not shown in the figure since they are well outside the observational range. The other two sets of models have been computed with the two new suggested values for $M_{up}$, as labelled in the figure.

Figures 2 and 3 show the same diagram for the rest of our galaxies. Only models with the standard value of $M_{up}=8\,M_\odot$ are shown. The general results of the models, as can be seen from the figures, points to a value of $\alpha_{RV}$ between 1.0 and 1.5. If the standard value of $M_{up}$ is adopted, the observations are better described by models with $\alpha_{RV}$ closer to 1.0, which corresponds to a primary fraction of nitrogen between 30 and 60%. For $M_{up} = 5$ or $6\,M_\odot$ a value of $\alpha_{RV}$ closer to 1.5 is needed in order to provide the same required fraction.

In the case of the Galaxy, the observed $^{12}C/^{13}C$ ratio provides a further constraint for the values of $\alpha_{RV}$ and $M_{up}$ since $^{13}C$ primary and secondary production depends on them in the same way as $^{14}N$. With $M_{up}=8\,M_\odot$ and $\alpha_{RV}=1.0$ the present $^{12}C/^{13}C$ ratio predicted for the solar neighbourhood is 170 by mass, while the corresponding observed value is 65 (Henkel et al. 1982, Henkel et al. 1985). Models with lower $M_{up}$ and $\alpha_{RV}=1.5$ provide values of this ratio between 48 and 99 in better agreement with the presently available data. Therefore, this latter combination provides the best fit to both the N/O and $^{12}C/^{13}C$ observations and leads us to conclude that a primary fraction of nitrogen between 30 and 60% can account for the observed data.

REFERENCES

Arnett, W.A., 1978, Astrophys.J. 219, 1008
Bertelli, G., Bressan, A., Chiosi, C., 1985, Astron.Astrophys. submitted

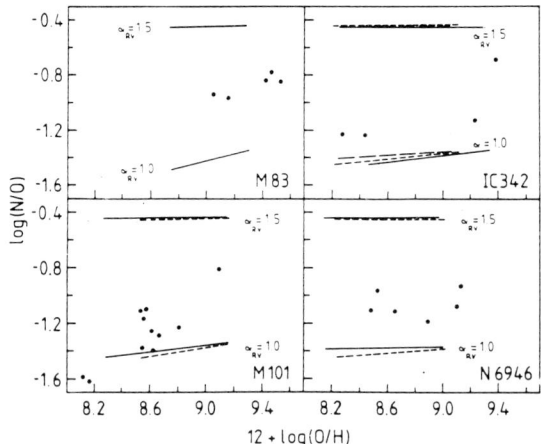

Fig.3. Logarithmic N/O vs O/H diagram for M83, M101, NGC6946 and IC342. Symbols are as in Fig.1.


Castellani, V., Chieffi, A., Pulone, L., Tornambe', A., 1985, Astrophys. J.Lett. submitted
Diaz, A.I., and Tosi, M., 1985, Astron.Astrophys. submitted
Henkel, C., Gusten, R., Gardner, F.F., 1985, Astron.Astrophys. $\underline{143}$, 148
Henkel, C., Wilson, T.L., Bieging, J., 1982, Astron.Astrophys. $\underline{109}$, 344
Maeder, A., 1981, Astron.Astrophys. $\underline{101}$, 385
Maeder, A., 1983, Astron.Astrophys. $\underline{120}$, 113
Pagel, B.E.J., and Edmunds, M.G., 1981, Ann.Rev.Astron.Astrophys. $\underline{19}$, 77
Renzini, A., Bernazzani, M., Buonanno, R., Corsi, C.E., 1985, Astrophys. J.Lett. submitted
Renzini, A., and Voli, M., 1981, Astron.Astrophys. $\underline{94}$, 175
Serrano, A., and Peimbert, M., 1983, Rev.Mex.Astron.Astrofis. $\underline{8}$, 177
Shaver, P.A., McGee, R.X., Newton, L.M., Danks, A.C., Pottasch, S.R., 1983, M.N.R.A.S. $\underline{204}$, 53
Tomkin, J., and Lambert, D.J., 1984, Astrophys.J. $\underline{279}$, 220


DISCUSSION

Mould: There is one constraint on primary nitrogen production available from observation of intermediate mass stars on the AGB. That is that this process is not so wonderfully successful that it converts carbon stars to M stars in the LMC. I don't know how to make this constraint quantitative yet, but it will need to be satisfied.

Tosi: What we infer from our study is that a primary fraction of nitrogen between 30 and 60% explains the observed properties of this element and that the hot-bottom burning can contribute this fraction. Whether or not this mechanism satisfies other observational constraints should be checked in a different framework.

EFFECTS OF METAL-DEPENDENT STELLAR MODELS ON THE YIELD OF NITROGEN


C. Forieri
International School of Advanced Studies
Strada Costiera 11
Trieste
Italia

ABSTRACT. The relation between nitrogen and oxygen is studied in the case of variable yields. The influence of the chemical composition on stellar evolution is taken into account. We found that new models mimick the primary origin of nitrogen.


1. INTRODUCTION

Nitrogen and oxygen are very abundant elements in the Universe. In fact both take part in the effective H-burning by means of the CNO-cycle and in particular oxygen is also the main product of the He-burning. The best information on their abundances comes from HII-regions that are taken as indicators of the interstellar medium composition.
    In a log O/H-log N/O diagram most of the data are confined between two 45° lines, as shown by Edmunds and Pagel (1981). Nevertheless a single galaxy shows a constant N/O ratio. In particular that is true also for the Galaxy. Data for galactic HII-regions from Shaver et al. (1983) are summarized in Figure 1. This rough constancy is very difficult to explain; for a pure secondary nitrogen -i.e. if oxygen and carbon seeds are required at the star formation to synthesized nitrogen as a by-product of the CNO-cycle- Serrano and Peimbert (1983) suggest the relation log N/O=log O/H + b. On the contrary the observations imply either a substantial production of primary nitrogen as a result from stellar evolutionary models of intermediate mass stars (Iben and Renzini, 1983, 1984) -i.e. nitrogen synthesized directly from hydrogen and helium- or a delay in the ejection of secondary nitrogen with respect of oxygen (Edmunds and Pagel, 1978, 1984), if the latter is synthesized mainly in massive stars, while the former comes from long-living low-mass stars.


2. VARIABLE YIELDS FOR Z-DEPENDENT MODELS

It is important to stress that also a time variation in the N yield

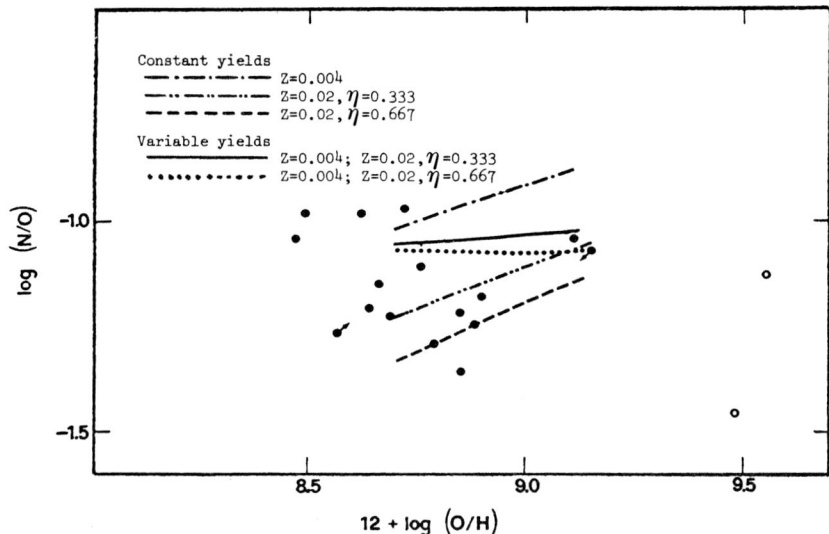

Fig.1. The N/O abundance ratio versus the total oxygen abundance; observations for galactic HII-regions (Shaver et al., 1983) and computed models

could influence the present N abundance. In principle the yields could depend on the initial metal content of the ejecting stars, even if the yields are usually considered constant -or only proportional to the primary abundances in the case of secondary elements- because of the lack of both complete grids of evolutionary models of stars with different metal contents and observational evidence of metal dependent nucleosynthesis. On the other hand it is worth mentioning the interesting feature of the models computed by Renzini and Voli (1981) for intermediate mass stars. The higher the metal content, the higher is also the quantity of secondary nitrogen synthesized by stars, but the efficiency of primary nitrogen production is shown to decrease for increasing initial metal content. This implies a variability of N yield that should modify its present abundance.

To investigate the effect of this dependence of nitrogen production on total metallicity, numerical computation of chemical evolution of the Galaxy are performed, using the code suggested by Chiosi and Matteucci (1982). We assume a linear dependence of primary nitrogen production on metal content, as the poor grid does not permit a more precise estimate.

## 3. RESULT

The same Figure 1 shows the theoretical relation between nitrogen and oxygen abundances that results from computations. Solid and dotted lines denote models with variable yields. Dotted dashed lines and dotted line denote models with constant yields. All these models reproduce the metallicity gradient in the galactic disk but only for the first two the ratio between nitrogen and oxygen remains almost constant along the galactic radius in the solar neighborhood, the exact value depending on the assumed specific models of the grid for the nucleosynthesis.

Models with constant yields do not work at all and predict an increasing N/O ratio in the solar neighborhood for increasing O abundance. On the contrary models with variable yields give a flat slope of the curve even if the predicted ratio is slightly too high. However the new rate for the reaction $^{12}C(\alpha,\gamma)^{16}O$ given by Kettner et al. (1982) implies a decrease in the N/O ratio. Moreover the new models of stellar evolution with overshooting by Bertelli et al. (1984) offer another possibility, being the mass range of intermediate mass stars different.

The full results of these computations will be published in a forthcoming paper (Forieri, 1985).

REFERENCES
Bertelli,G.,Bressan,A.G. and Chiosi,C.:1984, Astron.Astrophys.(in press)
Chiosi,C. and Matteucci,F.:1982, preprint.
Edmunds,M.G. and Pagel,B.E.J.:1978, MNRAS 185,779
Edmunds,M.G. and Pagel,B.E.J.:1984, in C.Chiosi and A.Renzini(eds.),
    Stellar Nucleosynthesis,D.Reidel Publ.Co.,Dordrecht, p.341.
Forieri,C.: 1986, Astrophys. and Space Science 114, 119.
Iben,I. and Renzini,A.:1983,Ann.Rev.Astron.Astrophys. 21,271.
Iben,I. and Renzini,A.:1984,Physics Rep. 105,329.
Kettner,K.U.,Becker,H.W.,Buchmann,L.,Clayton,D.D.,Macklin,R.L. and
    Ward,R.A.:1982,Astrophys.J. 257,821.
Pagel,B.E.J. and Edmunds,M.G.:1981,Ann.Rev.Astron.Astrophys. 19,77.
Renzini,A. and Voli,M.:1981,Astron.Astrophys.94,175.
Serrano,A. and Peimbert,M.:1983,Rev.Mex.Astron.Astrof. 8,117.
Shaver,P.A.,McGee,R.X.,Newton,L.M.,Danks,A.C. and Pottash,S.R.:1983,
    MNRAS 204,53.

DISCUSSION

Comte: From an unpublished study by L.Vigroux, G.Stasinśka and myself, in which we have re-reduced >70 observations of 55 irregular and blue compact galaxies (including some new results taken at ESO), we find a very good correlation between N/H and O/H extending from [O/H] ~ .08 [O/H] solar up to ~ solar value, and we got a far small dispersion. The data were all re-reduced according to the grid of ionization models by G.Stasinśka, in each case we have a good estimate of the effective temperature from the [OIII]4363Å line.

SURFACE BRIGHTNESS DISTRIBUTIONS IN TWO EDGE-ON SPIRAL GALAXIES

M.A.Shaw
Dept. of Astronomy, Univ. of Edinburgh, Edinburgh, U.K.

G.F.Gilmore
Institute of Astronomy, Madingley Road, Cambridge, U.K.

1. INTRODUCTION

Of considerable importance to modelling the formation and early evolution of galaxies is the relationship between the spheroids of spirals and S0´s, and those of ellipticals. In this context, the isolation of a thick disc in our own Galaxy (Gilmore et al. (1985)) and in other spirals (van der Kruit (1984); herafter vdK) - aside from their initial discovery in S0´s by Burstein (1979) - has added significance because the luminosity profiles of several spirals seem to be dominated by this component over considerable ranges of galacto-centric distance (R) and height (Z). On the other hand, "elliptical-like" spheroids, generally characterised by the de Vaucouleurs $r^{1/4}$ law, seem to be barely detectable in many galaxies in complete contradiction to our preconception that this model adequately describes all forms of galaxian spheroids.

Here we present analyses of the surface brightness distributions in the two well known nearby edge-on spirals NGC 4565 and NGC 891 (adopted distances of 10 and 9.5 Mpc respectivly). This is the first such an investigation of the NGC 4565 profiles presented by Jensen & Thuan (1982; herafter JT), while other recent data on this galaxy (van der Kruit & Searle 1981; herafter KS1) are used to supplement those of JT. The analysis of NGC 891 (from van der Kruit & Searle 1981; herafter KS2) is included both to extend the results of vdK and to complement those on NGC 4565.

2. THE ADOPTED MODELS

In attempting to describe the thin disc in the following analysis, we have made the, now usual, assumption of a locally isothermal and self-gravitating sheet. In its present form the model used is that first proposed (and more fully described) by KS1, reducing to the following expressions in terms of surface brightness

$$\mu(R,Z) = \begin{cases} \mu(0,0)\,[1+(R^2/2h_R^2)\ln(R/2h_R)]\,\mathrm{sech}^2(Z/Z_0) & \text{for } R \ll h_R \\ \mu(0,0)\,[(\pi R/2h_R)^{1/2}\exp-(R/h_R)]\,\mathrm{sech}^2(Z/Z_0) & \text{for } R \gg h_R \end{cases}$$

$\mu(0,0)$ mag arcsec$^{-2}$ (herafter "$\mu$") being the central surface brightness.

Implicit in the derivation of this model is the approximation of a constant scaleheight with R. Although it is not an immediately obvious result, KS1 have given some evidence to support its validity.

Modified forms of the above equations are also used to represent a possible thick disc (with a suitable choice for $h_R$ and scaleheight $(Z_1/2)$. Alternatively, the familiar exponential model is also applied

$$L(R,Z) = L_0 \exp-[(R/h_r) + (Z/h_z)]$$

$h_z$ being the thick disc scaleheight.

A spheroid component is represented either by an exponential similar to that of the thick disc, or by the more usual de Vaucouleurs $r^{1/4}$ law

$$\log(L/L_0) = -3.33[(\theta/\theta_E)^{1/4} -1]$$

where $\theta = (q^2R^2 + Z^2)^{1/2}$, q being the axis ratio, and $\theta_E$ is the effective radius.

## 3. RESULTS

### 3.1 NGC 4565

In this section we present the results of a full analysis of the existing B band data of JT for NGC 4565 – the running of single, two or three component fitting iterations being used in an attempt to describe the form of these observed profiles.

Major discrepencies, when compared to the actual data, are generated by the adoption of either single or two component model fits. For example, a two component thin disc and spheroid combination consistently underestimates the amount of light present in the galaxy at intermediate Z (typically ~2.4 to 5.1 kpc) by as much as 1.2 $\mu$ (amounting to ~30 times the quoted errors of JT). Similar underestimations are revealed, but in this case in the outer regions of each profile (Z in excess of 3.4 kpc), by the superposition of a thin disc and thick disc – the dicrepencies here running from 0.5 to 3.0 $\mu$ or ~3 to 20 times quoted errors. In neither of these cases do the discrepencies appear to be model dependant nor systematic in origin.

Observations are, on the other hand, very well described by a combination of all three of the components noted above – as figure 1 shows. For such a combination, the thin disc has a $(Z_0)$ scaleheight of 0.82 kpc comparing quite favourably with that found by KS1 and with the currently accepted value for our own Galaxy as deduced from star counts. The resultant thick disc scaleheights also prove to be independent of R, having values of 0.84 kpc for the exponential and 0.80 kpc

($Z_1/2$) for the $sech^2$ model forms - such values again in accord with the existing data. The spheroid, it turns out, is equally well described by exponential or $r^{1/4}$ laws. Use of the former gives rise to an unambiguous increase in scaleheight with R although this trend, rather than describing any fundamental property of the spheroid, merely seems to indicate the increasing dominance of this component over the other two as R increases. This is supported by the fact that at large Z ($\gtrsim 8.7$ kpc or $\sim 3\theta_E$), the ($h_z$) scaleheight of the spheroid approaches a limiting value of 3.1 kpc.

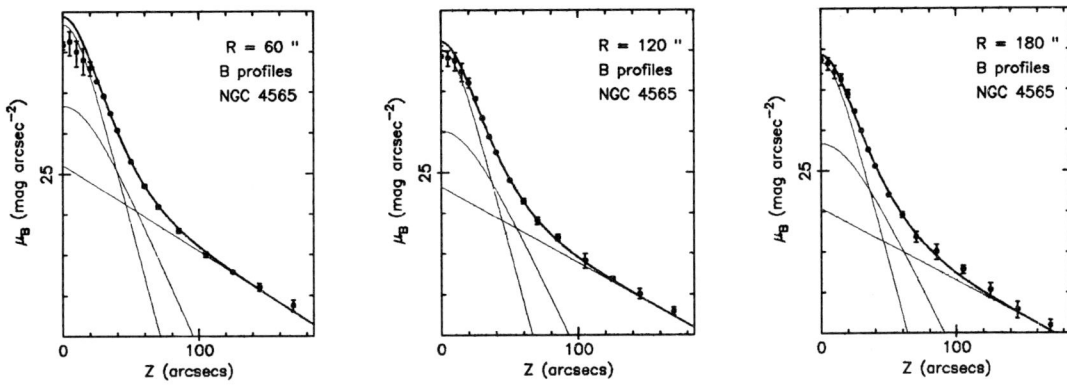

Figure 1. Selected surface brightness profiles of NGC 4565 with superimposed three component (thin disc, thick disc plus spheroid) fits and associated errors. R is the galactocentric distance.

Isophote shapes provide a more graphic representation of the above. We find clear evidence of global isophotal flattening with R for all models used, further deconvolution of these isophotes into their respective components indicating the presence of a very highly flattened thick disc at all R underlying a much rounder spheroid (albeit still a little flatter than one might expect from a preconception that it is of the same form as an elliptical galaxy) - see fig 3.

3.2 NGC 891

We now apply the same modelling procedure used above to the existing J($\lambda 4700$Å) and F($\lambda 6400$Å) band data on NGC 891 from KS2. The three examples shown in figure 2 clearly indicate how well described these profiles are by the combination of a thin disc and an exponential thick disc (the $sech^2$ form is unable to model the thick disc in this case). The best-fit thin disc scaleheights ($Z_0$) are 0.94 kpc in F and 0.99 kpc in J, while for the thick disc the scaleheight ($h_z$) is 1.92 kpc in both bands (it is again constant for all R). The appropriate central surface brightnesses of the two components are given in table 1.

The isophote shapes of this thick disc noticeably flatten beyond ~ 8.3 kpc as figure 3 shows. Interior to this the isophotes are moderately rounded and are consistent with ellipses having axis ratios of between 0.6 and 0.66 (compared to the 0.6 adopted by vdK and the range 0.6 to 0.7 quoted by Bahcall & Kylafis (1985)), so in this respect the thick disc seems a little different to that of NGC 4565. Inspection of the relative scaleheights in table 1 supports this result.

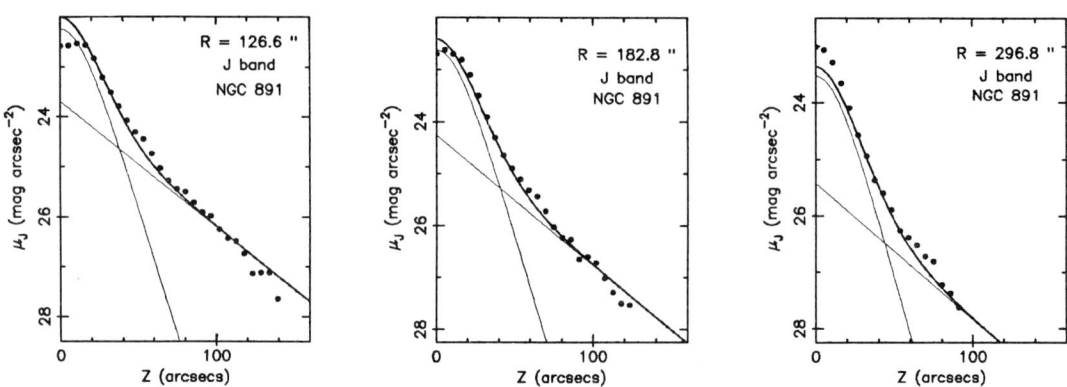

Figure 2. Selected J surface brightness profiles of NGC 891 with superimposed two component fits using a thin disc and exponential thick disc.

Hence, taking the above photometric data in isolation, we see no evidence for a spheroid in this galaxy down to a limiting surface brightness ~26.5 or $27_\mu$ in F and ~27.5 to $28^\mu$ in J. It is quite possible, of course, that a spheroid component does exist but that it is only apparent at levels fainter than this. Indeed, if this is the case, and if we can use the results of NGC 4565 (for which the spheroid only becomes prominent around $\mu_B$ ~$27_\mu$) as a guide, then we might expect such a component in NGC 891 also to appear only around 27 - 27.5 $\mu$ in J assuming the galaxies to be fairly similar. This is very close to the limit of the current photometry and so the failure to detect a spheroid is not really suprising.

Problems in analysing NGC 891 first become apparent only when one wishes to try and model all the profiles with one component alone or when one insists on adding to the models a spheroid brighter than ~$27\mu$. Use of just such a bright spheroid together with a thin disc, for example, generates disagreements very similar to those noted in the case of NGC 4565 between the predicted and observed surface brightness distributions. Here, as found previously, such discrepencies can only be eliminated by the addition of a third component - the thick disc - to the models.

# 4. POSSIBLE PROBLEMS

Some of the general problems associated with profile deconvolution of edge-on spirals have been addressed by, for example, Kormendy (1982) - two of these are of particular concern here.

The first is the effect that even a small amount of internal absorption due to gas and dust might have on the resultant parameters - in particular those of the thin disc. In the case of NGC 891, for which we have good multicolour data analysed with the above techniques, such absorption is the likely cause of differences in both thin disc and thick disc central surface brightnesses - and to a lesser degree the thin disc scaleheights - between the J and F bands. As far as the scaleheights are concerned, the greater problem of obscuration in J gives rise to artificially flattened profiles at small Z (artificially compared to F that is). To continue to fit the data in such a situation, the thin disc scaleheight in the J band must be correspondingly increased. The preferential light loss in J also results in the central surface brightnesses of both thin and thick discs appearing fainter there than in F.

The second major concern lies with the effect on the parameter values of galaxy inclination, although with an inclination $\approx 89°$ such

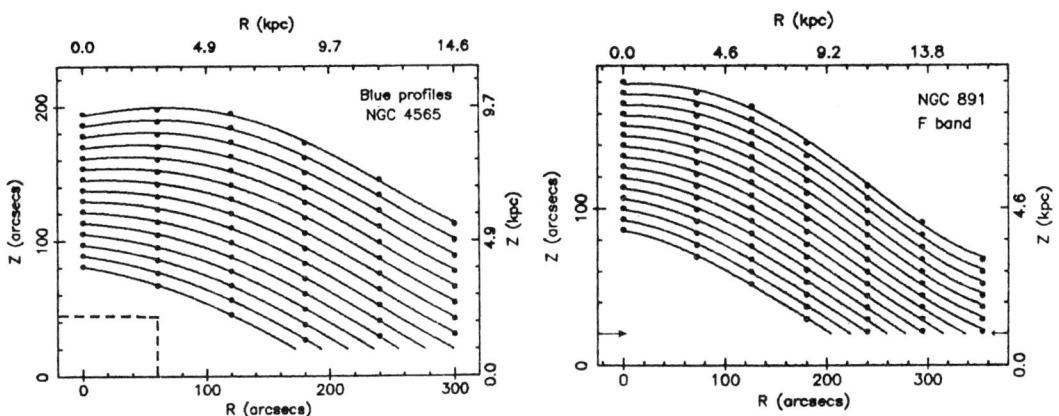

Figure 3. Isophote shapes of the spheroid of NGC 4565 (left) and the thick disc of NGC 891 (right) both derived using exponential models. Each curve corresponds to a specific surface brightness - in the case of NGC 4565 running from 26.2µ (bottom curve) to 29.0µ (top) and for NGC 891 from 24.2µ (bottom) to 27.0µ (top). The dashed box for NGC 4565 represents the area of significant contamination by the "peanut" bulge, while arrows on the NGC 891 plot marks the Z-height below which dust obscuration and thin disc contamination are likely to be a problem.

a difficulty is not as severe for NGC 891 as it is likely to be for NGC 4565 (for which $i$ is only $85° - 86°$). The effects on NGC 4565 will only become apparent in our analysis when we have completed a study of the red profiles of JT.

## 5. DISCUSSION

We see no evidence in the photometry of NGC 891 by KS2 (currently the best available on this galaxy) for the existence of a spheroid, the profiles being very well described by a locally isothermal thin disc in combination with an exponential thick disc, both possessing uniform scaleheights at all R. The constancy of the thick disc scaleheight in both J and F, together with only marginal evidence for a (J-F) colour gradient with R or Z in the KS2 data (despite strong evidence of (U´-F) gradients), argues in favour of a uniform age to this component. In agreement with the contour plots presented by KS2, the thick disc becomes progressivly flatter at large R.

Analysis of NGC 4565 suggests the presence of a thin disc resembling that of NGC 891 even to a similar scaleheight, and a thick disc again having a constant scaleheight with R. In addition, a spheroid (becoming progressively more dominant for $Z \gtrsim 8.7$ kpc) has had to be invoked because of the failure of single or two component modelling to reproduce adequately the observed profiles. The isophote maps indicate a moderate flattening of the spheroid, but a high degree to the thick disc.

Aside from the obvious advantage of being relatively nearby, another of the major motivations for studying these two galaxies in particular has been to investigate how they might resemble our own Galaxy were it viewed from a similar aspect, and a strong case has been recently been made for NGC 891 in this regard (vdK). The existence of a warp in the disc of NGC 4565 - a feature in common with our own Galaxy - together with the similarities of its parameters derived above to those of the Galaxy and NGC 891, argue that NGC 4565 is also a good candidate for such a comparison.

However, it has been stressed that additional problems of interpretation arise when modelling NGC 4565 because of the presence both of the warp and also the so-called "peanut" bulge. In particular, vdK has argued that residual light from the warp contaminates the outer spheroid sufficiently to lead to the observed flattening of the isophotes. Thus we have attempted to estimate the degree of importance of either to our conclusions. It seems that despite the obvious dominance by the bulge of the minor axis profile, it is not of fundamental significance to our results mainly because it becomes unimportant beyond $R \sim 3$ kpc (i.e. $\sim 1 \theta_E$) and $Z \sim 2$ kpc. In our analysis, however, further investigation of the bulge is hampered partly by dust obscuration but in particular by the inherent uncertainties resulting from subtraction of all the other components.

An investigation into the likely effect of the warp is best addressed using the KS1 data for NGC 4565 because it has not been smoothed over all the galaxy quadrants. A comparison of surface brightnesses at corresponding points in the (R,Z) plane over the whole galaxy (together with even a brief inspection of the contour plots presented by KS1 themselves) clearly shows this warp only to be present over regions fainter than $\mu_B \sim 27\,\mu$ and R in excess of ~23 kpc. However, flattening in the thick disc - and to a lesser extent in the spheroid - first becomes noticeable at much brighter magnitudes and much smaller R. It is also important to note that the same degree of global isophotal flattening is seen in all quadrants and is not appreciably different in that containing the strongest evidence for the warp. Even if it is both more extensive and more complex than is assumed - admittedly rather naively in this discussion - we suspect the warp may be too faint to have a significant effect on the above conclusions.

As for the importance of these isophote shapes, even modest deviations of such shapes from true ellipses in any spheroid are highly significant because, as pointed out by Binney & Petrou (1985), they arise from correspondingly much larger deviations in the underlying three dimensional structure of the galaxy. In analysing these shapes, therefore, it is interesting to note the recent study of variations in ellipticity for a large sample of E´s and SO´s presented by Michard

TABLE I - Derived and adopted parameters for NGC 4565 and NGC 891

| component | models | NGC 4565 | NGC 891 |
|---|---|---|---|
| thin disc | sech$^2$ | $\mu(0,0) = 21.12\ (\pm 0.30)$ mag arcsec$^{-2}$ | $\mu(0,0) = \begin{cases} 20.16\ (\pm 0.05)\ \text{mag arcsec}^{-2}\ \text{in F} \\ 21.30\ (\pm 0.30)\ \text{mag arcsec}^{-2}\ \text{in J} \end{cases}$ |
| | | $z_0 = 0.82\ (\pm 0.02)$ kpc | $z_0 = \begin{cases} 0.94\ (\pm 0.01)\ \text{kpc in F} \\ 0.99\ (\pm 0.03)\ \text{kpc in J} \end{cases}$ |
| | | $h_r = 5.5$ kpc | $h_r = 4.9$ kpc |
| thick disc | exp. | $\mu(0,0) = 21.44\ (\pm 0.20)$ mag arcsec$^{-2}$ | $\mu(0,0) = \begin{cases} 21.60\ (\pm 0.12)\ \text{mag arcsec}^{-2}\ \text{in F} \\ 22.40\ (\pm 0.19)\ \text{mag arcsec}^{-2}\ \text{in J} \end{cases}$ |
| | | $h_r = 5.5$ kpc | $h_r = 4.9$ kpc |
| | | $h_z = 0.84\ (\pm 0.09)$ kpc | $h_z = 1.92\ (\pm 0.14)$ kpc |
| | sech$^2$ | $\mu(0,0) = 23.10\ (\pm 0.40)$ mag arcsec$^{-2}$ | -------- |
| | | $h_r = 5.5$ kpc | |
| | | $z_1 = 1.60\ (\pm 0.19)$ kpc | |
| spheroid | r$^{1/4}$ law | $\mu_0 = 24.20\ (\pm 0.30)$ mag arcsec$^{-2}$ | -------- |
| | | $q = 0.7$ | |
| | | $\theta_e = 2.9$ kpc | |
| | exp. | $\mu(0,0) = 24.20\ (\pm 0.30)$ mag arcsec$^{-2}$ | -------- |
| | | $h_r = 5.5$ kpc | |
| | | $h_z = 3.1\ (\pm 0.2)$ kpc for $Z \gtrsim 180$ " | |

(1984). These results show that the shape of a galaxy flatter than E3 varies in a fashion which is qualitatively similar to the observed behaviour of the global (i.e. thick disc plus spheroid) isophotes of NGC 4565. Extending the conclusions of Michard, a natural explanation of this result is that E galaxies also have a two component (thick disc plus spheroid) structure as seen in NGC 4565 after thin disc subtraction. There exists, therefore, the interesting possibility that flattened (thick disc) structure is an integral part of a spheroidal galaxy, leading in turn to the possibility that thin and thick disc structure are unrelated in spirals.

6. ACKNOWLEDGEMENTS

We would like to thank in particular Piet van der Kruit for supplying his data on NGC 4565 and NGC 891 in such a usable form, and also Trinh Thuan for allowing the analysis of the NGC 4565 data. Tim Hawarden provided several very helpful comments on the manuscript. MAS acknowledges the support of an Science & Engineering Research Council studentship.

REFERENCES

Bahcall, J.N. and Kylafis, N.D.: 1985, Ap. J., 288, 252
Binney, J. and Petrou, M.: 1985, M N R A S., 214, 449
Burstein, D.: 1979, Ap. J., 234, 435
Gilmore, G.F., Reid, I.N. and Hewett, P.: 1985, M N R A S., 213, 257
Jensen, E.B. and Thuan, T.X.: 1982, Ap. J. Supp., 50, 421 (JT)
Kormendy, J.: 1982 in *Morphology and Dynamics of Galaxies, 12th Adv. course Swiss Soc. of Astron. & Astrophys., Geneva Obs. Publ.*
Kruit, P.C. van der: 1984, A. & A., 140, 470 (vdK)
Kruit, P.C. van der and Searle, L.: 1981, A. & A., 95, 105 (KS1)
Kruit, P.C. van der and Searle, L.: 1981, A. & A., 95, 116 (KS2)
Michard, R.: 1984, A. & A., 140, L39

# SUBJECT INDEX

| | |
|---|---|
| Abundance gradient | 451,463 |
| Active galactic nuclei | 382 |
| Age: | |
|     discrepancy | 243,251 |
|     limits | 164 |
|     MC clusters | 135,217 |
|     M giants | 157 |
|     molecular clouds | 35 |
|     OB associations in LMC | 75 |
|     simple stellar population (SSP) | 214 |
|     star clusters | 135,251 |
| Age-metallicity relation | 452 |
| Asymptotic giant branch (AGB) : | |
|     stars | 159,166,171,204,219,243,287 |
|     luminosity function | 244,254 |
| | |
| Baade's window | 145,161 |
| Biochemistry | 140 |
| Bolometric luminosity evolution | 91,203 |
| Broad band : | |
|     colors | 324 |
|     galaxy counts | 392 |
|     photometric comparisons | 394 |
| Bulge : | |
|     component stars | 137 |
|     M giants | 157,163 |
| | |
| Carbon stars | 135,171 |
| Cepheid stars | 255 |
| Chemical abundances | 67 |
| Chemical evolution of galaxies | 5,449 |
| Cloud evolution | 38 |
| Cluster galaxies | 375 |
| Cluster galaxy samples | 268 |
| C/M star ratios | 172 |
| Colors : | |
|     irregular galaxies | 81 |
|     3CR and FRHR galaxies | 403 |
| Colour-magnitude diagram : | |
|     M 31 bulge | 139 |
|     bulge stars | 165 |
|     E galaxies | 309 |
| Convective overshooting | 237,244,246,256 |
| CPC far-infrared maps | 129 |
| Cumulative metallicity distribution | 452 |

| | |
|---|---|
| Disk formation | 455 |
| Evolutionary flux | 197,207 |
| Evolutionary models : | |
|     physical fundaments | 244 |
|     self gravitating object | 49 |
| Evolutionary synthesis | 283 |
| Far infrared excess | 114 |
| Far IR emission | 63 |
| Far UV emission | 63 |
| Far UV excess | 360 |
| First ranked : | |
|     cluster galaxies | 270,396,403 |
|     high redshift galaxies (FRHR) | 403 |
| First stars | 15,50 |
| Fragmentation | 40,47 |
| Frequency-mass histograms | 52 |
| Fuel consumption theorem | 200 |
| Galactic bulge | 143 |
| Galactic bulge distance | 161 |
| Galaxies : | |
|     anemic | 412 |
|     blue compact | 57,88 |
|     disk | 411 |
|     dwarf | 87,136 |
|     dwarf spheroidal | 136 |
|     elliptical | 121,222,264,309,324,363 |
|     giant elliptical | 393,403 |
|     high redshift | 326,412,419 |
|     interacting | 57,100,103,127 |
|     irregular | 57,81,136,265 |
|     late type | 81,265 |
|     Local Group | 133,171 |
|     Markarian | 81,98 |
|     metal poor | 87 |
|     normal galaxies | 237,403 |
|     peculiar | 100,274 |
|     radio | 396,404 |
|     Seyfert | 106 |
|     S0 | 321,357,413 |
|     spiral | 62,103,127,265,411,469,477 |
| Galaxy : | |
|     count surveys | 393 |
|     encounters | 97 |
|     evolution | 3,49,65,391 |
|     formation | 6,15 |

# SUBJECT INDEX

| | |
|---|---|
| morphology | 5 |
| number counts | 439 |
| Galaxy-galaxy interactions | 103 |
| Gas | 4,67 |
| Gas mass | 81 |
| Globular clusters | 217,324 |
| Globular cluster giants | 163 |
| Gravitational instability | 47 |
| | |
| High redshift clusters | 378,411 |
| Holmberg effect | 97 |
| Horizontal branch (HB) | 135,138 |
| Horizontal branch stars | 224 |
| HR diagram morphology | 242 |
| Hydrogen mass | 81 |
| | |
| Infrared : | |
| colors | 403 |
| emission | 113,346 |
| galaxies | 111 |
| H-K colors | 403 |
| observations (bulge giants) | 149 |
| photometry | 148,404 |
| spectral energy distribution | 113 |
| Initial mass function (IMF) | 4,21,48,76,206,210,457 |
| Integrated spectra | 295 |
| Intermediate age : | |
| clusters | 250 |
| populations | 324 |
| Intermediate mass stars | 242,250 |
| IRAS : | |
| observations | 91,111,150,346 |
| far-infrared | 127 |
| selected galaxies | 93 |
| source counts | 95 |
| Isochrone synthesis | 351 |
| IUE : | |
| S0 galaxies spectra | 357 |
| elliptical galaxies en.distr. | 363 |
| | |
| K correction | 392,405 |
| | |
| Local Group galaxies | 133,171,184 |
| Look back time | 228,321,375 |
| Look back time method | 391 |
| Luminosity : | |
| irregular galaxies | 81 |

|  |  |
|---|---|
| M giants | 160 |
| Luminosity function : | |
|     AGB stars | 254 |
|     C stars | 178 |
|     main sequence | 134 |
|     M 33 | 183 |
|     nearby galaxies | 60,184 |
| Luminosity function evolution | 391 |
| Lyman continuum flux | 62 |
| | |
| Magellanic clouds : | |
|     clusters | 217 |
|     stellar populations | 133 |
|     very young associations | 75 |
| Mass function | 4,47,52,184 |
| Mass of gas | 67 |
| Massive stars | 239,249 |
| Mass loss | 76,150,160,166,237,244 |
| Metal abundance | 90 |
| Metal dependent stellar models | 473 |
| Metallicity gradient : | |
|     gas | 453 |
|     M giants | 159 |
|     M 31 | 138 |
| M giants : | |
|     bulge | 143,151,157 |
|     colors | 161 |
|     luminosity and mass | 160 |
| Models : | |
|     photometric evolution | 310 |
|     spectral evolution | 264 |
|     spectrophotometric evolution | 405,412 |
| Molecular cloud observations | 31 |
| Molecular flows | 36 |
| | |
| Neutral Hydrogen : | |
|     distribution (M 33) | 189 |
|     mass (Irregular galaxies) | 81 |
| Nitrogen | 469,473 |
| Nuclear activity | 103 |
| Nucleosynthesis | 461,470 |
| | |
| OB associations ( LMC ) | 75 |
| Open clusters and associations | 33 |
| Optical spectrum | 348 |
| Optical surveys of primeval galaxies | 419 |
| Overshooting | 237,256 |
| Oxygen | 469,473 |

# SUBJECT INDEX

| | |
|---|---|
| Photometric evolution of galaxies | 309 |
| Population phase transition | 221 |
| Population models | 158 |
| Population synthesis | 345 |
| Post AGB stars | 225,288 |
| Post star-burst galaxies | 379 |
| Present day star formation rate | 58,61 |
| Primeval galaxies | 27,419 |
| | |
| Radial distribution (M 33) | 183 |
| Radio galaxies | 396,404 |
| Radio-infrared correlation | 120 |
| Ram-pressure stripping | 411 |
| RR Lyrae stars | 135,137 |
| | |
| Selection effects | 404 |
| Simple stellar population (SSP) | 197 |
| Simultaneous photometer (SPIV) | 404 |
| Spectra : | |
|     galaxies | 378 |
|     old stellar generations | 294 |
|     young stellar generations | 290 |
| Spectral energy distribution (SED) | 212,223,266,301,396 |
| Spectral evolution | 195,263,395 |
| Spectral synthesis | 67,322 |
| Spectrophotometric models | 283,412 |
| Spectrophotometry : | |
|     clusters | 377 |
|     galaxies | 105,397 |
| Spheroidal population | 138 |
| Star-burst : | |
|     evolution | 84 |
|     galaxies | 118 |
| Star counts | 58 |
| Star clusters | 133,217,242 |
| Star formation | 38,108,357,406,412 |
| Star formation : | |
|     bursts | 79,90,358,407 |
|     efficiency | 35 |
|     history | 15,64,133,189,391 |
|     physical processes | 31,75 |
|     rate | 58,76,91,97,133,159,167,459 |
|     tracers | 57 |
| Stellar atmospheres theory | 288 |
| Stellar evolution clock | 197 |
| Stellar evolutionary models | 213,244 |
| Stellar evolution theory | 237,285 |
| Stellar populations : | |

|  |  |
|---|---|
| elliptical galaxies | 310 |
| galaxies | 133,195 |
| S0 galaxies | 321,357 |
| Stellar population synthesis | 357 |
| Subgiant branch | 134,204 |
| Super-metal-rich stars (SMR) | 157,159 |
| Surface brightness distribution | 477 |
| Synthesis library and techniques | 349 |
| | |
| Thick disk | 477 |
| Thin disk | 477 |
| 3C radio galaxies | 274,396,403 |
| | |
| UV energy distribution | 363 |
| UV light evolution | 228 |
| UV spectra | 369,407 |
| UV variations | 407 |
| | |
| Virgo cluster | 115 |
| V luminosity function<br>( nearby irregular galaxies ) | 60 |

RAYMOND H. FOGLER LIBRARY
DATE DUE

BOOKS ARE SUBJECT TO
RECALL AFTER TWO WEEKS

NOV 13